D1076809

QD
481
ATK

STEREOSELECTIVE
SYNTHESIS

WITHDRAWN FROM RADCLIFFE SCIENCE LIBRARY	
LEGAL DEPOSIT COPY	NO
LAST BODLEIAN COPY	NO
RECENT USAGE	NO

£45 BLA
303056170

STEREOSELECTIVE SYNTHESIS

Robert S. Atkinson
University of Leicester, Leicester, UK

JOHN WILEY & SONS
Chichester · New York · Brisbane · Toronto · Singapore

Copyright © 1995 by John Wiley & Sons Ltd,
Baffins Lane, Chichester,
West Sussex PO19 1UD, England

Telephone: National Chichester (01243) 779777
International +44 1243 779777

Reprinted March 1996
Reprinted January 1997
Reprinted February 1999
All rights reserved.

No part of this book may be reproduced by any means,
or transmitted, or translated into a machine language
without the written permission of the publisher.

Other Wiley Editorial Offices

John Wiley & Sons, Inc., 605 Third Avenue,
New York, NY 10158–0012, USA

Jacaranda Wiley Ltd, 33 Park Road, Milton,
Queensland 4064, Australia

John Wiley & Sons (Canada) Ltd, 22 Worcester Road,
Rexdale, Ontario M9W 1L1, Canada

John Wiley & Sons (SEA) Pte Ltd, 37 Jalan Pemimpin #05–04,
Block B, Union Industrial Building, Singapore 2057

Library of Congress Cataloging-in-Publication Data:

Atkinson, Robert S.
    Stereoselective synthesis / Robert S. Atkinson.
        p.   cm.
    Includes bibliographical references and index.
    ISBN 0-471-95250-8
    1. Stereochemistry.   2. Organic compounds—Synthesis.   I. Title
    QD481.A79   1995
    547′.2—dc20   94-20918
                                                                    CIP

British Library Cataloguing in Publication Data:

A catalogue record for this book is available from the British Library

ISBN 0 471 95250 8 (Cloth)
ISBN 0 471 95419 5 (Paper)

Typeset in 10/12 Times in Great Britain by Mackreth Media Services, Hemel Hempstead
Printed and bound in Great Britain by Antony Rowe Ltd, Chippenham, Wiltshire

CONTENTS

INTRODUCTION

The synthesis of new organic molecules and the improved synthesis of known ones will always be a major task for the professional organic chemist. The stereoisomerism which can arise even when only two appropriately substituted sp^3-hybridised carbons are contained in a molecule makes it inevitable that the synthesis of such a molecule will require the exercise of stereocontrol.

The concepts of conformational analysis, stereoelectronic control and reaction mechanism have been vital for the development of methods for stereoselective synthesis. Progress in this area has been stimulated by developments in a number of other areas since the 1950s, including: spectroscopic methods, in particular nuclear magnetic resonance spectroscopy; advances in chromatographic techniques for separating mixtures; the availability of rapid X-ray crystal structure determinations and the use of computers in molecular modelling and molecular mechanics calculations. In addition, there is the ongoing discovery of new reagents and reactions that bring about functional group interconversions stereo-, regio- and chemoselectively.

The discovery of the Sharpless–Katsuki enantioselective epoxidation of allylic alcohols in the early 1980s led to the development of other 'chemzymes'—reagents which resemble enzymes in bringing about functional group interconversions enantioselectively. A complementary development has been the use—one might say abuse—of enzymes to bring about functional group interconversions on substrates other than their natural ones. Genetic engineering techniques are now making available modified enzymes which will be better suited to this task.

These developments in the area of enantioselective synthesis have been given added impetus in many countries by the imminence of legislation which will require pharmaceuticals whose components contain chiral centres to be sold in enantiopure form.

The teaching of synthesis at undergraduate level is given structure by the application of retrosynthetic analysis to the target molecule in question. At this level, however, the emphasis usually is on functional group interconversion and chemoselectivity. Real synthesis design, of course, will normally require that the problem of stereostructure in the target molecule be addressed. Although the methods of retrosynthetic analysis can still be applied, in many cases this analysis will be dominated by the need to accommodate the stereostructure of the target.

Undergraduates are often introduced to stereoselective synthesis in courses which examine the synthesis of natural products. They may also encounter courses which are based on the application of selected reagents or reactions to stereoselective synthesis. It is hoped that this book can complement such courses by explaining in more detail the basis for, and the background to, the stereoselectivity of the reactions employed. Alternatively, the book may be used as the basis of a course which examines the various ways in which simple molecules can be synthesised stereoselectively. Such a course could provide an ideal illustration of the ways in which factors such as conformational analysis, steric and stereoelectronic effects and orbital symmetry combine to bring about stereoselectivity.

An important part of this book is the classification of stereoselective reactions as Type 0, I, II, III or IV. Although this classification is new, it is for the most part a shorthand for classes of stereoselective reaction already recognised in the literature. It does at the least have the merit of being considerably less cumbersome to use. Thus a Type II/III$_{r.c.}$ reaction is one in which chiral centres are formed as a result of simple diastereoselectivity and reagent-controlled asymmetric induction. Type 0 reactions are those in which no additional chiral centres are formed and no bonds to existing chiral centres are broken except where they are converted into achiral centres.

To design a synthesis requires more than a knowledge of the sum of its parts. The individual functional group conversions and transformations that it incorporates may be trivial but directing these towards a target molecule with appropriate choice of starting material, reagent, attention to functional group compatibility *and* stereochemistry provides a test of the student's grasp of organic chemistry as a whole that cannot be surpassed.

One problem in writing a book such as this is that of nomenclature. Wherever available, IUPAC recommended terms have been used (one exception is the abbreviation enant. for enantiomeric). The terms chiral and prochiral have been preferred over stereogenic and prostereogenic, although both are currently widely used. *Cis* and *trans* refer to a spatial relationship of substituents in a substrate or product; *syn* and *anti* refer to a spatial relationship of substituents in a reaction.

A distinction is drawn between the *stereostructure* of a *compound* (a description of the relative (or absolute) configuration of its chiral elements/double bonds) and the *stereochemistry* of a *reaction* (its course described by reference to the stereostructure of the reactants and products). I have found it necessary to introduce two new terms that I believe will prove helpful heuristically, namely *inherent diastereoselectivity* (for the most part synonymous with diastereospecificity) and *occasional diastereoselectivity*.

I would like to thank my co-workers and colleagues in Leicester for their help in various ways in the writing of this book: Emma Barker, Tom Claxton,

Ann Crane, Paul Cullis, Andrew Ellis, Paul Jenkins, Ray Kemmitt, Pavel Kocovsky, John Malpass, Ian Lochrie, Bernard Rawlings and Paul Williams. I am particularly grateful to Martin Harger who read through the manuscript and made innumerable valuable suggestions. My thanks are also due to Brian Judkins (Glaxo) and Nick Lawrence (UMIST) for their help and to John Wiley & Sons who have taken considerable pains to produce the diagrams to as high a specification as possible. Finally, I would like to thank my wife Lois without whose help this book would not have been written.

ABBREVIATIONS

AIBN	α-azo-*iso*butyronitrile
BINAP	[2,2′-bis(diphenylphosphino)-1,1′-binaphthyl]; see Chapter 14, compound (**11**)
Bn	benzyl
Bu	butyl
BuLi	*n*-butyllithium
Bz	benzoyl
CAN	cerium (IV) ammonium nitrate
COD	cyclooctadien(yl)
DBU	1,8-diazabicyclo[5.4.0]undeca-7-ene
DEAD	diethyl azodicarboxylate
DHAP	dihydroxyacetone phosphate
DIBAL	diisobutylaluminium hydride
DMAP	4-dimethylaminopyridine
DME	dimethoxyethane
DMF	dimethylformamide
DMSO	dimethyl sulphoxide
enant.	enantiomer(ic)
FGI	functional group interconversion
HLADH	horse liver alcohol dehydrogenase
HMPA	hexamethylphosphoramide
HOMO	highest occupied molecular orbital
imidaz.	imidazole
LDA	lithium diisopropylamide
LUMO	lowest unoccupied molecular orbital
mCPBA	*m*-chloroperoxybenzoic acid
MsCl	methanesulphonyl chloride
NBS	*N*-bromosuccinimide
Nu	nucleophile
PCC	pyridinium chlorochromate
PLE	pig liver esterase
PPL	porcine pancreas lipase
pyr.	pyridine
RAMA	rabbit muscle aldolase
R-Ni	Raney nickel
TBAF	tetrabutylammonium fluoride
Tf	trifluoromethanesulphonyl
TFA	trifluoroacetic acid
THF	tetrahydrofuran
THP	tetrahydropyran(yl)
TMEDA	*N,N,N,N*-tetramethylethylenediamine
TMS	tetramethylsilyl

Ts	toluene-*p*-sulphonyl
TsCl	toluene-*p*-sulphonyl chloride
TSG	transition state geometry
Type 0	see Chapter 3
Type I	see Chapter 4
Type II	see Chapters 6–8
Type III	See Chapters 9–16
Type IV	see Chapter 2
Type II/III	see Chapter 13

1 SELECTIVITY IN ORGANIC SYNTHESIS: STEREOCHEMICAL VOCABULARY

Organic synthesis requires the exercise of control in the way that organic molecules react. The consequences of imperfect control are a reduced yield of product and the necessity for its separation from by-products. Since this separation is often expensive in time, materials and resources, the benefits of complete control are clear.

One can identify four major types of control or selectivity which may be required in the reactions of organic molecules:

chemoselectivity
regioselectivity
diastereoselectivity $\left.\right\}$ stereoselectivity
enantioselectivity

In this book we shall be concerned for the most part with stereoselectivity, but the problems of chemoselectivity and regioselectivity will inevitably and persistently intrude.

Chemoselectivity is the preferential reaction of one functional group over another under the reaction conditions employed. Thus, using sodium tetrahydroborate, a ketone can usually be reduced to a secondary alcohol in the presence of an ester (Scheme 1).

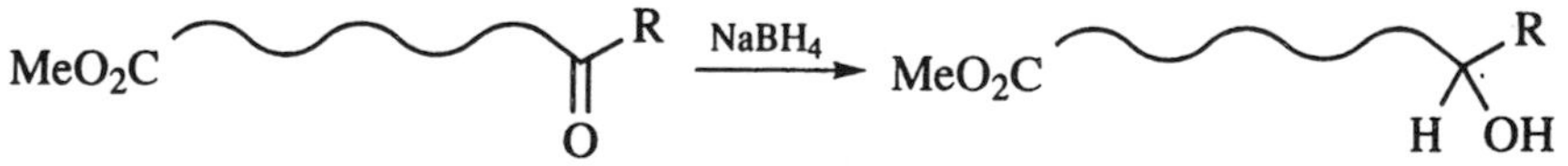

$$MeO_2C \overset{\text{NaBH}_4}{\longrightarrow} MeO_2C$$

Scheme 1

Direct reduction of an ester in the presence of a ketone, however, is not so straightforward and is usually accomplished using a *protecting group* for the ketone (Scheme 2).

Protecting groups, such as the acetal in Scheme 2, are available for a

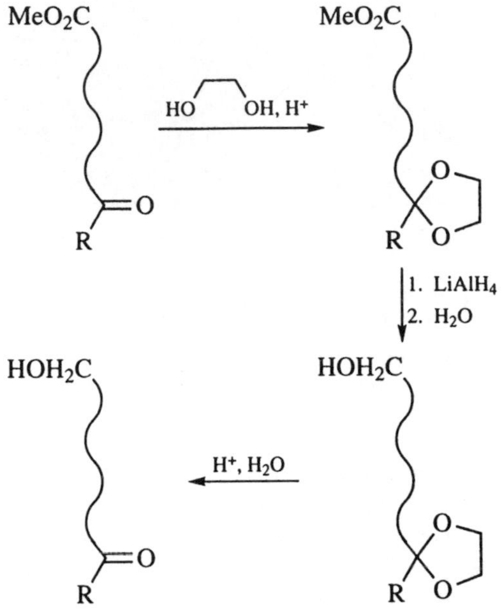

Scheme 2

variety of functional groups and their widespread use in synthesis is a measure of how imperfect is our control of chemoselectivity. Our recourse to the use of protecting groups will, of course, also require the input of additional time, materials and resources.

Regioselectivity is the preferential formation of one *isomer* of the product in a reaction in which one (or less commonly two) other isomer(s) may also be formed. A familiar example is the ionic addition of hydrogen bromide to an alkene such as 2-methylpropene (Scheme 3), which proceeds with complete *regioselectivity* to give *tert*-butyl bromide.

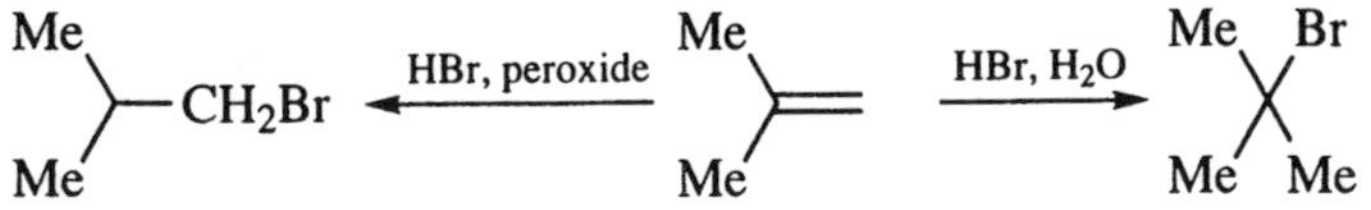

Scheme 3

Which *regioisomer* is formed in a reaction may depend critically on the conditions under which it is carried out: the regioselective formation of isobutyl bromide in Scheme 3 can be brought about by homolytic addition of hydrogen bromide using the reagents indicated.

The different regioselectivities obtaining in these hydrogen bromide additions are, of course, the result of two different mechanisms for addition but, more often, regioisomers are formed by similar mechanisms (Scheme 4).

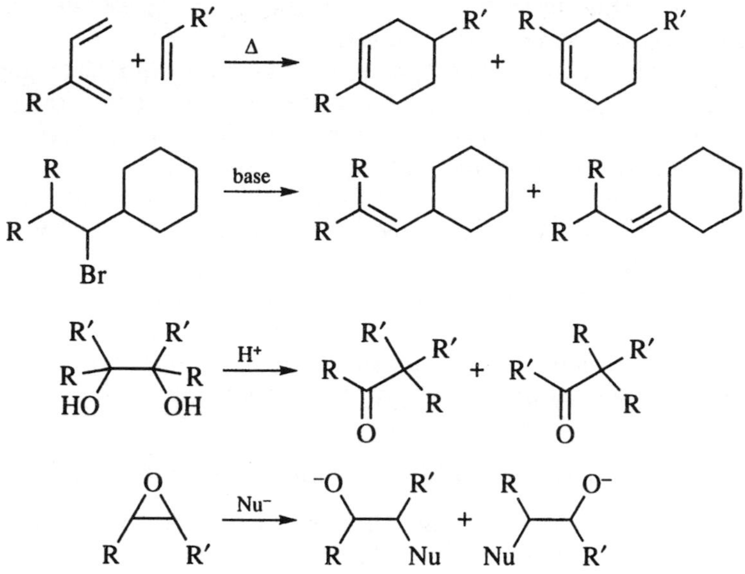

Scheme 4

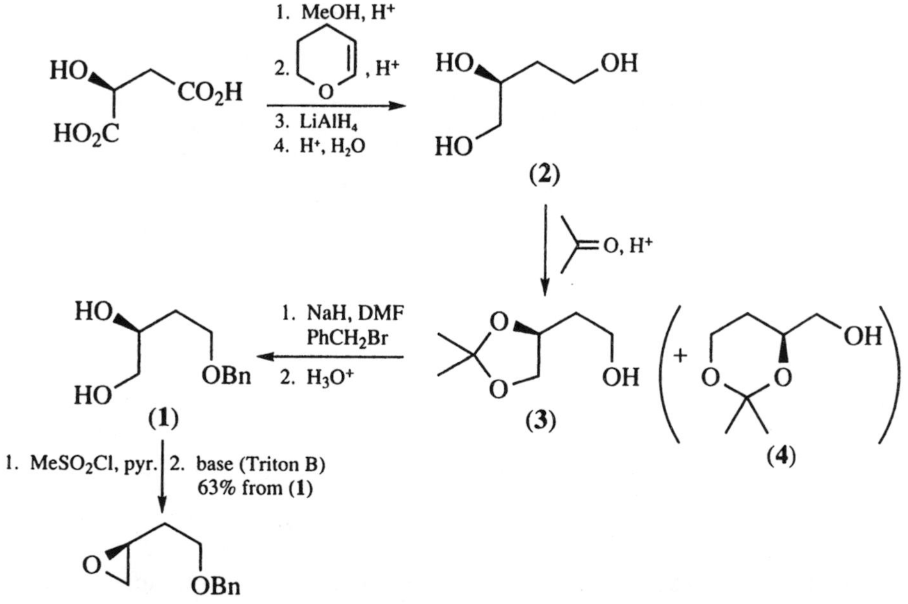

Scheme 5

Control of regioselectivity in these cases is less easy although not impervious to changes in reaction conditions.

If a molecule contains two (or more) identical but distinguishable functional groups, preferential reaction of one of them requires the exercise of regioselectivity. Protection of one or more of these groups is often used, allowing the unprotected one to be manipulated as desired.

In Scheme 5,[1] mesylation of (1) is regioselective as a result of the greater reactivity of the primary over the secondary hydroxyl group. Differentiation between the primary hydroxyl groups of the triol (2), however, requires regioselective formation of the five-membered ring acetal (3) as indicated [this five-membered ring acetal was freed from the ~10% of six-membered ring acetal (4), which is also formed, by crystallisation of the 3,5-dinitrobenzoate of (3) followed by hydrolysis back to the alcohol[2]].

Stereoselectivity: enantioselectivity and diastereoselectivity

Stereoisomers are isomers which have the same atoms and bonds in common but different arrangements in space. Thus the acyclic molecules (6), $_{enant.}$(6), (7) and $_{enant.}$(7) are stereoisomers and, provided that rotation around the central C—C bond is rapid (as ordinarily it is), these are the only isolable stereoisomers having structure (5).

For each of these stereoisomers there are an infinite number of non-isolable *conformational isomers* corresponding to incremental degrees of rotation around the central C—C bonds in these molecules. However, for the case of (6), for example, there are two particular conformations, (6′) and (6″), which, together with (6), are both more important than all others and also more readily identifiable since they have a staggering of their bonds which minimises the torsional strain that is present in other conformations. In solution at room temperature, therefore, (6), (6′) and (6″) will be in dynamic equilibrium although it is unlikely that they will contribute equally to this equilibrium.

In the same way that stereoisomers are a special sort of isomer, *enantiomers* are a special sort of stereoisomer where the two molecules in question have a *mirror-image* or *enantiomeric* relationship.

Thus (6) and $_{\text{enant.}}$(6) are enantiomers, as are (7) and $_{\text{enant.}}$(7). The relationship between (6) and (7) [$_{\text{enant.}}$(7)] is a diastereoisomeric one: *diastereoisomers are stereoisomers which do not have an enantiomeric relationship*. As we progress, therefore, from isomer to diastereoisomer and to enantiomer, the kinship between two molecules becomes increasingly closer.

Diastereoisomers can only arise in a chiral molecule when it contains at least two chiral elements (see below), as in (5); a molecule such as (8) can exist in only two (enantiomeric) forms, (9) and $_{\text{enant.}}$(9). Some achiral molecules can exist as diastereoisomers: the *E/Z* stereoisomers of alkenes are not enantiomers and are therefore diastereoisomers.

A reaction which proceeds stereoselectively is one which results in the formation of an excess of one stereoisomer of the product over others. Since the term stereoisomers embraces enantiomers and diastereoisomers, stereoselectivity can be sub-divided into *enantioselectivity* and *diastereoselectivity*.

Enantioselectivity in a reaction is either the preferential formation of one enantiomer of the product over the other or the preferential reaction of one enantiomer of the (usually racemic) starting material over the other. The latter is also known as an asymmetric transformation or de-racemisation.

Diastereoselectivity is the preferential formation in a reaction of one diastereoisomer of the product over others. Note that the product in a diastereoselective reaction may be produced either as a single enantiomer or as a racemate, or indeed may contain any excess of one enantiomer over the other between these extremes. Thus a reaction which produces a mixture of only (6) and $_{\text{enant.}}$(6) by *anti*-addition of the elements of a–z across a configured double bond (Scheme 6) is completely diastereoselective, but the product is bound to be racemic since attack in either of the two senses indicated in Scheme 6 is equally likely.

On the other hand, the completely diastereoselective addition of z$^-$ to the C=y′ bond of the single enantiomer (10) in the sense indicated will produce only a single enantiomer (6) (Scheme 7). Although a single enantiomer is formed in this reaction, one would not describe it as enantioselective since there is no expectation that the enantiomer of (6) could have been formed.

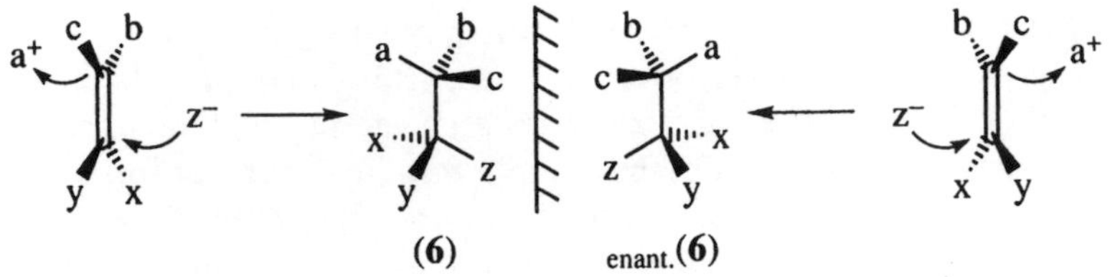

(6)   enant.**(6)**

Scheme 6

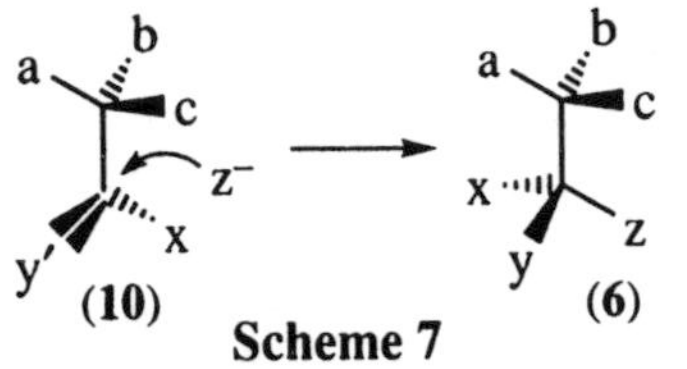

(10)   **Scheme 7**   **(6)**

Stereoselectivity, enantioselectivity and diastereoselectivity may be described as complete, high (say $\geq 90\%$), moderate or low.

Enantioselectivity is normally quoted as the enantiomeric excess (e.e.), which is the mole fraction of the major enantiomer expressed as a percentage, i.e. with an excess of the R form:

$$\text{e.e.} = \frac{\text{mole fraction } R - \text{mole fraction } S}{\text{mole fraction } R + \text{mole fraction } S} \times 100\% = \frac{[\alpha]_{\text{obs.}}}{[\alpha]_{\text{max.}}} \times 100\%$$

$[\alpha]$ = optical rotation

Similarly, diastereoselectivity is usually given as the diastereoisomeric excess (d.e.) which is the mole fraction of the major diastereoisomer D_1 in a mixture of two diastereoisomers D_1 and D_2, again usually expressed as a percentage:

$$\text{d.e.} = \frac{\text{mole fraction } D_1 - \text{mole fraction } D_2}{\text{mole fraction } D_1 + \text{mole fraction } D_2} \times 100\%$$

Sometimes diastereoselectivity may be quoted as the ratio of major to minor diastereoisomers (d.r.)

Stereospecificity is a term reserved for the circumstance in which stereo-differentiated reactants give stereo-differentiated products and/or show different reactivity.[3] Thus the S_N2 reaction is stereospecific (Scheme 8).

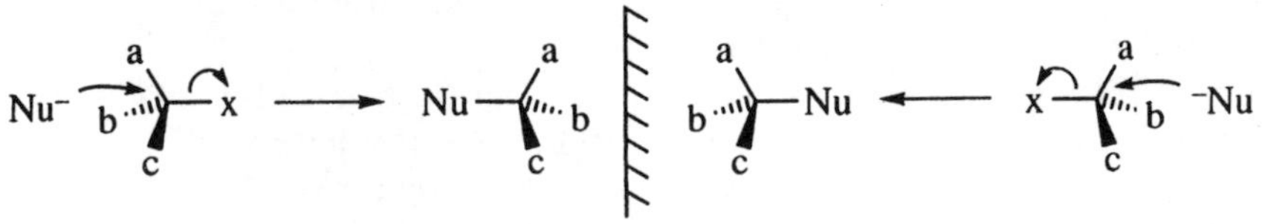

Scheme 8

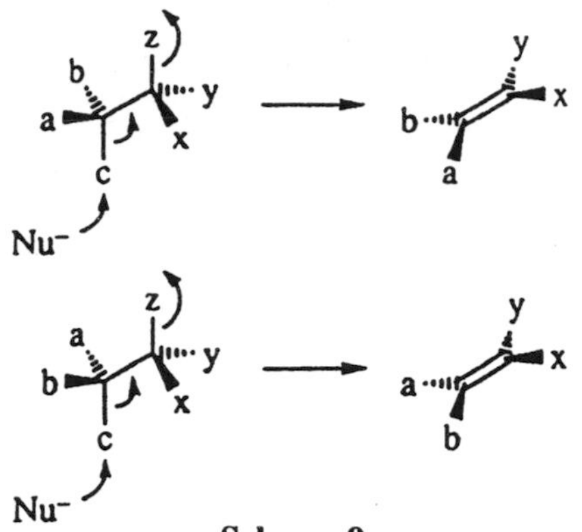

Scheme 9

In this case, each enantiomer of the starting material gives a single enantiomer of the product with inversion of configuration. Although the reaction would still be stereospecific even if a racemate of the starting material were used, it is only by using single enantiomers of the substrate that the specificity can be demonstrated.

Likewise, the *E*2 elimination is diastereospecific (Scheme 9). In this case, each diastereoisomer of the starting material gives a different diastereoisomer (double bond isomer) of the product. Although the starting materials in Scheme 9 are drawn in only one enantiomeric form, the diastereospecificity of the reaction would still hold even if racemic starting material were used in each case. It is unfortunate that whilst regiospecificity and enantiospecificity are used to indicate complete regioselectivity[4] and enantioselectivity, stereospecificity is not synonymous with complete stereoselectivity. At the present time, the correct use of the term stereospecificity is more often honoured in the breach rather than the observance. To avoid the confusion associated with the use of this term we shall also refer to a reaction showing diastereospecificity as *inherently diastereoselective* (see Chapter 6).

Other stereochemical vocabulary

CHIRALITY: CHIRAL ELEMENTS

A molecule which is non-superimposable upon its mirror image is *chiral*. Its *chirality* arises from the absence of some symmetry elements. The chiral element is most usually a *chiral centre* or, as it is sometimes called, a *stereogenic centre*[5] in the molecule which is an atom (usually carbon) bearing four different substituents. The *configuration* at this chiral centre is the arrangement in space of the substituents responsible for its chirality. The two configurations possible for a molecule containing a single chiral centre are described as *R* or *S* using the Cahn–Ingold–Prelog convention.[6]

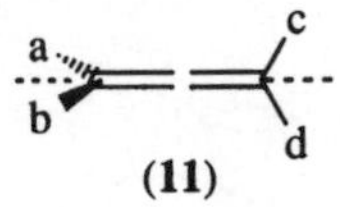

(11)

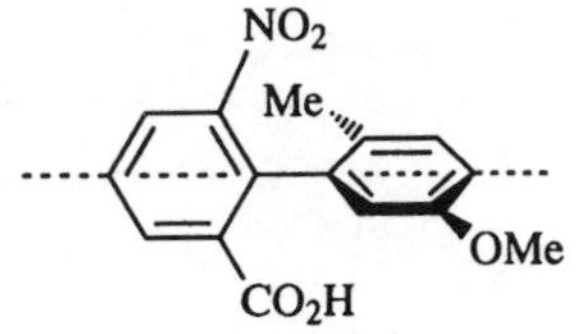

(12)

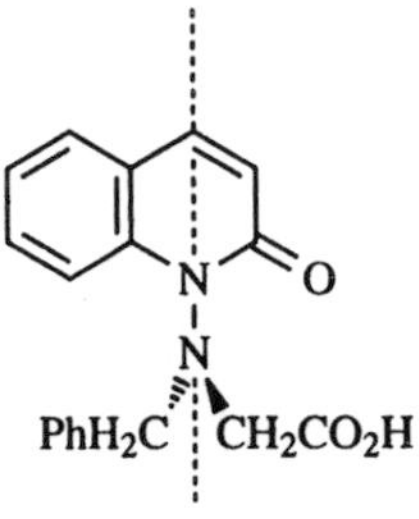

(13) 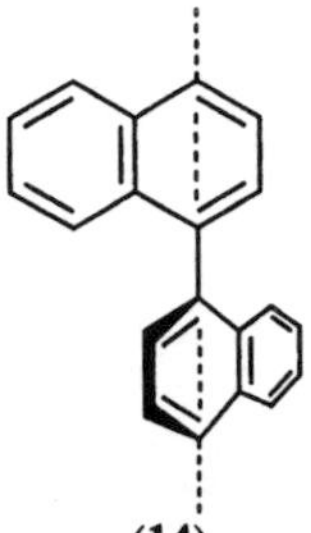   (14)

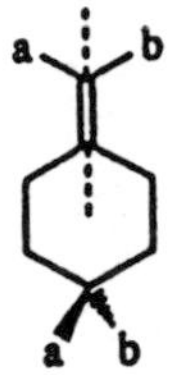

(15) 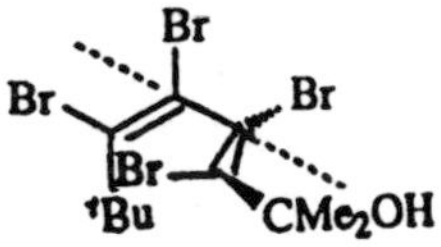   (16)

The chirality of a molecule may be alternatively ascribed to the presence of a *chiral axis*. This is a single or double bond within the molecule, rotation around which would convert one enantiomer into the other. Maintenance of enantiopurity in such molecules, therefore, depends upon the barrier to such a rotation being sufficiently high. Examples of molecules containing chiral axes are (11)–(16) with the chiral axes shown as dashed lines. Descriptions of their absolute configurations require conventions similar to those used for chiral centres.[7]

(17) 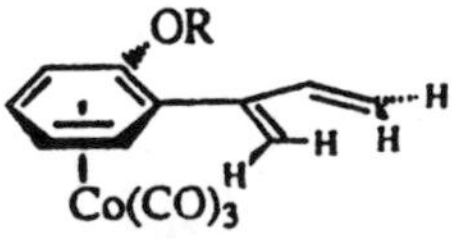   (18)

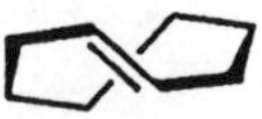

(19) 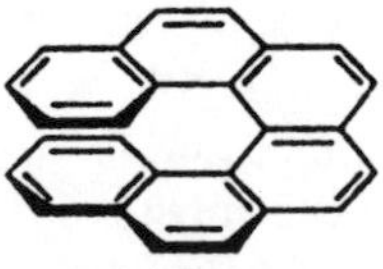   (20)

The presence of chiral centres or axes is not mandatory for a molecule to be non-superimposable upon its mirror image and chirality may arise from the absence of a plane, centre and alternating axis of symmetry, e.g. as in **(17)–(20)**.

In this book we shall be dealing overwhelmingly with synthesis of target molecules having one or more chiral centres. These target molecules may be required in a single enantiomeric form (enantiopure) or it may be sufficient to synthesise them with less than complete exclusion of the other enantiomer and, most commonly, in racemic form where both enantiomers are present in equal amounts. Thus a chiral compound is not necessarily an optically active one, though it has the potential to become one by resolution; a compound which is superimposable on its mirror image is *achiral*.

PROCHIRALITY: ENANTIOFACES

A *prochiral double bond* in an addition reaction is one in which one or two chiral sp³ centres are formed and the addition product itself is chiral. The prochiral double bond in the reaction may be contained in a molecule which is either chiral or achiral.

If the prochiral double bond is contained in an achiral molecule and the reagent is achiral, the resulting product will inevitably be racemic since the two transition states involved are enantiomeric and therefore equal in energy. Thus, in Scheme 10 the two products obtained in each reaction are enantiomers. Since in each case attack of the reagent on one face of the double bond gives one particular enantiomer and attack on the other face gives its mirror image, the double bond is said to have *enantiofaces* or *enantiotopic* faces.

In its reaction with a prochiral double bond, an achiral reagent cannot distinguish between enantiofaces and the product, therefore, is always racemic.

Enantiofaces can be identified by an adaptation of the Cahn–Ingold–Prelog rule for assigning absolute configuration to chiral centres.[8]

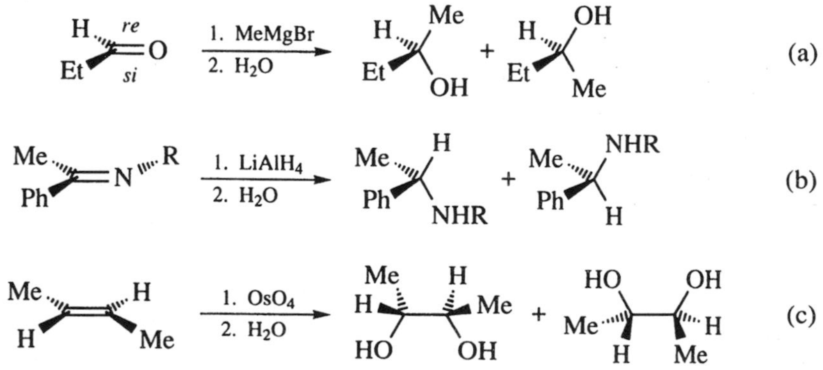

Scheme 10

Thus the substituents are assigned priority in the usual way when one face or other is viewed. A clockwise ordering of the substituents on the double bond defines the face viewed as *re*, an anticlockwise ordering as *si*. The *re* and *si* faces of propanal, therefore, are as designated in Scheme 10(a).

Prochirality is not an inherent property of the double bond alone but only arises in its combination with the reagent. Thus bromination of (*E*)-but-2-ene results in a single *meso*-dibromide (Scheme 11) and so the double bond in this reaction has identical or *homotopic* faces rather than enantiofaces which the same alkene has in its hydroxylation with osmium tetroxide [Scheme 10(c)].

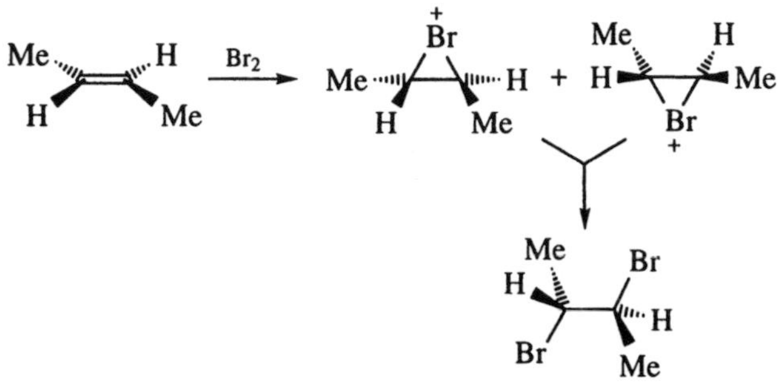

Scheme 11

Unlike the carbonyl group, a carbon–carbon double bond may be *configured* and exist in (diastereoisomeric) *Z*- and *E*-forms as in the case of but-2-ene above. A configured prochiral double bond can give rise to two contiguous chiral centres from attack on its two enantiofaces [Scheme 10(c)] whereas a non-configured prochiral double bond can give rise to only one. [(Scheme 10(a)].

The carbon–nitrogen double bond may also be configured as in Scheme 10(b) but the addition product does not normally have an assignable configuration at the pyramidal amine nitrogen since it is undergoing rapid inversion of configuration at room temperature. If this inversion at nitrogen is sufficiently retarded, as is the case when the latter is part of an oxaziridine ring, then here also two chiral centres are produced (Scheme 12).[9]

The possession of enantiofaces is not restricted to isolated double bonds;

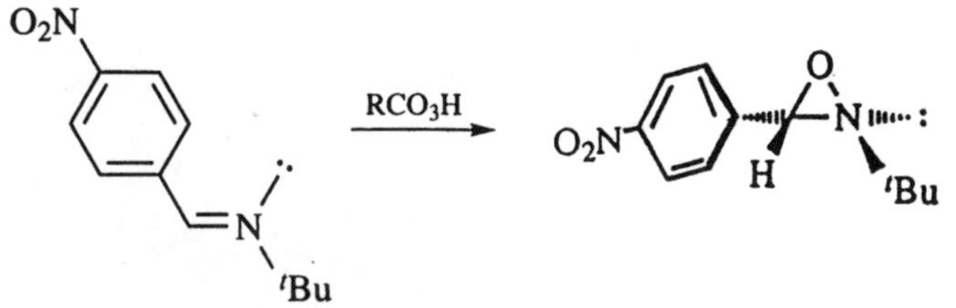

Scheme 12

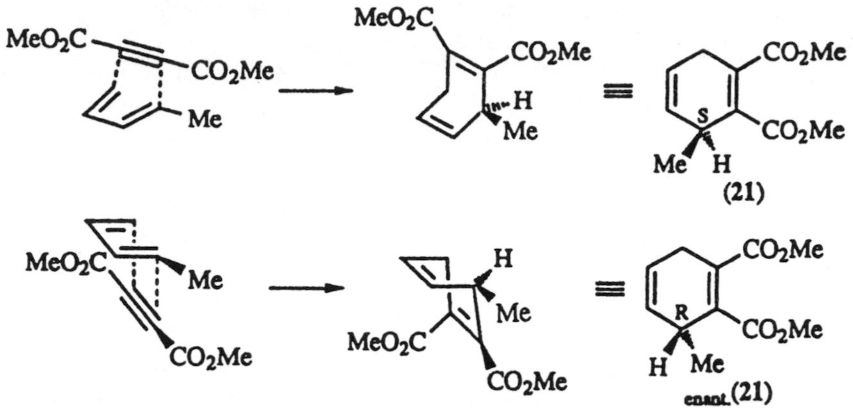

Scheme 13

the diene in Scheme 13 undergoes Diels–Alder addition at both enantiofaces to give the racemic cyclohexene derivative (21).

PROCHIRAL (PROSTEREOGENIC) CENTRES: ENANTIOTOPIC ATOMS OR GROUPS

Prochirality is a description which is also applied to groups such as the methylene group in ethanol: if each of the component C–H bonds of this methylene group is formally substituted in turn by, e.g., deuterium, as in Scheme 14, enantiomers are produced.

The prochiral methylene group in ethanol therefore contains two *enantiotopic* hydrogens. In fact, two identical substituents of any kind on an sp^3-hybridised atom (usually carbon) which bears two additional different substituents will be enantiotopic. Thus the two carboxymethylene groups in (22) (Scheme 15) are enantiotopic and monoesterification will give a racemic product. *The two transition states leading to the two enantiomeric products are equal in energy, as was the case for attack on enantiofaces of prochiral double bonds above* (Schemes 10–13). Identification of the individual enantiotopic atoms or groups uses an obvious extension of the Cahn–Ingold–Prelog

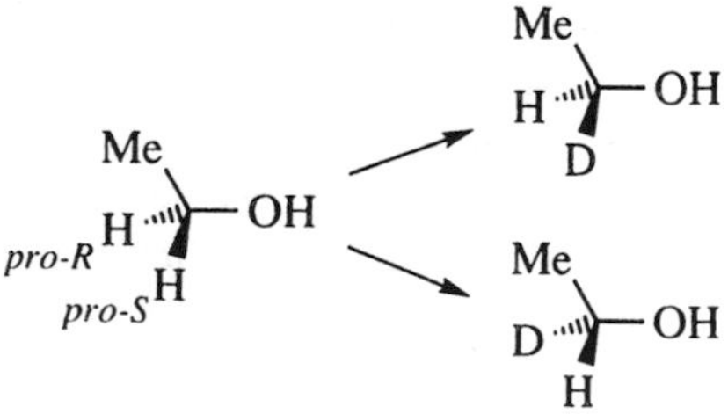

Scheme 14

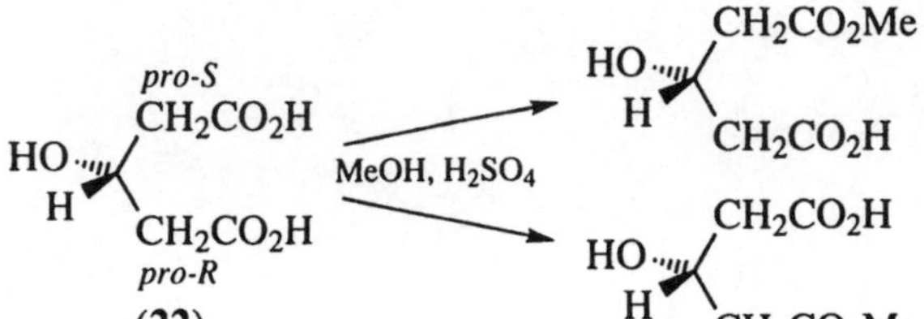

Scheme 15

convention:[8] one or other of the two atoms or groups is assigned priority over the other (e.g. by substituting D for H or ^{18}O for ^{16}O), but the priority of the remaining two substituents is not affected. The resulting configuration (*R* or *S*) defines the atom or group promoted as *pro-R* or *pro-S*, respectively. Thus the two methylene hydrogens in ethanol (Scheme 14) and the two carboxymethylene groups in (**22**) (Scheme 15) have the *pro-R* and *pro-S* designations shown.

Two enantiotopic atoms or groups in a molecule are not necessarily substituents on the same carbon atom: the two hydrogen atoms in the *meso*-diacid (**23**) are also enantiotopic since their individual replacement by, e.g., deuterium again produces molecules having an enantiomeric relationship (Scheme 16).

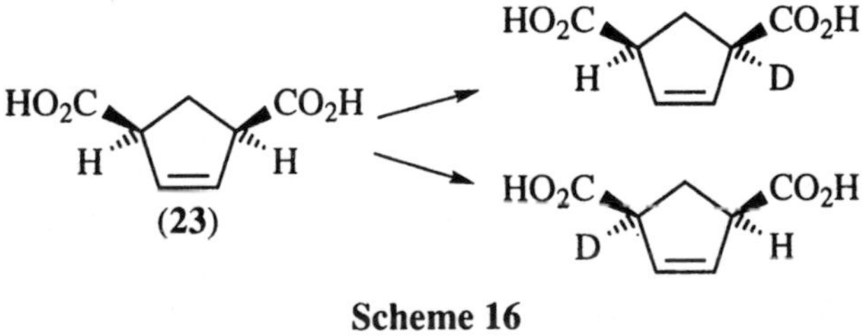

Scheme 16

PROCHIRALITY: DIASTEREOFACES

Consider now the reaction of a single enantiomer (**24**) of an alkene having a chiral (stereogenic) centre directly attached to the prochiral double bond (Scheme 17) with an achiral epoxidising agent [O].

The relationship between epoxides (**25**) and (**26**) is now a diastereoisomeric one and, in principle, *attack on the two faces of the alkene may take place at different rates:* the reaction may proceed diastereoselectively. This is because the two transition states leading to epoxides (**25**) and (**26**) are also diastereoisomeric and in contrast to the enantiomeric transition states referred to above, *not* necessarily equal in energy. The alkene (**24**) in this reaction is described as having *diastereofaces* or diastereotopic faces.

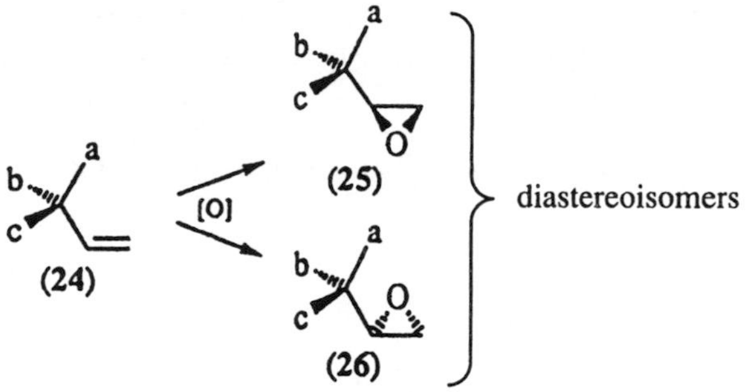

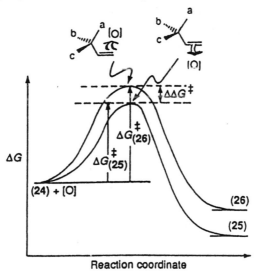

Scheme 17

It is helpful to picture the two diastereoisomeric transition states leading to epoxides **(25)** and **(26)** with the aid of an energy profile diagram (Figure 1), which is a graphical representation of the (free) energy changes which accompany the progress of the reaction (the reaction coordinate).

Figure 1

In a kinetically controlled reaction the rate constants for formation of epoxides **(25)** and **(26)** are related to the activation energies $\Delta G^{\ddagger}$ **(25)** and $\Delta G^{\ddagger}$ **(26)** according to equation 1 and hence the difference in activation energies $\Delta\Delta G^{\ddagger}$ is given by equation 2.

$$-\Delta G^{\ddagger} = RT\ln k \tag{1}$$

$$\Delta\Delta G^{\ddagger} = [\Delta G^{\ddagger}\,(26) - \Delta G^{\ddagger}\,(25)] = RT\ln(k_1/k_2) \tag{2}$$

where k_1 and k_2 are the rate constants for formation of the transition state complexes leading to (25) and (26), respectively.

For epoxide (25) to be formed 100 times faster than (26) at room temperature (300 K):

$$\Delta\Delta G^{\ddagger} = 8.314 \times 300 \times 2.303 \log_{10} 100 \text{ J mol}^{-1}$$

$$= 11.5 \text{ kJ mol}^{-1} \text{ or } 2.74 \text{ kcal mol}^{-1}$$

For most practical purposes, a 100:1 ratio of epoxide (25) to epoxide (26) would be as good as a completely diastereoselective reaction, but although 11.5 kJ mol^{-1} is a small amount of energy (the energy barrier to rotation around the C–C bond in ethane is ~12.5 kJ mol^{-1}), it might still correspond to ~10% of the total activation energy for the epoxidation.

A diastereoselectivity of higher than 100:1 would be normal for a naturally occurring enzyme-mediated reaction, but would be less common for one in which enzymes were not involved. A $\Delta\Delta G^{\ddagger}$ value of 7.3 kJ mol^{-1} (1.75 kcal mol^{-1}) would be accessible for a larger number of diastereoselective reactions. This $\Delta\Delta G^{\ddagger}$ value corresponds to a 95:5 ratio of diastereoisomers and although generalisations here abound with exceptions, this is usually a high enough ratio for such a mixture to be used in the next stage of a synthesis without purification should the latter prove troublesome. The underlying assumption here is that the unwanted diastereoisomer or its transformation products would be removed at this next or some later stage of the synthesis, most expediently by crystallisation of the major product.

EFFECT OF LOWERING TEMPERATURE ON DIASTEREOSELECTIVITY

The effect of lowering the temperature on the diastereoselectivity of a reaction can be illustrated by reference to Figure 1. At higher temperatures, the reactants will be able to surmount either the barrier leading to (25) or that leading to (26), but as the temperature is lowered, a greater proportion will be able to surmount only the barrier leading to (25).

Applying the Arrhenius equation and assuming that A is identical for the formation of (25) and (26):

$$k_1 = A \exp \left[-\Delta G^{\ddagger}_{(25)}/RT \right]$$
$$k_2 = A \exp \left[-\Delta G^{\ddagger}_{(26)}/RT \right]$$

Hence

$$k_1/k_2 = \frac{A \exp \left[-\Delta G^{\ddagger}_{(25)}/RT \right]}{A \exp \left[-\Delta G^{\ddagger}_{(26)}/RT \right]} = \exp \left[-\Delta G^{\ddagger}_{(25)} + \Delta G^{\ddagger}_{(26)} \right]/RT$$

k_1/k_2 is a measure of the diastereoselectivity and since $\Delta G^{\ddagger}_{(26)} > \Delta G^{\ddagger}_{(25)}$, the term $\exp \left[-\Delta G^{\ddagger}_{(25)} + \Delta G^{\ddagger}_{(26)} \right]/RT$ is maximised by small values of T.

In general, therefore, the likelihood of diastereoselectivity is increased as the temperature at which the reaction is carried out is lowered.

PROCHIRALITY: DIASTEREOTOPIC ATOMS OR GROUPS

Whereas the two hydrogens in the prochiral methylene group of ethanol are enantiotopic (Scheme 14), the two hydrogens in the prochiral methylene group of (27) are *diastereotopic* since replacement of each of them in turn by, for example, deuterium gives two molecules, (28) and (29), which are diastereoisomers (Scheme 18).

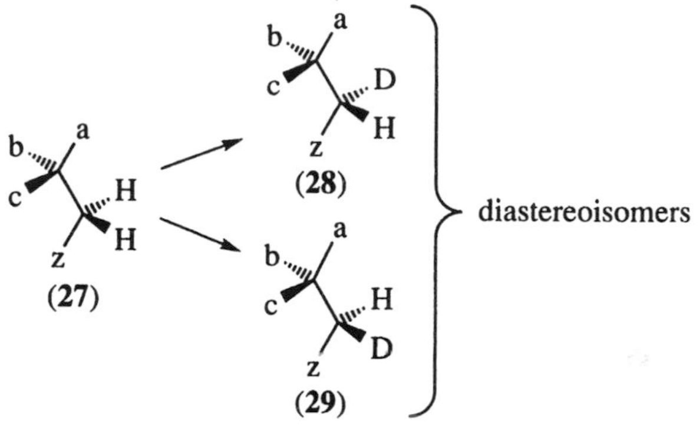

Scheme 18

Reaction of each of these diastereotopic protons by whatever reagent (including an electrophile E^+, a nucleophile Nu^- or a radical $R^·$) can, in principle, take place at different rates since the corresponding transition states will not necessarily be equal in energy. The origin of any diastereoselectivity in such a reaction would be analogous to that which might obtain in epoxidation at the two diastereofaces of the alkene (24) in Scheme 17 and a free-energy profile diagram analogous to that in Figure 1 can be drawn.

This term diastereotopic can be applied to any pair of identical substituents on an sp^3-hybridised carbon bearing two different additional substitutents at least one of which contains a chiral centre. Thus the two methyl groups in (30) and the two phenyl groups in (31) are diastereotopic.

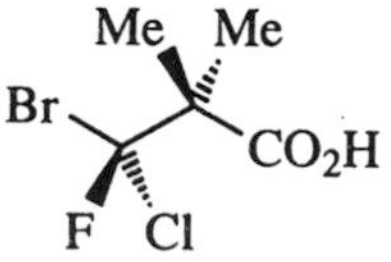

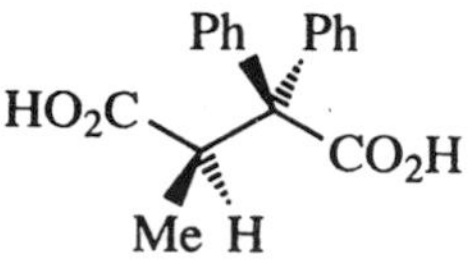

There is one particular substitution pattern which gives rise to compounds which contain diastereotopic groups or atoms but no chiral centres. We have seen previously that the diacid (22) has two enantiotopic carboxymethylene groups (Scheme 15); the two hydrogens within each methylene group, however, are diastereotopic although the molecule contains no chiral centre. Hence replacement in turn of each of the hydrogens in one of the methylenes by deuterium (cf. Scheme 18) gives two molecules which are clearly diastereoisomeric (Scheme 19). In this by no means uncommon situation

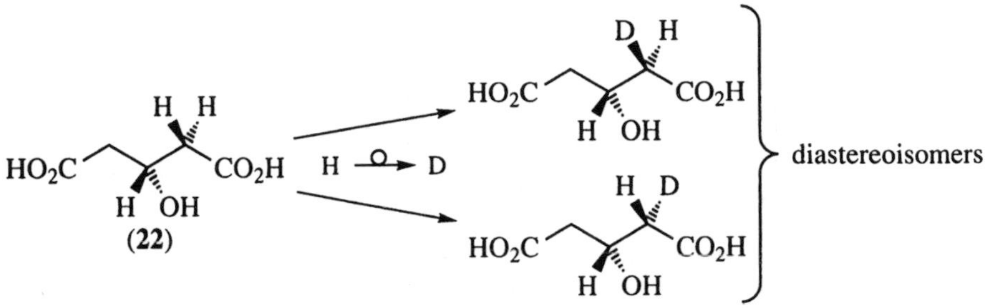

Scheme 19

a molecule contains two prochiral centres, each of which is contained within one of two enantiotopic groups: the two pairs of diastereotopic atoms or groups comprise two pairs of equivalent (homotopic) atoms or groups, e.g. $2 \times R_A$ and $2 \times R_B$ in (32).

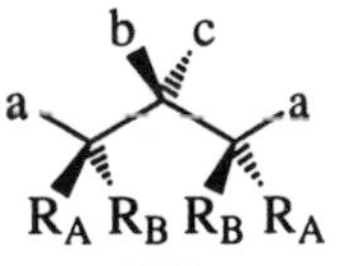

Diastereotopic atoms or groups also are not necessarily bound to the same atom; the two methyl groups in (33) are diastereotopic since their individual replacement by, for instance, ethyl groups (Scheme 20) gives products having a diastereoisomeric relationship.

We have illustrated the concepts of diastereofaces and diastereotopic groups with reference to compounds in which a *chiral centre* is *adjacent* to the prochiral double bond or prochiral centre (Schemes 17 and 18). However, any chiral element, e.g. a chiral axis, could play the same role as the chiral centre. Furthermore, in principle, any number of atoms may separate the chiral element from the prochiral double bond or prochiral centre.

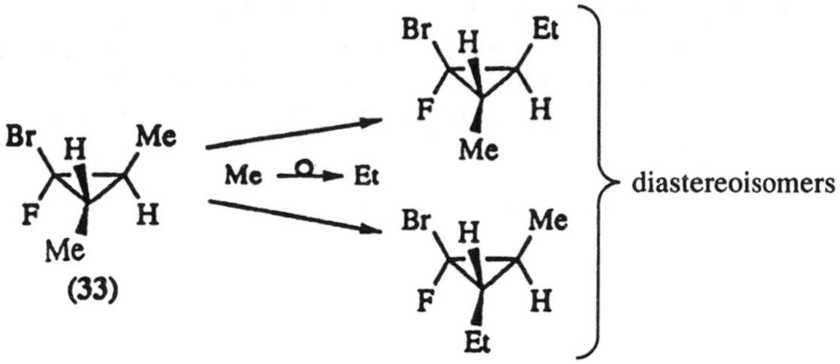

Scheme 20

HOMOTOPIC FACES AND GROUPS

A double bond which is not prochiral in a reaction has homotopic faces: the products obtained from reaction at either face are identical.

This term can also be applied to particular faces of different double bonds within the same molecule. Thus, the arrowed faces of diene (34) are homotopic since, for example, the same product (a single enantiomer) is obtained from Michael addition to either of them (Scheme 21).

Similarly, homotopic groups are those whose individual replacement by, say, deuterium gives identical products. Thus the three hydrogen atoms of

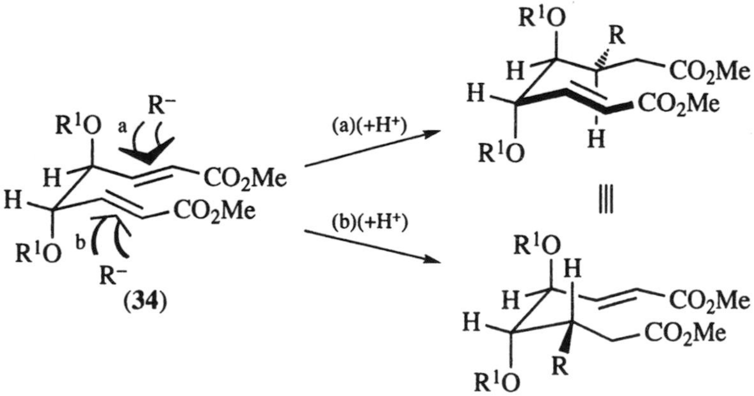

Scheme 21

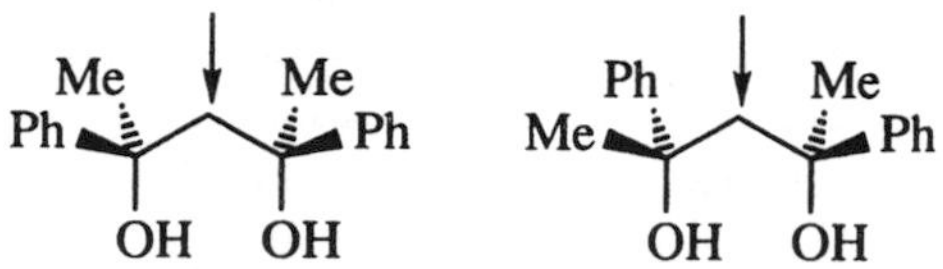

the methyl group of ethanol are homotopic. In propane, the two methyl groups are homotopic, the six hydrogens within these methyl groups are homotopic and the two methylene protons are also homotopic; the two arrowed diastereotopic protons in (35) become homotopic in (36).

REACTION OF DIASTEREOFACES: ASYMMETRIC INDUCTION

It has been pointed out above that, in principle, an achiral reagent can distinguish between two diastereofaces of a double bond. If epoxidation of the prochiral double bond in (24) (Scheme 22) proceeds to give an excess of (25) over (26) (or vice versa), the reaction is said to proceed diastereoselectively. It may also be said to be proceeding with *asymmetric induction*: the existing chiral centre in (24)—the *parent* chiral centre—*induces* the preferential formation of one of the two possible configurations for the *newly created* chiral centre.

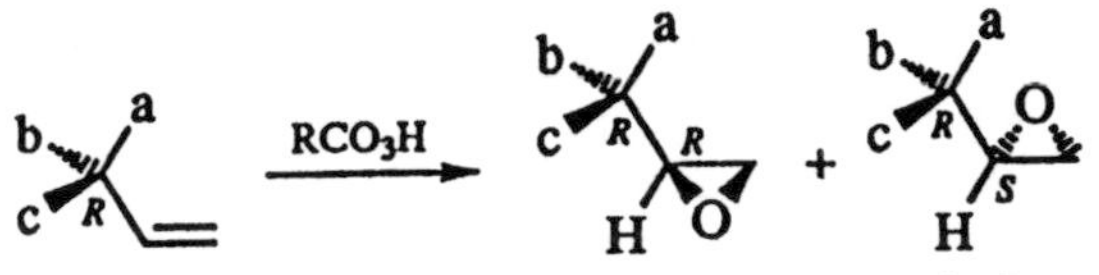

Scheme 22

Suppose the parent chiral centre in (24) is R and the induced configuration in (25) is R, then that in (26) is S. A *completely* diastereoselective epoxidation of (24) (Scheme 22) would give either the single enantiopure diastereoisomer (25) or the single enantiopure diastereoisomer (26). Since we define diastereoselectivity without reference to the enantiomeric purity of the starting material, epoxidation of a mixture of (24) and its enantiomer to give either racemic (25) or racemic (26) would also be completely diastereoselective.

The *sense* of asymmetric induction in Scheme 22 is a description of the relationship between the configuration of the parent and the newly created chiral centres. Thus the sense of induction in Scheme 22 leading from (R)-(24) to (R,R)-(25) is opposite to that leading from (R)-(24) to (R,S)-(26). However, since the sense of induction leading from (R)-(24) to (R,R)-(25) is the same as that leading from (S)-(24) to (S,S)-(25), the application of this concept of sense of induction to the epoxidation of *racemic* (24) is shown in Scheme 23. The formation of an excess of (25) over (26) (or vice versa) would constitute a case of 1,2-asymmetric induction where the inducing chiral centre is assigned the 1-position and the newly created chiral centre is numbered accordingly.

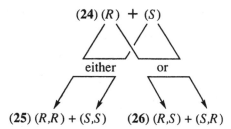

Scheme 23. A completely diastereoselective epoxidation of (**24**) and its enantiomer showing opposite senses of 1,2-asymmetric induction.

The means by which the chiral centre in (**24**) can direct epoxidation in a diastereoselective epoxidation to one face rather than the other will be discussed later but, for the present, it is important to recognise that a prochiral double bond has diastereofaces even when separated from the chiral centre (or chiral element) by any number (n) of atoms. In principle, $1,(n+2)$ asymmetric induction is possible in such compounds, although intuitively we would expect this to become less likely at larger values of n in an acyclic substrate.

REACTION OF ENANTIOFACES: INTERMOLECULAR ASYMMETRIC INDUCTION

Enantiofaces react at the same rate with achiral reagents since the two transition state energies involved are identical (see above). For one enantioface to react in preference to the other they must be converted into diastereofaces. One way in which this can be accomplished is by attachment of a chiral auxiliary (see below). However, it is not mandatory that the two faces of the alkene are diastereofaces from the outset of the reaction. Figure 2 represents the energy profile diagram for regiospecific *syn*-addition of an enantiopure reagent x–y* (with y* as the chiral element) to a prochiral but achiral alkene (**37**).

It is clear that the two addition products (**38**) and (**39**) are diastereoisomers since they have the same configuration for y* but different configurations for the created chiral centres. It is also clear that, as in the epoxidation in Figure 1, asymmetric induction takes place *in the transition state for the addition*: the two transition states are diastereoisomeric or, in other words, the two enantiofaces of the alkene (**37**) become diastereofaces in the transition state. If these diastereoisomeric transition states differ sufficiently in energy, then (**38**) will be produced exclusively with complete asymmetric induction [*exclusive* formation of one of two stereoisomers requires $\Delta\Delta G^{\ddagger} = \infty$; for practical purposes, we shall consider $\geqslant 99\%$ (e.e. or d.e.) as 'exclusive'].

An example of the type of induction shown in Figure 2 is the hydroboration of alkene (**41**) by dialkylborane (**40**) (Scheme 24).[10] In this

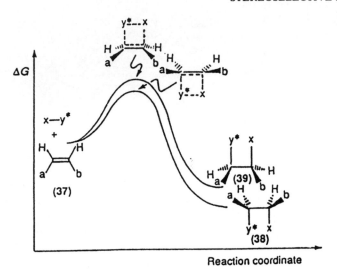

Figure 2

case, asymmetric induction is brought about by the enantiopure isopinocampheyl substituents on the boron atom. The absolute configuration of the newly created chiral centre in the borane (**42**) was determined by exchange of the boron by OH with retention of configuration at carbon (see Chapter 4), in the usual way, to yield the *R* enantiomer of the alcohol in 60% enantiomeric excess.

Hence, asymmetric induction in addition reactions to double bonds does not require the inducing (parent) chiral centre to be present within the same molecule.

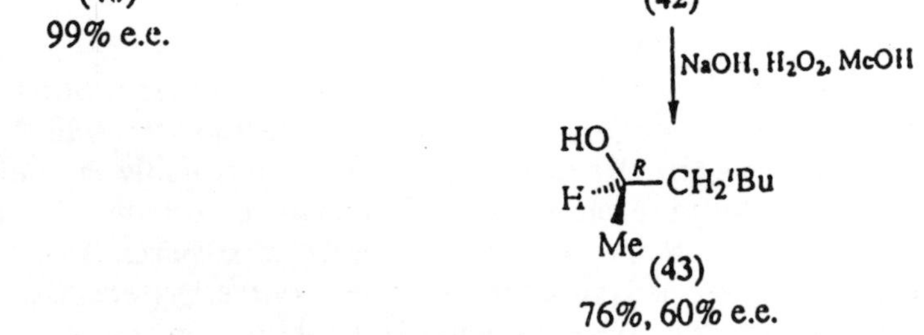

Scheme 24

There is, phenomenologically, little difference between induction brought about by a parent chiral centre which is fully covalently bonded to the developing chiral centre (cf. Figure 1) and that brought about by the chiral centre in the enantiopure reagent x–y* in Figure 2 where one of the bonds separating this parent chiral centre from the developing chiral centre is only partially formed in the transition state.

It is but a small step to see that the chiral element only needs to be *associated with the prochiral alkene in the transition state* to render the faces diastereotopic and to bring about asymmetric induction; *it need not be retained in the product at all.* For example, the chiral element could be contained in the solvent or in a catalyst. The energy profile diagram for addition of an achiral reagent x–y to a prochiral alkene in the presence of an enantiopure solvent S* would be represented by Figure 3.

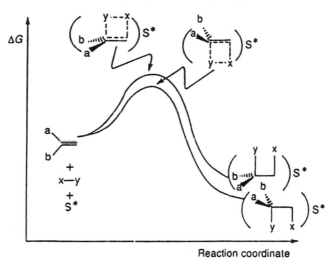

Figure 3

In Figure 3, the involvement of one or more enantiopure solvent molecules in two otherwise enantiomeric transition states renders these diastereoisomeric. Photolysis of acetophenone in the enantiopure solvent in Scheme 25 gives a mixture of diastereoisomeric 1,2-diols (pinacols) in which **(44)** is enantioenriched (8.3% e.e.).[11] The two radical pairs, through which dimerisation of the first-formed ketyl radical **(45)** to give **(44)** proceeds, become diastereoisomeric in their association with the solvent and radical dimerisation is (slightly) preferred via one of them.

Reactions involving high levels of asymmetric induction which proceed via a single transition state with or without retention of the chiral element in the product (Figures 2 and 3) are fewer in number than those where there is involvement of one or more (usually non-isolable) intermediates. Thus in the

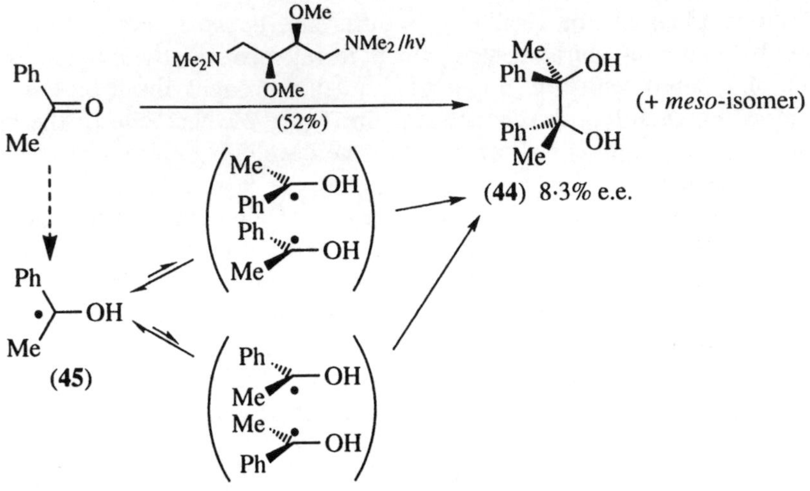

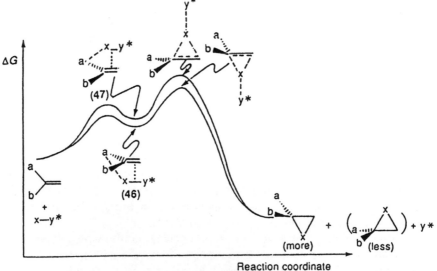

Scheme 25

profile shown in Figure 4, complete or partial bonding between the reagent x–y* and both substituent a and one or other face of the prochiral alkene occurs leading to intermediates (**46**) and (**47**) from which elimination of y* can occur in a subsequent product-forming step. As shown in Figure 4, the rates of formation of (**46**) and (**47**), which are themselves diastereoisomers, may differ since diastereoisomeric transition states are also involved in *their* formation (which may or may not be reversible). Even if (**46**) and (**47**) become identical (complete or partial bonding to substituent *a* only) the

Figure 4

starting alkene now has diastereofaces at which (intramolecular) reaction is more likely to take place at different rates than direct (intermolecular) reaction between the alkene and x–y*.

ASYMMETRIC INDUCTION IN REACTION OF ENANTIOTOPIC ATOMS OR GROUPS

Energy profile diagrams analogous to those above for attack on one enantioface by a chiral enantiopure reagent (Figures 2 and 4) can be drawn to show selective attack on one of two enantiotopic atoms or groups.

Thus the profile in Figure 5, showing selective functionalisation of one of the enantiotopic groups b using x–y*, is analogous to that in Figure 4. In both reactions, the chiral element y* in the reagent is not found in the product.

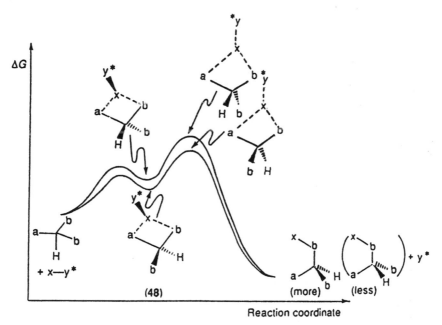

Figure 5

Figures 4 and 5 resemble closely the energy profile diagrams which illustrate the way in which many enzymes function. Enzymes are chiral catalysts of the greatest importance for regio- and stereoselective reactions in living cells and, increasingly, for reactions in the laboratory (see Chapter 16). For example, the conversion of glycerol into its monophosphate enantiomer shown (Scheme 26) is an important metabolic process catalysed by glycerol kinase in the presence of adenosine triphosphate (ATP).

Thus the two enantiotopic hydroxymethylene groups of glycerol are

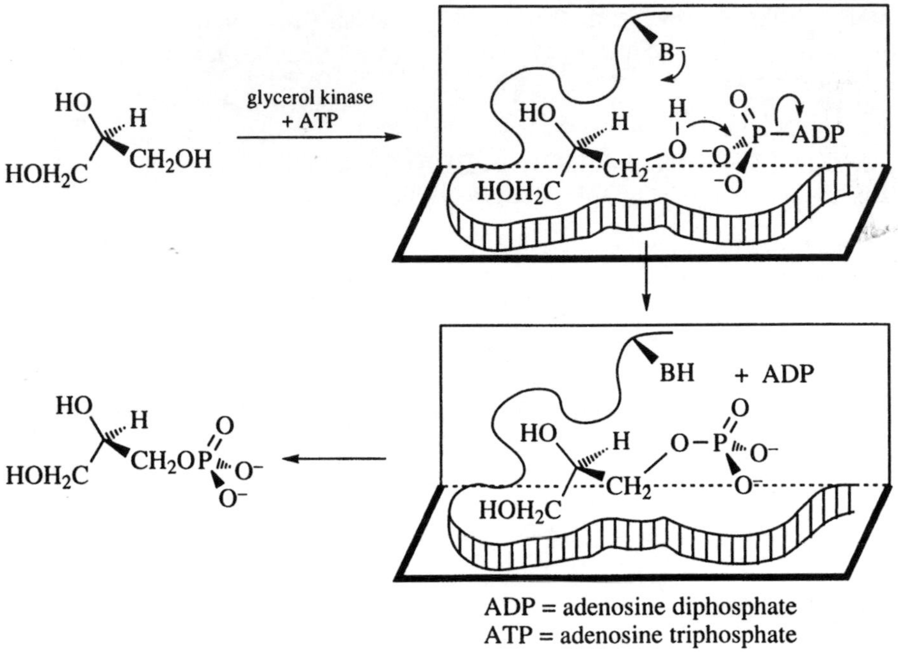

ADP = adenosine diphosphate
ATP = adenosine triphosphate

Scheme 26

distinguished (become diastereotopic) when bound to the enzyme surface [cf. (**48**) in Figure 5] and one of them (the *pro-R*) is positioned such that it can rob ATP (also bound to the enzyme surface) of one of its phosphoryl groups in the product-forming step.

DISTINGUISHING BETWEEN ENANTIOMERS

Distinguishing between two enantiofaces or two enantiotopic atoms or groups in a reaction requires the presence of a chiral element in the reagent or reaction ensemble (see above). Similarly, enantiomers can only be distinguished chemically by reaction with a chiral reagent. Complete *kinetic resolution* is the reaction of only one enantiomer of a racemate leaving the unreacted enantiomer in enantiopure form (see Chapter 15).

Diastereoisomers, on the other hand, can be distinguished by achiral reagents. A good example is the chromatographic separation of a mixture of diastereoisomers which can be accomplished as a result of the unequal energies of interaction of the individual diastereoisomers with the achiral chromatographic absorbent. Chromatographic separation of a racemate requires a chiral absorbent which must be used in enantiopure form.

CHEMZYMES

We have seen that stereoselective reactions are of two types depending upon whether the parent chiral element(s) in the reaction ensemble are retained in the product (Figures 1 and 2) or not (Figures 3–5).

Of course, the chiral elements of the enantiopure reagent which are found in the product (Figure 2) in diastereoselective reactions of the first kind can always be removed in a subsequent reaction to release the newly created chiral centre as in the hydroboration–oxidation procedure in Scheme 24. Alternatively, it may be that all of the chiral elements of the enantiopure reagent are required in the product and both the reagent and the stereoselective reaction involved were selected with this in mind.

Highly enantioselective but non-enzymic reactions of the second type, in which the parent chiral elements in the reaction ensemble are not found in the product, are especially valuable but still relatively few in number and type (see Chapters 12 and 14). The Sharpless–Katsuki epoxidation reaction[12] converts a prochiral allylic alcohol into the corresponding epoxide with high enantioselectivity using the cocktail shown (Scheme 27).

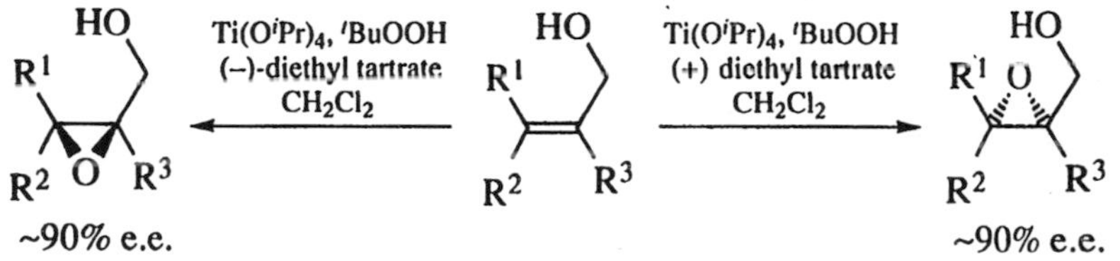

Scheme 27

Thus, which enantiomeric epoxide is formed in this reaction is determined by which enantiomer of diethyl tartrate is selected. A fuller discussion of this reaction is given in Chapter 14, but for the present it should be noted that the energy profile diagram for the reaction is most probably of the 'enzyme' type shown in Figure 4: the enantiotopic faces of the allylic alcohol are rendered diastereotopic in a complex which is believed to include two molecules each of diethyl tartrate and titanium isopropoxide and one each of allylic alcohol and tert-butyl hydroperoxide.[13] Intramolecular delivery of the oxygen from the hydroperoxide takes place in a second product-forming step.

The term 'chemzyme' has been suggested for reagents which are non-enzymic but bring about high levels of enantioselectivity approaching those which are characteristic of enzyme-mediated reactions.[14]

Enzymes are almost invariably completely enantioselective in the transformations they bring about on their natural substrates; although chemzymic reactions are not so completely enantioselective they are invariably more catholic in the variety of substrates to which they can be applied.

CHIRAL AUXILIARIES

Any attempt to alkylate the amide (49) (Scheme 28) with an achiral alkylating agent R^2—X will inevitably lead to a racemate since the two faces of the enolate (50) are enantiotopic. It cannot be assumed that enolate (50) will be the only one formed by removal of a proton from (49), but if any enolate (51) is formed it will also undergo alkylation from both enantiofaces with equal facility. To distinguish between the two enantiofaces of enolate (50), their conversion to diastereofaces is required.

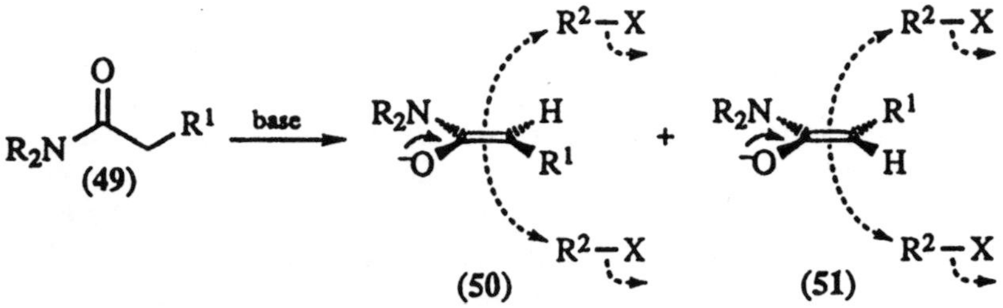

Scheme 28

Reaction of the acid chloride (52) with the lithium salt of enantiopure oxazolidone (53) (Scheme 29) gives the amide derivative (54). Proton removal from (54) using lithium diisopropylamide (LDA) gives only the (Z)-enolate (55), in which the lithium is believed to be coordinated to two oxygen atoms.[15]

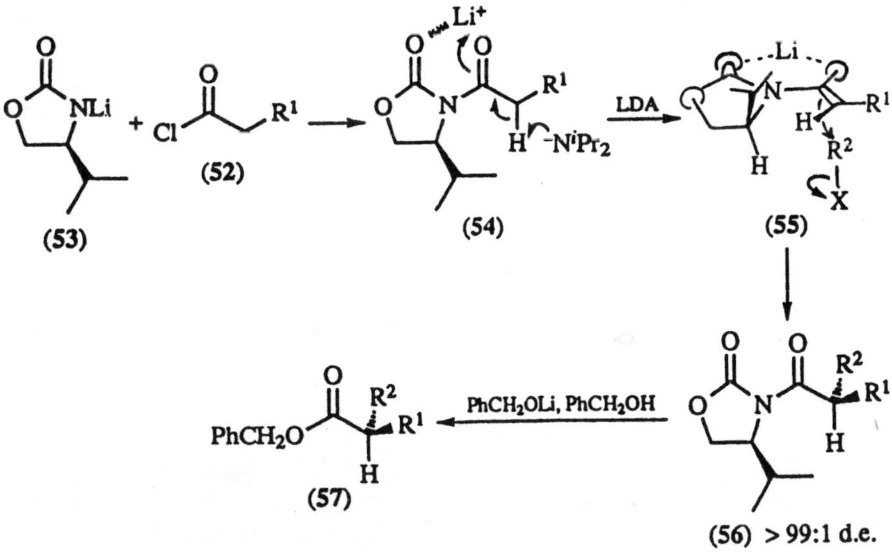

Scheme 29

The two faces of this enolate (55) are now diastereotopic and alkylation takes place from the less hindered face as shown to give (56) with high diastereoselectivity. Consequently, cleavage of the oxazolidone with benzyl alcoholate liberates the ester (57), which is obtained in high enantiomeric excess. The *chiral auxiliary* is that part of the derivative [the Evans' oxazolidone in (54)] which arises from the enantiopure reagent employed.

The most effective chiral auxiliaries, of course, are those which direct attack of the achiral reagent exclusively on to one diastereoface. Usually this is brought about by a steric effect arising from the shape of the auxiliary which screens one diastereoface and exposes the other, as is the case in attack on enolate (55).

Auxiliaries which are popular are those with C_{2v} symmetry.[16] In Scheme 30, the bottom face of the enolate (58) is screened by the 2-substituent on the pyrrolidine. The C_{2v} symmetry of this auxiliary means that rotation around the C—N bond through 180° as shown reproduces the same screening of this bottom face (by the 5-substituent).[17]

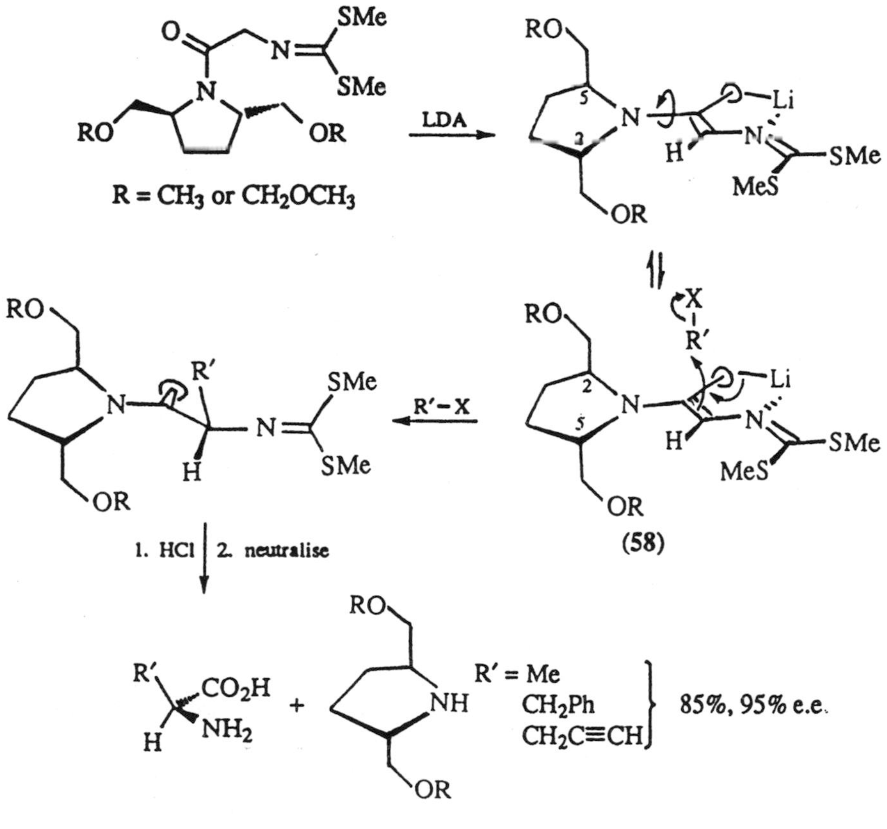

Scheme 30

In some cases, the origin of the diastereoselection may be apparently electronic in origin as in the alkylation of lithium enolate (59) bearing the sultam (60) (the Oppolzer chiral auxiliary) (Scheme 31).[18] The lone pair of electrons of the pyramidal nitrogen atom on the enolate (59) is believed to direct attack of the alkylating agent to the underside of the double bond.

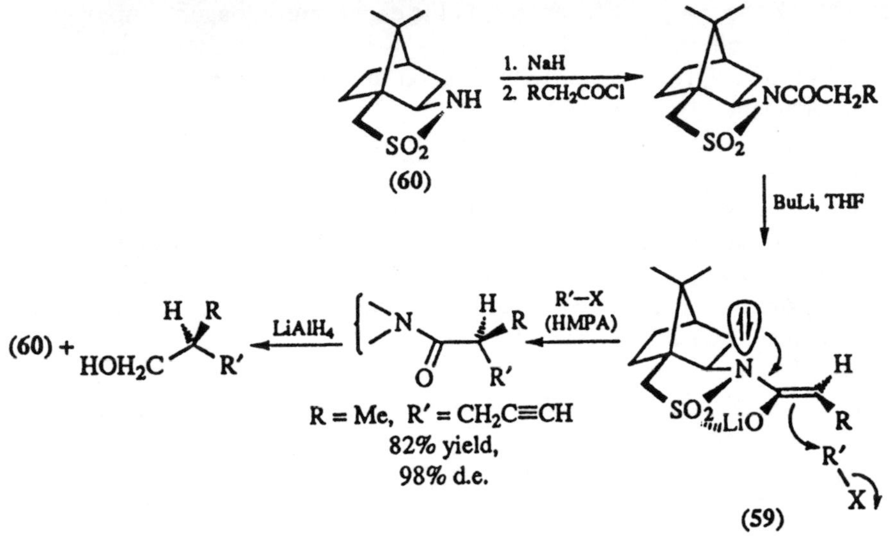

Scheme 31

However, since the configuration of this nitrogen pyramid is dictated by the nature of the bicyclic system to which is is attached, it is from the chiral framework of the latter that the facial discrimination in the enolate originates, with the nitrogen pyramid acting as a relay.

Ideally, chiral auxiliaries should be cheap and readily available in both pure enantiomeric forms. Most of them are derived from naturally occurring 'chiral pool' materials (see Chapter 3) including terpenes [(60), from camphor] and amino acids [(53) from valine]. Menthol (61) is a readily available terpenoid which has been used as an auxiliary on the C–5 position of unsaturated γ-lactone (62) (Scheme 32) such that Michael attack *anti* to oxygen at C–5 followed by reductive cleavage gives an enantiopure 2-substituted 1,4-diol.[19]

It is also desirable, particularly in large-scale work, that methods for removal of the chiral auxiliary should be non-destructive so that it can be recovered and recycled.

An auxiliary not derived directly from chiral pool materials, which has been developed by Davies and co-workers,[20] is the air-stable iron complex (63). This compound has been shown to have the conformation shown from an X-ray crystal structure determination and it is believed to retain this conformation in solution. It has also been obtained in both enantiopure forms.

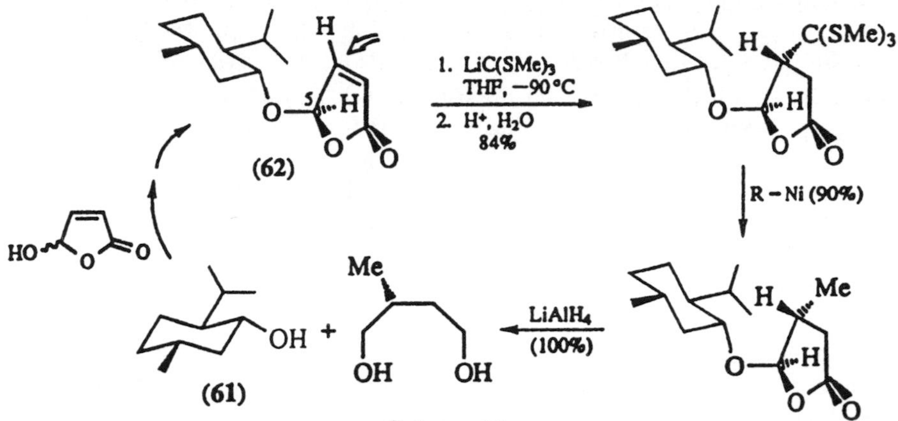

Scheme 32

Scheme 33 exemplifies the use of (**63**) in the synthesis of (S,S)-captopril (**65**), an anti-hypertensive drug.[20] The enolate configuration and accessible diastereoface of (**64**) mean that the newly created chiral centre has the configuration shown. Bromine brings about cleavage of the iron–carbonyl bond and allows recovery of the auxiliary.

Although the use of chiral auxiliaries requires more labour and lacks the directness of chemzyme- and enzyme-mediated conversions, it can often lead

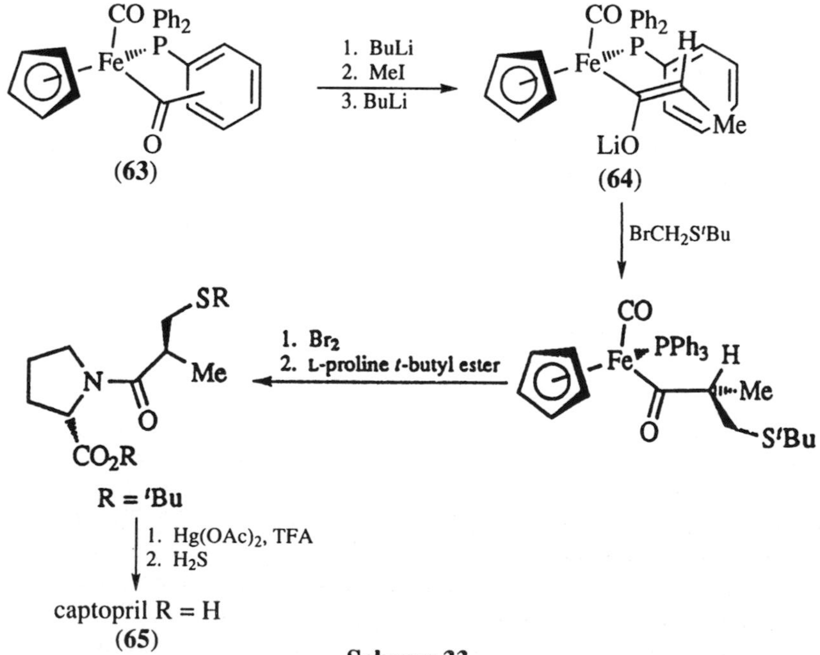

Scheme 33

to enhanced enantiopurity in the isolated (auxiliary-freed) enantiomer. This is because the diastereoisomeric products which arise from attack on the two diastereofaces of the auxiliary-linked substrate are isolable and can, in principle, be purified, most expediently by crystallisation. Cleavage of the auxiliary from the purified diastereoisomer will then yield a single enantiomer of the product.

Chiral auxiliaries are more often used as above to direct reaction of an achiral reagent on to one diastereoface, but they may also be used in a similar way to allow selective reaction of one of a pair of enantiotopic atoms or groups [see Scheme 14, Chapter 16].

There are obvious similarities between the mode of action of chiral auxiliaries on the one hand and enzymes/chemzymes on the other. Both involve the conversion of enantiotopic faces, atoms or groups within a molecule into diastereotopic faces, atoms or groups by the temporary attachment of an enantiopure unit. The advantage with the use of enzymes/chemzymes is that attachment of the chiral unit and its uncoupling take place spontaneously in the enantioselective conversion.

The dearth of readily available enzymes and chemzymes for many routine functional group interconversions means that chiral auxiliaries will continue to be useful for the forseeable future when enantioselective versions of these interconversions are required. Just as the widespread use of protecting groups is a measure of our inability to exercise chemoselective control, so the similar use made of chiral auxiliaries is an indication of the dearth of methods which are available for exercising complete enantioselective control in a single reaction.

IDENTIFICATION OF DIASTEREOTOPIC AND ENANTIOTOPIC ATOMS OR GROUPS BY N.M.R. SPECTROSCOPY

In the n.m.r. spectra of two diastereoisomeric compounds, none of the corresponding nuclei are, in principle, magnetically equivalent. Likewise, diastereotopic atoms or groups within a single molecule are intrinsically non-equivalent and, in principle, distinguishable by n.m.r. spectroscopy, always assuming that there are nuclei in these atoms or groups which have magnetic moments are are, therefore, n.m.r. 'visible.'

In practice, with modern high-field n.m.r. spectrometers, diastereoisomers can ususally be distinguished by the non-identical signals from at least some of the corresponding nuclei unless the two chiral centres are remote from each other (say separated by more than four atoms in an acyclic system).

Diastereotopic nuclei in a single molecule which are most likely, in practice, to be identifiably non-equivalent are those which are as close as possible to the chiral element and ideally adjacent to it. Whether other diastereotopic nuclei display non-equivalence depends on a number of factors besides their distance from the chiral element including; the connectivity between them and the chiral element, the nature of the nuclei and, of course, the resolving power of the instrument.

The chiral centre in (66) means that the protons in the adjacent methylene group are diastereotopic. The non-equivalence of the CH_2 protons is evident from inspection of the three staggered Newman projections (66a), (66b) and (66c) for this molecule. It is clear that the environments of H_A and H_B are different in all these three conformations (and in others where staggering is absent). The intrinsic non-equivalence of H_A and H_B cannot, therefore, be removed by rapid rotation around the C—C bond indicated in (66) (or around any other bond).

Enantiomers are not distinguishable directly by n.m.r. spectroscopy in achiral solvents. For such a distinction to be made and, in particular, to allow

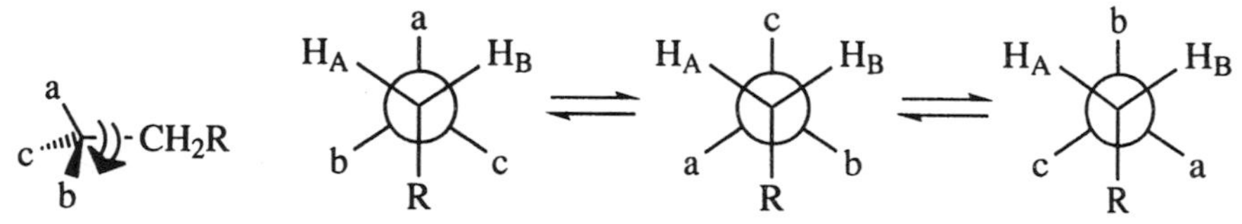

the excess of one enantiomer over another (enantiomeric excess) to be quantified, the enantiomeric relationship must be converted into a diastereoisomeric one. This can be accomplished by derivatisation using an enantiopure reagent, by using an enantiopure shift reagent or by measuring the spectrum in an enantiopure solvent. One of the most commonly used derivatives of chiral secondary alcohols or amines is the corresponding ester (67) or amide (68) (Scheme 34) with α-methoxy-α-trifluoromethylphenylacetic acid (MTPA) (Mosher's acid).[21]

Although the ratio of, e.g., diastereoisomeric MTPA esters (67) can be determined using [1]H n.m.r., it is advantageous to use the singlet ^{19}F signals of

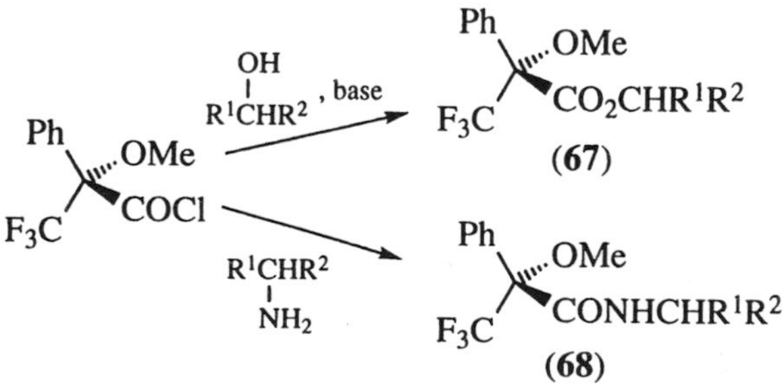

Scheme 34

the trifluoromethyl groups since these are likely to be more clearly separated and less likely to overlap with other signals than comparable signals in the ^{1}H spectrum. Similar advantages derive from the use of phosphorus-containing derivatising agents and examination of the diastereoisomers formed by ^{31}P n.m.r.

An interesting alternative to the use of an enantiopure derivatising agent is the use of, e.g., the achiral O,O-diphenylphosphorodithioic acid (69) (1 equiv.) (Scheme 35) and the chiral alcohol (2 equiv.). For example, in CDCl$_3$ solution at 20 °C, racemic menthol (70) gives a mixture of *meso*- (72) and (±)-derivatives (71) in equal amounts.[22]

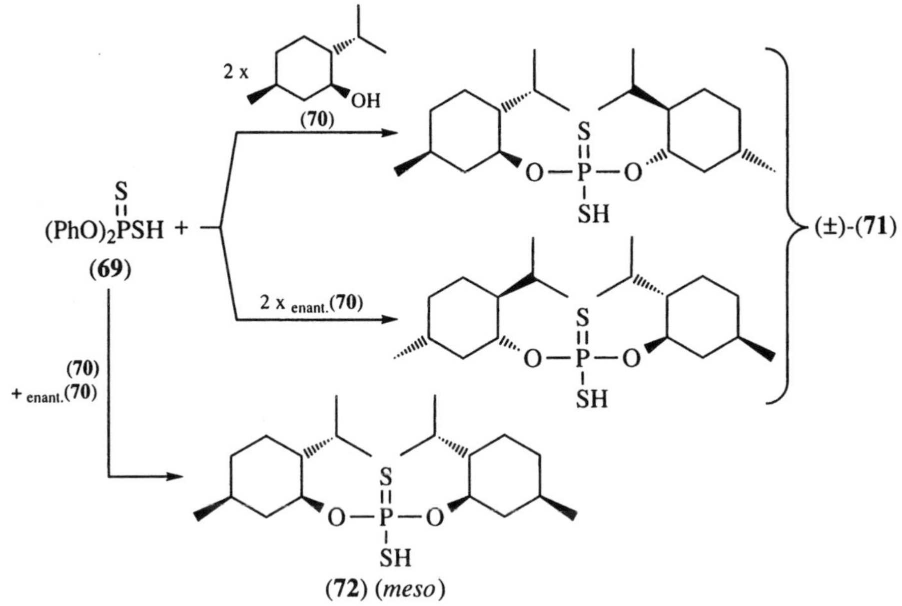

Scheme 35

The two enantiomers of (71) have the same ^{31}P chemical shift, which, however, is different from that of the *meso*-isomer (72). If the menthol (70) is enantiopure, the ^{31}P signal for the *meso*-isomer will be absent.

In this alternative, the alcohol is, in effect, used as its own derivatising agent: the two diastereoisomers which are formed (cf. the normal derivatisation in Scheme 34) are the *meso*- and (±)-forms.

Note that in (71) and (72), the tetrahedral phosphorus is not a configurationally defined chiral centre since tautomerism by exchange of the thiol hydrogen between the two sulphur atoms is rapid.

Chiral shift reagents often used are europium, praseodymium, and other rare-earth complexes of camphor derivatives, e.g. (73).[23] These (enantiopure) chiral reagents form diastereoisomeric complexes with individual

(73)

(74)

Ar = phenyl, α-naphthyl, 9-anthracenyl

(facam)₃M; R = CF₃
(hfc)₃M; R = C₃H₇ } M = Eu, Pr, etc.

Eu(dcm)₃   dcm =

(75)

enantiomers which are fluxional and the signals observed in the n.m.r. spectra are time-averaged ones between those in complexed and uncomplexed enantiomers. Similarly, the use of (enantiopure) chiral solvents, e.g. **(74)**, involves fluxional complexes between the solvent and the individual enantiomers which establishes the distinguishing diastereoisomeric relationship between them.[24]

Enantiotopic atoms or groups may be distinguished by n.m.r. using the same means as those for distinguishing between enantiomers. Tris(*d,d*-dicampholylmethanato)europium(III) **(75)** is a chiral shift reagent recomended for this purpose.[23]

THE IMPORTANCE OF PURE DIASTEREOISOMERS AND ENANTIOMERS

The synthesis of any product capable of existing as diastereoisomers will ideally require the synthesis of that product as a single diastereoisomer. In general, the intermediates in such a synthesis will also be required as single diastereoisomers.

Because of their similarity in structure, diastereoisomers often have very similar chemical and physical properties. Consequently, separation of diastereoisomer mixtures will usually be more difficult than separation of mixtures of constitutional isomers. It is for this reason that a synthesis which leads to a single diastereoisomer and reactions which are completely diastereoselective are most valuable.

Enantiopure materials have long been important for the synthesis of natural products since the latter almost always are found in nature as single enantiomers. However, the importance of methods which lead to a (diastereoisomerically pure) product as a single *enantiomer* as opposed to a racemate is increasing rapidly. Many drugs and medicines have active components whose molecules are chiral. It seems likely that in many countries in the near future, legislation will be enacted which will require these components to be used clinically as single enantiomers. This is because in many cases these drugs, in their mode of action, interact with enzyme systems and receptors and very likely one enantiomer of the drug will be more active than the other. In some cases one enantiomer may be completely inactive; in others this enantiomer may be active in another deleterious sense, leading to undesirable side-effects of the drug.[25]

Pheromones are organic compounds, or mixtures of compounds, by which many animals and especially insects communicate with one another. They are usually low molecular mass and volatile materials whose synthesis is of importance for their use in lures and traps as a method of insect pest control. Again, where chiral centre(s) are present in the pheromone, it is invariably only one enantiomer that has maximal biological activity. Sometimes the presence of small amounts ($\sim$1%) of the other enantiomer can negate the effect of the otherwise active enantiomer.

Herbicides, insecticides and fungicides also act in many cases via their interference with enzyme systems of plants, insects and fungi. It seems likely that here also these biocides, when used in chiral form, will eventually be used as single enantiomers and thus, by elimination of the redundant enantiomer, lead to a reduction in the possibility of environmental damage.[26]

It is clear that the devising and development of methods for the synthesis of single stereoisomers is, and will continue to be in the years ahead, one of the major challenges facing both commercial and academic research workers.

References

1. M. Majewski, D.L.J. Clive and P.C. Anderson, *Tetrahedron Lett.*, 1984, **25**, 2101.
2. A.I. Meyers and J.P. Lawson, *Tetrahedron Lett.*, 1982, **23**, 4883.
3. H.E. Zimmerman, L. Singer, and B.S. Thyagarajan, *J. Am. Chem. Soc.*, 1959, **81**, 108: E.L. Eliel, in *Stereochemistry of Carbon Compounds*, McGraw-Hill, New York, 1962, p. 436.
4. See, however, H.L. Goering and V.D. Singleton, *J. Org. Chem.*, 1983, **48**, 1531.
5. K. Mislow and J. Siegel, *J. Am. Chem. Soc.*, 1984, **106**, 3319.
6. IUPAC, *Pure Appl. Chem.*, 1976, **45**, 11.
7. R.S. Cahn, C.K. Ingold and V. Prelog, *Angew. Chem., Int. Ed. Engl.*, 1966, **5**, 385.
8. E.L. Eliel, *J. Chem. Educ.*, 1971, **48**, 163.
9. D.R. Boyd, R. Spratt and D.M. Jerina, *J. Chem. Soc.C*, 1969, 2650.
10. H.C. Brown, M.C. Desai and P.K. Jadhav, *J. Org. Chem.*, 1982, **47**, 5065.
11. D. Seebach, H.A. Oei and H. Daum, *Chem. Ber.*, 1977, **110**, 2316; D. Seebach, H.-O. Kalinowski, W. Langer, G. Crass and E.-M. Wilka, *Org. Synth.*, 1983, **61**, 24, and references cited therein.
12. T. Katsuki and K.B. Sharpless, *J. Am. Chem. Soc.*, 1980, **102**, 5974; M.G. Finn and K.B. Sharpless, *Asymmetric Synthesis*, Academic Press, New York, 1985, Vol. 5, pp. 247–308.
13. M.G. Finn and K.B. Sharpless, *J. Am. Chem. Soc.*, 1991, **113**, 113.
14. E.J. Corey, C.-P. Chen and G.A. Reichard, *Tetrahedron Lett.*, 1989, **30**, 5547.
15. D.A. Evans, M.D. Ennis and D.J. Mathre, *J. Am. Chem. Soc.*, 1982, **104**, 1737.

16. Review: J.K. Whitesell, *Chem. Rev.*, 1989, **89**, 1581.
17. S. Ikegami, H. Uchiyama, T. Hayama, T. Kasuki and M. Yamaguchi, *Tetrahedron*, 1988, **44**, 5333.
18. W. Oppolzer, R. Moretti and S. Thomi, *Tetrahedron Lett.*, 1989, **30**, 5603.
19. J.F.G.A. Jansen and B.L. Feringa, *Tetrahedron Lett.*, 1989, **30**, 5481.
20. G. Bashiardes and S.G. Davies, *Tetrahedron Lett.*, 1987, **28**, 5563; S.G. Davies, *Chem. Br.*, 1989, **25**, 268.
21. J.A. Dale, D.L. Dull, and H.S. Mosher, *J. Org. Chem.*, 1969, **34**, 2543; J.A. Dale and H.S. Mosher, *J. Am. Chem. Soc.*, 1973, **95**, 512; G.R. Sullivan, J.A. Dale and H.S. Mosher, *J. Org. Chem.*, 1973, **38**, 2143.
22. B. Feringa, *Chem. Commun.*, 1987, 695.
23. H.L. Goering, J.N. Eikenberry, G.S. Koermer and C.J. Lattimer, *J. Am. Chem. Soc.*, 1974, **96**, 1493; M.D. McCreary, D.W. Lewis, D.L. Wernick and G.M. Whitesides, *J. Am. Chem. Soc.*, 1974, **96**, 1038.
24. W.H. Pirkle, R.L. Muntz and I.C. Paul, *J. Am. Chem. Soc.*, 1971, **93**, 2817; W.H. Pirkle, T.G. Burlingame and S.D. Beare, *Tetrahedron Lett.*, 1968, 5849.
25. A.N. Collins, G.N. Sheldrake and J. Crosby, *Chirality in Industry*, Wiley, Chichester, 1992.
26. G.M. Ramos Tombo and D. Bellus *Angew. Chem., Int. Ed. Engl.*, 1991, **30**, 1193.

2 CLASSIFICATION OF STEREOCHEMICAL REACTIONS USED IN THIS BOOK

Stereoselective synthesis of molecules having two chiral centres

Molecules (1), (2) and (3) represent three common ways in which two chiral centres may be linked together. The connecting chains p, q and r may be assumed to consist of any number and variety of atoms, including zero in which case the two chiral centres are directly bonded together.

(1)   (2)   (3)

A problem in the synthesis of (1), (2) or (3) is, of course, to ensure that the two chiral centres have the correct *relative configuration* if the molecule is racemic, or the correct *absolute configuration* if the target molecule is enantiopure. Whether the target molecule is required in enantiopure form or not can profoundly influence the route which is devised for its synthesis.

In particular, therefore, it is syntheses which, for the case of, e.g., (1), maximise the yield and minimise the yield of its diastereoisomer (1′), which will be of greatest value. A reaction in which comparable yields of (1) and (1′) are produced is not necessarily of little value but the isolation of (1) from the mixture will often pose a separation problem which is circumvented by using a reaction which produces (1) completely stereoselectively.

(1)   (1′)

There are four different methods which can be applied to the synthesis of, e.g., (1), which seek to minimise the amount of (1′) produced.

METHOD A

The first method (A) is to find a starting material, or starting materials, which have chiral centres in (1) already in place. By 'in place' we mean that, ideally, the two chiral centres are separated by a chain of the appropriate length and composition. Some functional group manipulation within this connecting chain then brings about its identity to p in the product (1).

For example, suppose the required molecule was the enantiopure tetraol (4). Mannitol (5) (Scheme 1) could serve as a suitable starting material if the chiral centres at positions 3 and 4 could be transformed into CH_2 groups. This has been accomplished as shown.[1]

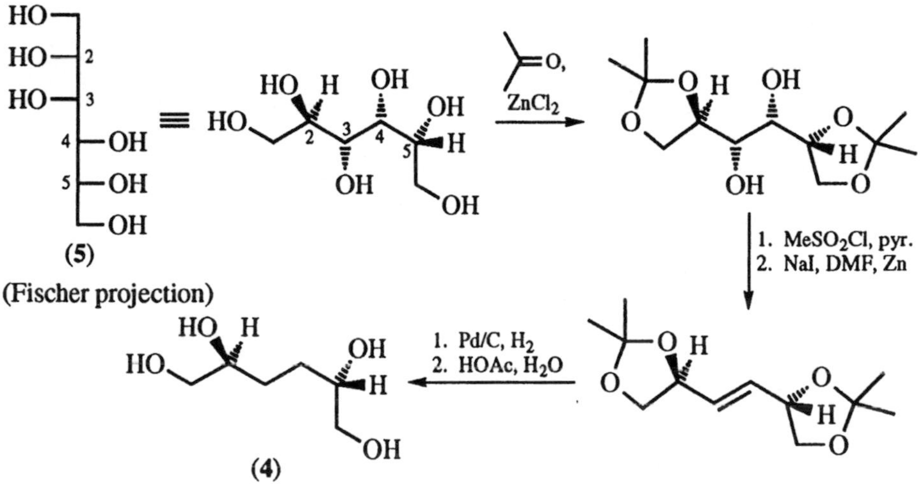

Scheme 1

In this synthesis, the bonds to the chiral centres at positions 2 and 5 remain undisturbed and therefore the configurations at these centres in the product can confidently be assumed to be the same as those in mannitol. Mannitol is a readily available starting material, the absolute configuration of whose chiral centres is known to be as depicted in the Fischer projection formula shown.

An alternative version of method A for the synthesis of (1) would be to link two fragments, each containing a chiral centre of the required *absolute* configuration so as to form the required connecting chain p directly, or with the application of appropriate functional group interconversion (FGI) (Scheme 2). The best known example of this alternative is the formation of peptides by the reaction of two appropriately protected enantiopure amino acids (Scheme 3). Here the connecting linkage p—the amide bond—is formed directly as required but the free amino and carboxylic acid groups of the dipeptide usually require liberation from the protecting groups in which they were clad to ensure the regiospecificity and heterocoupling of the reaction.

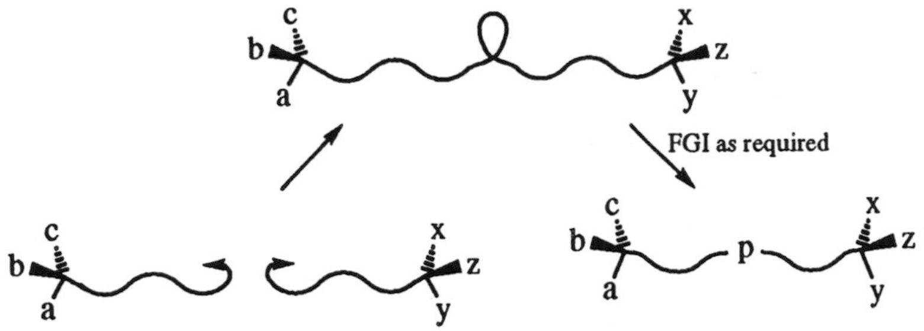

Scheme 2

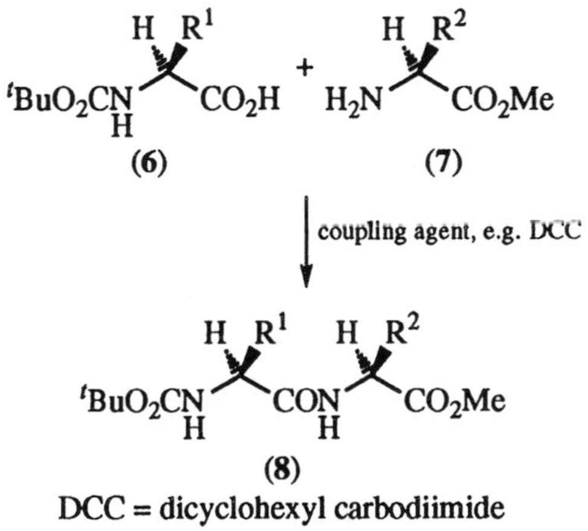

DCC = dicyclohexyl carbodiimide

Scheme 3

This route in Scheme 3 cannot be applied to the synthesis of, e.g., racemic **(8)** by linking the two fragments **(6)** and **(7)** in racemic form since, in general, both (racemic) diastereoisomers of **(8)** will be formed ($S,R + R,S$ as well as $S,S + R,R$).

METHOD B

Method B for the synthesis of **(1)** also uses a starting material with two chiral centres already present but, unlike method A, involves the breaking and making of bonds at one or both of these chiral centres independently by reactions that are completely and reliably stereoselective. This includes reactions in which the bond making/bond breaking takes place with inversion (S_N2) [Scheme 4(a)] or retention of configuration [Scheme 4(b)].

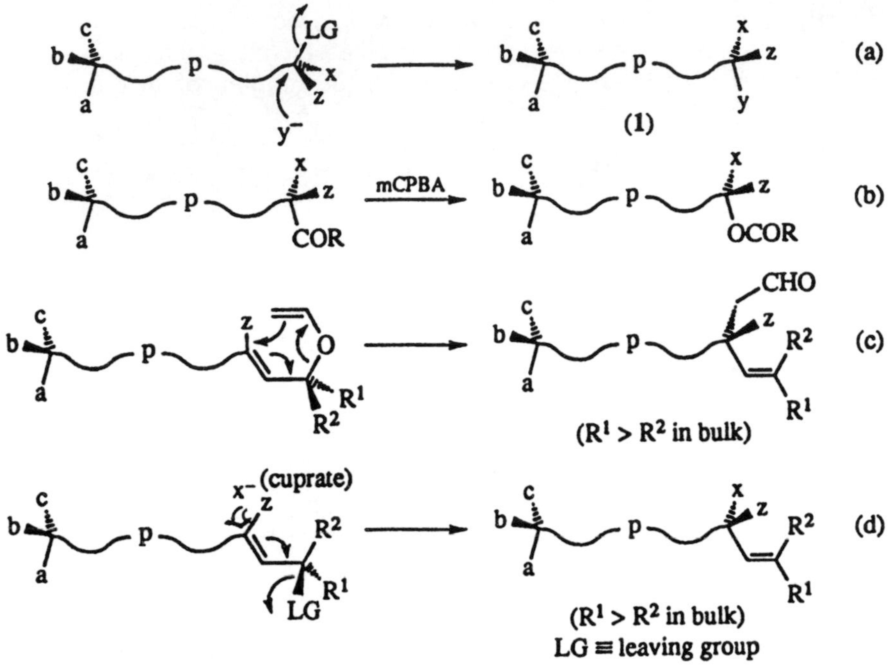

Scheme 4

Related reactions include those in which a chiral centre is transferred, also completely stereoselectively, from one position in the molecule to another with concomitant shift of a configured double bond as in many Claisen rearrangements [Scheme 4(c)] or S_N2' reactions [Scheme 4(d)].

METHOD C

The third method for synthesis of (1) is to use a reaction in which both chiral centres are created, usually simultaneously, from achiral but prochiral precursors and achiral reagents. Diastereoselectivity arises in this method from the stereocontrol inherent in the reaction.

For acyclic molecules such as (1), this method includes reactions which fashion adjacent chiral centres from a configured carbon–carbon double bond: bromination, hydroboration and hydroxylation are examples (Scheme 5).

In contrast to the products from application of methods A or B, those from method C are bound to be racemic since there is no chiral element in the reaction ensemble.

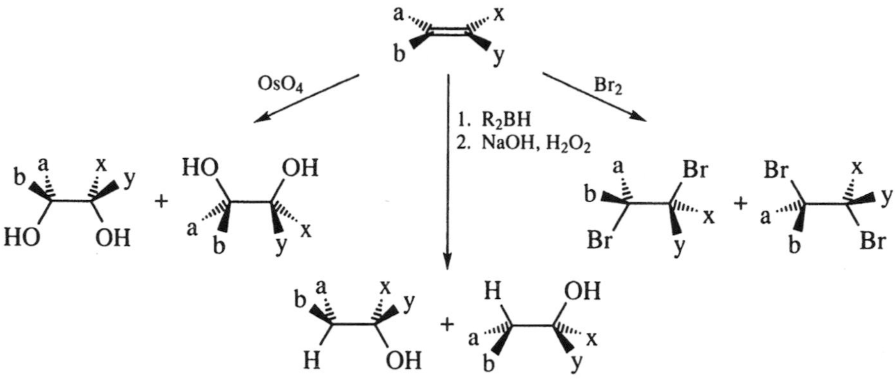

Scheme 5

METHOD D

The fourth method for the synthesis of (**1**) is to use a parent chiral centre, C_{abc}, in a precursor (**1**), e.g. (**9**) (Scheme 6) to *control* the configuration of the newly created chiral centre, C_{xyz}, by directing attack on z on only one diastereoface of the prochiral double bond $C=y'$. This attack by z may, in principle, take place by any mechanism including electrophilic, radical, nucleophilic or pericyclic attack on the $C=y'$ double bond.

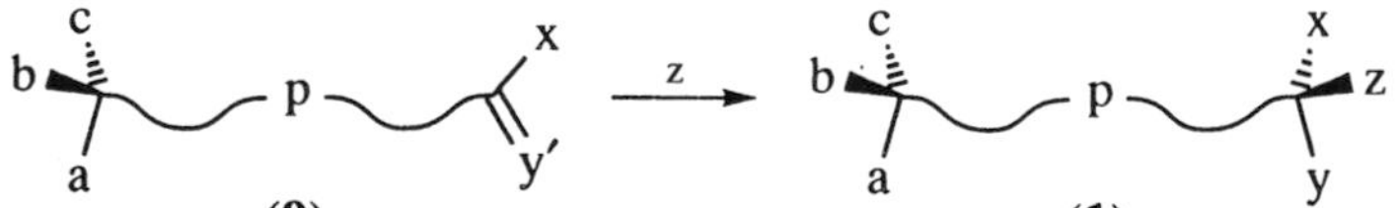

Scheme 6

A variant on method D to that given in Scheme 6 is one in which the parent chiral centre C_{abc} in a precursor of (**1**), e.g. (**10**), controls which one of the two diastereotopic groups or atoms (say x) undergoes reaction and conversion into z (Scheme 7).

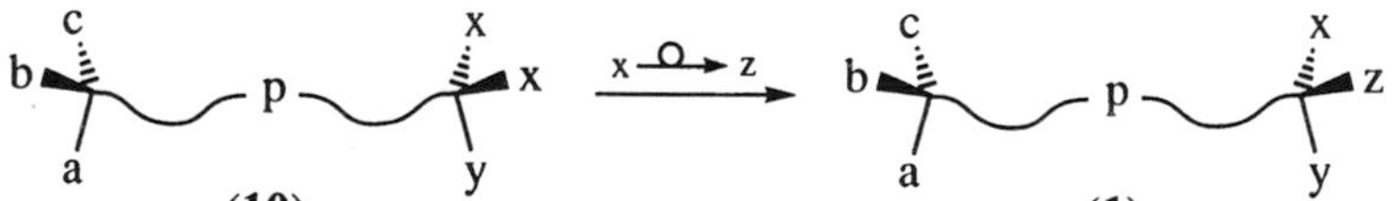

Scheme 7

In both Schemes 6 and 7, the parent chiral centre C_{abc} and prochiral element are contained within the same molecule but this is not necessarily the case. Scheme 8 illustrates the attack of an enantiopure reagent on the prochiral

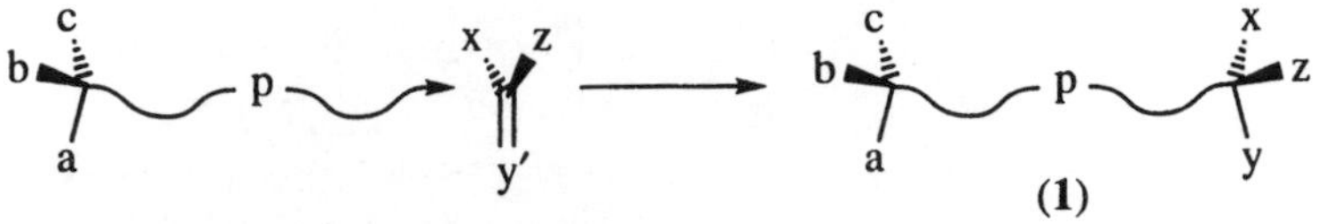

(1)

Scheme 8

double bond $C{=}y'$ where the prochiral element and the parent chiral centre are contained in different molecules; again the parent chiral centre C_{abc} in the reagent determines which face of the prochiral double bond is attacked.

Method D can also be applied to the synthesis of target molecules in racemic form by using racemic starting materials in Schemes 6–8.

The means by which the existing or parent chiral centre or centres can induce a particular configuration of newly created chiral centre(s) is one of the major concerns of this book (see Chapters 9–16).

Classification of stereoselective reactions

Methods A–D for the stereoselective synthesis of (1) outlined above may be used as a basis for classification of stereoselective reactions. This classification is based on the number of chiral elements created in the product in comparison with the number present in the starting material. The chiral elements will be assumed to be chiral centres since this is the case for the great majority of stereoselective reactions.

TYPE 0 REACTIONS

Type 0 reactions are those in which no stereoselectivity is required or obtained. In stereoselective synthesis they are carried out on substrates having one or more chiral centres. No new chiral centres are created and all bonds remain unbroken to the chiral centres which survive the reaction. Type 0 reactions will be considered in Chapter 3.

TYPE I REACTIONS

Type I reactions involve reaction with either complete inversion or retention

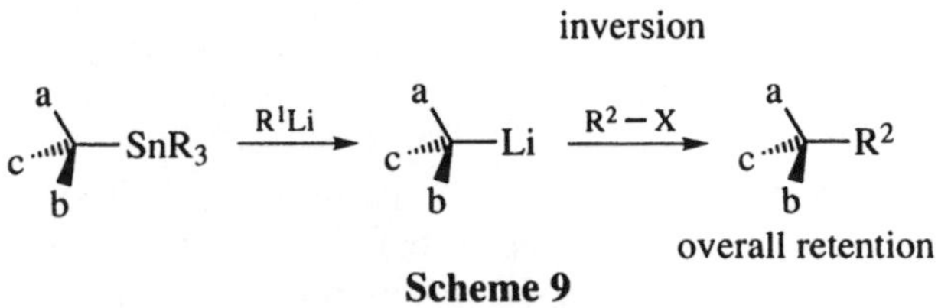

inversion

overall retention

Scheme 9

of configuration at an existing chiral centre; there is no increase in the number of chiral centres.

Examples of Type I reactions are given in Scheme 9 and these reactions will be covered in Chapter 4. Related simple chirality transfer reactions will be considered in Chapter 5.

TYPE II REACTIONS

Type II reactions involve the diastereoselective formation of a product containing at least two chiral centres from the reaction of one or more prochiral double bonds contained in one or more achiral starting materials using achiral reagents.

Type II reactions include many cycloadditions, electrocyclic reactions and sigmatropic rearrangements in which the starting materials are achiral (and no reagents are required), and also additions to double bonds in which two chiral centres are created (see Scheme 5).

Since Type II reactions use achiral starting materials and reagents, the products are bound to be racemic. Examples of Type II reactions are given in Scheme 10 and will be considered in more detail in Chapters 6–8.

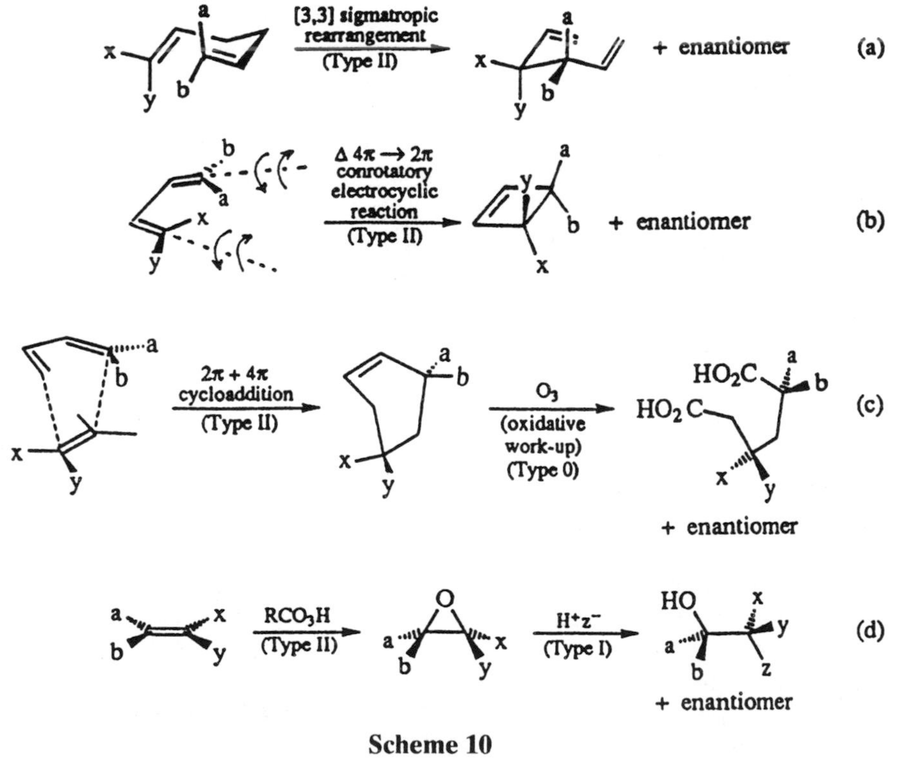

Scheme 10

TYPE III REACTIONS

Type III reactions involve the stereoselective formation of one or more additional chiral centres under the influence of one or more existing chiral centres in the starting material, the reagents or some part of the reaction ensemble. All Type II reactions have a Type III component when one of the reactants contains a chiral centre.

The major part of this book will be concerned with Type III reactions and, in particular, the means by which the existing or parent chiral centre or centres control the relative or absolute configuration of the newly created chiral centre(s). Examples of Type III reactions are given in Chapter 1 (e.g. Schemes 24, 27 and 29–33). This class and its sub-categories will be considered in Chapters 9–16.

TYPE IV REACTIONS

Type IV reactions are those which require the exercise of diastereocontrol but do not involve the formation of tetrahedral (chiral) centres or other chiral elements.

Type IV reactions involve, for the most part, formation of configured carbon–carbon double bonds by methods which include:

(i)   elimination from a substrate containing two contiguous chiral centres;
(ii)   addition to alkynes or allenes;
(iii)   substitution on a (usually already configured) double bond;
(iv)   sigmatropic rearrangement in which formation of configured double bond(s) accompanies the loss of chiral centres;
(v)   union of two functional groups with direct formation of the double bond, e.g. the Wittig reaction.
(vi)   addition–elimination using a molecule containing a double bond, e.g. ketone → enol derivative (Chapter 8).

Because of limitations of space, Type IV reactions will not be considered in detail in this book.

We have introduced this classification by reference to the synthesis of an acyclic molecule (**1**), but it can also be applied to the synthesis of cyclic (**2**) and bicyclic (**3**) target molecules. Thus, a Type II reaction as applied to the synthesis of the monocyclic (**2**) is exemplified by the Diels–Alder reaction in Scheme 10(c) and by the epoxidation in Scheme 10(d).

These monocyclic products are a valuable source of diastereoisomerically pure acyclic compounds by ring-cleavage reactions using Type 0 reactions as in Scheme 10(c), particularly since such 1,3-substituted products are not generally accessible by Type II reactions from acyclic starting materials. Likewise, regiospecific and completely stereoselective ring opening of epoxides (Type I) as in Scheme 10(d) provides a versatile means of preparing representatives of (**1**) with $p = O$.

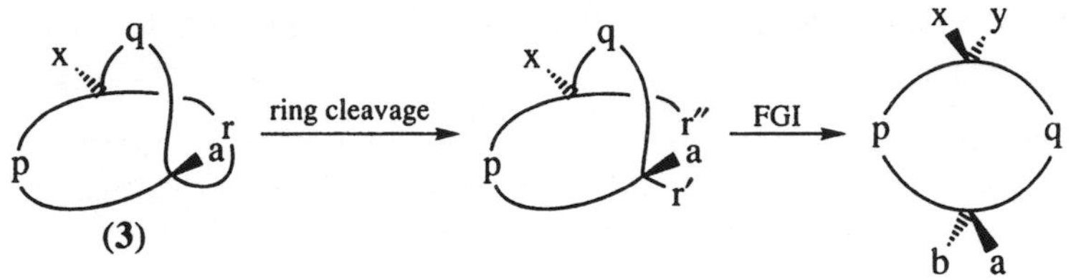

Scheme 11

Many bicyclic systems (**3**) are available directly by Type II cycloaddition reactions. A versatile means for the synthesis of monocyclic (**2**) using Type 0 reactions is the cleavage of e.g., the r bridge of (**3**) (Scheme 11) with additional FGI as appropriate for conversion into b and y. This will result in a defined relative configuration at the two chiral centres in (**2**) in which b and y are *cis* (see Chapter 3).

The above classification of reactions used in stereoselective synthesis contrasts with the normal classification of reactions which is based on mechanism. Thus, depending on the substitution, a [3,3] sigmatropic rearrangement may belong to Type II, III or IV (Scheme 12), although all these transformations are believed to proceed via similar mechanisms. On the other hand, there are reactions of diverse mechanistic types which can be grouped together in terms of the common stereochemical changes involved.

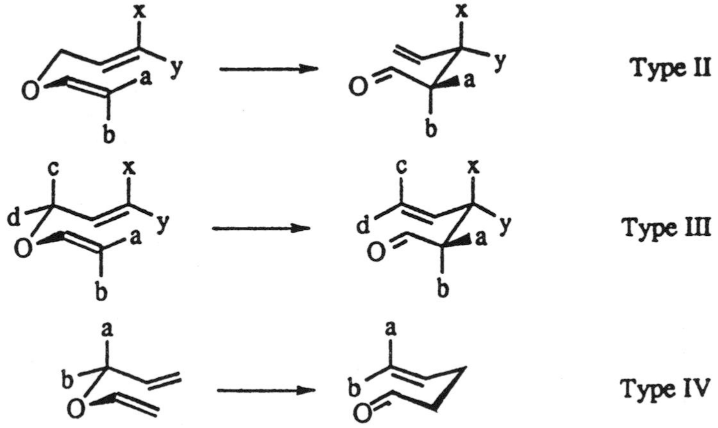

Scheme 12

In this book, it is expedient to use the classification given above which focuses on the stereochemical changes involved in a reaction; understanding these changes will necessarily involve consideration of the mechanism by which the individual reactions are thought to proceed. Further sub-division, particularly for Type III reactions, will be necessary (Chapters 9–16).

Stereoselective synthesis of molecules having three or more chiral centres

The classification outlined above can also be applied to the reactions used in the synthesis of molecules containing three or more chiral centres, but as the number increases the likelihood is that the synthesis will require more than a single step.

The more widespread use of Type 0 reactions alone for the synthesis of target molecules in this category is limited by the availability of diastereoisomerically pure starting materials having three or more chiral centres whether enantiopure or racemic (see Chapter 3).

Although it is possible, using Type II reactions, to generate three or more chiral centres in a single reaction, notably in the Diels–Alder and other cycloaddition reactions, *by itself* this is not a general method for synthesis of target molecules of this complexity; the product, moreover, will necessarily be racemic.

In practice, the synthesis of a target molecule containing three chiral centres will commonly use a starting material having two chiral centres, prepared using Type 0 or I reactions and introduce the third chiral centre using Type III. Alternatively, two or more chiral centres may be generated by a cycloaddition reaction (Type II) before the third centre is introduced using again Type III. For target molecules containing four or more chiral centres, the use of one or more Type III reactions is usually unavoidable.

Stereoselective synthesis of molecules having a single chiral centre

Stereoselectivity in the synthesis of a molecule containing a single chiral centre means the synthesis of that molecule as a single enantiomer. Type II reactions are not applicable in this case since they can only result in racemic products. Of course, resolution of racemates will always be a viable way of obtaining single enantiomers but the outcome of an untried resolution can rarely be predicted with confidence. Just as a completely diastereoselective reaction avoids the necessity for separation of diastereoisomers, so a completely enantioselective reaction avoids the necessity for resolution.

The increasingly widespread use of enzymic or chemzymic methods for the conversion of prochiral double bonds or centres into chiral centres has been referred to previously. These methods are a variation on Type III reactions in which the existing parent chiral centres are contained in the reagent but do not appear in the product (see Chapter 1, Figures 3–5, and Chapters 14 and 16).

Type 0 reactions provide an invaluable method for the synthesis of molecules containing a single chiral centre, particularly when used in combination with Type I and simple chirality transfer reactions (Chapters 4 and 5).

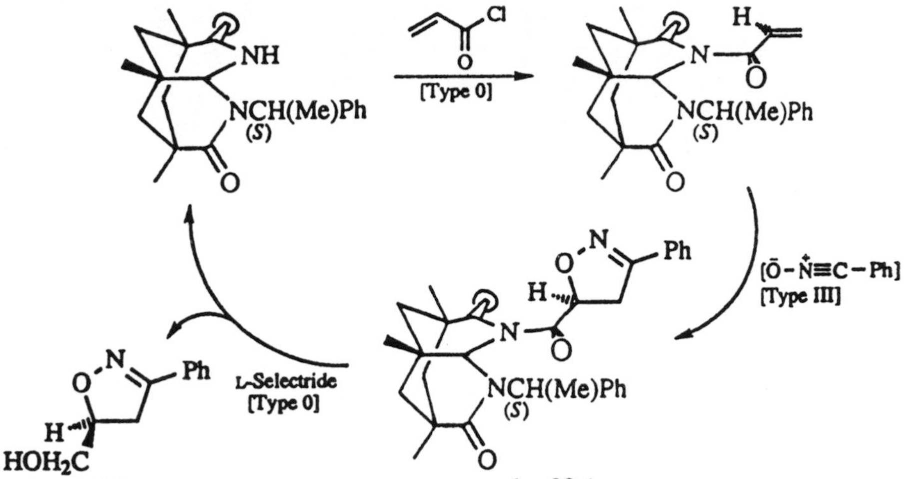

(11)

d.r. 99:1

$$\text{L-Selectride} = Li(C_2H_5\overset{\displaystyle CH_3}{\overset{|}{C}}H-)_3BH$$

Scheme 13

Sequential use of Type 0 and III reactions can be applied to synthesise molecules in this category as in the use of chiral auxiliaries. Thus, in Scheme 13, the attachment of the auxiliary containing one or more chiral centres to the prochiral element (Type 0), followed by diastereoselective cycloaddition (Type III), followed by cleavage of the auxiliary (Type 0) gives a single enantiomer of the product (**11**).[2]

References

1. C.C. Deane and T.D. Inch, *Chem. Commun.*, 1969, 813.
2. D.P. Curran, K.-S. Jeong, T.A. Heffner and J. Rebek, *J. Am. Chem. Soc.*, 1989, **111**, 9238.

3 TYPE 0 REACTIONS: REACTIONS IN WHICH NO NEW CHIRAL CENTRES ARE CREATED

Since, by definition, Type 0 reactions do not require stereoselectivity, it might be assumed that a synthesis consisting of only Type 0 reactions would not constitute a stereoselective one. However, *stereoselective synthesis* is now widely taken to be the synthesis of a single stereoisomer of the product by whatever means which include only Type 0 reactions if the chiral centres of the product are already in place in the starting material (it could be said that the selection comes in the choice of stereoisomer of the starting material). The Type 0 reactions necessary in this case will convert functional groups in the starting material into those required in the product. They may also include the removal of superfluous chiral centres.

The application of Type 0 reactions to a limited number of readily available enantiopure starting materials having one, two or more chiral centres is the means by which a much greater number of enantiopure products are made available (see below). These enantiopure products will often in turn be used for further transformations or they may be synthetic targets themselves.

Syntheses consisting of only Type 0 reactions will require a knowledge of the availability of starting materials having chiral centres in place with the correct relative or absolute configuration for those required in the products. For example, (S)-3-hydroxypiperidine (1) has been prepared from the readily available butyrolactone (2),[1] [readily obtained by a Type I reaction from (S)-glutamic acid (Scheme 1)[2]].

Similarly the δ-lactone (3), which is an insect pheromone, has been prepared (Scheme 2) from the enantiopure epoxide (4) as shown by Type 0 reactions only.[3]

Epoxide (4) is a useful chiral building block which is available by a chemo-enzymatic route (a route which includes at least one enzyme transformation) from the ketone (5) by reduction using bakers' yeast (Scheme 3). Although

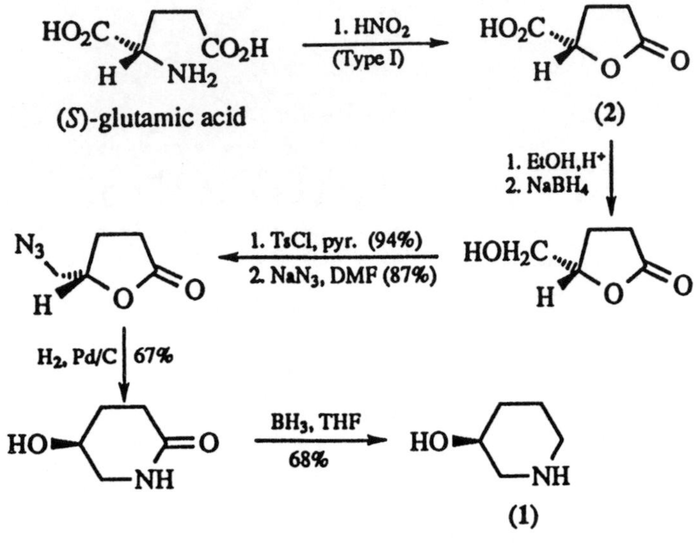

Scheme 1

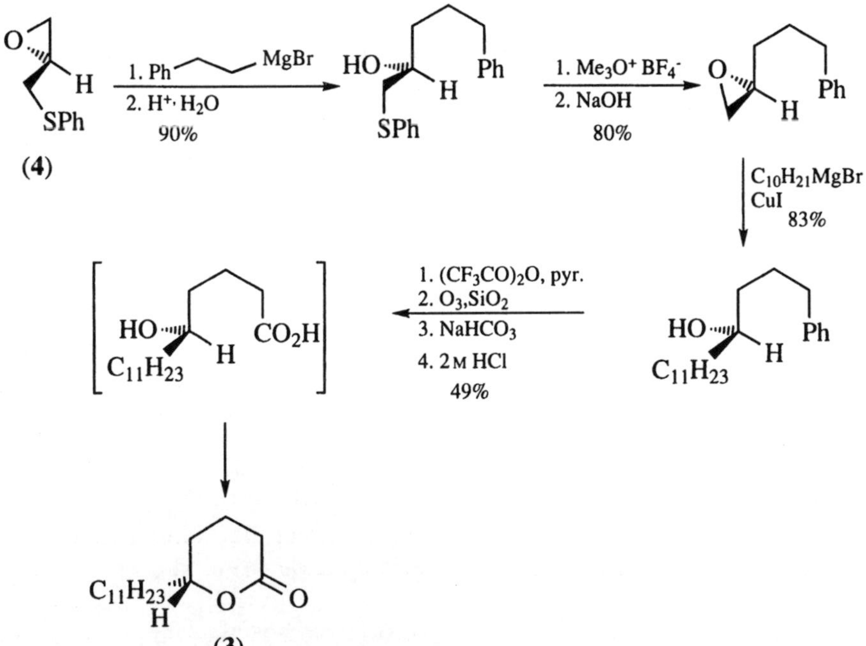

Scheme 2

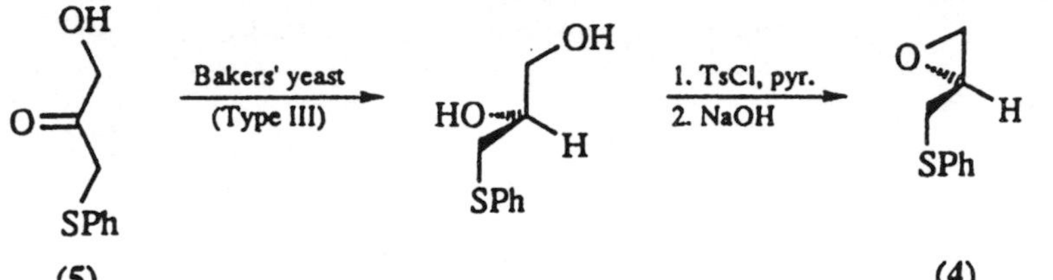

Scheme 3

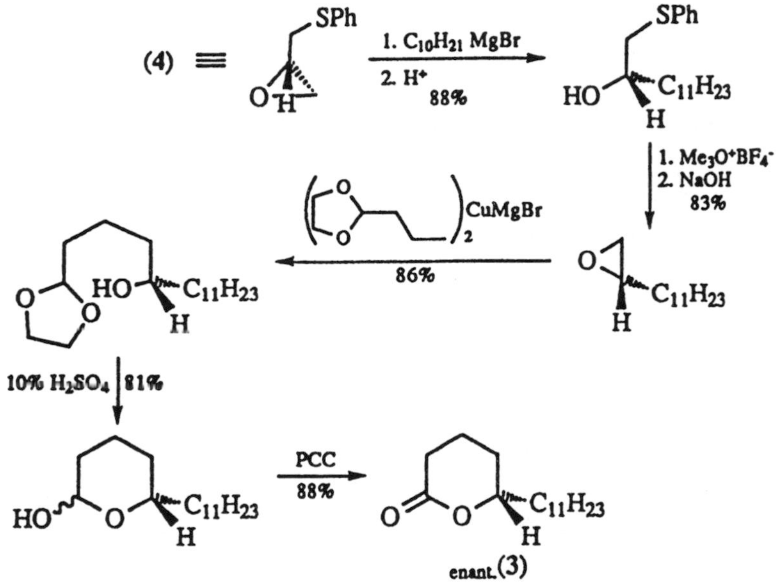

Scheme 4

the enantiomeric excess of (4) is only 79%, it was obtained enantiopure by crystallisation from benzene.

Scheme 4 illustrates how, by a nice choice of functional group manipulation (Type 0 reactions) the enantiomer of (3) [= $_{enant.}$(3)] in Scheme 2 can be prepared.[3] A device such as this may be of real value when only a single enantiomer of the starting material is available, which is often the case in enzymically prepared starting materials.

Table 1 shows a selection of commercially available starting materials which contain a single chiral centre and which are available as single enantiomers, most at less than £5 per gram. They include amino acids, α-hydroxy acids, amines, alcohols and a number of terpenes. Most common amino acids are available in both enantiomeric forms, although the S-enantiomer is usually much cheaper than the R-form. These compounds in Table 1 are invaluable starting materials for synthesis of other enantiopure products.[4]

Table 1. Readily available enantiopure materials having one chiral centre (unless otherwise indicated, both enantiomers are available).

α -Amino acids

$$\underset{NH_2CHCO_2H}{\overset{R}{|}}$$

R =   Me   CH$_2$Ph   CH$_2$CO$_2$H   (CH$_2$)$_2$CO$_2$H

    alanine   phenylalanine   aspartic acid   glutamic acid

    (CH$_2$)$_2$CONH$_2$ CHMe$_2$ CMe$_2$SH

    glutamine   valine   penicillamine
    (S)

R =   CH$_2$CHMe$_2$ CH$_2$OH   CH$_2$SH   (CH$_2$)$_2$SMe   (CH$_2$)$_3$NH$_2$

    leucine   serine   cysteine   methionine   ornithine

    (CH$_2$)$_4$NH$_2$ CH$_2$CONH$_2$

    lysine   asparagine

R =   (CH$_2$)$_3$NHČNH$_2$ (NH)   (CH$_2$)$_3$NHCONH$_2$ histidine

    arginine   citrulline
    (S)

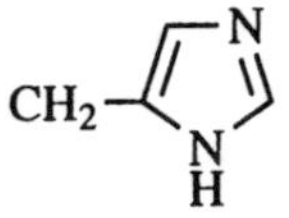

R =

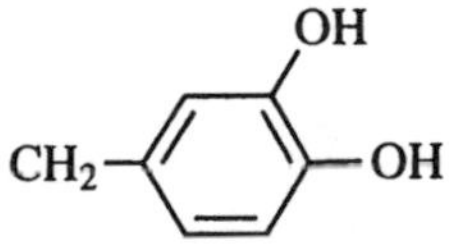

    dihydroxyphenylalanine   tyrosine
    (S)

R =   (CH$_2$)$_2$CH$_3$

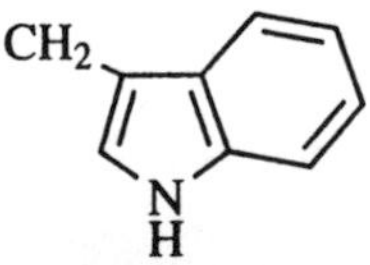

    tryptophan   norvaline

and

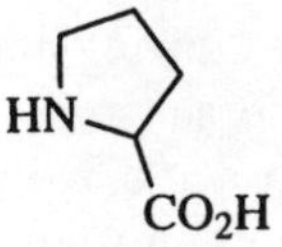

    proline   pyroglutamic   thiazolidine-4-carboxylic
                  acid   acid
                  (S)   (R)

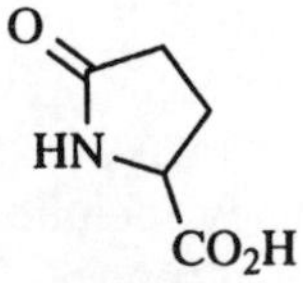

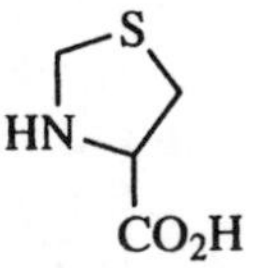

Table 1 *(continued)*

α -Hydroxy acids

$$\overset{R}{\underset{|}{HOCHCO_2H}}$$

R =	Me	Ph	CH_2CO_2H
	lactic acid	mandelic acid	malic acid

and

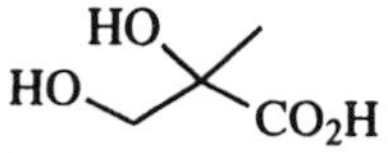

(as methyl ester)

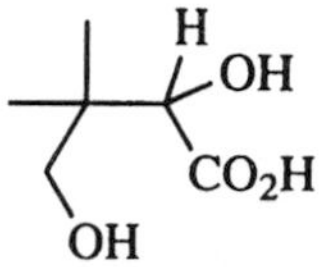

(as lactone)
(R)

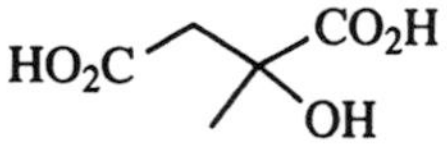

citramalic acid

β -Hydroxy acids

$$\overset{Me}{\underset{|}{HOCH_2CHCO_2H}}$$
(as methyl ester)

Amines

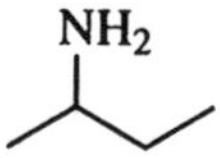

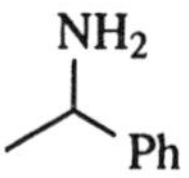

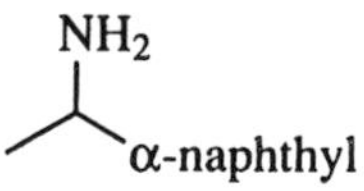

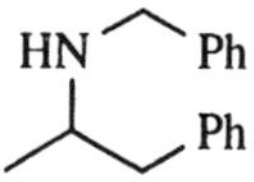

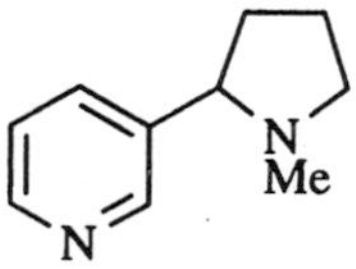

nicotine
(S)

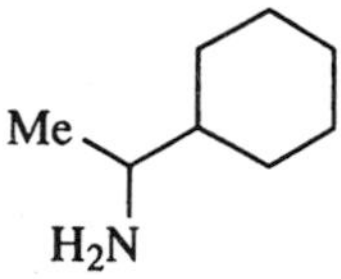

(continued)

Table 1 *(continued)*

Alcohols

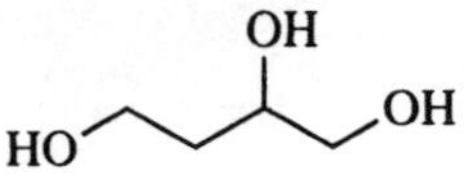

(S) 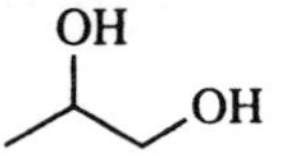　　　　　　(S) 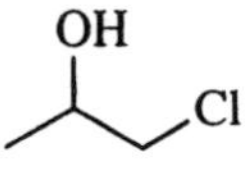　　　　　　(S)

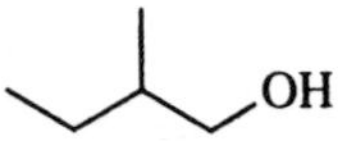

(S) 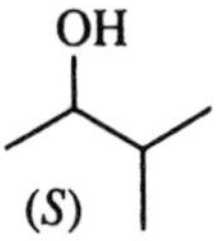　　　(S) 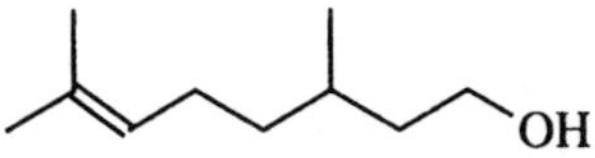　　　citronellol

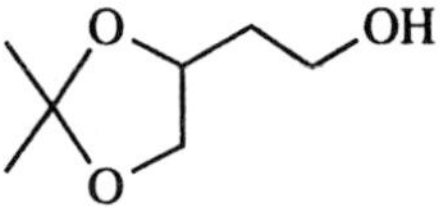

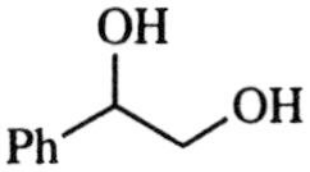

Amino alcohols

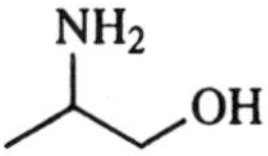

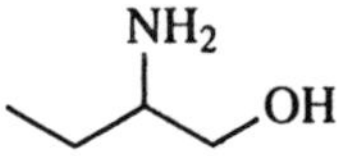

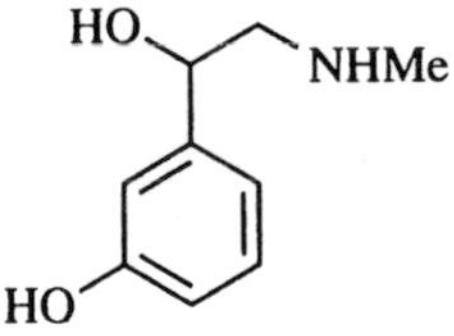

(R)

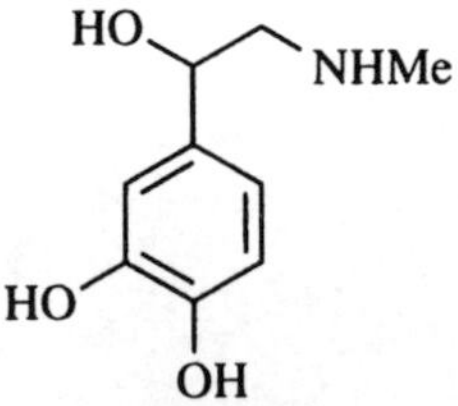

epinephrine
(R)　　　　　　prolinol

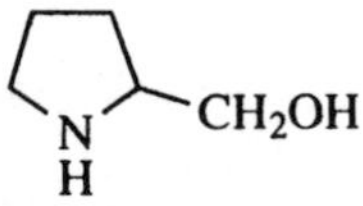

(continued)

Table 1 *(continued)*

Terpenes

limonene

perillaldehyde
(also alcohol)
(*S*)

pulegone

(*R*)

α-terpineol
$[\alpha]_D -30°$

carvone

terpinene-4-ol
$[\alpha]_D +29°$

Many of these materials are available as single enantiomers by virtue of the fact that they are products of enzyme-mediated processes. The harnessing of enzyme systems as in Scheme 3 to bring about enantioselective conversions is becoming increasingly important as a source of enantiopure starting materials (see Chapter 16); it can be anticipated that the number of compounds in Table 1 will be increased in years to come by the availability of single enantiomers prepared by enzymatic and chemo-enzymatic methods.

Table 2 shows a selection of commercially available starting materials containing more than one chiral centre which are available as single enantiomers. These include a number of bifunctional compounds, more terpenes and a number of sugars and their derivatives.

It was inevitable that the availability of naturally occurring sugars with their plenitude of chiral centres would mean that they would find use in synthesis employing for the most part Type 0 reactions.[5]

exo-Brevicomin (**6**) is an aggregation pheromone of the Western pine beetle. Retrosynthesis (Scheme 5) by disconnection of the acetal and comparison of the derived diol enantiomer (**7**) with D-glucose (Fischer projections) shows that the chiral centres at positions 3 and 4 in the sugar need to be retained and a number of Type 0 reactions carried out including the addition of three carbons to C-1 and elimination of chiral centres at C-2 and C-5.

Table 2. Readily available enantiopure materials having two or more chiral centres

Amino acids

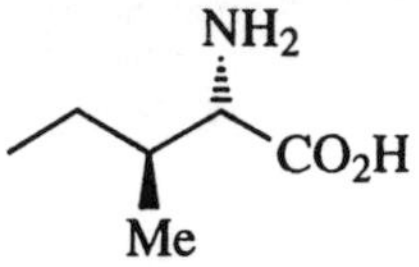

(2S,3S)-isoleucine
(also enant.)

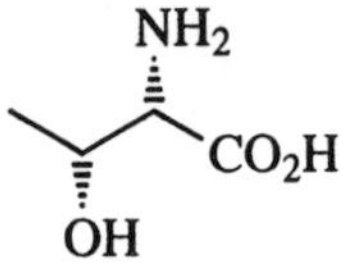

threonine
(also enant.)

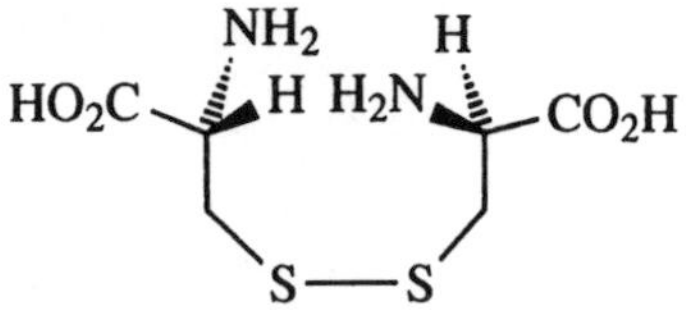

(R,R)-cystine
(also enant.)

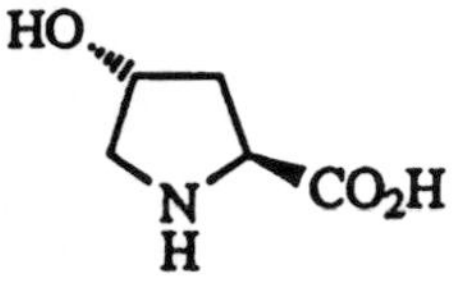

4-hydroxyproline
(also enant.)

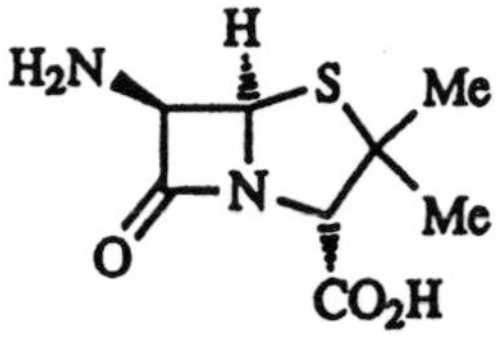

6-aminopenicillanic acid

Amines and amino alcohols

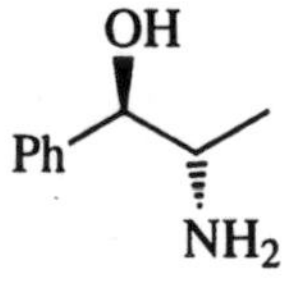

norephedrine
(also enant.)

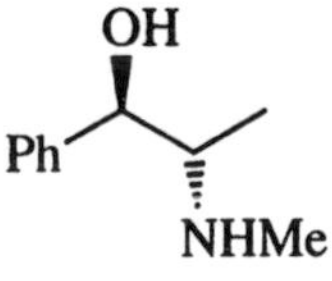

ephedrine
(also enant.)

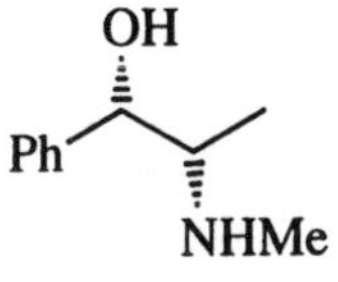

pseudoephedrine

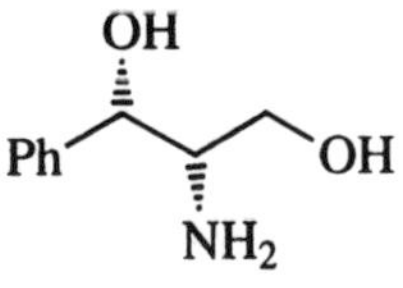

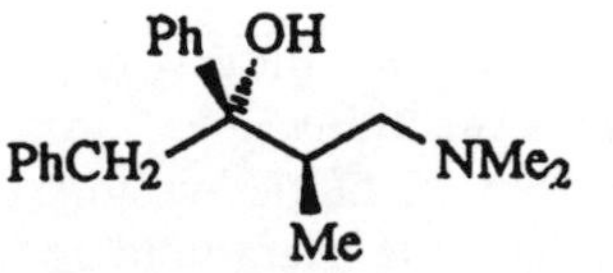

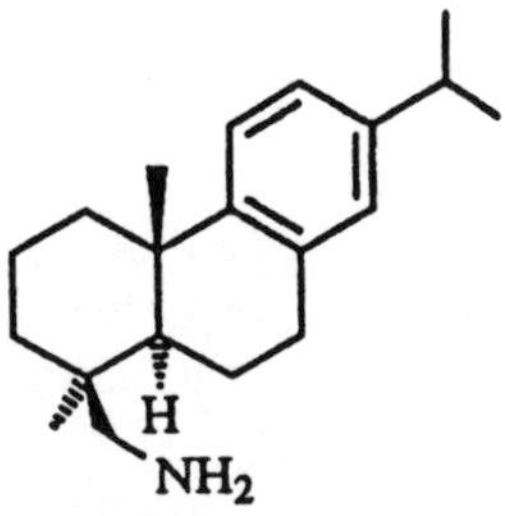

dehydroabietylamine

Table 2 (*continued*)

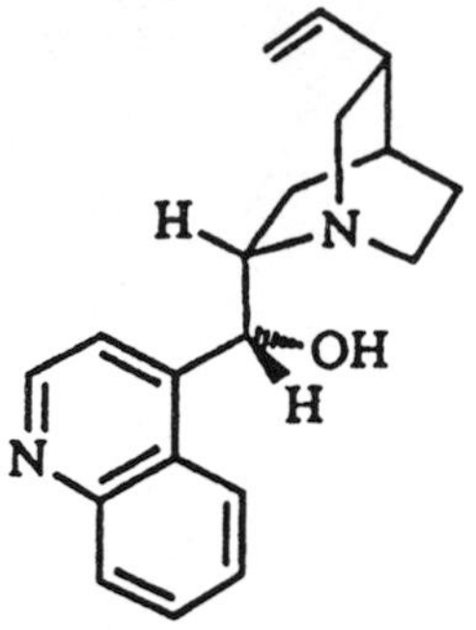

cinchonidine 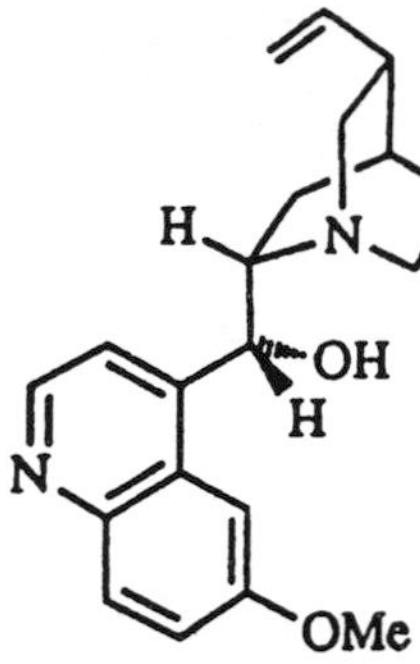   quinine 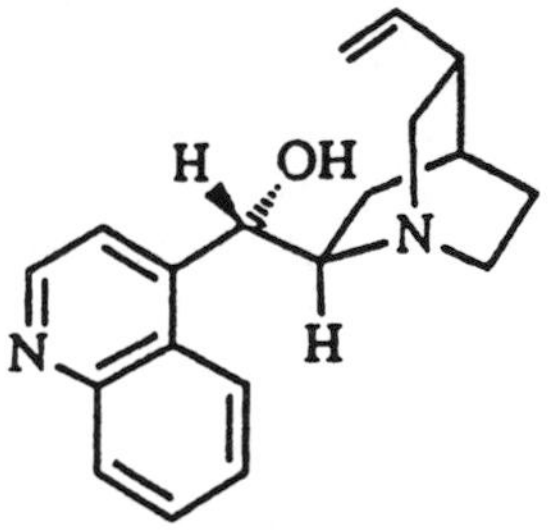   cinchonine

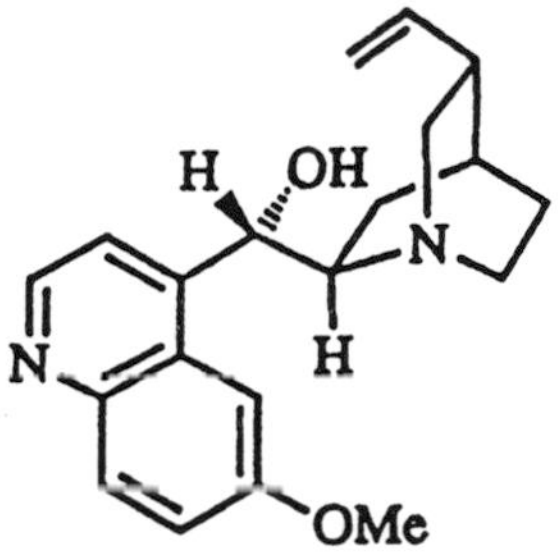

quinidine 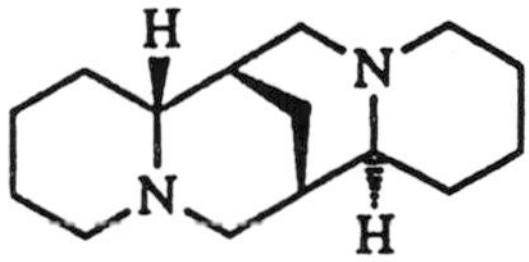   sparteine

Acids

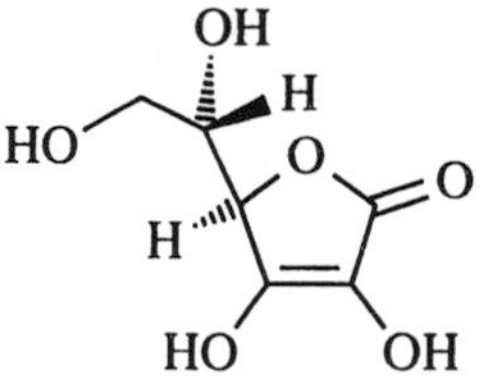

ascorbic   isoascorbic
acid   acid

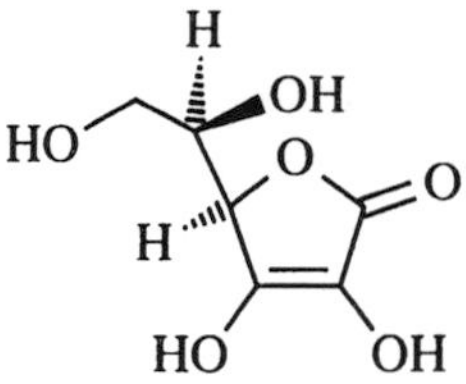

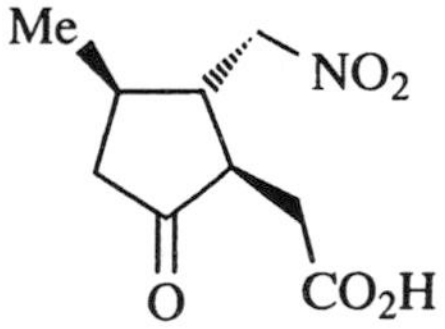

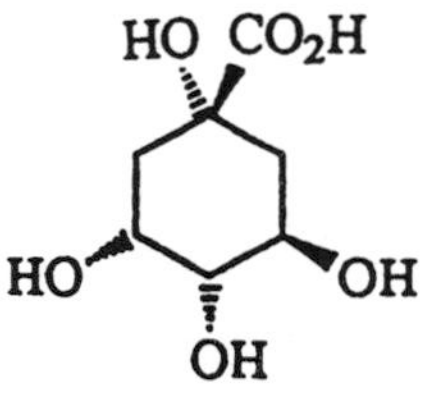

quinic acid 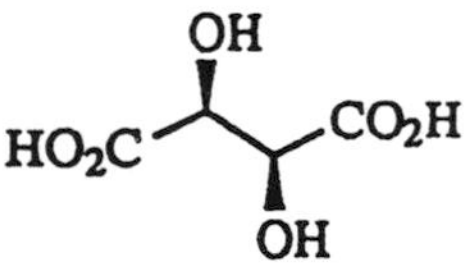   tartaric acid   pantothenic acid
                     (also enant.)

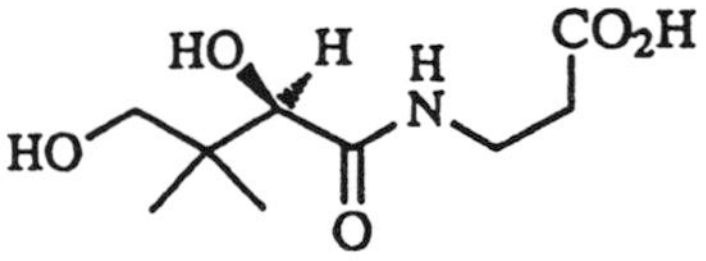

(*continued*)

Table 2 (*continued*)

Terpenes

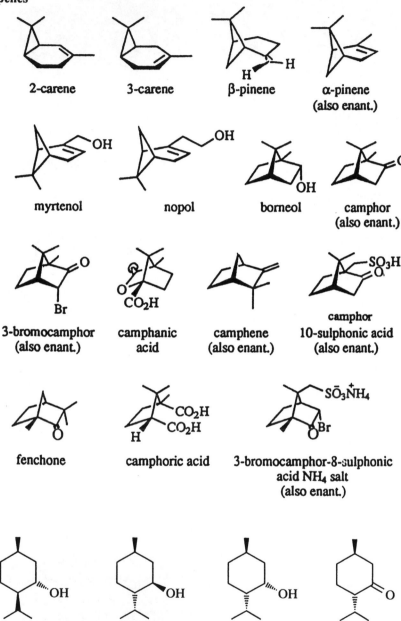

2-carene

3-carene

β-pinene

α-pinene
(also enant.)

myrtenol

nopol

borneol

camphor
(also enant.)

3-bromocamphor
(also enant.)

camphanic
acid

camphene
(also enant.)

camphor
10-sulphonic acid
(also enant.)

fenchone

camphoric acid

3-bromocamphor-8-sulphonic
acid NH₄ salt
(also enant.)

isomenthol

menthol
(also enant.)

neomenthol

menthone

Table 2 (*continued*)

Sugars

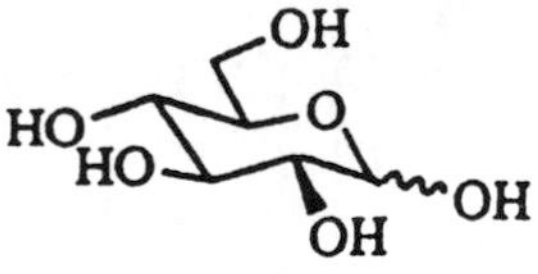

glucose

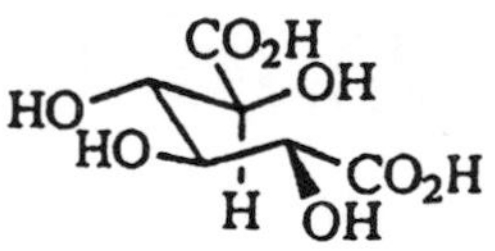

saccharinic acid

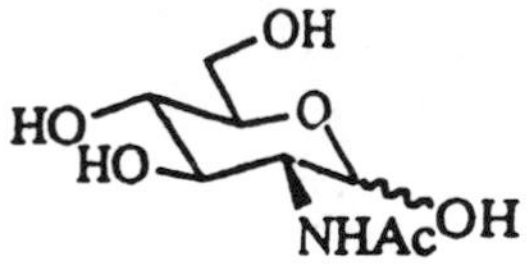

N-acetylglucosamine

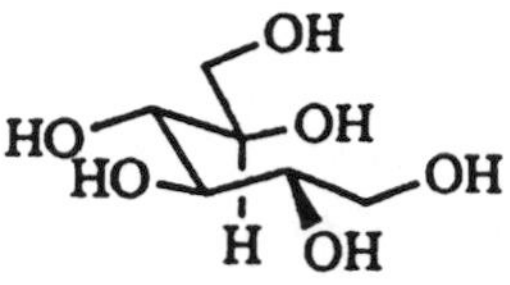

sorbitol

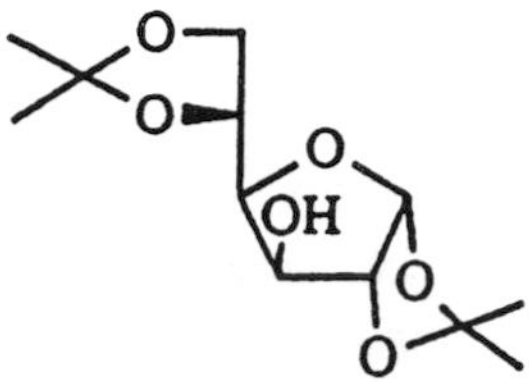

diacetone glucose

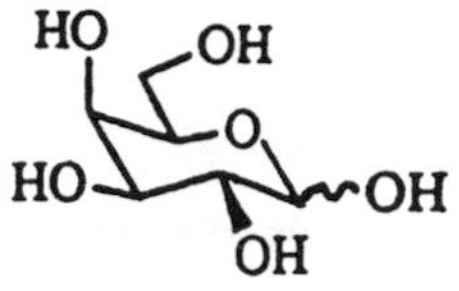

galactose

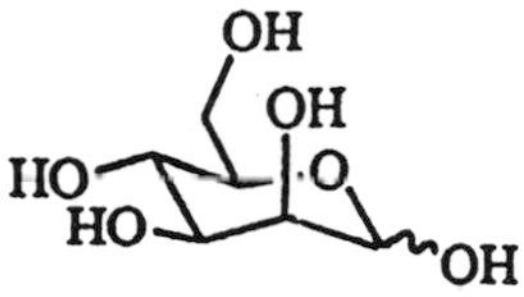

mannose

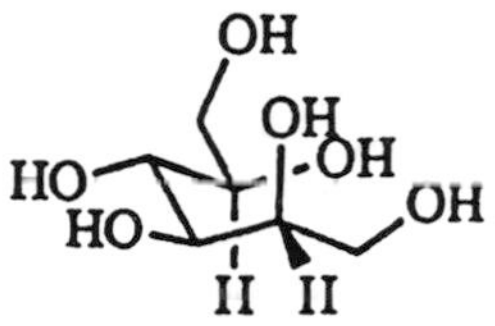

mannitol

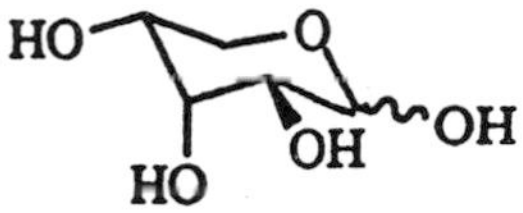

ribose

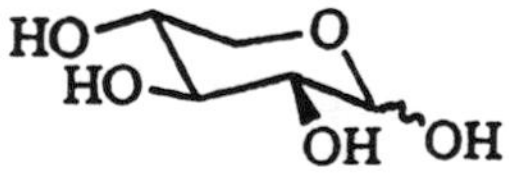

xylose

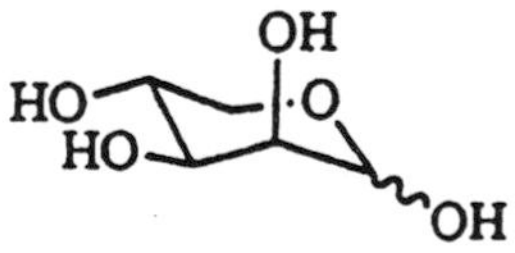

arabinose
(also enant.)

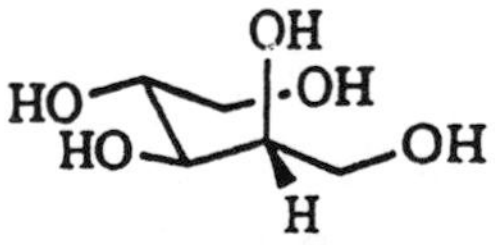

arabitol
(also enant.)

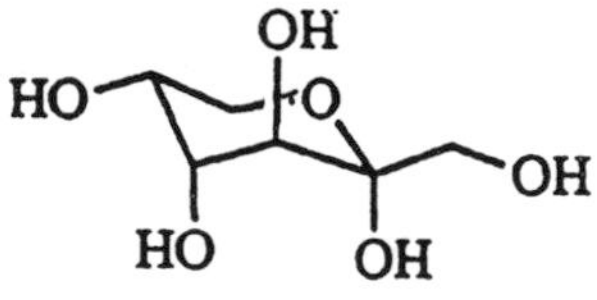

fructose

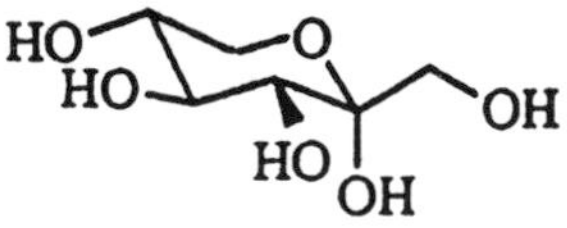

sorbose

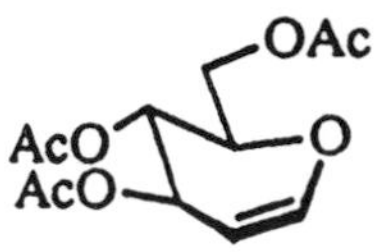

tri-*O*-acetylglucal

Scheme 6 shows how this was achieved by Sherk and Fraser-Reid[6] with the chiral centres in diacetone glucose which must be retained, shown starred. The specific rotation of the product was +80.7°, in good agreement with that reported in the literature (+84.1°) for the material produced by the insect.

Synthesis using sugars is invariably an exercise in protecting group manipulation; in Scheme 6, for example, no less than five steps are concerned

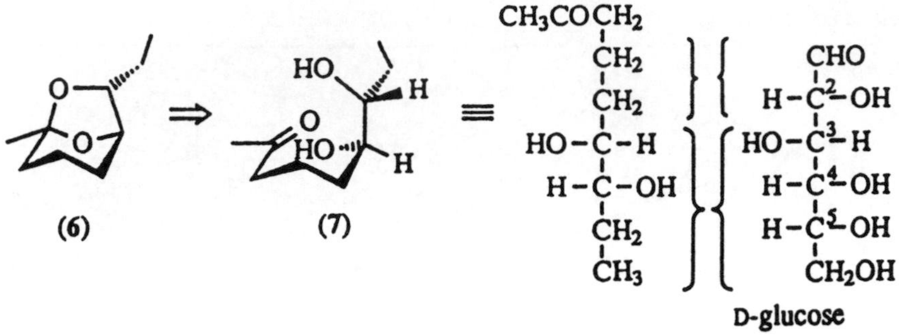

$$\text{(6)} \Rightarrow \text{(7)} \equiv$$

$$\begin{array}{c} CH_3COCH_2 \\ | \\ CH_2 \\ | \\ CH_2 \\ | \\ HO-C-H \\ | \\ H-C-OH \\ | \\ CH_2 \\ | \\ CH_3 \end{array} \qquad \begin{array}{c} CHO \\ | \\ H-C^2-OH \\ | \\ HO-C^3-H \\ | \\ H-C^4-OH \\ | \\ H-C^5-OH \\ | \\ CH_2OH \end{array}$$

D-glucose

Scheme 5

with protection or deprotection of specific hydroxyl groups. It will probably be some time before the regiocontrol of reactions in sugar chemistry has developed sufficiently to allow individual functionalisation of each of the hydroxyl groups in glucose without recourse to protection or to enzymic methods.

Certain reactions become important in Type 0 reactions for removal of unwanted chiral centres. The Barton–McCombie method,[7] used for conversion of (8) into (9) in Scheme 6, is one such method, particularly when sugars are used; it replaces a hydroxyl group by hydrogen via the radical-mediated route shown and is therefore safe from the intervention of any carbocation rearrangements which may accompany other methods of elimination.

Readily available starting materials enantio- and diastereoisomerically pure

What constitutes a readily available starting material? Just because a compound is commercially available this does not mean it is readily available for use as a starting material; its cost may be prohibitive for the amount that is required even for the most elastic of budgets.

The Aldrich Chemical catalogue lists over 20 000 compounds but less than 5% are offered as single enantiomers; the cost of the majority of these is such that their use as starting materials for a synthesis in which several grams of the target molecule are required cannot normally be contemplated.

Those commercially available enantiopure compounds which are cheap enough to be routinely used as starting materials are few in number; most of them are shown in Tables 1 and 2 or are simple derivatives of these.

Most of the materials in Table 1 are *chiral pool*-derived, that is, they are products of enzyme-mediated processes in plants, fungi or other microorganisms, and are relatively easy to isolate in gram quantities. The largest class is the amino acids, most of which are available as both pure

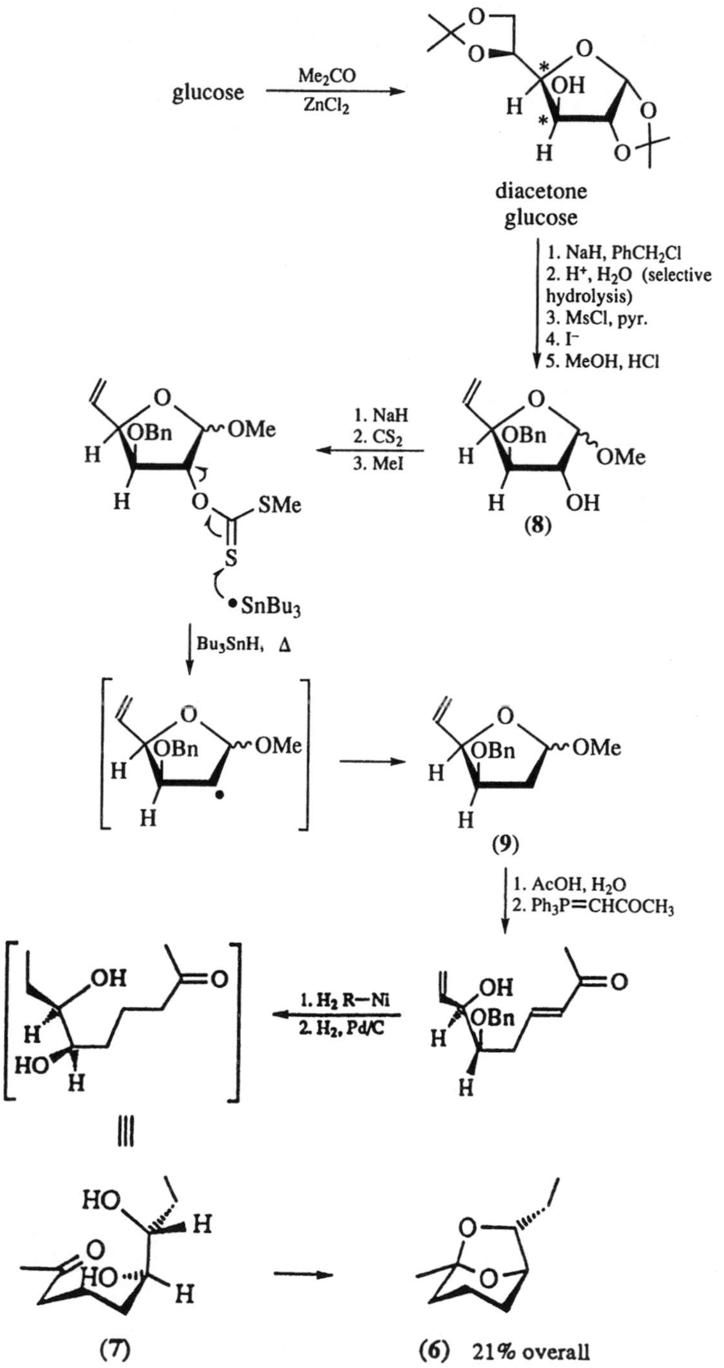
glucose
Me2CO
ZnCl2
diacetone glucose
1. NaH, PhCH2Cl
2. H+, H2O (selective hydrolysis)
3. MsCl, pyr.
4. I−
5. MeOH, HCl
1. NaH
2. CS2
3. MeI
(8)
SMe
•SnBu3
Bu3SnH, Δ
OMe
(9)
1. AcOH, H2O
2. Ph3P=CHCOCH3
1. H2 R—Ni
2. H2, Pd/C
(7)
(6) 21% overall

Scheme 6

enantiomers. Table 1 contains a limited number of terpenes (C_{10}), whose structural variations are limited. Both enantiomers of limonene and carvone are available.

Similarly, most of the compounds in Table 2 are chiral pool-derived. Of these, the largest group is the sugars, most of which are available in only one enantiomeric form. Some terpenes in Table 2 are available as both pure enantiomers (camphor, menthol, α-pinene), but again there is not a great variety in structural type (not surprisingly since they are all derived by similar biogenetic routes).

It is certain that, because of the importance of enantiopure compounds, e.g. for the manufacture of medicines referred to in Chapter 1, the number and variety of available inexpensive enantiopure compounds will continue to increase. In fact, a number of compounds find themselves in Tables 1 and 2 not because they occur plentifully in nature but because they are products of commercial fermentation processes, e.g. 6-aminopenicillanic acid.

Although Tables 1 and 2 are not comprehensive, they do give an indication of the limited number and variety of compounds available at reasonable cost. However, these 'feedstock' enantiopure compounds are of great importance for the preparation of a much larger and considerably more diverse range of

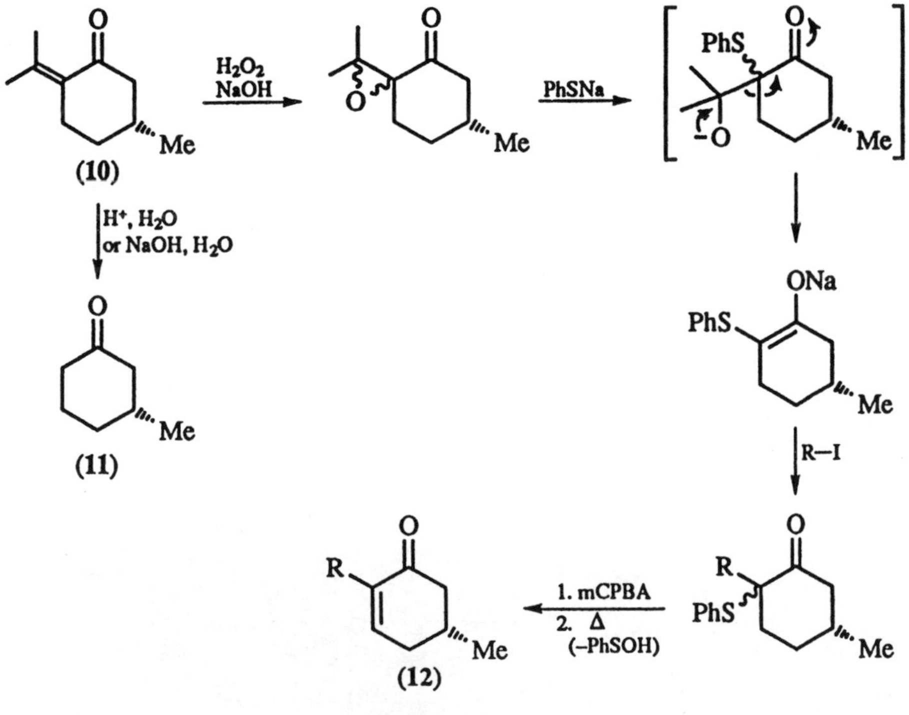

Scheme 7

enantiopure compounds, which are not commercially available. Awareness of these and of methods for their synthesis can make them readily available in a well-found laboratory. We shall refer to these compounds as second-rank starting materials.

Take, for example, the use of pulegone (10) (Scheme 7): it is relatively inexpensive and is transformed into a number of second-rank starting materials, including (R)-3-methylcyclohexanone (11) by retro-aldolisation and (2-substituted)-5-methylcyclohexenones (12) by a modification of this reaction.[8]

Compound (11) (which is commercially available) and compound (12; R=H) (which is not) will always be important starting materials so long as organic synthesis is practised.

Pulegone has also been transformed into (R)-(+)-3-methylcyclopentanone (13) (commercially available) and trans-pulegenic acid (14) (Scheme 8), both important enantiopure building blocks.[9]

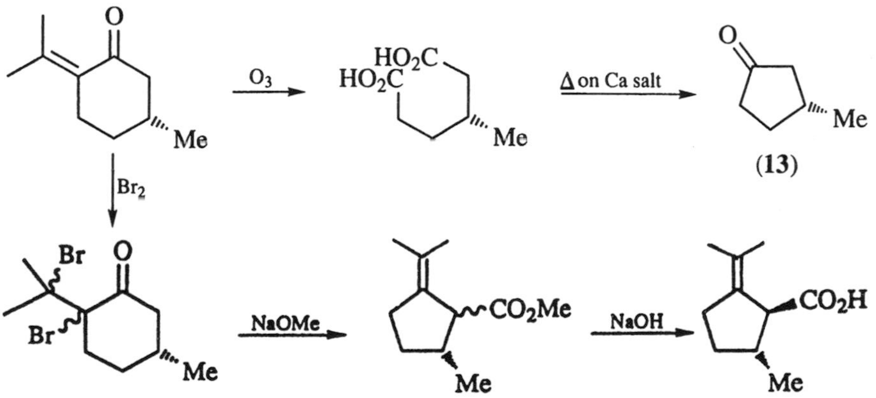

Scheme 8

The Favorskii rearrangement of pulegone dibromide (15) gives a mixture of cis- and trans-esters (16), but hydrolysis yields exclusively trans-pulegenic acid. This is a result of base-catalysed equilibration of the cis- and trans-esters and a retarded rate of hydrolysis of the sterically hindered cis-isomer.

Finally, pulegone can be converted into (R)-citronellic acid (17) as shown in Scheme 9. Citronellic acid is itself a naturally occurring material but is available in higher enantiomeric purity from pulegone.[10]

Many compounds in Tables 1 and 2 have, like pulegone, a number of readily available, but not necessarily commercially available, derivatives. Apart from the disadvantage that for many of the compounds only one enantiomer is available (like pulegone), a further limitation is the lack of variety in their substitution, e.g. the prevalence of hydroxyl groups in sugars

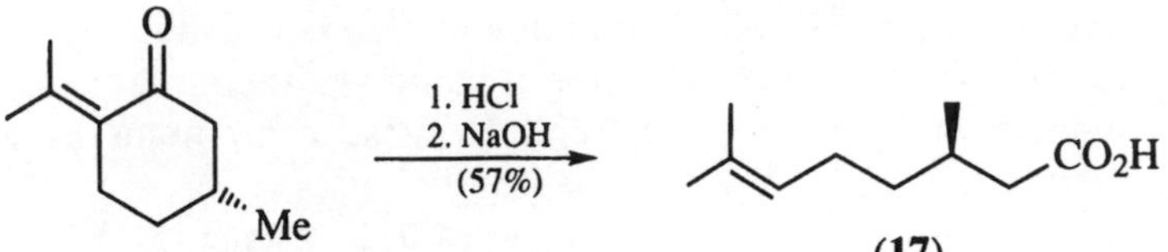

Scheme 9

and the *gem*-dimethyl substitution of many of the terpenes in Table 2. This *gem*-dimethyl group in terpenes is not always wanted, but is not easy to remove or modify chemically. However, camphor (available as both enantiomers) can be transformed by bromination first into 3-bromocamphor **(18)** and then into 3,9-dibromocamphor **(19)** (Scheme 10). Zinc–acid reduction of the labile 3-*endo*-bromine then gives 9-bromocamphor **(20)** (Scheme 10) with the *gem*-dimethyl group functionalised.[11]

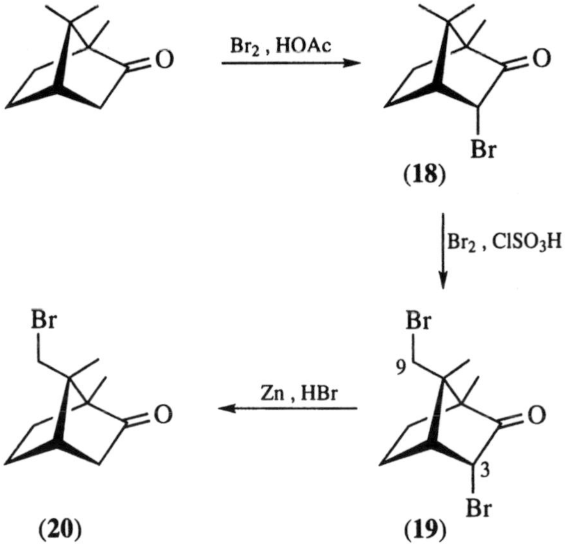

Scheme 10

The extraordinary regioselectivity of this (Type III) 9-methyl bromination of 3-bromocamphor **(18)** is a consequence of the carbocation-mediated and highly disciplined rearrangements peculiar to the bicyclo[2.2.1]heptane system; the mechanism of formation of **(19)** from **(18)** is believed to be that shown in Scheme 11.[12]

Compound **(21)**, which contains two of the three chiral centres present in 9-bromocamphor **(20)**, was used in a synthesis of a steroid; it was obtained from **(20)** as shown in Scheme 12.[13]

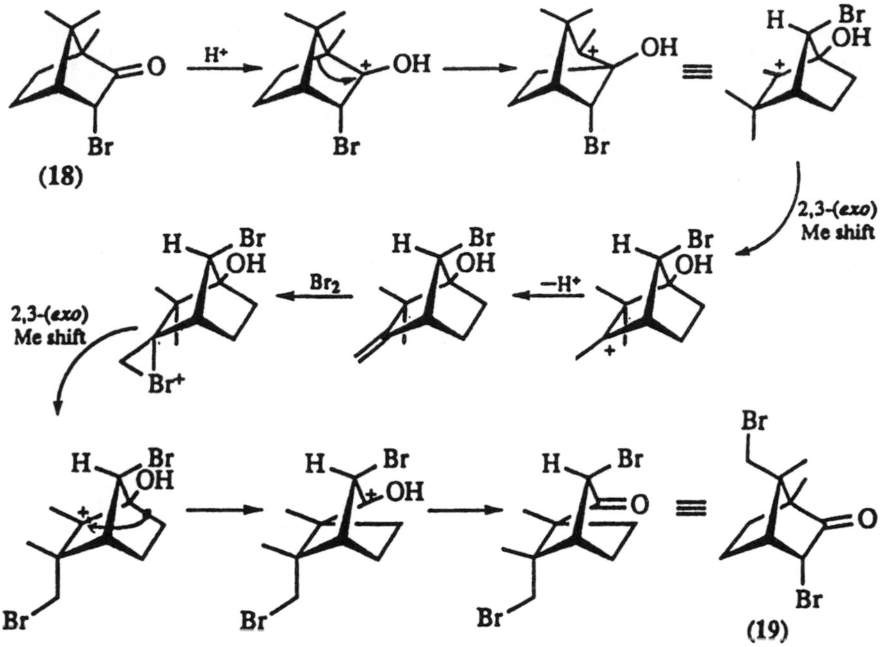

Scheme 11

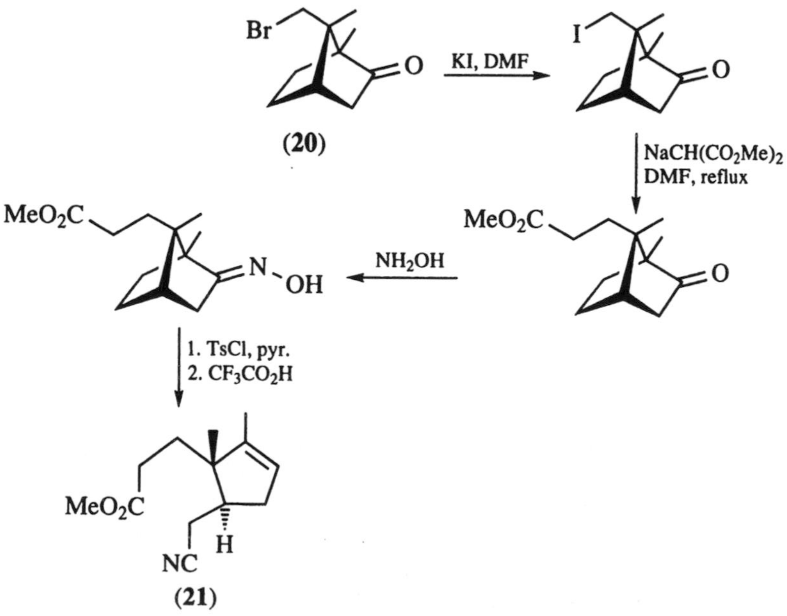

Scheme 12

Just as remarkable as the regioselective formation of the 9-bromocamphor derivative (19) from (18) is the complementary formation of an 8-bromocamphor derivative (22) by the route shown in Scheme 13.[12] Both of these derivatives of camphor and others have provided the chiral centres required for a variety of natural product syntheses.[14]

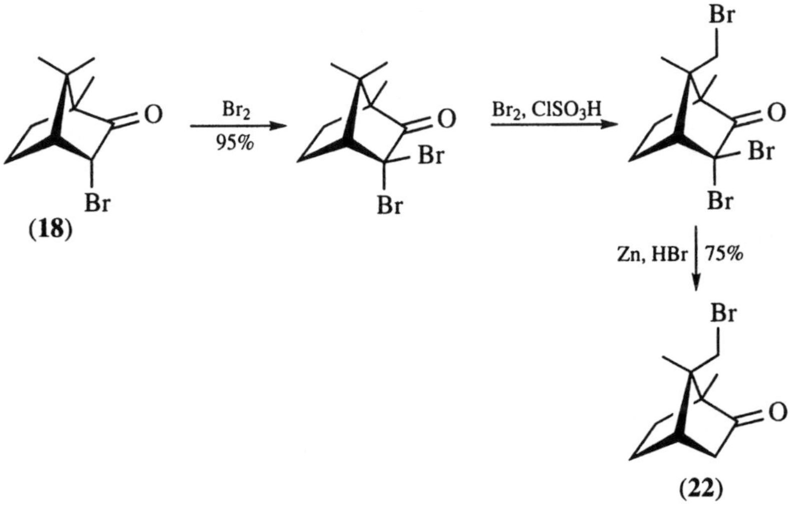

Scheme 13

Neither (20) nor (22) is commercially available but both are in practice now readily available. Awareness of their existence would be unlikely without some acquaintance with the chemistry of camphor.

Synthesis of single diastereoisomers using Type 0 reactions

Although synthesis involving only Type 0 reactions is usually directed at target molecules as single enantiomers, this is by no means always the case. For example, the lactone (23) (Scheme 14) was prepared in racemic form as an intermediate in the synthesis of (±)-compactin, a compound of interest in the study of cholesterol biosynthesis inhibition.[15] Retrosynthesis on lactone (23) (Scheme 14a) leads to the all-cis-trihydroxycyclohexane (24).

The clever parts of this retrosynthesis and the derived synthesis (Scheme 14b) are (i) the recognition of (24) as a second-rank starting material and (ii) the symmetry of the intermediate ketone (25) which obviates the need for regiocontrol in the Baeyer–Villiger reaction.

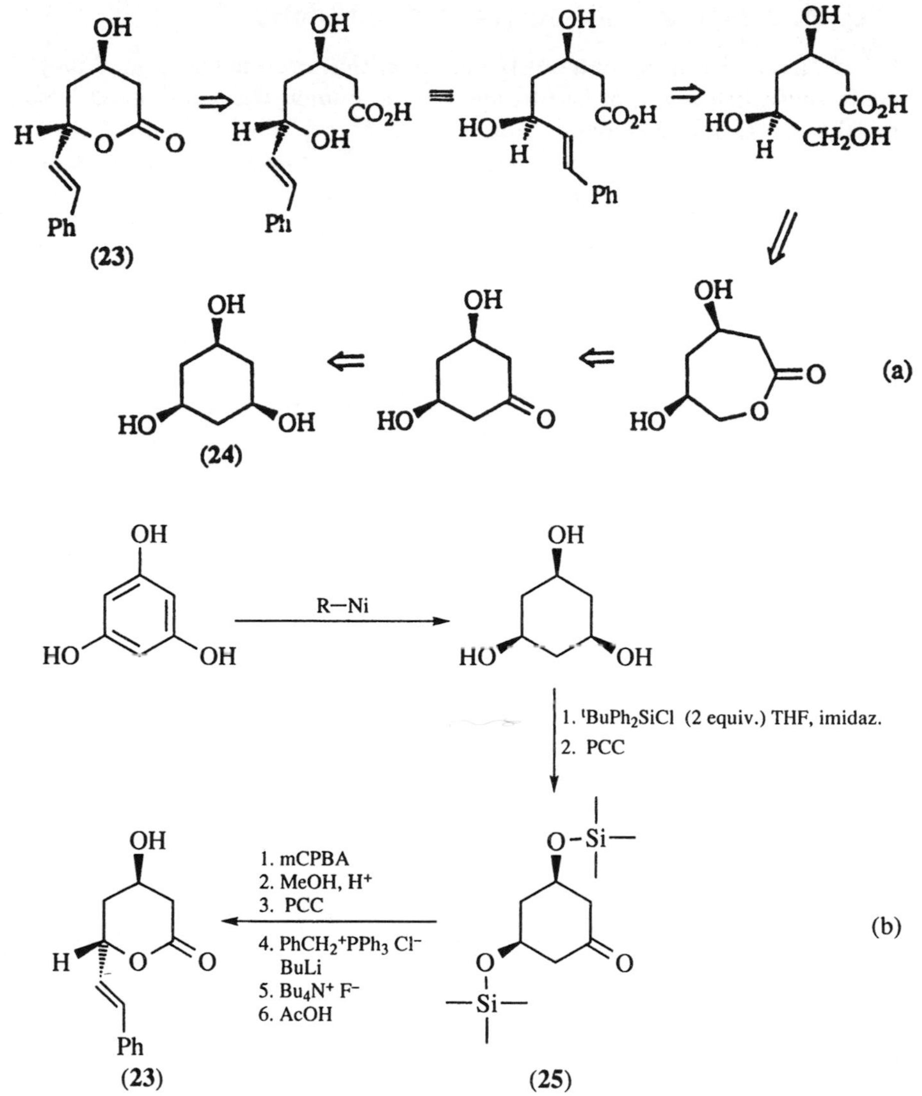

Scheme 14

The number of racemic compounds containing two or more chiral centres which are commercially available as single diastereoisomers is not large by comparison with the number of enantiopure compounds in Tables 1 and 2. Many compounds containing two or more chiral centres which are available as single diastereoisomers in racemic form are obtained from bicyclic compounds by ring cleavage (see below).

Bicyclic compounds as sources of chiral centres

It has already been pointed out (Chapter 2) that even unsubstituted bicyclic compounds can serve as useful sources of monocyclic compounds whose relative or absolute configuration at two chiral centres is defined. This requires that one of the bridges in the bicyclic compound can be cleaved without affecting the configuration at either chiral centre originating from the bridgehead positions. Thus ozonolysis of bicyclo[2.2.1]heptene with oxidative work-up gives *cis*-1,3-cyclopentanedicarboxylic acid (Scheme 15).

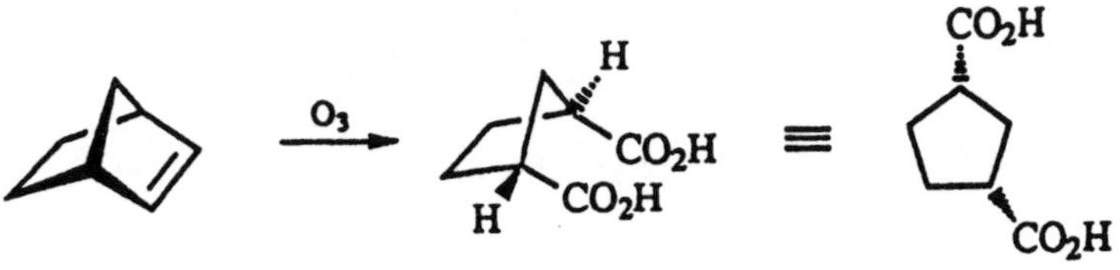

Scheme 15

Since bicyclo[2.2.1]heptene derivatives are readily available by cycloaddition reaction of cyclopentadiene (see Chapter 6), their ring cleavage as in Scheme 15 makes available *cis*-1,3-disubstituted cyclopentanes. In fact, for the synthesis of a compound having any two *cis*-1,3-disposed groups on a five-membered ring, reconnection to the bicyclo[2.2.1] skeleton should always be considered.

The Schreiber modification[16] of the ozonolysis reaction is relevant here since it produces two carbons from cleavage of a symmetrically substituted double bond which are chemo-differentiated (Scheme 16).

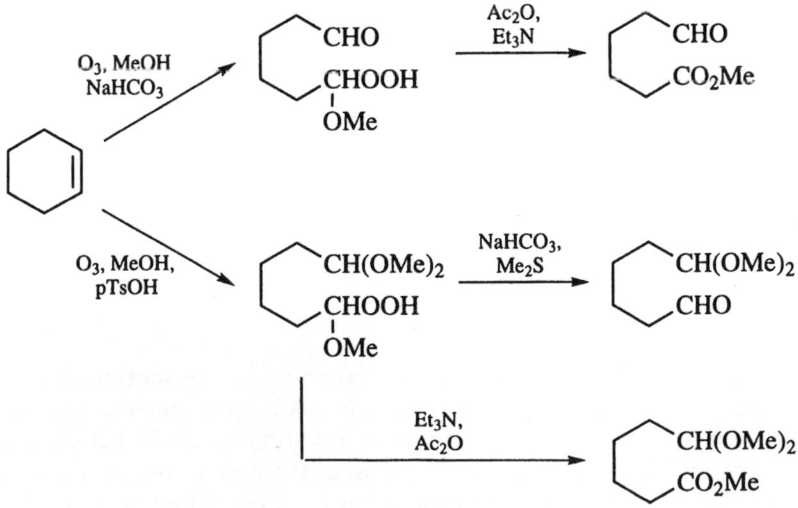

Scheme 16

The tactic of reconnection to a bicyclo[2.2.1]heptene may still be applicable even though one of the *cis*-1,3-substituents in the target molecule is not linked by a C—C bond. Thus, completely stereoselective conversion of C—C → C—N with retention of configuration at carbon, shown retrosynthetically in Scheme 17, can be accomplished using the Hoffmann degradation (a Type I reaction; see Chapter 4).

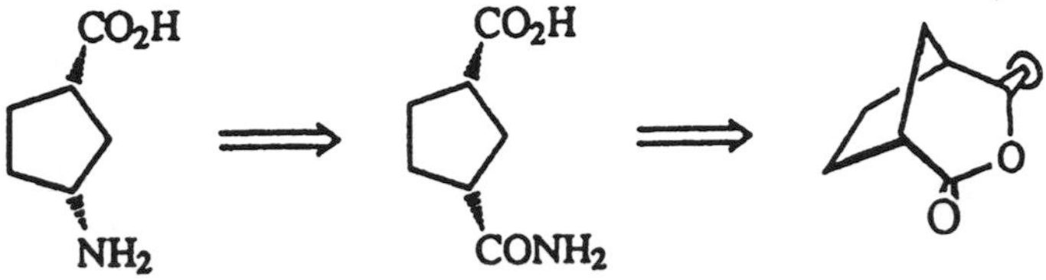

Scheme 17

Synthesis of single stereoisomers of monocyclic compounds by ring cleavage of bicyclic compounds is, of course, a method far wider in scope than is indicated in Scheme 15. Thus, substituted bicyclo[2.2.1]heptenes can be prepared diastereoselectively by Diels–Alder reactions, leading, after ring cleavage to *cis*-1,3-disubstituted cyclopentanes bearing additional chiral centres (see Chapter 6). Also, Diels–Alder cycloaddition, followed by double bond cleavage, can be carried out using dienes contained in rings other than cyclopentadiene, e.g. cyclohexadienes and cycloheptadienes, leading to *cis*-1,3-disubstituted cyclohexanes or cycloheptanes.

Neither is this route limited to the synthesis of *cis*-1,3-substituted rings, since these may be less stable than the corresponding *trans*-1,3-stereoisomers and the presence of carbonyl groups resulting from, e.g., ozonolysis may allow easy base-catalysed thermodynamic equilibration (see Chapter 9).

Finally, the diene may be contained in a heterocyclic ring (e.g. furan) and/or the dienophile may contain heteroatoms, leading to a variety of heteroatom-substituted bicyclic compounds which can serve as progenitors of *cis*(or *trans*)-1,3-disubstituted (heteroatom-containing) rings.

Readily available fissionable bicyclic compounds

The use made of bicyclic compounds for the preparation of disubstituted monocyclic compounds of defined relative configuration by Type 0 reactions (e.g. Scheme 15) will depend upon their availability and on the ease with which one ring can be cleaved. A large number of bicyclic compounds are readily available by a variety of Type II cycloadditions including the Diels–Alder reaction (see Chapter 6).

In bicyclic compounds of type (27), the configuration of one of the chiral centres in (26) is inverted as a result of which one of the bridgehead substituents a projects towards the interior of the bicyclic system. Such compounds are not readily available and are unknown with p+q+r ≤ 9. Exceptionally, however, when one of the bridges p, q or r contains zero atoms, this gives rise to ring-fused bicyclic systems which are available with *cis*- or *trans*-ring junctions, e.g. (28) and (29), respectively.

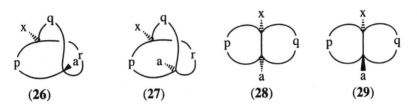

   (26)   (27)   (28)   (29)

Ring cleavage of (28) or (29) at the p or q bridge will produce two adjacent chiral centres in the monocyclic products.

Fissionable bridges

Any reaction which leads to bond breaking can be harnessed to bring about cleavage of a ring provided that the atoms of the functional group involved can be incorporated into a ring. Lactones are particularly useful fissionable rings because they are cleaved under mild conditions to give two chemo-differentiated functional groups. Lactones are often made available by Baeyer–Villiger reaction on a ketone (Type I) as in Scheme 18 using camphor.

Scheme 18

Alternatively, controlled reduction of lactones (DIBAL) gives acetals which are generally in equilibrium with a sufficient concentration of the corresponding aldehyde–alcohol to allow reaction of the latter in, e.g., a Wittig reaction (Scheme 18).

In Scheme 19, nucleophilic attack on the lactone ring of (30) by the lithiomethyl phosphonate followed by oxidation and intramolecular Horner–Emmons reaction is a route to the enantiopure cyclopentenone derivative (31), i.e. a carbocyclic ring can be reconstituted from a lactone.[17]

Bridges in (26) consisting of oxygen or nitrogen atoms (O—O, N—O, N—N) are easily cleaved by reductive methods (Scheme 20). As indicated in Scheme 20, singlet oxygen O=O, appropriately substituted nitroso (RN=O) and azo (RN=NR) compounds are reactive dienophiles in the Diels–Alder reaction.[18–20]

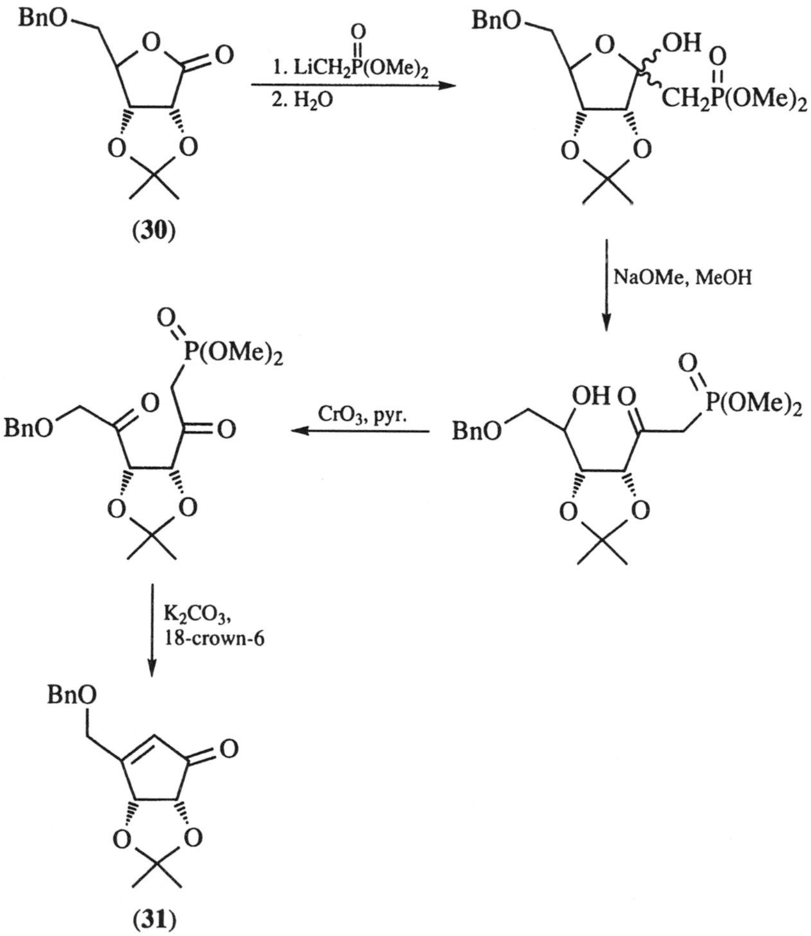

Scheme 19

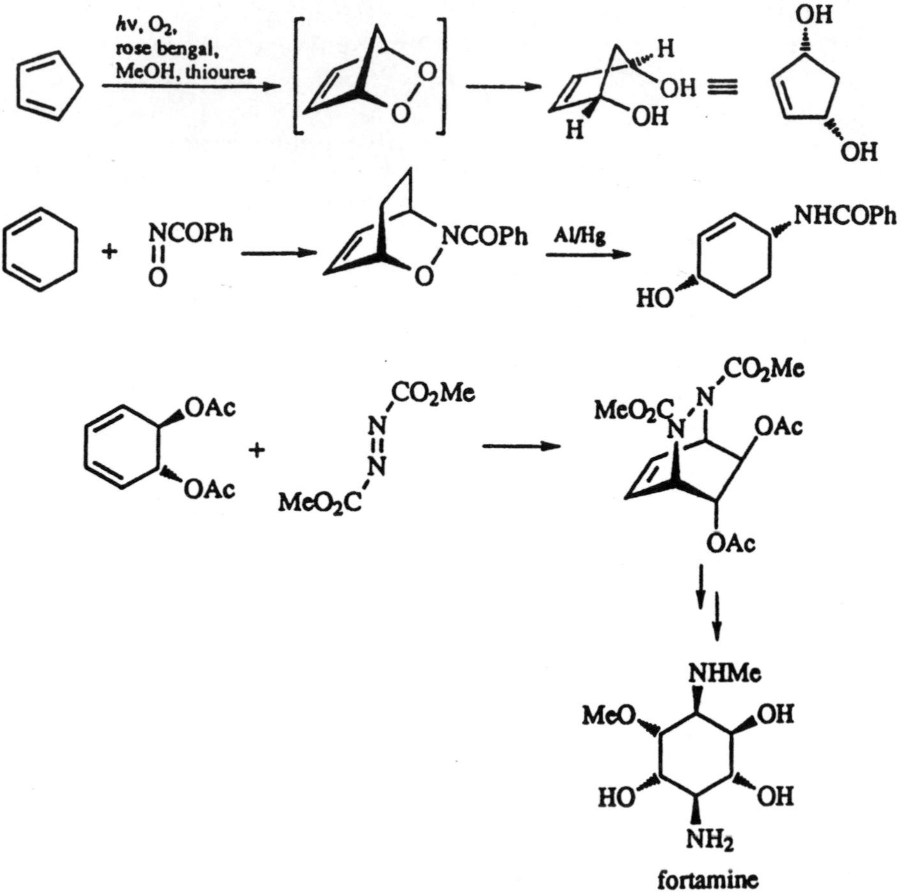

Scheme 20

A ketone-containing bridge may be directly oxidised to two carboxylic acids. Often, however, it may be expedient to carry out some preliminary FGI to render this bridge more easily or efficiently broken or to ensure that the functional groups arising from the cleavage are appropriate for the further transformations which are required. Thus a ketone may be converted into its silyl enol ether and then cleaved more easily by, for example, ozonolysis (Scheme 21) even in the presence of a lactone.[21]

Although we have illustrated the use of bicyclic compounds in the preparation of *mono*cyclic compounds having two chiral centres of defined relative configuration, the presence of two fissionable bridges in the bicyclic compound will allow the synthesis of acyclic compounds having two such chiral centres as in Scheme 22. In this case, the racemic bicyclic system (prepared by an intramolecular 1,3-dipolar cycloaddition; see Chapter 6) was

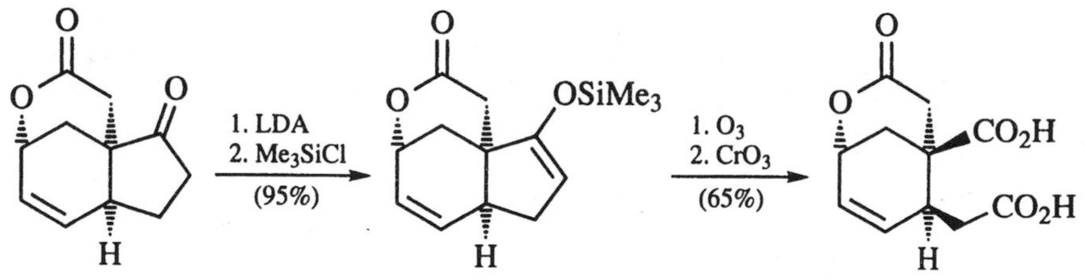

Scheme 21

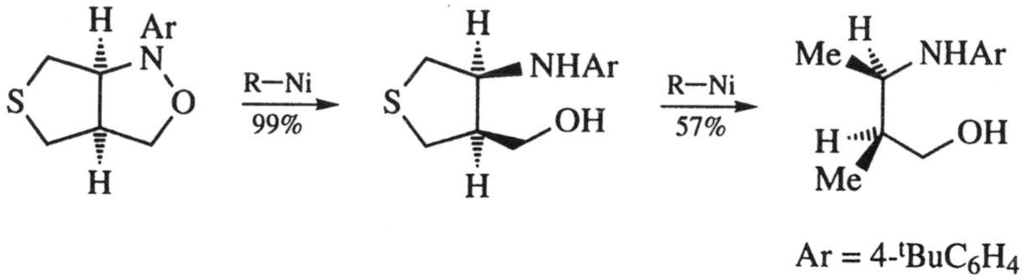

Ar = 4-^{t}BuC$_6$H$_4$

Scheme 22

first reductively cleaved at the N—O bond and subsequently the sulphide-containing ring gave rise to two methyl groups on further reduction with Raney nickel.[22]

We have restricted this discussion so far to the application of bicyclic compounds in Type 0 synthesis where two chiral centres at bridgehead or ring junction positions are retained in the product. However, the presence of additional chiral centres on the bicyclic skeleton may allow alternative cleavage reactions to be explored which result in destruction of one (or both) chiral centres at the bridgehead or ring junction and yet allow retrieval of two or more chiral centres from the bicyclic starting material in the monocyclic product. These additional chiral centres may be present either as a result of substituents on the partners in the reaction giving rise to the bicyclic compound (e.g. a cycloaddition) or they may be introduced into the bicyclic system by Type III reactions for which many bicyclic compounds are particularly well suited (Chapter 10). In Scheme 23, the sulphone-substituted 7-oxabicyclo[2.2.1]heptene (**32**), prepared from the ketone (**33**) as shown, suffers cleavage of its oxygen bridge on treatment with methyllithium. Addition of the methyl anion is completely stereoselective (Type III) and although one of the chiral centres at the bridgehead is lost, two chiral centres are retained in the product (**34**).[23]

The ketone (**33**) is obtained by a sequence which starts with a

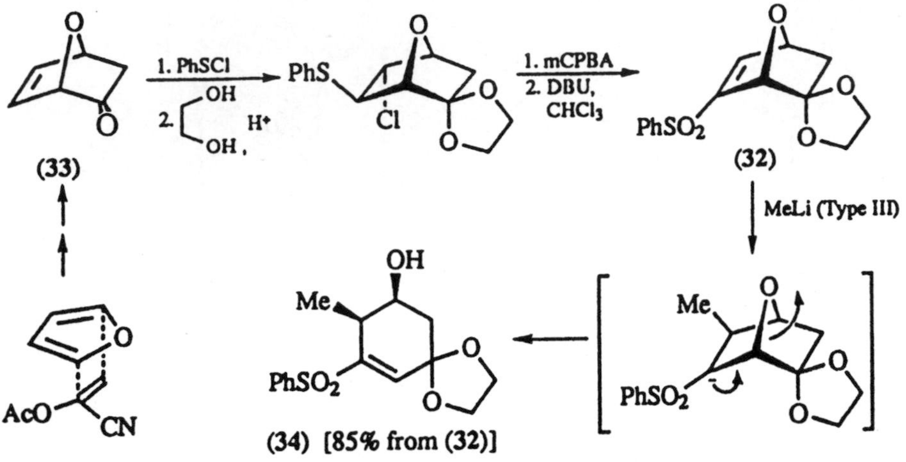

Scheme 23

Diels–Alder reaction between furan and α-acetoxyacrylonitrile; it has been resolved by Black and Vogel[24] via brucine complexes of the cyanohydrin derivatives.

Readily available fissionable tricyclic compounds

The use of tricyclic compounds for the synthesis of bicyclic/monocyclic compounds is widespread. However, the small number of commercially available tricyclic compounds of any complexity means that assembly of the required tricyclic compound will almost always be necessary with substitution on one or more bridges appropriate for their cleavage as required.

Some derivatives of adamantane serve as precursors for 3,7-disubstituted bicyclo[3.3.1]nonanes (Scheme 24).[25]

The known tricyclic noradamantanone (**35**) (Scheme 25) has been used to synthesise the azabicyclo[4.3.0]nonane (**36**), which was shown to be identical with one of the products resulting from solvolysis of the azatwistane derivative (**37**).[26]

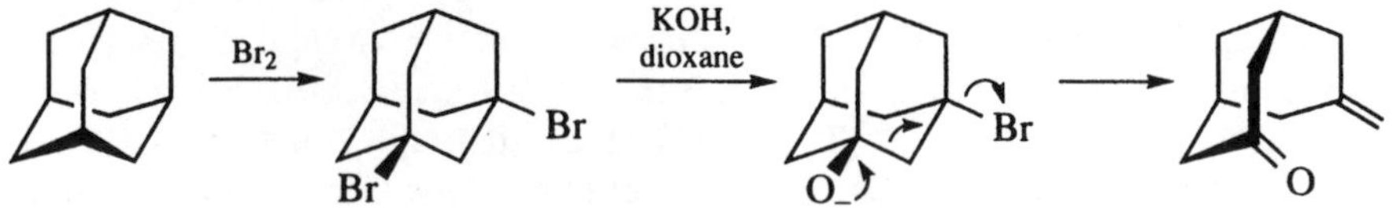

Scheme 24

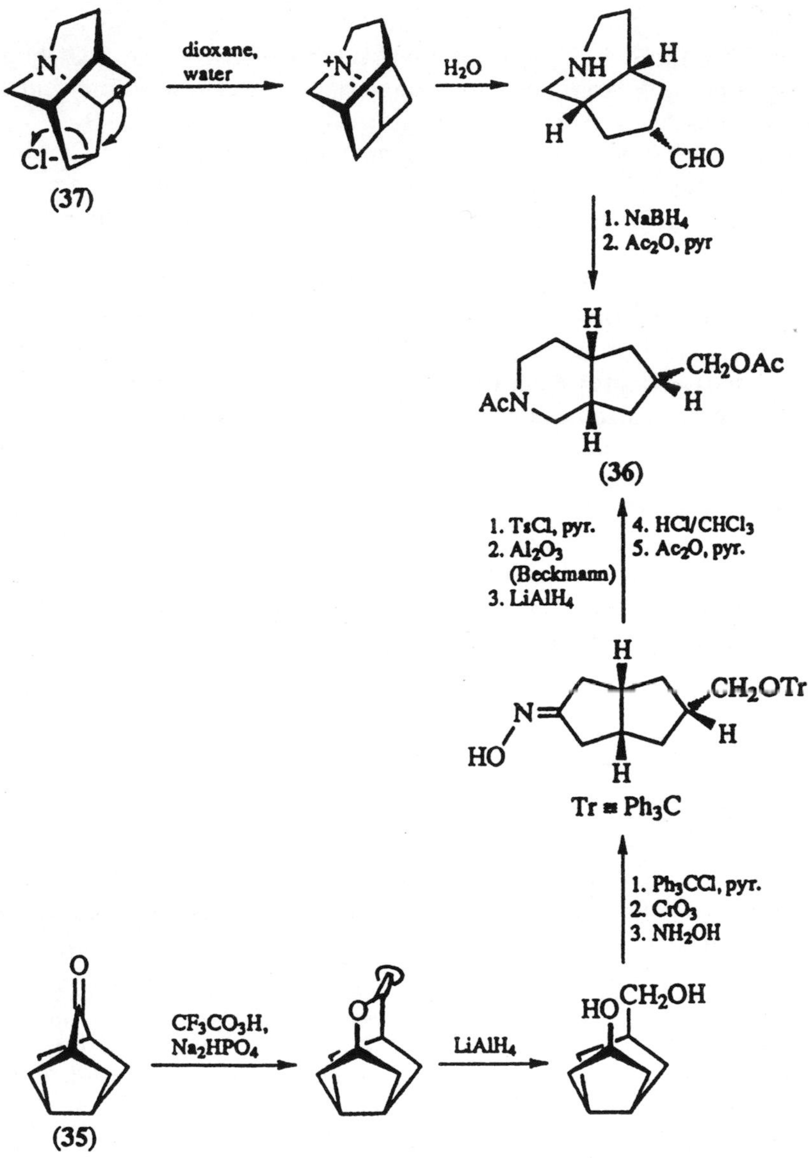

Scheme 25

As always in synthesis involving only Type 0 reactions, the clever part of the design is awareness of the availability of the tricyclic starting material (**35**) together with a recognition of a viable pathway linking it to the target.

Summary

Synthesis using only Type 0 reactions relies heavily on the availability and recognition of appropriate starting materials in which the chiral centre(s) in the target molecule must already be present.

Although the most useful starting materials containing two chiral centres in Type 0 synthesis are those which are single enantiomers, racemic compounds may also be of use in the synthesis of target molecules in racemic form.

The number of cheap enantiopure compounds which are commercially available is small and their variety is limited. However, in the hands of a competent chemist working in a well-found laboratory this small number of compounds can provide access to a much larger number of second-rank starting materials, most of which are not commercially available.

Bicyclic compounds are fertile sources of monocyclic and acyclic molecules having at least two chiral centres of defined relative or absolute configuration. The two chiral centres present at the bridgehead or ring junction positions of these bicyclic compounds are conserved in Type 0 reactions which open one or both rings.

Awareness of the existence of second-rank starting materials and retrieval of methods for their preparation can be accomplished by chemical literature searching methods including computer-assisted searches. Ultimately, however, there is no substitute for familiarity with as wide a range of organic chemistry as possible and particularly with the chemistry of the compounds in Tables 1 and 2.

References

1. R.K. Olsen, K.L. Bhat, R.B. Wardle, W.J. Hennen and G.D. Kini, *J. Org. Chem.*, 1985, **50**, 896.
2. M. Taniguchi, K. Koga and S. Yamada, *Tetrahedron*, 1974, **30**, 3547.
3. T. Fujisawa, T. Itoh, M. Nakai and T. Sato, *Tetrahedron Lett.*, 1985, **26**, 771.
4. G.M. Coppola and H.F. Schuster, *Asymmetric Synthesis: Construction of Chiral Molecules Using Amino Acids*, Wiley-Interscience, New York, 1987; T.L. Ho, *Enantioselective Synthesis, Natural Products from Chiral Terpenes*, Wiley, New York, 1992.
5. S. Hanessian, *Total Synthesis of Natural Products: the Chiron Approach*, Pergamon Press, Oxford, 1983.
6. A.E. Sherk and B. Fraser-Reid, *J. Org. Chem.*, 1982, **47**, 932.
7. D.H.R. Barton and S.W. McCombie, *J. Chem. Soc., Perkin Trans. 1*, 1975, 1574.
8. D. Caine, K. Procter and R.A. Cassell, *J. Org. Chem.*, 1984, **49**, 2647.
9. J. Wolinsky and D. Chan, *J. Org. Chem.*, 1965, **30**, 41.
10. C.G. Overberger and J.K. Weise, *J. Am. Chem. Soc.*, 1968, **90**, 3525.
11. E.J. Corey, S.W. Chow and R. A. Scherrer, *J. Am. Chem. Soc.*, 1957, **79**, 5773; W.L. Meyer, A.P. Lobo and R.N. McCarty, *J. Org. Chem.*, 1967, **32**, 1754.
12. C.R. Eck, R.W. Mills and T. Money, *J. Chem. Soc., Perkin Trans. 1*, 1975, 251.
13. R.V. Stevens and F.C.A. Gaeta, *J. Am. Chem. Soc.*, 1977, **99**, 6105.

14. T. Money, *Nat. Prod. Rep.*, 1985, **2**, 253.
15. K. Prasad and O. Repic, *Tetrahedron Lett.*, 1984, **25**, 2435.
16. S.L. Schreiber, R.E. Claus and J. Reagen, *Tetrahedron Lett.*, 1982, **23**, 3867.
17. V.E. Marquez, M.-I. Lim, C.K.-H. Tseng, A. Markovac, M.A. Priest, M.S. Khan and B. Kaskar, *J. Org. Chem.*, 1988, **53**, 5709.
18. C. Kaneko, A. Sugimoto and S. Tanaka, *Synthesis*, 1974, 876.
19. G.E. Keck and S.A. Fleming, *Tetrahedron Lett.*, 1978, 4763.
20. C.H. Kuo and N.L. Wendler, *Tetrahedron Lett.*, 1984, **25**, 2291.
21. T.V. Lee and J. Toczek, *Tetrahedron Lett.*, 1985, **26**, 473.
22. H.G. Aurich and K.-D. Möbus, *Tetrahedron Lett.*, 1988, **29**, 5755.
23. J.L. Acena, O. Arjona, R.F. de la Pradilla, J. Plumet and A. Viso, *J. Org. Chem.*, 1992, **57**, 1945, and references cited therein.
24. K.A. Black and P. Vogel, *Helv. Chim. Acta*, 1984, **67**, 1612.
25. A.R. Gagneux and R. Meier, *Tetrahedron Lett.*, 1969, 1365.
26. W. Holick, E.F. Jenny and K. Heusler, *Tetrahedron Lett.*, 1973, 3421.

4 TYPE I REACTIONS: THOSE PROCEEDING WITH EITHER INVERSION OR WITH RETENTION OF CONFIGURATION AT A SINGLE CHIRAL CENTRE

In Type I reactions, at least one bond to a chiral centre is broken and a new bond is formed completely stereoselectively.

Inversion of configuration

Scheme 1 illustrates the inversion of configuration which characterises an S_N2 reaction. The bond changes which accompany the S_N2 reaction take place *concertedly*: as the Nu—C bond is forming so the C—X bond is breaking and a single transition state (**1**) separates the starting materials and products.

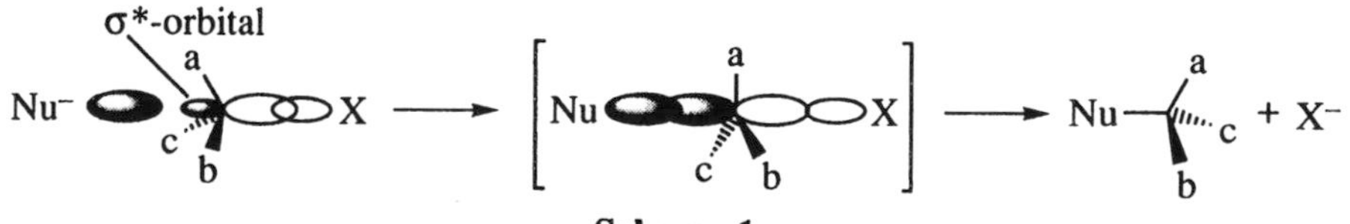

Scheme 1

Stereoelectronic control

In concerted reactions such as the S_N2 type, the orbitals which are involved in making new bonds have a defined spatial relationship to those orbitals which arise from the breaking of existing bonds. Thus, in the S_N2 reaction (Scheme 1), that part of the σ^* orbital shown is directed at an angle of 180° to orbitals comprising the C—X σ bond undergoing cleavage. Bond formation, therefore, which occurs by overlap of this empty σ* orbital with the filled orbital of the nucleophile, takes place with inversion of configuration as illustrated.

When, as in the S_N2 reaction, a specific spatial relationship exists between bonds made and broken, the reaction is said to proceed under stereoelectronic control.

Transition-state geometry (TSG)

In Scheme 1, the transition state is assumed to be (1) in which C—a, C—b and C—c bonds lie in a plane and the Nu—C and C—X bonds are colinear. In all probability this *transition-state geometry* (TSG) is an idealised representation for most S_N2 reactions. If, for example, C—X bond breaking runs ahead of Nu—C bond making, or vice versa, then this will result in a transition state in which C—a, C—b and C—c bonds deviate from coplanarity.

Limitations of the S_N2 reaction in stereoselective synthesis

Since the S_N2 reaction is sensitive to steric effects, ideally this reaction requires attack by a non-bulky and highly polarisable nucleophile on a sterically unencumbered sp³-hybridised (carbon) atom bearing a good leaving group.

The most widely used leaving groups are halides and OZ groups which are esters of the hydroxyl group with strong acids, e.g. tosylates (Z = OSO_2-*p*-tolyl), mesylates (Z = OSO_2Me), triflates (Z = OSO_2CF_3). The best nucleophiles are those bearing a negative charge including CN^-, N_3^-, RS^-, I^-, etc.

The stereoelectronic control present in the S_N2 results in inversion of configuration in the product. This complete stereoselectivity is impervious to the presence of other chiral centres elsewhere in the molecule. A comparison between the stereochemistry of S_N2 and S_N1 substitutions here is instructive. The S_N1 reaction can occasionally also be highly stereoselective (Scheme 2) if the carbocation through which it proceeds is captured by a nucleophile predominantly from one face.[1] However, if the latter does occur it is invariably the result of the molecular environment in which the carbocation is generated as in ion-pair formation in Scheme 2: unlike the S_N2 reaction, the S_N1 reaction is not an inherently stereoselective (stereospecific) reaction.

In spite of the stereochemical reliability of the S_N2 reaction there are a number of factors which limit the usefulness of *inter*molecular S_N2 reactions in stereoselective synthesis. First, attack at tertiary centres is so sterically retarded that reaction, if it occurs at all, results in elimination and alkene formation rather than substitution (Scheme 3).

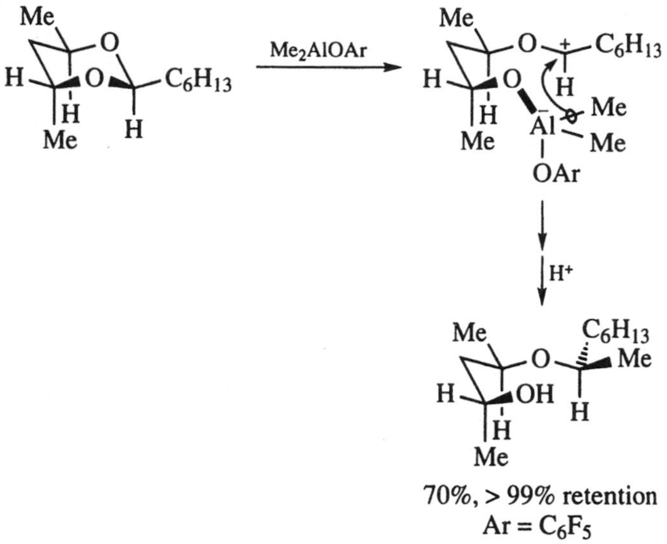

70%, > 99% retention
Ar = C$_6$F$_5$

Scheme 2

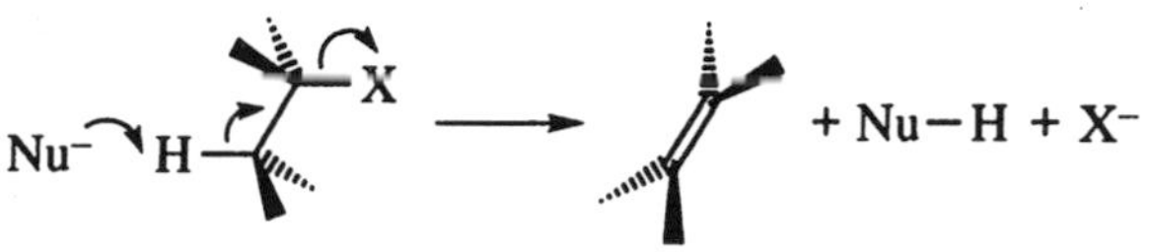

Scheme 3

Second, although primary carbon centres are well suited for S_N2 substitution, inversion of configuration is only apparent when deuterium (or tritium) substitution is present (Scheme 4).

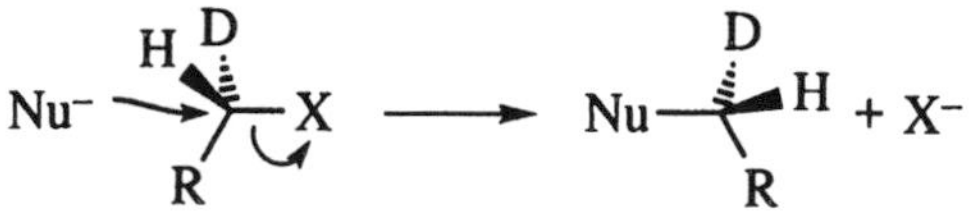

Scheme 4

Third, attempted S_N2 substitution at secondary centres is also often blighted by competitive elimination especially when the nucleophile has any basic character. Completely stereoselective formation of carbon–carbon bonds by intermolecular S_N2 substitution at secondary carbon centres by many carbon-centred nucleophiles is unreliable or low-yielding, although it is successful in the reaction in Scheme 5, one of the early steps in the synthesis of sarracenin (2),[2] a constituent of the roots and leaves of the golden trumpet, *Sarracenia flava.*

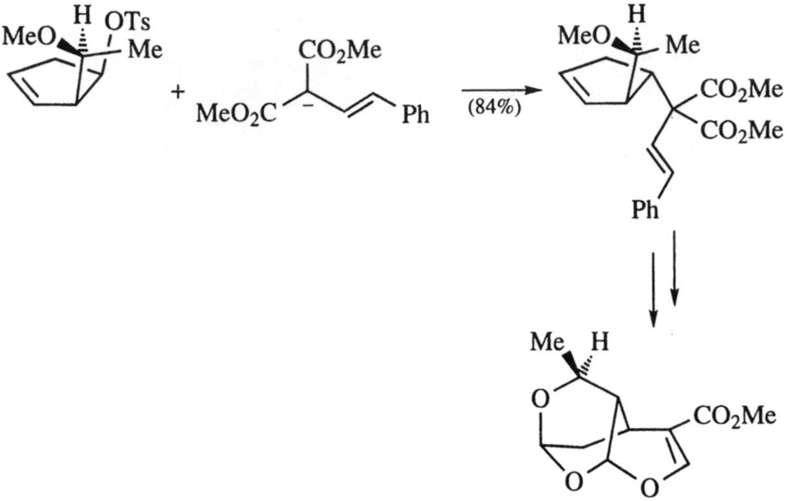

Scheme 5

In epoxides, the angle between the two exocyclic bonds is widened as a consequence of the higher s-character which these bonds contain (see p. 104). This increased s-character is itself a consequence of the higher p-character which is required for formation of the ring bonds to accommodate the geometry of the small ring (Figure 1). Widening of the bond angles as in Figure 1 allows easier ingress of the nucleophile. Complexation of the epoxide oxygen can assist in the ring opening such that the substitution has both 'push' from the nucleophile and 'pull' from the leaving group [Scheme 6(a)].[3]

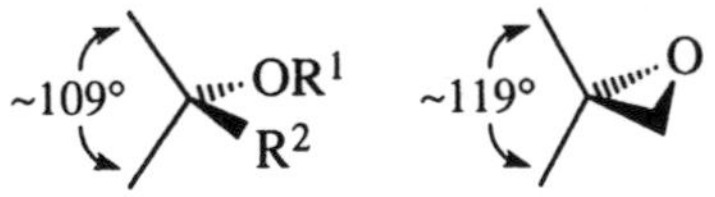

Figure 1

The diminished bascity and increased nucleophilicity of carbanions when complexed with copper (cuprates) facilitates substitution at secondary centres [Scheme 6(b)][4] and particularly those on epoxide rings [Scheme 6(c)].[5]

In Scheme 6(c), the epoxide is opened at the less hindered position highly regio-selectivity (11:1) and also completely stereoselectively. However, as the example in Scheme 6(d) shows, epoxide ring opening by nucleophiles does not necessarily occur at the sterically less hindered position.[6] In this case attack of the nucleophile is believed to take place initially by addition to the silicon substituent and is thence directed to the more hindered position on the ring.[7]

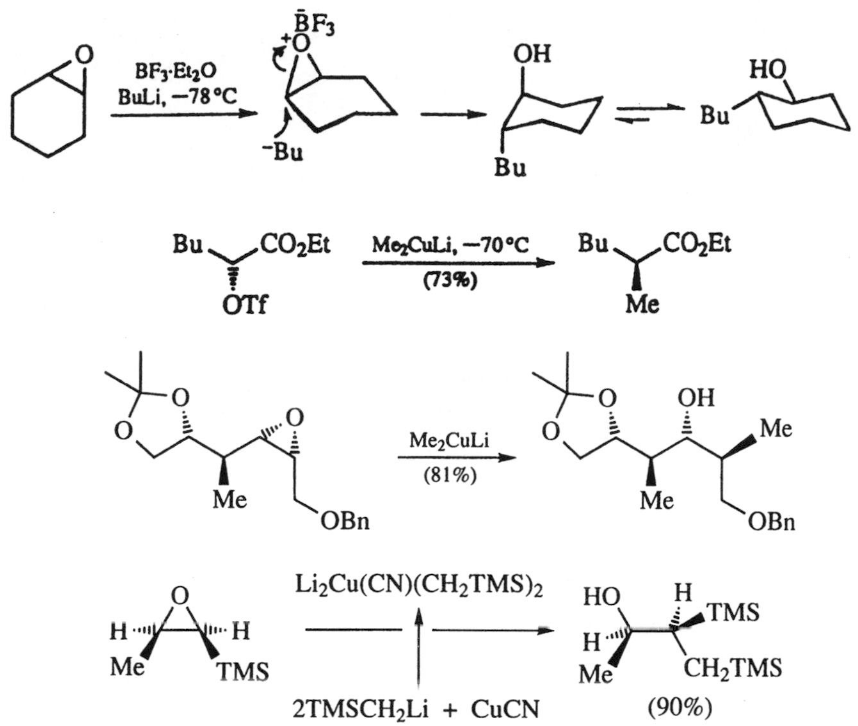

Scheme 6

S_N2 substitution at secondary centres is also facilitated when steric congestion in the transition state is reduced in other ways. This is the case when one or more of the substituents on C_α in the substrate is sp^2- or sp-hybridised, i.e. a vinyl or ethynyl group. Likewise, a heteroatom substituent (O, N, S) is smaller by virtue of the lone pair(s) (rather than substituents) that it bears. Unfortunately, these unsaturated substituents and heteroatoms may also facilitate S_N1 substitution (C$^+$ stabilisation) with its only occasionally complete stereoselectivity.

Ring opening of the enantiopure epoxide (**3**) (Scheme 7) with azide ion followed by reaction of (**5**) and (**6**) with triphenylphosphine (Staudinger reaction) results in the formation of aziridine (**4**).[8] In spite of the fact that the ring opening is not regiospecific, only a single enantiopure aziridine is obtained.

The mechanism in Scheme 7 [shown for one regioisomer (**5**) only] accounts for this because each chiral centre in the epoxide undergoes inversion of configuration in the conversion to (**4**). In the formation of (**4**) via the other regioisomer (**6**), the order in which the two inversions occur at the two chiral centres is reversed.

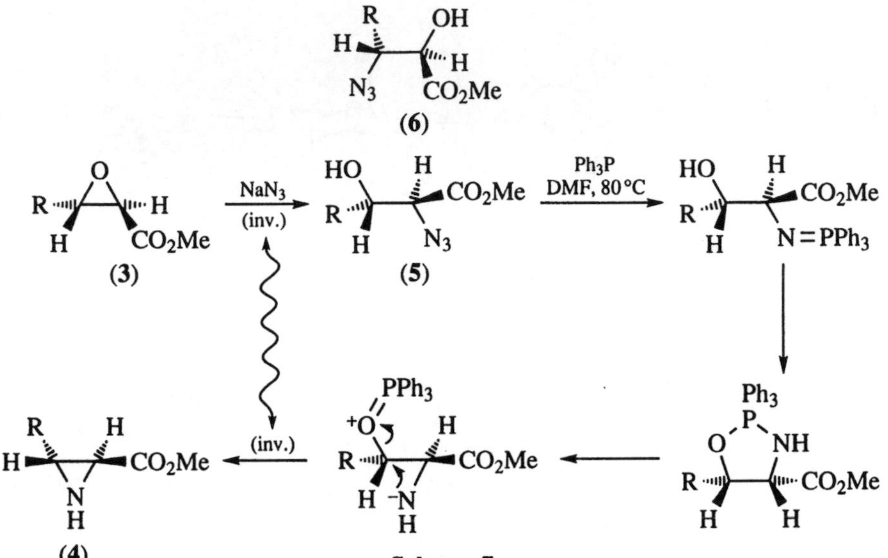

Scheme 7

Like carbanions, oxyanions ^-OR are also basic and the classical method for replacement of one C—O bond by another O—C bond with inversion, e.g. (7) → (8) via (9) (Scheme 8), is invariably accompanied by some elimination as a result of the basic nature of the oxygen-centred nucleophile (hydroxide).[9] This transformation can now be accomplished in better yield by making use of the Mitsunobu reaction followed by hydrolysis of (10) (Scheme 8).

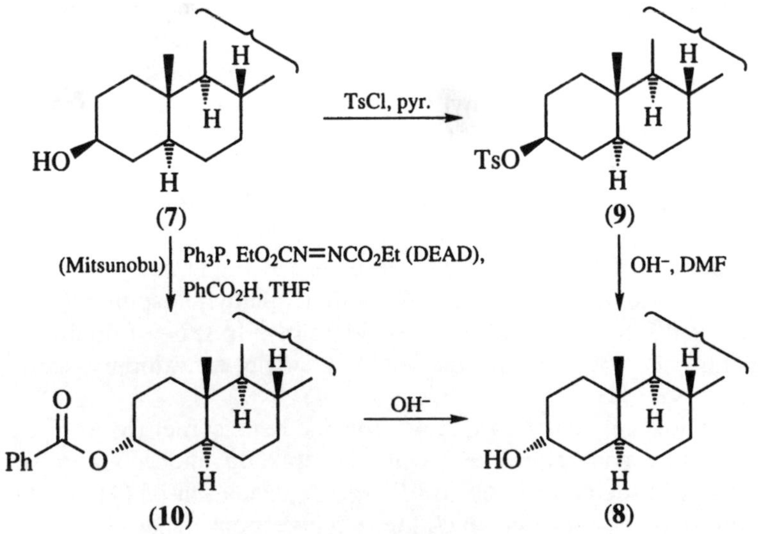

Scheme 8

The Mitsunobu reaction is believed to proceed via the mechanism shown in Scheme 9 and is now widely used for converting a secondary alcohol directly into its inverted ester with the nature of the ester being determined by the acid which is added.[10] This reaction can also be used (Scheme 10) for the construction of large-ring compounds as in the closure of (11) to the ring system (12) of gleosporone, a fungal germination self-inhibitor.[11]

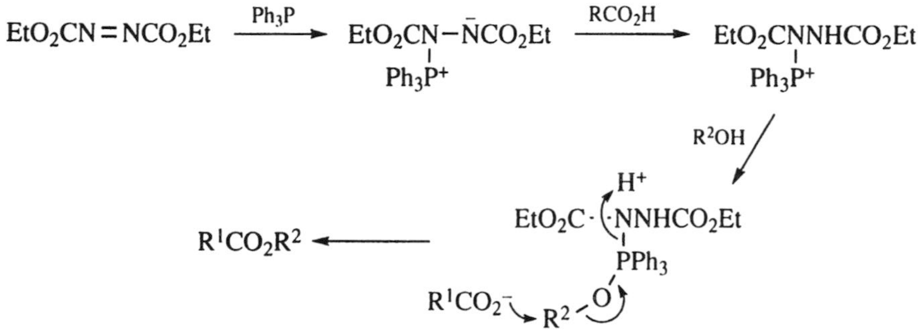

Scheme 9

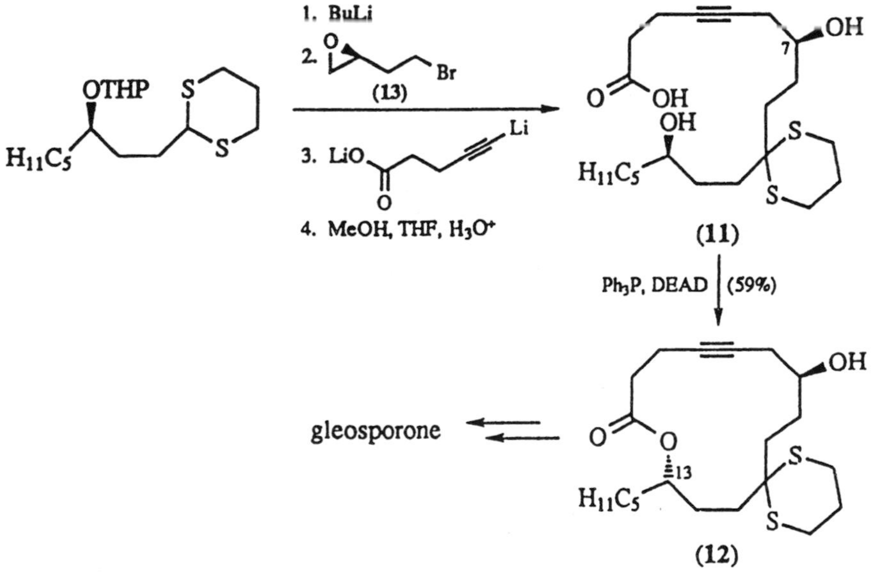

Scheme 10

Enantiopure bromoepoxide (13), used in the assembly of (11), was prepared[12] from the diester of the readily available (S)-malic acid by the route shown in Scheme 11, which uses exclusively Type 0 reactions.

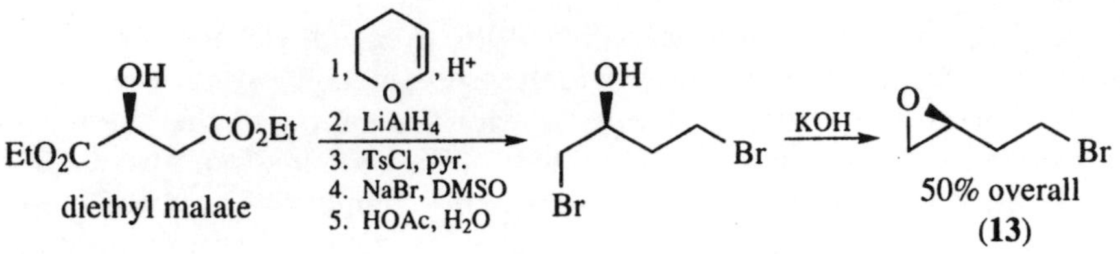

Scheme 11

The intramolecular Mitsunobu reaction in Scheme 10 occurs with inversion of configuration at C-13. Ring closure to form an eight-membered ring with the C-7 hydroxyl does not compete, presumably because of the strain imparted by the triple bond which this ring and the transition state leading to it would include.

Other methods for C—O→O—C conversion with inversion of configuration which minimise yields of alkene by-products are shown in Scheme 12:[13] some of

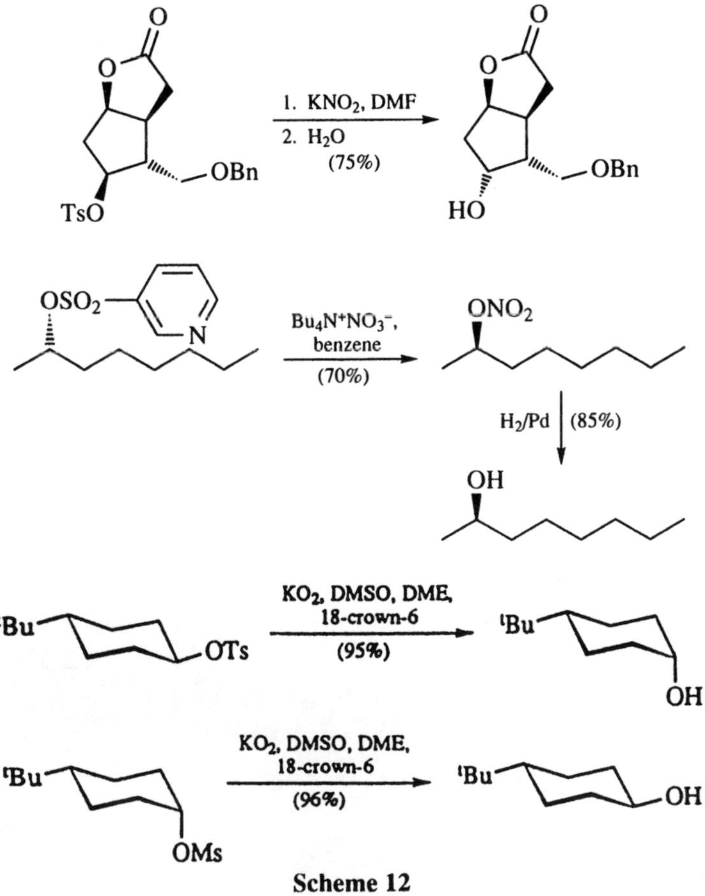

Scheme 12

these methods were developed for prostaglandin synthesis, where elimination to give alkenes was a particular problem with existing methods.

The factors referred to above which limit the use of intermolecular S_N2 reactions in stereoselective synthesis are less serious in their intramolecular versions, particularly when three-, five- or six-membered rings are being formed (Schemes 13 and 14).[14,15]

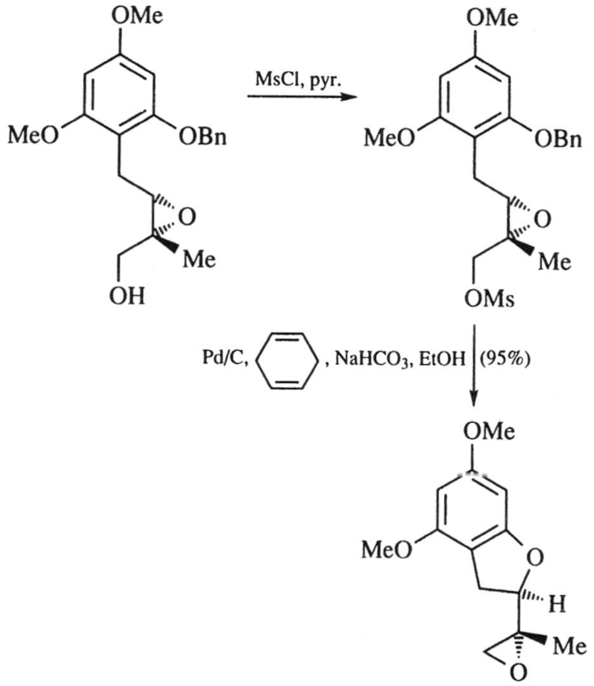

Scheme 13

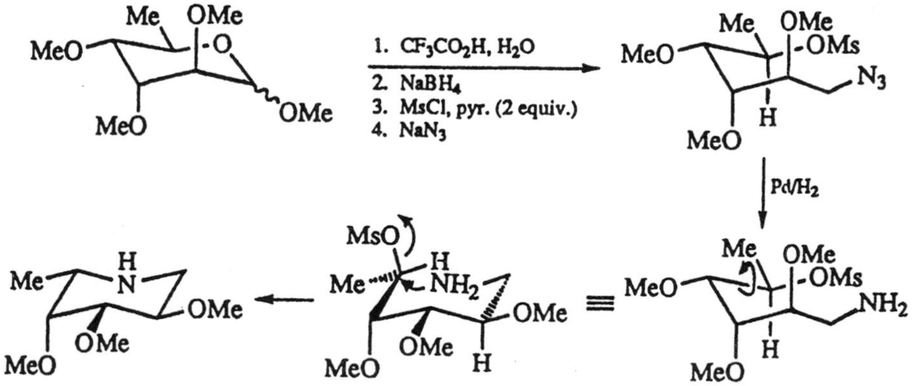

Scheme 14

Appropriately sited carbon nucleophiles can also react in an intramolecular S_N2-type reaction (Scheme 15) at secondary centres.[16]

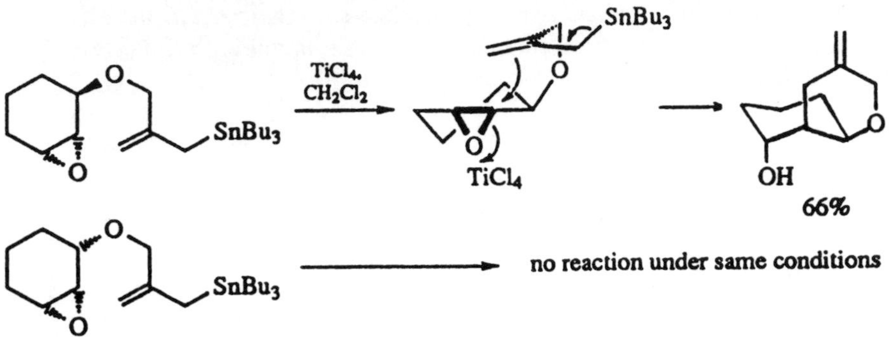

Scheme 15

The mechanism of the reaction in Scheme 16 (which was supported by studies using ^{18}O labelling) shows that an intramolecular S_N2 reaction can take place even at a tertiary centre.[17]

Intramolecular epoxide ring opening by the hydroxyl in an allylic alcohol epoxide with the formation of a new epoxide is known as the Payne rearrangement and is a common feature of epoxide chemistry.

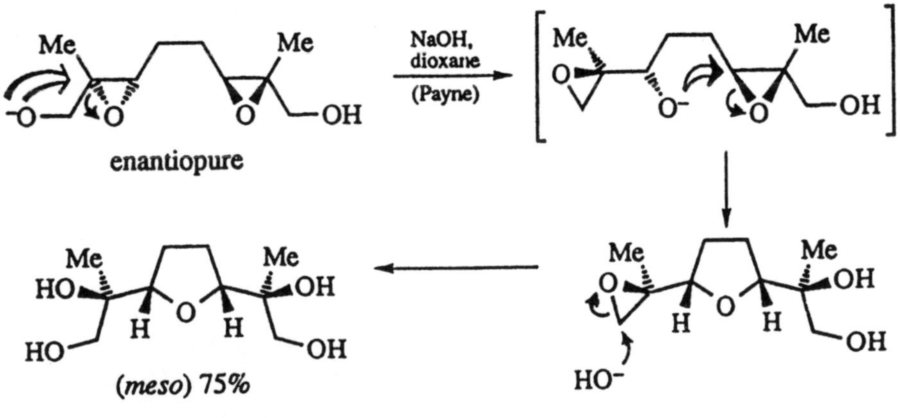

Scheme 16

Thus, treatment of the enantiopure acid in Scheme 17 with sodium ethoxide followed by acetylation gave diacetate (**14**), whose relative and absolute configurations suggest that a Payne rearrangement is involved and inversion at both adjacent chiral centres has occurred.[18]

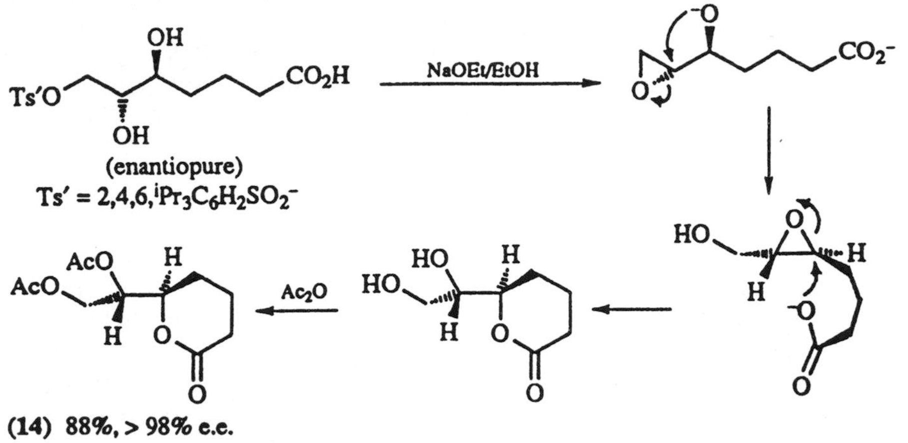

Scheme 17

Double inversion = retention

The conversion of (S)-glutamic acid into the butyrolactone acid (Scheme 18) proceeds with retention of configuration which is the result of double inversion:[19] an intramolecular S_N2 reaction with inversion and formation of an unstable α-lactone is followed by a second intramolecular substitution with inversion. α-Lactones are unusual in undergoing ring opening as shown rather than by attack at the carbonyl group, which is the case with five- and six-membered ring lactones.

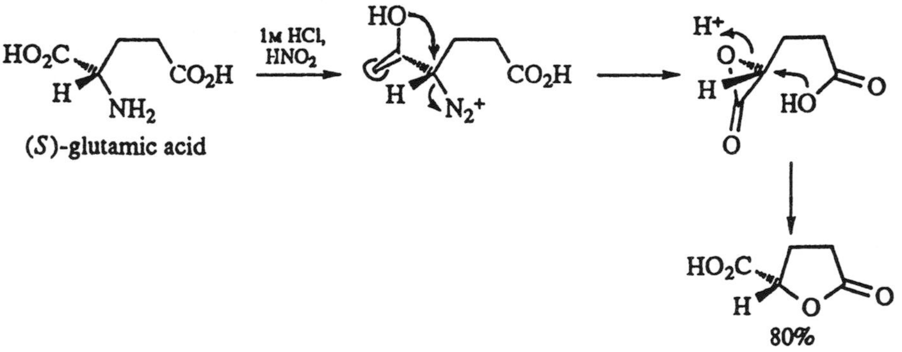

Scheme 18

Other amino acids are converted into α-hydroxy acids using nitrous acid with retention of configuration: the intermediate α-lactone in these cases is ring opened by intermolecular attack of water. In Scheme 19 the required

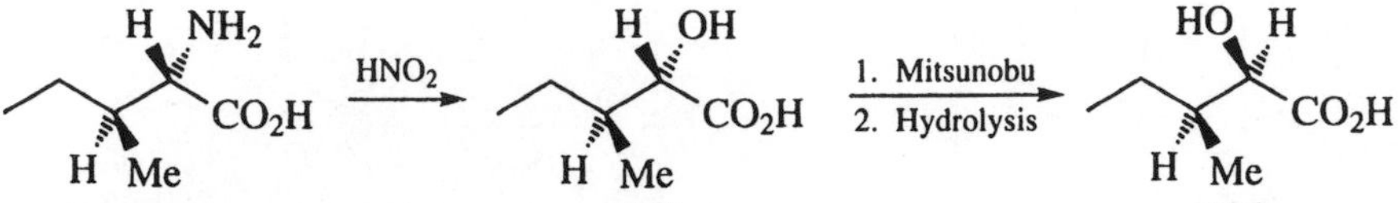

Scheme 19

product (16) corresponding to replacement of NH_2 by OH with inversion of configuration was actually obtained by Mitsunobu inversion on the product (15) followed by hydrolysis, i.e. with three inversions and overall just inversion.[20]

Double inversion commonly results from neighbouring group participation (NGP) of which α- and then γ-lactone formation in Scheme 18 is a particular example.[21] Lactonisation of (17) in Scheme 20 results in substitution of iodide with retention of configuration and is the result of intervention of an aziridinium ion (18).[22]

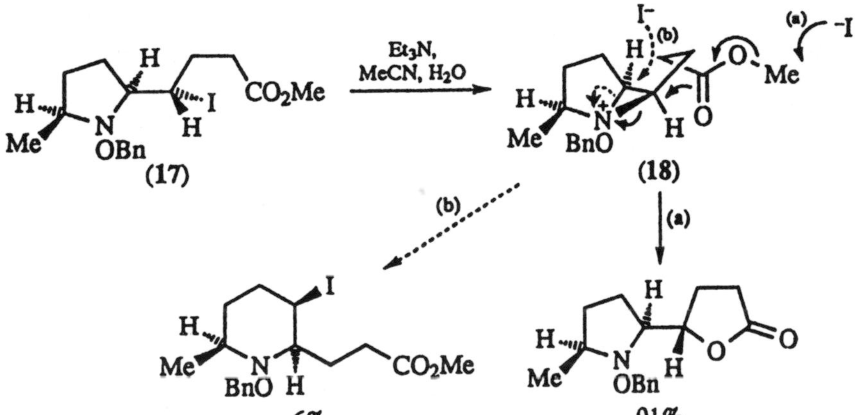

Scheme 20

Cyclisation of the enantiopure chromium-complexed benzylic alcohol (19) with acid (Scheme 21) takes place with retention of configuration because loss of the hydroxyl group is directed by participation from chromium with inversion and Friedel–Crafts cyclisation takes place with a second inversion.[23] Decomplexation of (20) with oxygen/sunlight gives the corresponding chromium-free enantiopure benzazepane (21). Cyclisation of (19) lacking the chromium tricarbonyl ligand gave rise to this product (21) with only 6% e.e.

Substitution with retention is also found in the acetolysis of exo-2-norbornyl tosylate (Scheme 22). The first inversion is the result of an intramolecular nucleophilic displacement mediated by partial delocalisation of the C_1—C_6 bond with the formation of a non-classical carbocation (22).

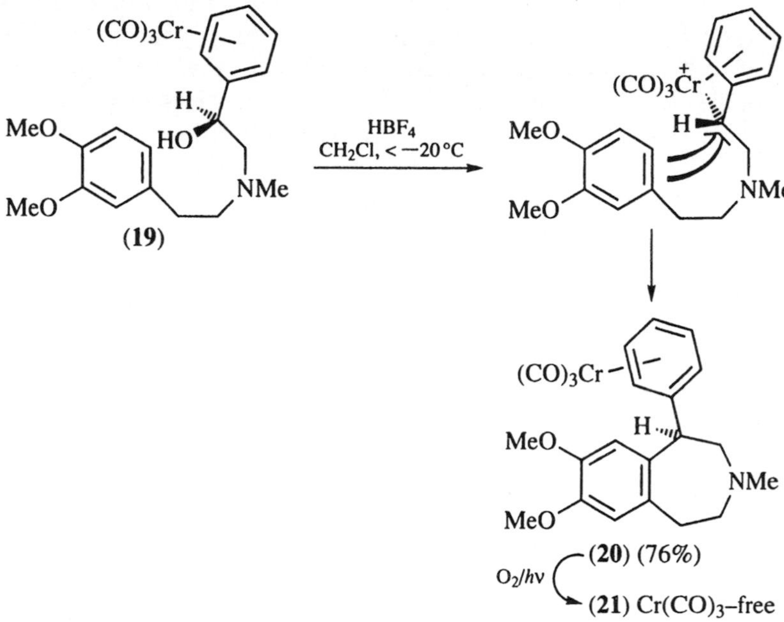

Scheme 21

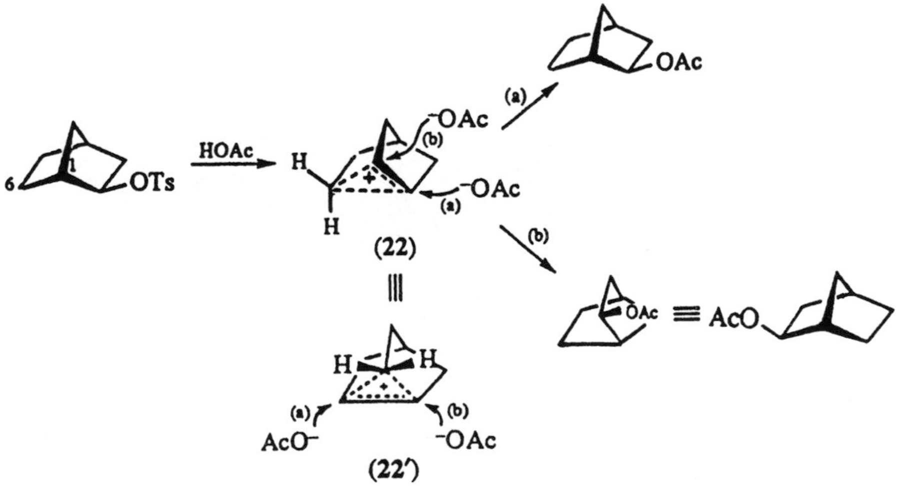

Scheme 22

The second substitution path (a) takes place by attack on this non-classical ion (22) by the solvent acetic acid with the partially formed bond acting as a leaving group. However, in this reaction the intermediate (22) is symmetrical (see 22′): attack as in (b) is equally probable and the overall result is that the product is racemic.[24]

Net retention of configuration can also be accomplished at allylic secondary carbon centres by carbon nucleophiles using palladium chemistry (Scheme 23).[25] The π-allyl system here must be sterically or otherwise biased to ensure that attack at one terminus of an unsymmetrical allyl system takes place (Scheme 24(a)).[26] Retention here is the result of two substitutions and although neither is an S_N2 reaction, the overall stereochemistry is that which would obtain if this were the case (Scheme 24).

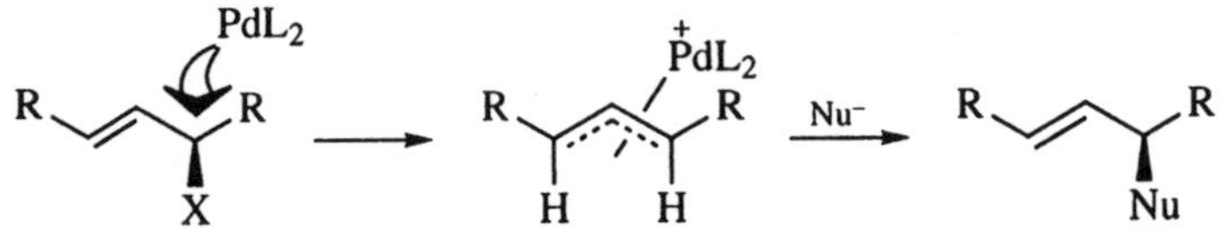

Scheme 23

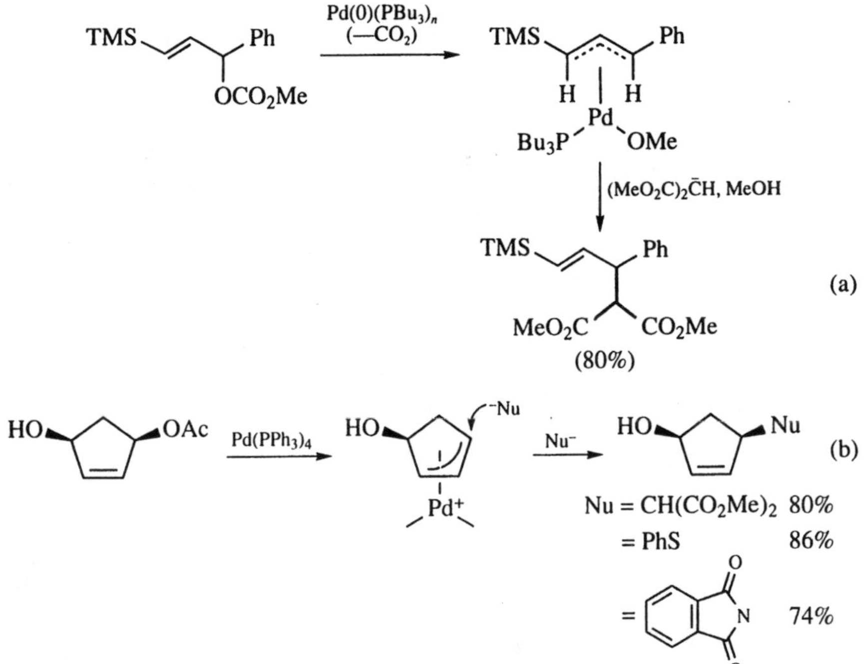

(a)

$Nu = CH(CO_2Me)_2$ 80%

$= PhS$ 86%

$= $ 74%

Scheme 24

1,2-Rearrangements with inversion of configuration at the migration terminus

The migration of a substituent from one carbon to an adjacent one is frequently accompanied by stereochemical changes. For example, the acid-catalysed rearrangement of the diol (23) in Scheme 25 is a Type I reaction with a net loss of one chiral centre.

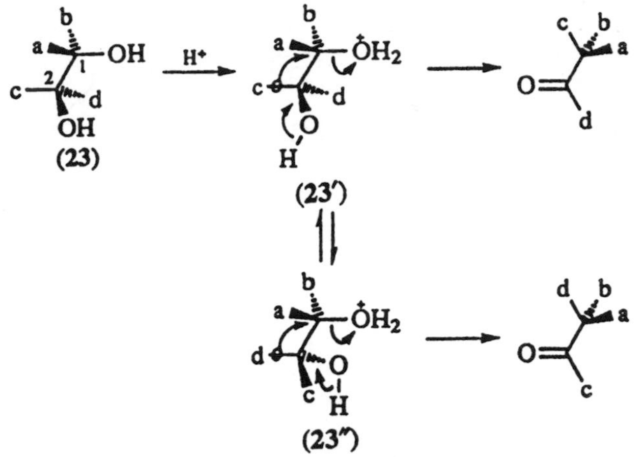

Scheme 25

The stereoelectronic requirement for this rearrangement is an antiperiplanar disposition of the bonds to migrating group and leaving group as shown in either (**23'**) or (**23''**). It should be clear, therefore, that this rearrangement is just a special case of (intramolecular) S_N2-substitution with the migrating c—C or d—C bond acting as the nucleophile.

In practice, pinacol rearrangement of acyclic diols as in Scheme 25, although completely stereoselective, is seldom regioselective. As shown in Scheme 25, a small rotation around the C—C bond in (**23'**) produces (**23''**) in which migration of d is stereoelectronically preferred. Moreover, protonation and ionisation of the other hydroxyl accompanied by migration of a or b can lead to other regioisomers.

Regioselectivity may be accomplished if migration of c is easier than d—if it has a higher migratory aptitude. Such is the case when c is a vinyl group and d an alkyl group as in Scheme 26.[27]

In Scheme 26, the problem of regioselectivity in the loss of a particular hydroxyl group is solved by selective mesylation of the secondary hydroxyl; this then becomes the better leaving group. Exclusive migration of the vinyl group occurs with retention of double bond configuration in the alkene and inversion of configuration at the migration terminus. Note that in this sequence it is not necessary that the alkyne addition to the carbonyl group of **24** is completely stereoselective because when the vinyl group migrates the chirality at the migration origin is eliminated.

The migration of c (or d) in the rearrangement in Scheme 25 is prompted and assisted by the 'pull' of the developing carbocation at C-1 but the migration may be assisted as much by a 'push' from a substituent on the migration origin. Conversion of the epoxide (**26**) into the ketone (**27**)

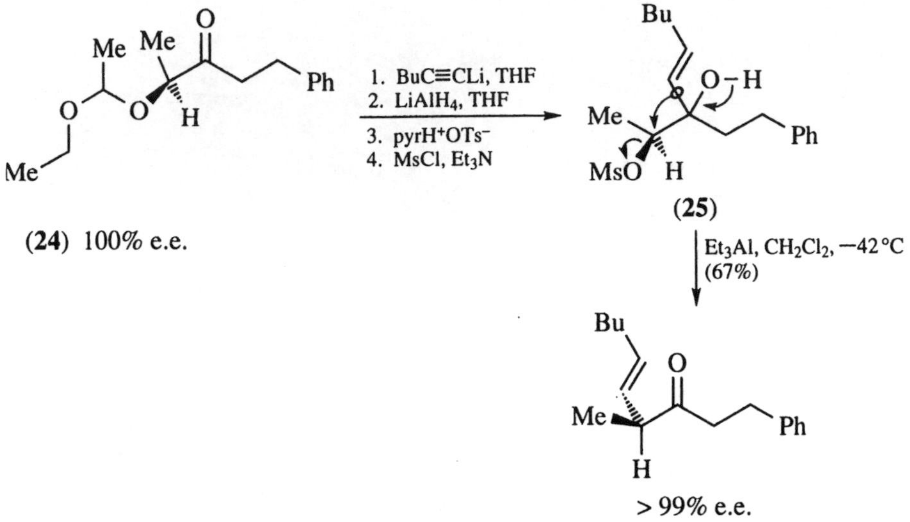

Scheme 26

(Scheme 27) is part of a synthesis of the antifungal metabolite avenaciolide.[28] The driving force for this Type I conversion includes the formation of the strong Si—Cl bond and the conversion of the resulting alkoxide anion into a carbonyl group along with relief of ring strain from opening of the epoxide.

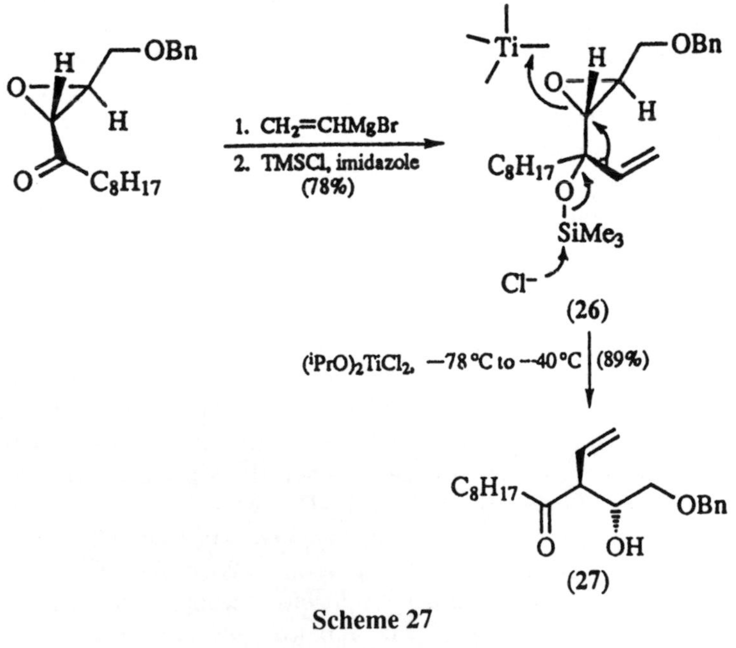

Scheme 27

Commercially available enantiopure compounds containing a C—O bond at a chiral centre are more abundant than analogues containing C—C bonds. It is noteworthy, therefore, that the overall conversions in Schemes 26 and 27 fashion C—C bonds at the expense of C—O bonds and this is accomplished diastereoselectively in spite of the fact that the substrates undergoing reaction are not necessarily single diastereoisomers.

It has been noted previously that intermolecular S_N2 substitution using carbon-centred nucleophiles at secondary and tertiary centres is not generally synthetically useful. Schemes 26 and 27 provide at least a partial solution to this problem; the nucleophile is added initially to a carbonyl group flanking the chiral centre and this is followed by intramolecular transfer to the chiral centre.

The 1,2-shift in Scheme 28 is a (Wagner–Meerwein) rearrangement in which a group c migrates with its σ-bonded pair of electrons to an electron-deficient centre. In general, a new more stable carbocation (**28**) is formed from that (**29**) formally derived by loss of a leaving group X. It is likely that the migration of c is already underway as X departs and that the carbocation (**29**) is not 'fully developed.' A stable product is formed from (**28**) either by loss of a positively charged fragment (usually a proton) or by reaction with a nucleophile. Alternatively, further rearrangement can occur and this daughter carbocation can be stabilised in either of the above ways. In any event, further stereo- or regiocontrol may be required if a single product is to be formed.

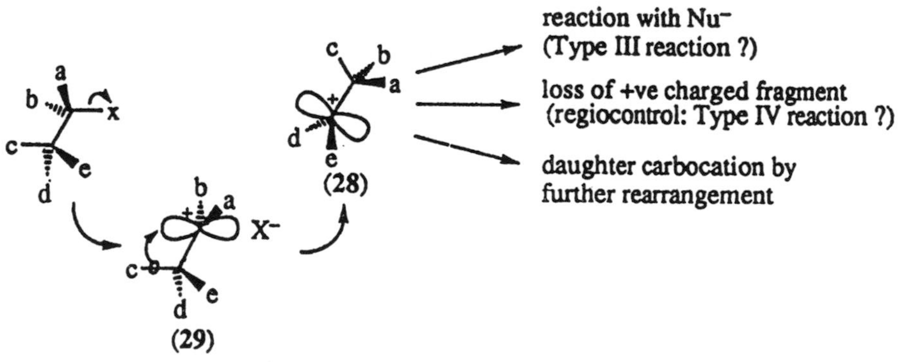

reaction with Nu⁻
(Type III reaction ?)

loss of +ve charged fragment
(regiocontrol: Type IV reaction ?)

daughter carbocation by
further rearrangement

Scheme 28

Not surprisingly, the stereoelectronic requirement in these Wagner–Meerwein rearrangements is just the same as that obtaining in Schemes 25–27 with the migrating groups acting as the nucleophile in an intramolecular S_N2-like substitution of the leaving group X.

The most common and useful 1,2-shifts of this type are those that are constrained to take place in conformationally restricted molecules such that, stereoelectronically, one group is better disposed for migration over others, even though their inherent migratory aptitudes might be similar. Thus, in the rearrangement of the bicyclo[4.4.0]decane skeleton of (30) to the bicyclo[5.3.0]decane skeleton of (31) (Scheme 29), the C—C bond common to both rings has the correct orientation for S_N2-type displacement of the tosyloxy group and generates a teriary carbocation as a result.[29]

Note that in this example a high yield of double bond isomer [(31)] is

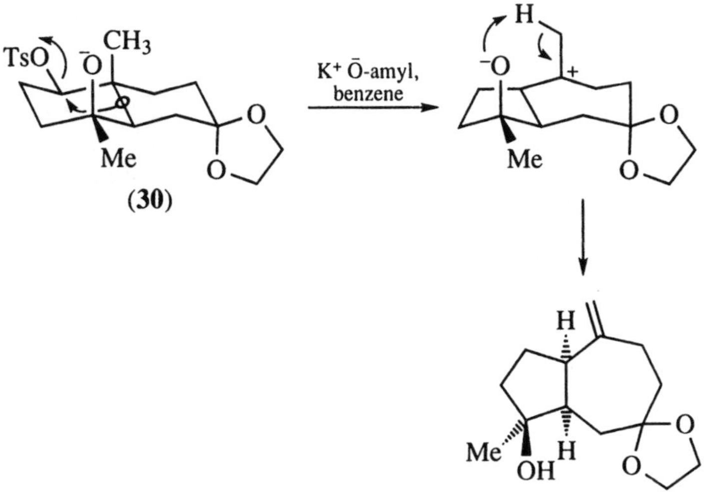

(30)

K⁺ Ō-amyl, benzene

Me

(31) 90%

Scheme 29

obtained (regiocontrol), which is attributed to the intramolecular removal of the proton as shown.

Wagner–Meerwein rearrangements dominate the carbocation-mediated chemistry of strained bicyclic systems and, in particular, that of substituted bicyclo[2.2.1]heptanes including camphor (see Chapter 3, Scheme 11).

Neighbouring group participation (NGP) can also result in rearrangement as in the conversion of (32) to the morphinane ring system (33) in Scheme 30; here there is NGP by the nitrogen in formation of the aziridinium ion and NGP by the double bond in formation of the additional ring and in both steps there is inversion of configuration.[30] Sulphur readily enters into NGP and migration of the sulphur substituent is often the result (Scheme 31), particularly when a more stable carbocation is generated thereby.[31]

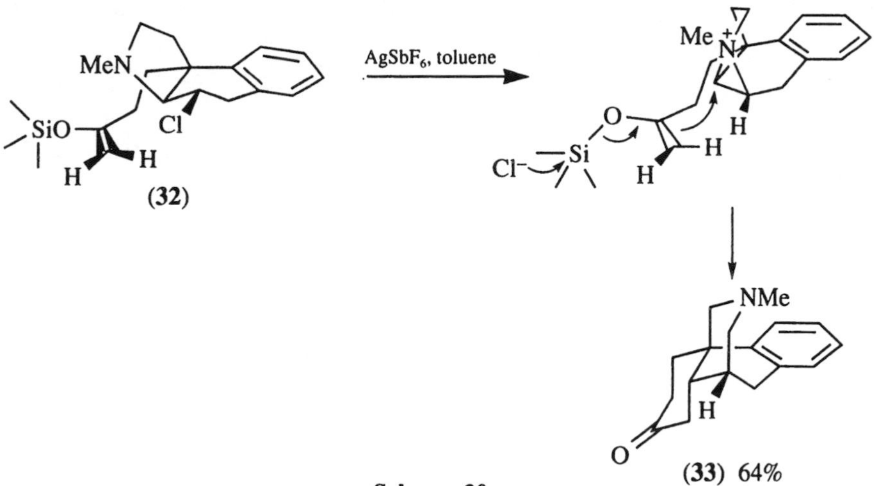

Scheme 30

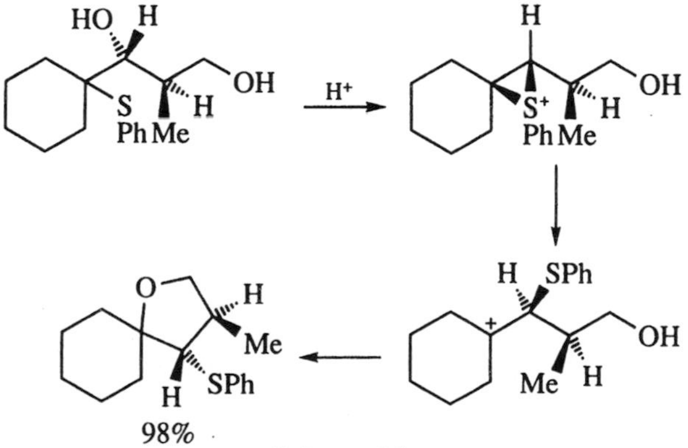

Scheme 31

1,2-Rearrangements involving retention of configuration in the migrating group

1,2-Rearrangements are not limited to those in which carbon is the migration terminus. The ease with which the O—O bond is broken, for example, facilitates migration from carbon to oxygen in the Baeyer–Villiger [Scheme 32(a)]. It is likely that this reaction includes an S_N2 attack on oxygen by the migrating group R^2 as in (34) but this cannot be proved from inversion of configuration since (bivalent) oxygen cannot be a chiral centre. However, if the migrating group R^2 is chiral then the rearrangement will be a Type I reaction since a bond to a chiral centre is broken and a new one is made [Scheme 32(b)].

HOOKE

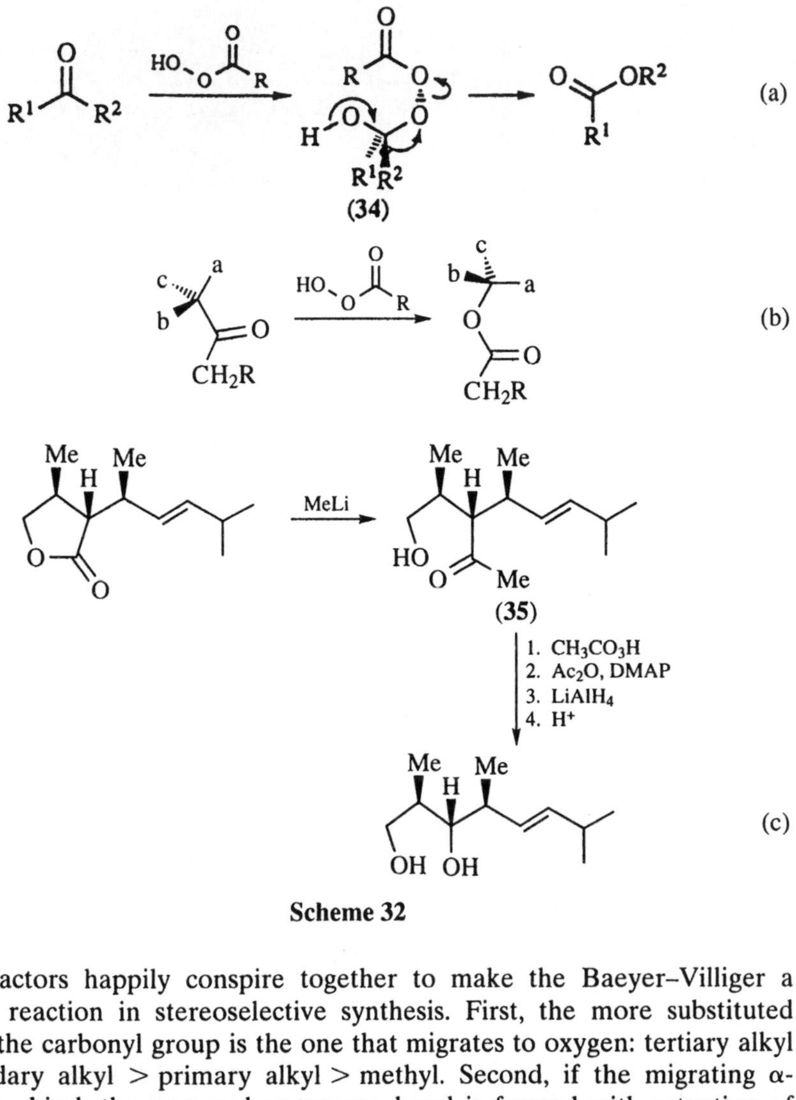

Scheme 32

Two factors happily conspire together to make the Baeyer–Villiger a valuable reaction in stereoselective synthesis. First, the more substituted bond to the carbonyl group is the one that migrates to oxygen: tertiary alkyl > secondary alkyl > primary alkyl > methyl. Second, if the migrating α-carbon is chiral, the new carbon oxygen bond is formed with retention of configuration [Scheme 32(b)].

The reluctance of methyl to migrate in the Baeyer–Villiger reaction is used in the Criegee sequence for conversion of a γ-lactone into a 1,3-diol [Scheme 32(c)]. Peroxyacetic acid brings about the interpolation of an oxygen into the more substituted bond of the ketone (35) and lithium aluminium hydride and then acid convert the mixed acetates and acetal intermediates into the diol product; migration of the secondary alkyl group occurs with retention of configuration.[32]

Hydroboration of an alkene followed by transformation of the carbon–boron bond into a carbon–oxygen bond is an invaluable route to alcohols (Scheme 33).[33] This reaction also uses the ease of cleavage of the O—O bond in (36) and migration of a substituted carbon atom with retention of its configuration; note that ⁻OH attacks boron and not carbon in the final step.

Regioselectivity in the migration to oxygen is usually not important here because either the other substituents on boron do not migrate competitively (R = OR') or all of the carbon–boron bonds (R = alkyl) can be transformed into carbon–oxygen bonds.

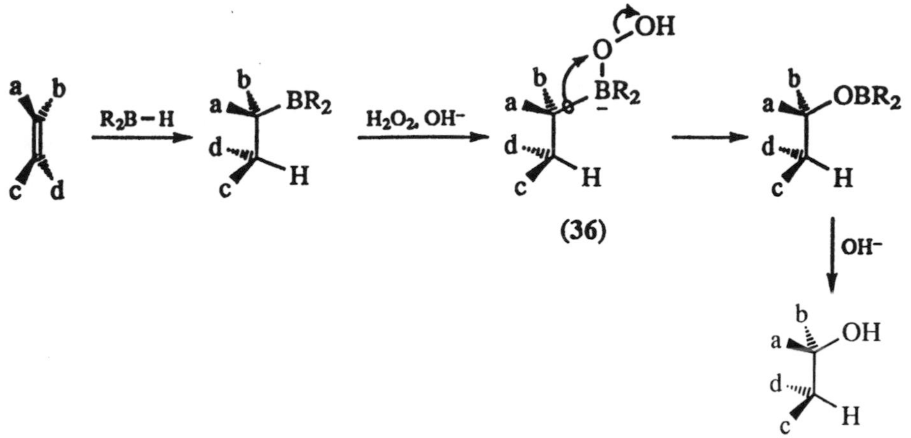

Scheme 33

Similarly, replacement of carbon–silicon bonds by carbon–oxygen bonds can now be reliably accomplished with retention of configuration at carbon (Scheme 34).[34] Silicon is prone to attack by a peroxy anion only when it is activated by a σ electron-withdrawing substituent (O, F, etc.). In Scheme 34(b), the function of the fluoroboric acid is to introduce such an activating fluorine by preliminary cleavage of the phenyl–silicon bond making use of the ability of silicon to stabilise a carbocation β to it. Oxidative cleavage of the C—Si bond can be accomplished without the need for isolation of the fluorosilane.

Again, these rearrangements are triggered by weakness of the O—O bond and mechanisms resembling that in Scheme 34(c) are operative.

Migration from carbon to nitrogen occurs in the Hofmann, Schmidt, Curtius and Lossen rearrangements, all of which proceed via an intermediate isocyanate formed with retention of configuration in the migrating group (Scheme 35).[35]

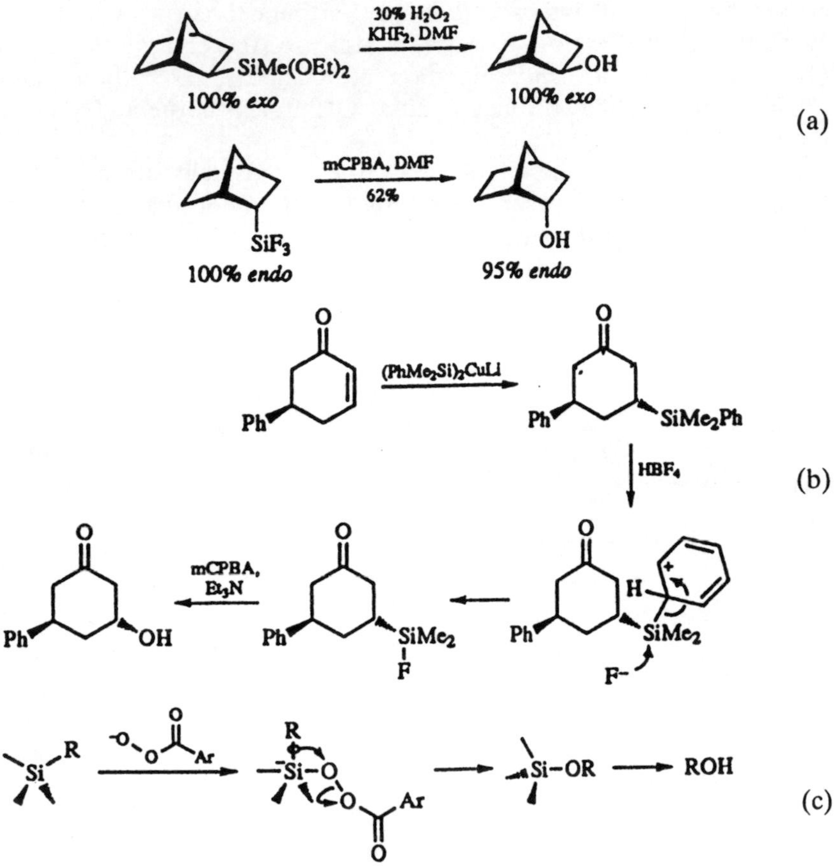

(a)

(b)

(c)

Scheme 34

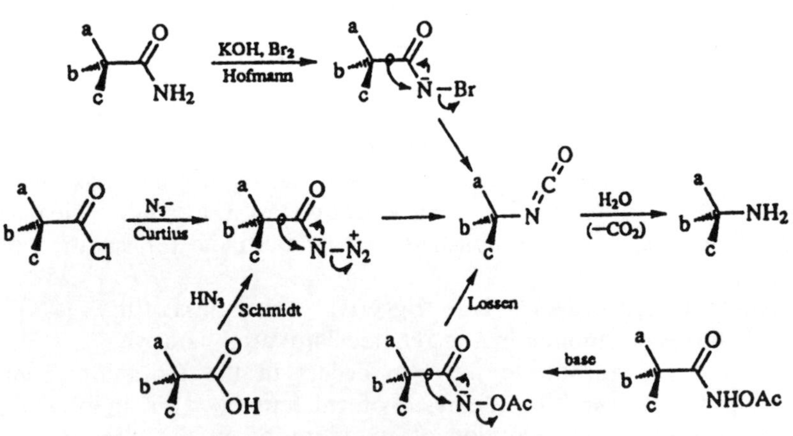

Scheme 35

Substitution with retention of configuration via configurationally stable carbanions

In stabilised carbanions, e.g. enolates, the carbanion is planar and achiral. Simple carbanions are usually tetrahedral but have low barriers to inversion. Consequently, the formation of a carbanion from an enantiopure precursor [Scheme 36(a)], followed by reaction with an electrophile, will in general give a racemic product because the carbanion is not configurationally stable.

(a)

enantiopure   racemic

E+ = electrophile

(b)

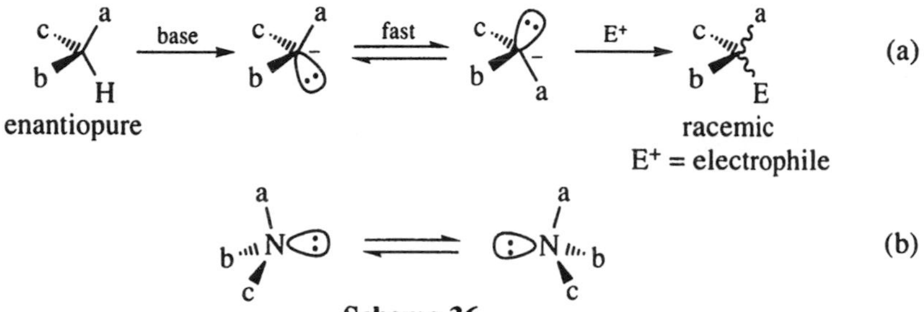

Scheme 36

The situation here is analogous to that of tetrahedral sp³-hybridised (pyramidal) trivalent nitrogen, where inversion is so fast as to preclude the possibility of isolating enantiomers [Scheme 36(b)].

One way in which the inversion barrier in sp³-hybridised amines can be raised is by having one or more of the substituents a, b or c as heteroatoms, e.g. O, N or halogen. The presence of non-bonding electron pairs on the heteroatom and/or its σ electron-withdrawing character raises the energy of the transition state for inversion relative to the ground state as compared with simple trialkylamines.[36]

It was shown by Still and Sreekumar[37] that the same inversion-retarding effect of heteroatom substitution could be applied to carbanions. Thus α-alkoxystannane (37) (Scheme 37) underwent Sn–Li exchange with

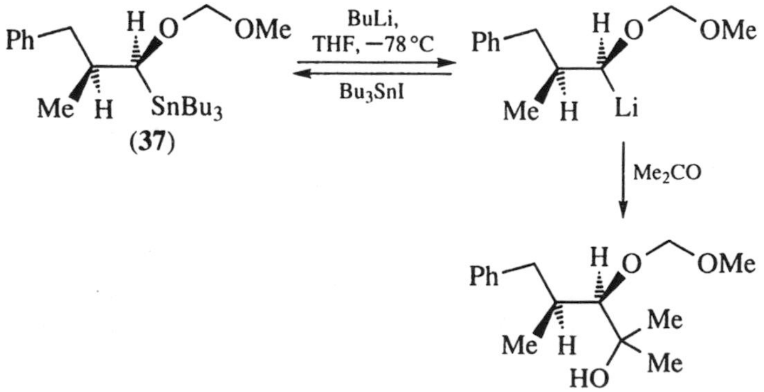

Scheme 37

butyllithium followed by reaction with tributyltin iodide with net retention of configuration. The lithiated intermediate also reacted with retention of configuration with other electrophiles.

Since 1980, enhanced pyramidal configurational stability has been demonstrated for carbanions bearing nitrogen, bromine, sulphur or selenium as substituents.[38]

The work of Chan and Chong[39] has shown that α-hydroxystannane (**39**) of ~98% e.e. (Scheme 38), prepared by enantioselective reduction of the α-stannyl ketone (**38**) using (S)-BINAL-H (see Chapter 12), is converted into the α-hydroxy acid derivative (**40**) with complete retention of configuration. The absolute configuration of this product was confirmed by relating it to (S)-valine via replacement of the amino group in the latter with retention of configuration (cf. Scheme 18).

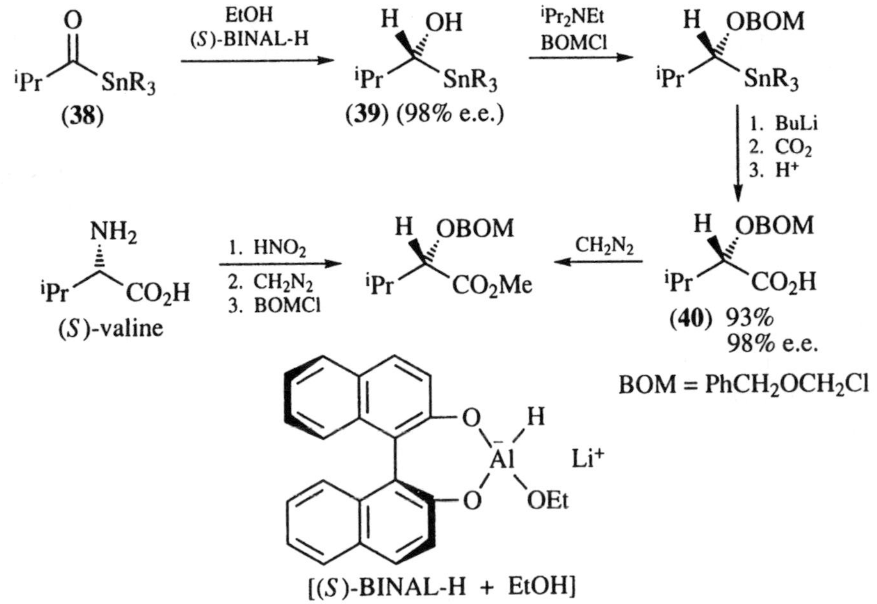

Scheme 38

From what has been said previously regarding the limitations of the S_N2 reaction, it might appear that with abcX as a secondary centre there is little prospect of being able to carry out the transformation shown in Scheme 39(a) to form two adjacent (carbon) chiral centres of defined absolute configuration.

However, work by Matteson and co-workers[40] has shown that secondary α-chloroboronates undergo substitution by nucleophiles with clean inversion of configuration [Scheme 39(b)] via rearrangement of intermediate boronate anions (**41**) (another example of S_N2 substitution via intramolecular

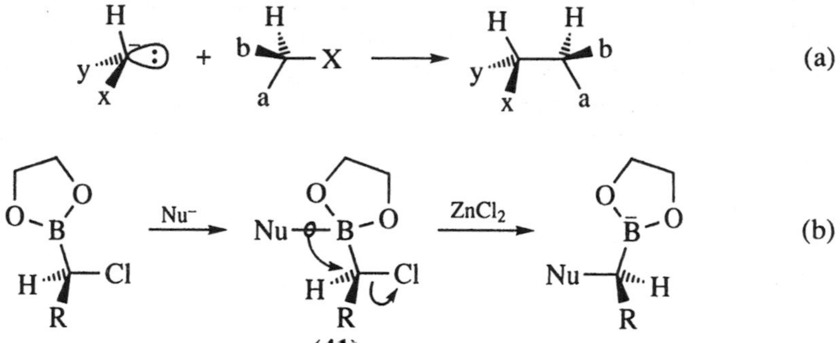

(a)

(b)

(41)

Scheme 39

rearrangement). They have shown that the transformation in Scheme 39(a) is feasible by reacting together single enantiomers of α-alkoxycarbanion (**42**) and α-chloroboronate (**43**) as shown in Scheme 40. In this scheme, the α-alkoxycarbanion (**42**) was itself prepared from the same α-chloroboronate (**43**) by substitution with tributyltin (with inversion) followed by replacement of boron by hydroxyl (with retention of configuration), hydroxyl protection and carbanion formation.

The scope of this unprecedented synthesis of a molecule containing two adjacent chiral centres is considerable, particularly since the configurations of the centres can be selected by appropriate choice of the enantiomer of the starting materials.

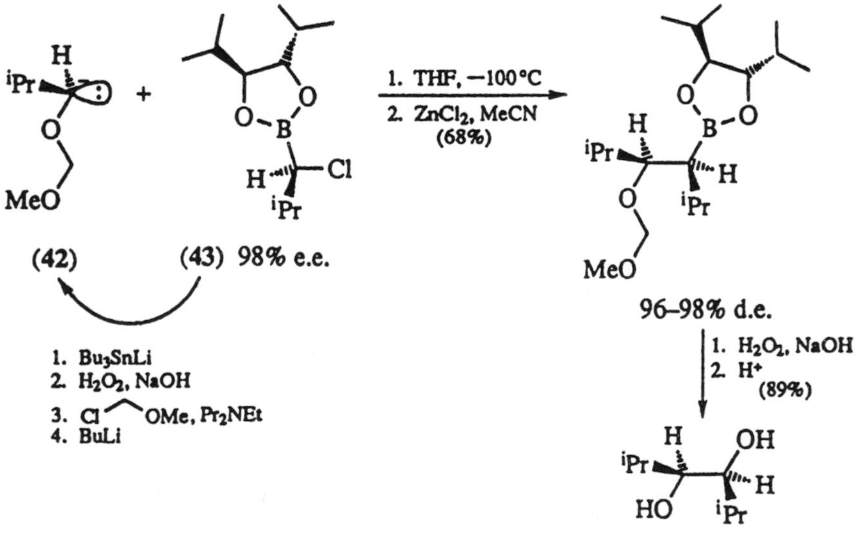

Scheme 40

Carbanions which are generated on three-membered rings also have enhanced configurational stability. Three-membered rings have considerable angle strain, having formally ring bond angles of ~60°. The hybridisation of their ring atoms provides a greater degree of p-character in the overlapping hybrid orbitals forming the bent 'banana' bonds of the ring (Figure 2); this allows for a narrower angle than the ~109° between normal sp³-hybrid orbitals. (As a corollary, the external bonds on the three-membered ring have increased s-character and the angle between them is widened; see Figure 1.)

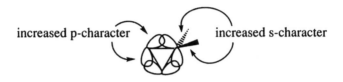

increased p-character   increased s-character

Figure 2

At the transition state (**44**) for inversion of the carbanion (Scheme 41), the ring bonds at this carbon are the result of overlap of (formally) sp²-hybrid orbitals having an increased angle of 120° between them. The net result is that the energy of the transition state (**44**) (Scheme 41) for this carbanion inversion is raised relative to that of a carbanion in an acyclic substrate which can accommodate normal sp²-bond angles without the same increase in strain. The same rationale accounts for the increased inversion barriers at nitrogen in aziridines by comparison with cyclic amines.

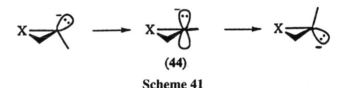

(**44**)

Scheme 41

Scheme 42 gives examples of carbanion generation and reaction with an electrophile with retention of configuration in a cyclopropane,[41] an epoxide[42] and an aziridine.[43] In Scheme 42(c), the optical rotation of the deuterated product was almost identical with that of the starting material, showing that carbanion and nitrogen inversion had not occurred.

Finally, the reaction of many metal–carbon bonds, with transfer of a ligand from the metal to carbon, takes place with retention of configuration (see Chapter 5). Transmetallation reactions, in which one metal–carbon bond is replaced by another metal–carbon bond, also usually proceed with retention of configuration at carbon.

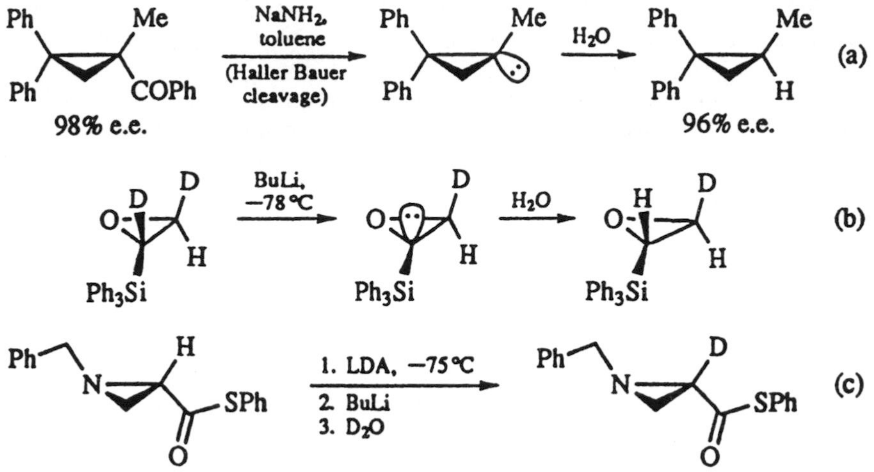

Scheme 42

Summary

There are limitations to the use of S_N2 reactions in stereoselective synthesis. Tertiary substrates do not react intermolecularly and the reactions of secondary substrates are unsatisfactory with nucleophiles of even modest bulk, including most carbanions. In intramolecular reactions, however, S_N2 substitution is practicable, sometimes even at tertiary centres.

Double inversion (= retention) of configuration is common in many reactions in which a neighbouring group participates. Net retention of configuration can also be accomplished using palladium chemistry.

1,2-Migration of carbon–carbon bonds to sp^3-hybridised centres with loss of a leaving group commonly proceeds with inversion at the migration terminus with the migrating bond behaving as the nucleophile in an S_N2-type substitution.

Concerted 1,2-migration with retention of configuration in the migrating sp^3 carbon takes place in the Baeyer–Villiger and related reactions where migration of a carbon–carbon, carbon–boron or carbon–silicon bond takes place to give a carbon–oxygen bond. Similar 1,2-migrations from carbon to nitrogen with retention of configuration in the migrating group take place in the Hofmann, Curtius and related reactions.

Pyramidal carbanions substituted with oxygen, nitrogen or halogens have retarded rates of inversion and, at sufficiently low temperatures, the carbanion is configurationally stable and can be generated and reacted with retention of configuration. Carbanions generated on three-membered rings are also configurationally stable at lower temperatures.

References

1. K. Ishihara, N. Hanaki and H. Yamamoto, *J. Am. Chem. Soc.*, 1991, **113**, 7074.
2. T.R. Hoye and W.S. Richardson, *J. Org. Chem.*, 1989, **54**, 688.
3. M.J. Eis, J.E. Wrobel and B. Ganem, *J. Am. Chem. Soc.*, 1984, **106**, 3693.
4. Y. Petit, C. Sanner and M. Larchevêque, *Tetrahedron Lett.*, 1990, **31**, 2149.
5. H. Toshima, S. Yoshida, T. Suzuki, S. Nishiyama and S. Yamamura, *Tetrahedron Lett.*, 1989, **30**, 6721.
6. J.A. Soderquist and B. Santiago, *Tetrahedron Lett.*, 1989, **30**, 5693.
7. J.J. Eisch and J.E. Galle, *J. Org. Chem.*, 1976, **41**, 2615.
8. J. Legters, L. Thijs and B. Zwanenburg, *Tetrahedron Lett.*, 1989, **30**, 4881.
9. A.J. Bose, B. Lal, W.A. Hoffman and M.S. Manhas, *Tetrahedron Lett.*, 1973, 1619.
10. O. Mitsunobu, *Synthesis*, 1981, 1.
11. G. Adam, R. Zibuck and D. Seebach, *J. Am. Chem. Soc.*, 1987, **109**, 6176.
12. B. Seuring and D. Seebach, *Helv. Chim. Acta*, 1977, **60**, 1175.
13. B. Radüchel, *Synthesis*, 1980, 292: G. Cainelli, F. Manescalchi, G. Martelli, M. Panunzio and L. Plessi, *Tetrahedron Lett.*, 1985, **26**, 3369; W.H. Kruizinga, B. Strijtveen and R.M. Kellogg, *J. Org. Chem.*, 1981, **46**, 4321; E.J. Corey, K.C. Nicolaou, M. Shibasaki, Y. Michida and C.S. Shiner, *Tetrahedron Lett.*, 1975, 3183.
14. K.S. Reddy, O.-H. Ko, D. Ho, P.E. Persons and J.M. Cassady, *Tetrahedron Lett.*, 1987, **28**, 3075.
15. G.W.J. Fleet, A.N. Shaw, S.V. Evans and L.E. Fellows, *Chem. Commun.*, 1985, 841.
16. G.A. Molander and S.W. Andrews, *J. Org. Chem.*, 1989, **54**, 3114.
17. T.R. Hoye and S.A. Jenkins, *J. Am. Chem. Soc.*, 1987, **109**, 6196.
18. P. Prasit and J. Rokach, *J. Org. Chem.*, 1988, **53**, 4421.
19. S. Iwaki, S. Marumo, T. Saito, M. Yamada and K. Katagiri, *J. Am. Chem. Soc.*, 1974, **96**, 7842.
20. L. Crombie and S.R.M. Jarrett, *Tetrahedron Lett.*, 1989, **30**, 4303; H. Irie, K. Matsumoto, T. Kitigawa, Y. Zhang, T. Ueno, T. Nakashima and H. Fukami, *Chem. Pharm. Bull.*, 1987, **35**, 2598.
21. D.R. Williams, M.H. Osterhout and J.M. McGill, *Tetrahedron Lett.*, 1989, **30**, 1331.
22. B. Capon, *Q. Rev. Chem. Soc.*, 1964, **18**, 45; B. Capon and S.P. McManus, Neighbouring Group Participation, Plenum, New York, 1976.
23. S.J. Coote, S.G. Davies, D. Middlemiss and A. Naylor, *Tetrahedron Lett.*, 1989, **30**, 3581.
24. G.D. Sargent, *Q. Rev. Chem. Soc.*, 1966, **20**, 301.
25. T. Hayashi, T. Hagihara, M. Konishi and M. Kumada, *J. Am. Chem. Soc.*, 1983, **105**, 7767. B.M. Trost, *Acc. Chem. Res.*, 1980, **13**, 385; J. Tsuji, *Pure Appl. Chem.*, 1982, **54**, 197.
26. J. Tsuji, M. Yuhara, M. Minato, H. Yamada, F. Sato and Y. Kobayashi, *Tetrahedron Lett.*, 1988, **29**, 343: D.R. Deardoff, R.G. Linde, A.M. Martin and M.J. Shulman, *J. Org. Chem.*, 1989, **54**, 2759.
27. K. Suzuki, E. Katayama and G. Tsuchihashi, *Tetrahedron Lett.*, 1984, **25**, 1817.
28. K. Suzuki, M. Miyazawa, M. Shimazaki and G. Tsuchihashi, *Tetrahedron Lett.*, 1986, **27**, 6237.
29. J.B.P.A. Wijnberg, L.H.D. Jenniskens, G.A. Brunekreef and A. de Groot, *J. Org. Chem.*, 1990, **55**, 941.
30. C.A. Broka and J.F. Gerlits, *J. Org. Chem.*, 1988, **53**, 2144.
31. V.K. Aggarwal and S. Warren, *Tetrahedron Lett.*, 1986, **27**, 101.

32. F.E. Ziegler and R.T. Wester, *Tetrahedron Lett.*, 1986, **27**, 1225.
33. H.C. Brown, G.W. Kramer, A.B. Levy and M.M. Midland, *Organic Syntheses via Boranes*, Wiley, New York, 1975.
34. K. Tamao, N. Ishida, T. Tanaka and M. Kumada, *Organometallics*, 1983, **2**, 1694; K. Tamao and N. Ishida, *J. Organomet. Chem.*, 1984, **269**, C37; I. Fleming, R. Henning and H. Plaut, *Chem. Commun.*, 1984, 29; K. Tamao, T. Kakui, M. Akita, T. Iwahara, R. Kanatani, J. Yoshida and M. Kumada, *Tetrahedron*, 1983, **39**, 983.
35. W. Lwowski, *Nitrenes*, Wiley–Interscience, New York, 1970, p.217.
36. W.B. Jennings and D.R. Boyd, in *Cyclic Organonitrogen Stereodynamics*, eds; J.B. Lambert and Y. Takeuchi, VCH, New York, 1992.
37. W.C. Still and C. Sreekumar, *J. Am. Chem. Soc.*, 1980, **102**, 1201; J.S. Sawyer, A. Kucerovy, T.L. Macdonald and G.J. McGarvey, *J. Am. Chem. Soc.*, 1988, **110**, 842.
38. J.M. Chong and S.B. Park, *J. Org. Chem.*, 1992, **57**, 2220, and references cited therein.
39. P.C.-M. Chan and J.M. Chong, *Tetrahedron Lett.*, 1990, **31**, 1985; P.C.-M. Chan and J.M. Chong, *J. Org. Chem.*, 1988, **53**, 5584.
40. D.S. Matteson, P.B. Tripathy, A. Sarker and K.M. Sadhu, *J. Am. Chem. Soc.*, 1989, **111**, 4399; D.S. Matteson, *Acc. Chem. Res.*, 1988, **21**, 294; D.S. Matteson *et al.*, *Pure Appl. Chem.*, 1985, **57**, 1741.
41. F.J. Impastato and H.M. Walborsky, *J. Am. Chem. Soc.*, 1962, **84**, 4838.
42. J.J. Eisch and J.E. Galle, *J. Am. Chem. Soc.*, 1976, **98**, 4646.
43. R. Häner, B. Olano and D. Seebach, *Helv. Chim. Acta*, 1987, **70**, 1676.

5 SIMPLE CHIRALITY TRANSFER REACTIONS: THOSE IN WHICH A SINGLE CHIRAL CENTRE IS TRANSFERRED WITH CONCOMITANT MIGRATION OF ONE OR MORE DOUBLE BONDS

Type I reactions discussed in the previous chapter are those in which either inversion or retention of configuration takes place at a chiral centre.

It is appropriate to consider in this chapter reactions which involve simple transfer of chirality or, more precisely, those which take place with the loss of the existing chiral centre, the migration of at least one configured double bond and the creation of a new chiral centre elsewhere in the molecule. These reactions are in fact substrate-controlled Type III reactions (see Chapter 9), but in those cases in which a single chiral centre is lost and gained as above there is obvious complementarity to Type I reactions discussed in Chapter 4.

1,3-Simple chirality transfer via the S_E2' reaction

The reaction of allylsilanes with electrophiles is an S_E2' reaction (Scheme 1): loss of the silyl group and reaction with electrophiles takes place in an *anti* fashion as the examples in Scheme 2 show, i.e. with *1,3-transfer of chirality*.[1,2] Note that in example (a), reaction takes place from only one of the two conformations shown in Scheme 1 i.e. with a=H, b=Ph (see below).

A small number of examples are known in which the inherent *anti*-bias of the S_E2' is overridden by other factors and *syn* attack of the electrophile and loss of the trialkylsilyl group is the stereochemical outcome.

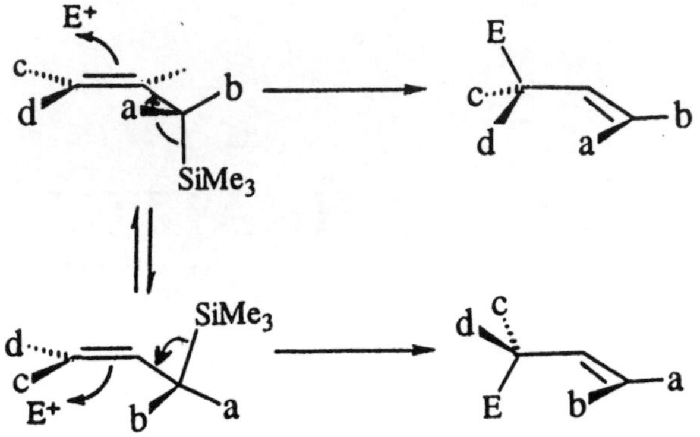

Scheme 1

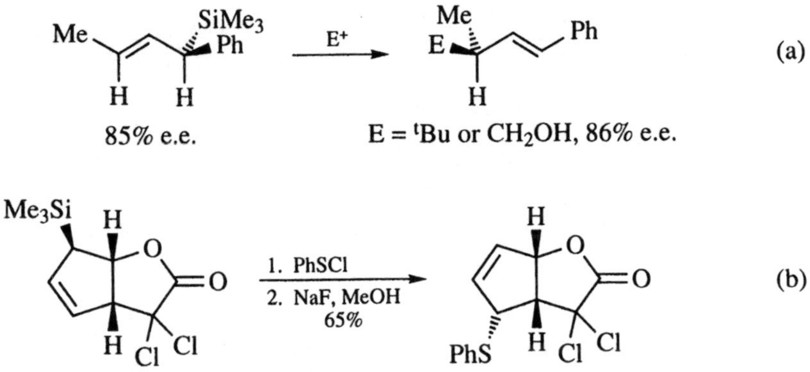

E = ^{t}Bu or CH$_2$OH, 86% e.e.

(a)

(b)

Scheme 2

Allylstannanes react with electrophiles in an S_E2' reaction with the same *anti* preference as allylsilanes.

1,3-Simple chirality transfer via the S_N2' reaction

An S_N2' reaction is the nucleophilic attack at the terminal sp^2 centre of an allyl system with migration of the double bond and expulsion of a leaving group from the allylic position (Scheme 3). With appropriate substitution on the allyl system, this reaction also results in 1,3-transfer of chirality.

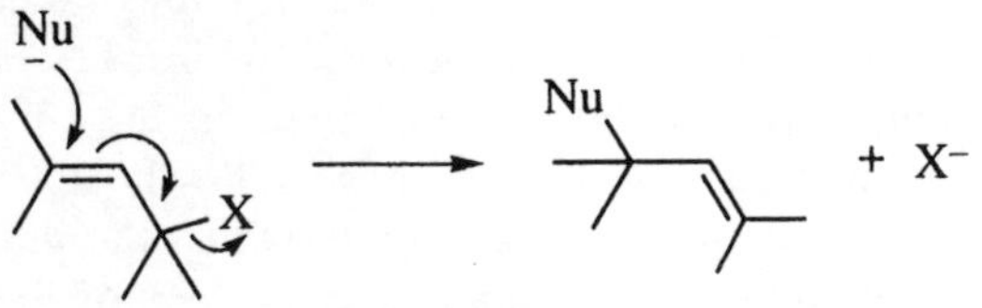

Scheme 3

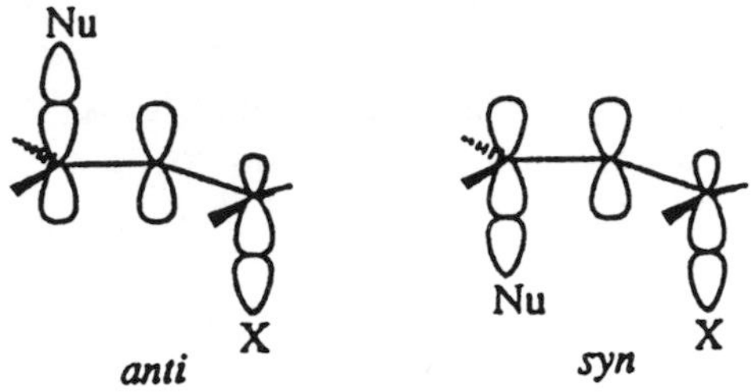

Nu

anti X   syn Nu X

Scheme 4

The stereoelectronic requirement for the S_N2' reaction arises from the necessity for the developing p-orbital formed from the breaking C—X bond to overlap with the adjacent p-orbital of the existing π-bond to form the new π-bond in the product. As indicated in Scheme 4, this stereoelectronic requirement can be satisfied by attack of the nucleophile in either of two orientations relative to the leaving group (*syn* or *anti*). In an enantiopure acyclic substrate having a double bond of defined configuration, there are two conformations from which both *syn* and *anti* reaction can proceed (Scheme 5); note that the products from each of these conformations, e.g. (**1**) and (**2**), have opposite configurations at both the chiral centre and the double bond.

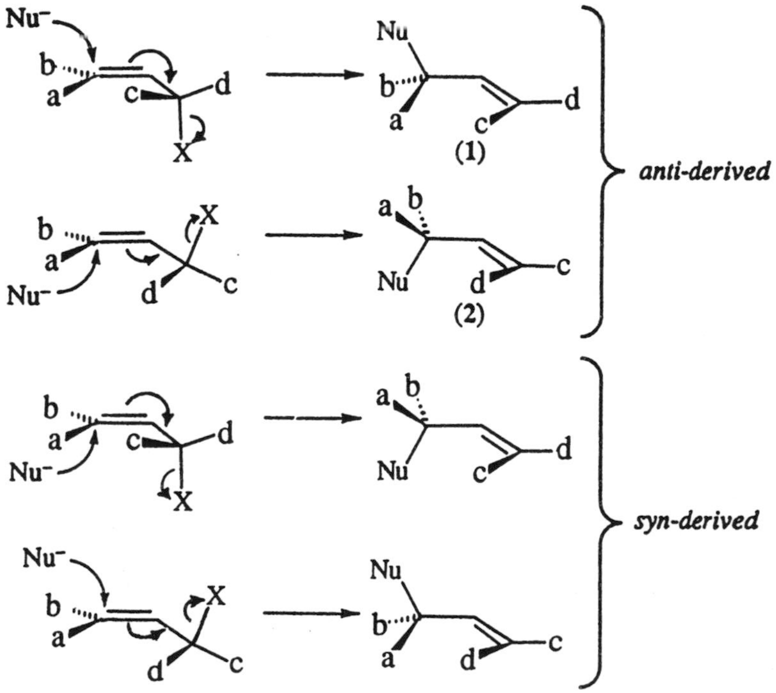

Nu⁻ b a c d X → Nu b a c d (**1**)

b a X d c Nu⁻ → a b Nu d c (**2**) ⎫ *anti*-derived

b a c d Nu⁻ X → a b Nu c d

Nu⁻ b a d X c → Nu b a d c ⎫ *syn*-derived

Scheme 5

In general, the S_N2' reaction is not highly stereoselective: attack of the nucleophile may be preferentially *syn* or *anti* to some degree depending on the nucleophile, the leaving group and whether the substitution is inter- or intramolecular.[3] However, the application of copper and palladium chemistry to the S_N2' reaction has greatly increased the reliability of its stereochemical outcome.[4]

The S_N2' reaction in Scheme 6 using lithium dimethylcuprate was examined with a view to the synthesis of enantiopure subunits of polypropionate-derived natural products.[5] Although the reaction has high *anti* stereoselectivity, two diastereoisomers are still obtained (albeit in disparate amounts) because two conformations of the starting material react competitively (cf. Scheme 5). The local steric effect indicated in (4) probably accounts for the preferential reaction via (3); chelation, as in (5), may also help to stabilise the transition state for (3).

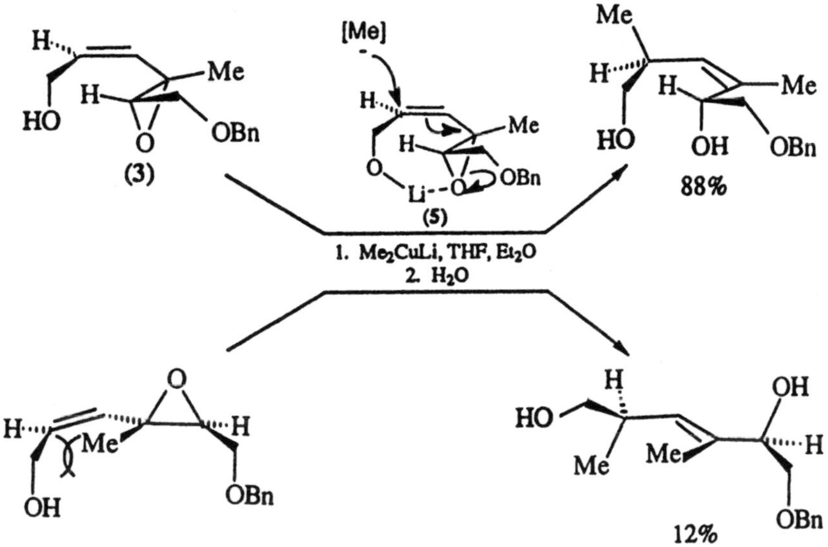

Scheme 6

In favourable cases, cuprate-mediated S_N2' reactions on acyclic substrates can be highly stereoselective (Scheme 7),[6] but this is more likely when one of the options of *anti* reaction is removed by conformational restraints, e.g. when the allylic system is incorporated in a ring (Scheme 8).[7]

In these reactions, a π-complex of the alkene with copper (6) is believed to be converted into a σ-complex (7) with loss of the leaving group followed by metal-to-carbon transfer of the ligand R (with retention of configuration at carbon) (Scheme 9). Control of the stereochemistry, therefore, is brought about by the π-complex formation which, in an otherwise unbiased five- or six-membered ring system, will be from the side opposite to the acetoxy group and hence lead to the *anti* mode of substitution. However, when this face is

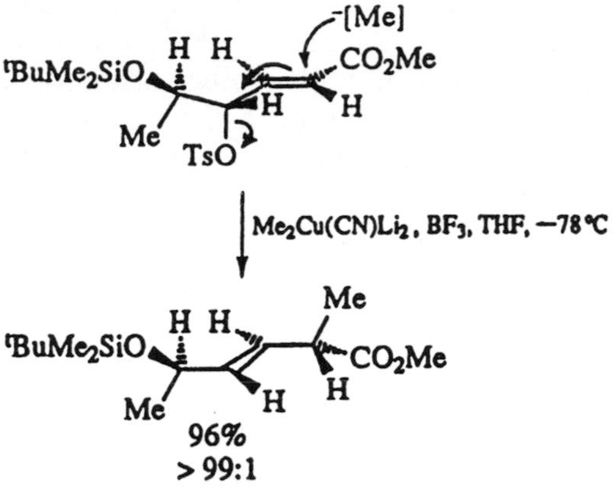

Scheme 7

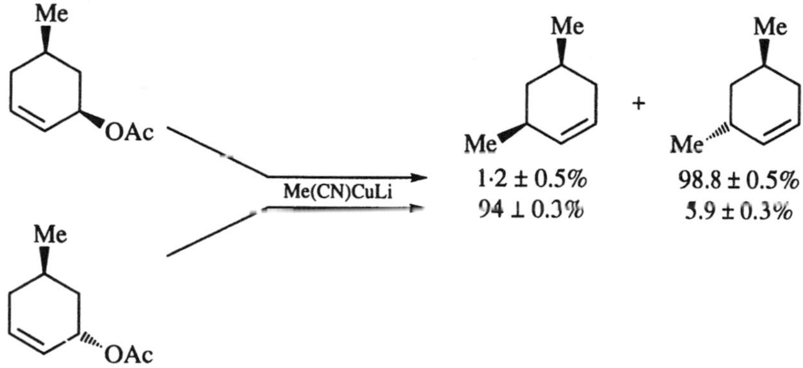

Scheme 8

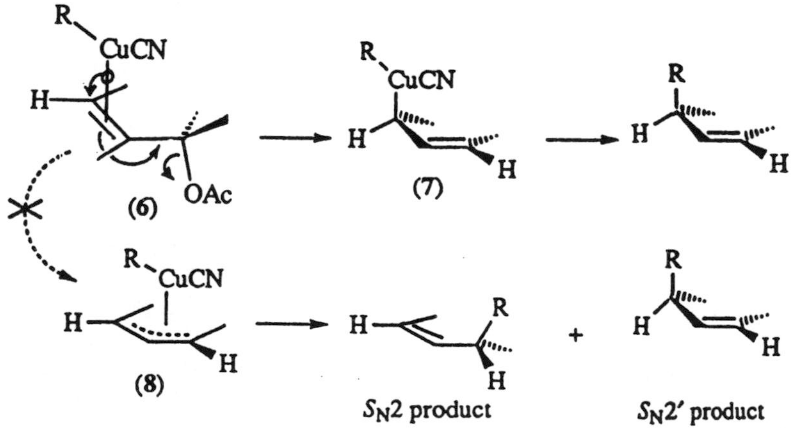

Scheme 9

hindered, even the copper-catalysed S_N2' reaction can revert to the *syn* mode.

It is noteworthy that in Scheme 9, the formation of a π-allyl complex (**8**) is apparently not favoured when the other ligand on copper is cyanide, since the product from transfer of R to the carbon which originally bore the acetoxy group—an S_N2 reaction—is at best a very minor one. However, using a dialkylcuprate, R_2CuLi, this is the major pathway and the regiospecificity of substitution of the S_N2' reaction is lost.

It has been suggested that the tendency for cuprates to bring about S_N2' reactions with *anti* stereochemistry can be ascribed to the ability of the filled and diffuse d-orbital of the nucleophilic copper atom to interact with both π*- and σ*-orbitals of the substrate in a bidentate fashion (**9**).[8]

Exceptionally, *syn* addition in these cuprate-mediated S_N2' substitutions can be effected by coordination of the cuprate to the leaving group, thus directing addition of the copper to the *syn* face of the allyl group as in Scheme 10.[9]

By contrast, S_N2' reactions which are brought about by involvement of palladium invariably proceed with *anti* formation of π-allyl complexes followed by *anti* attack of the nucleophile leading to an overall *syn* stereochemistry (Scheme 11).[10] Palladium is generally used in only catalytic amounts. Whereas in copper-mediated S_N2' reactions the cuprate is the

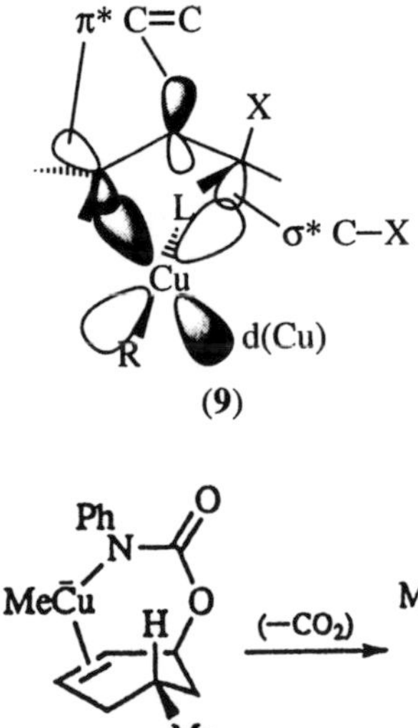

(**9**)

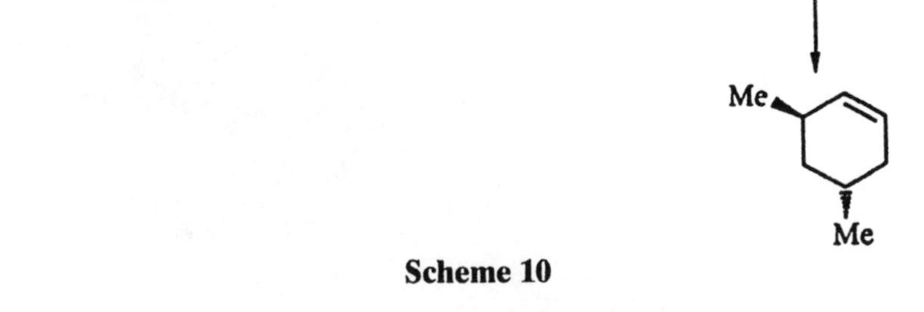

Scheme 10

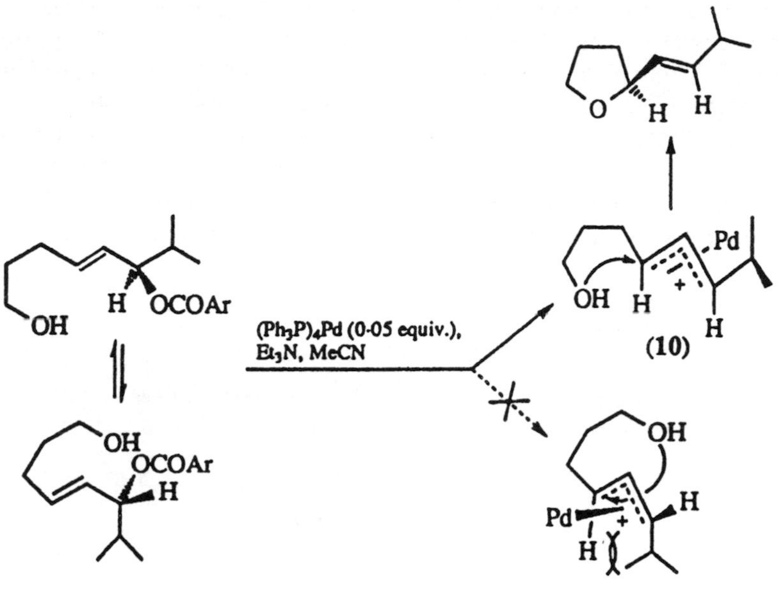

Scheme 11

source of the nucleophile, in palladium-catalysed S_N2' reactions the nucleophile comes from elsewhere.

The nucleophile in Scheme 11 attacks intramolecularly to form a five-membered ring; there can be no attack at the other terminus of the system in (10) because this would require a *trans*-double bond to be contained in a seven-membered ring.[11] The alternative transition state (11) for five-membered ring formation is disfavoured because of $A_{1,3}$-strain (see Chapter 9).

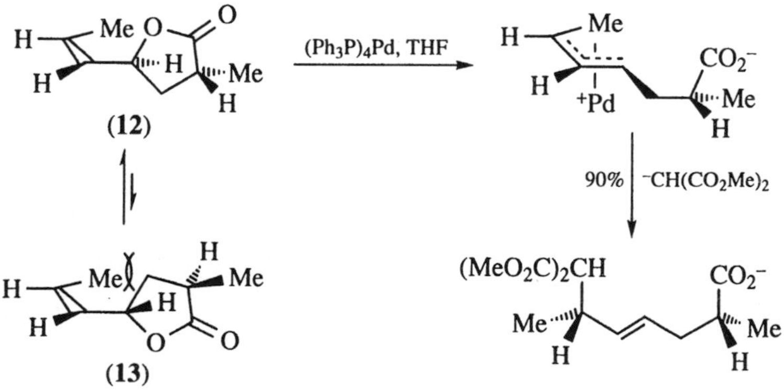

Scheme 12

In these palladium-catalysed reactions, complete stereo- and regioselectivity using intermolecular attack on acyclic substrates by nucleophiles has also been achieved using the lactone (12). The complete stereoselectivity here reflects the lower concentration of the alternative conformer (13) from which ionisation could occur ($A_{1,3}$-strain, see Chapter 9) and/or its sluggish reaction with the bulky (complexed) palladium.[12] The *regioselectivity* in this reaction is thought to be the result of charge repulsion between the incoming nucleophile and the carboxylate anion.

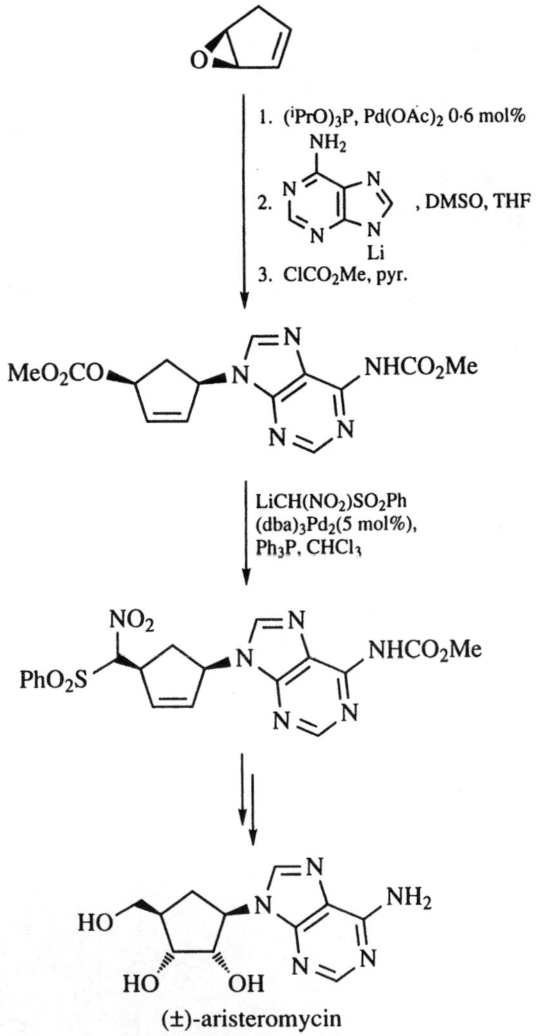

(±)-aristeromycin

dba = (PhCH=CH)$_2$C=O

Scheme 13

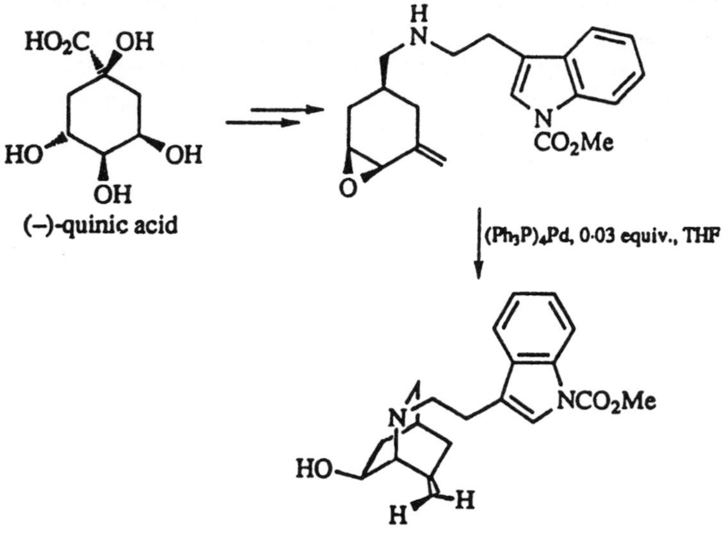

Scheme 14

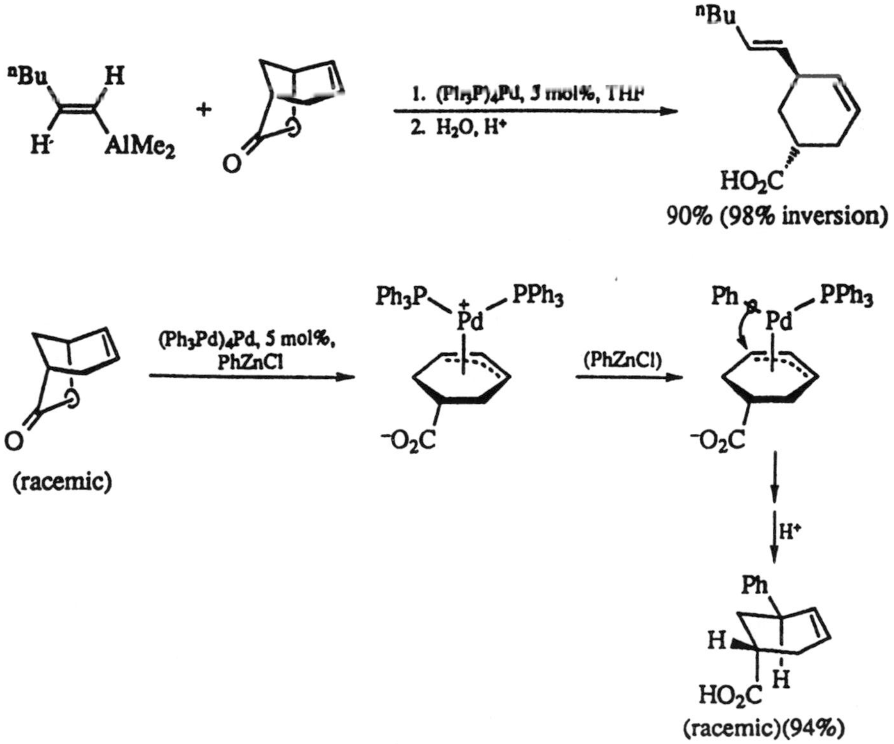

90% (98% inversion)

(racemic)(94%)

Scheme 15

As with cuprates, however, reaction of allylic systems contained within rings is generally more likely to lead to stereoselectivity as in Trost's (±)-aristeromycin synthesis (Scheme 13)[13] and isoquinuclidine synthesis[14] (Scheme 14). In the latter case the π-allylpalladium species is trapped intramolecularly by a nitrogen nucleophile with net retention of configuration at the epoxide carbon atom, i.e. overall this is not equivalent to an S_N2' reaction.

Whereas attack of a nucleophile on the π-allylpalladium intermediate is usually *anti*, attack involving aryl or vinyl zinc halides or vinylalanes is *syn* (Scheme 15); in these cases the nucleophile is transferred first to the palladium (transmetallation) and thence to the *syn* face (Scheme 15).[15]

Simple chirality transfer via concerted sigmatropic rearrangements

In a sigmatropic rearrangement, the breaking of an existing σ-bond is accompanied by the making of a new σ-bond elsewhere in the same molecule; consequential shifting of one or more double bonds is generally required.

Stereoelectronic control in these sigmatropic rearrangements is a consequence of transition-state geometries (TSGs) in which, as always, overlap of the appropriate interacting orbitals (including those from any intervening π-bonds) is maximised.

Sigmatropic rearrangements are one of the important sub-classes of pericyclic reaction first recognised by Woodward and Hoffman.[16] Pericyclic reactions are single-step reactions which proceed via cyclic transition states in which bonds are being made and broken concertedly. Concertedness in these pericyclic reactions is 'allowed' only when the orbitals which overlap have the correct symmetry. If the symmetry is not appropriate, the reaction may still proceed but via a *non-concerted* pathway involving more than a single transition state. *Concertedness in these reactions is often accompanied by high stereoselectivity*; non-concerted reactions are less likely to be highly stereoselective.

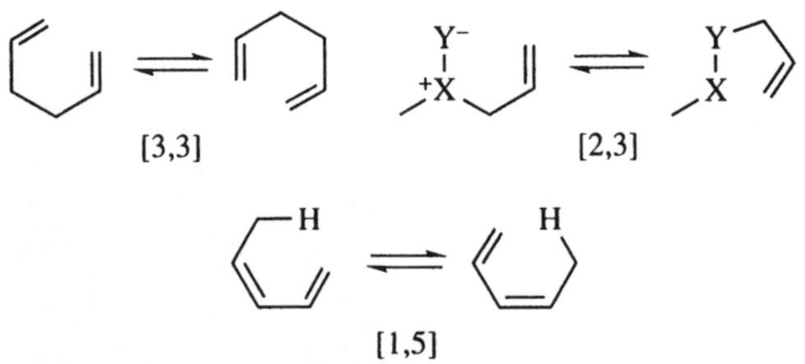

Scheme 16

A feature of pericyclic reactions is that their stereochemical course can usually be predicted when the demands of stereoelectronic control (including correct symmetry of the overlapping orbitals) are considered. Such predictions can be made by using rules devised by Woodward and Hoffman or by using the frontier molecular orbital modification (FMO) of those rules devised by Fukui.[17]

The more important sigmatropic rearrangements in simple chirality transfers are [2,3], [3,3] and [1,5] (Scheme 16).

1,3-Simple chirality transfer by [2,3] sigmatropic rearrangement[18]

IN ACYCLIC SYSTEMS

Probably the most widely used examples of this reaction in stereoselective synthesis are the sulphoxide–sulphenate ester, the [2,3] Wittig, the allyl sulphonium ylide–allyl sulphide and the [2,3] Stevens rearrangements (Scheme 17). Both the Wittig and Stevens rearrangements are prefaced by [2,3] to distinguish them from their [1,2] (non-concerted) variants. The sulphoxide–sulphenate ester interconversion is unusual in being reversible (see below).

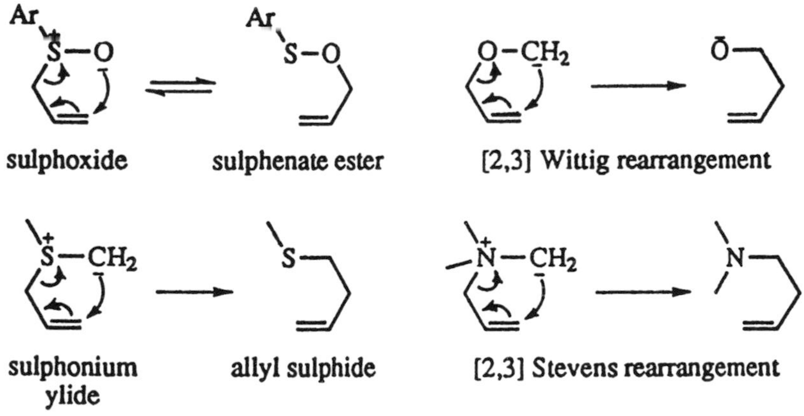

sulphoxide	sulphenate ester	[2,3] Wittig rearrangement
sulphonium ylide	allyl sulphide	[2,3] Stevens rearrangement

Scheme 17

Depending on the substitution present, the [2,3] sigmatropic rearrangement may involve the net creation of chiral centres in Type II or III reactions as illustrated for the Wittig rearrangement [Schemes 18(b) and (c)]. Here, however, we shall consider only those rearrangements in which chirality is transferred, i.e. there is no net gain in the number of chiral centres. For this to be so the double bond must be configured and both it and the existing chiral centre must be present in the pericyclic array [Scheme 18(a)].

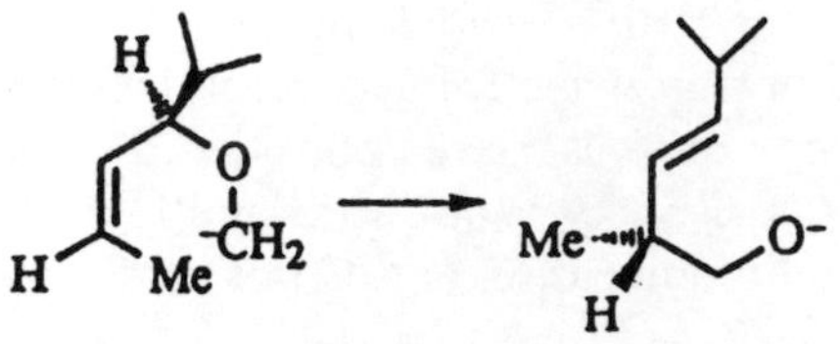

(a) simple chirality transfer (Type III) 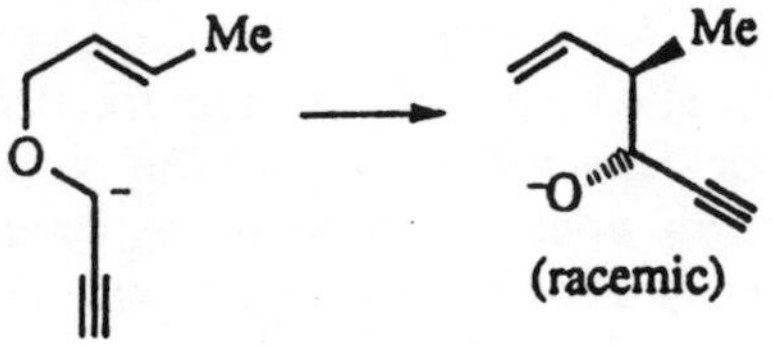   (b) Type II

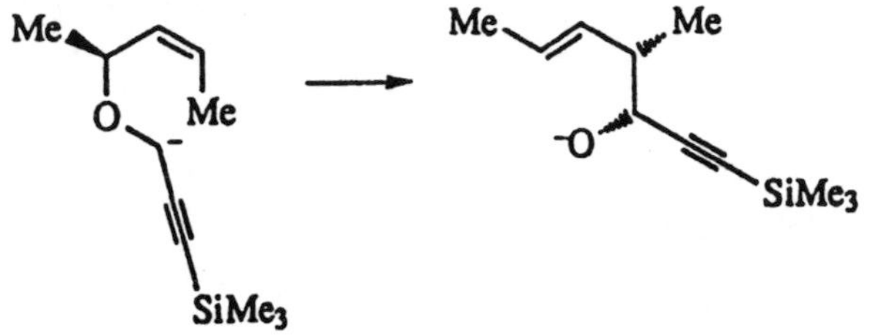

(c) Type III

Scheme 18

In [2,3] rearrangements involving simple chirality transfer there are only two positions for the chiral centre: either at the allylic (i) or homoallylic (ii) position (Scheme 19).

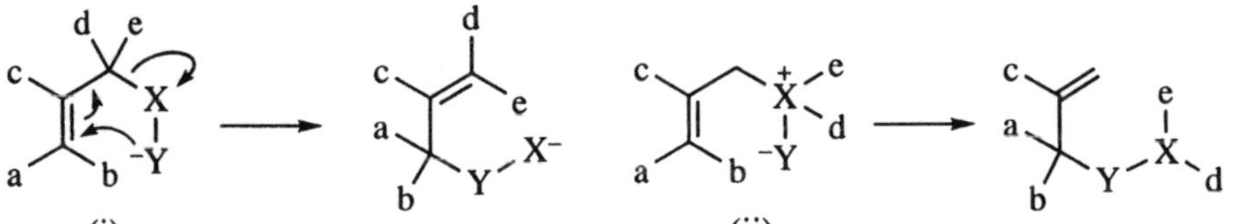

(i)   (ii)

Scheme 19

Maximised overlap of the orbitals involved in the [2,3] rearrangement can be accommodated as indicated in Scheme 20.

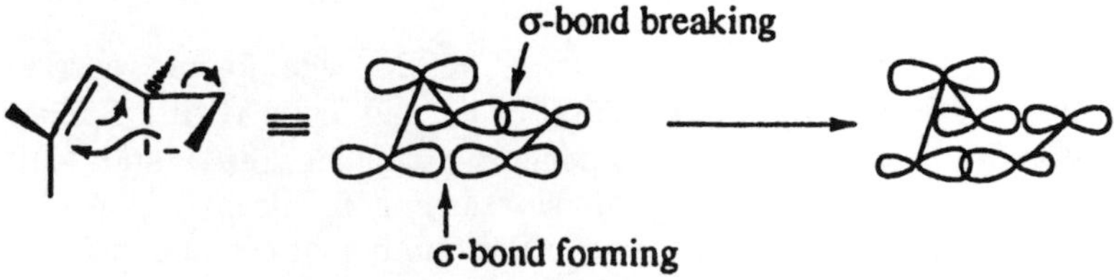

σ-bond breaking

σ-bond forming

Scheme 20

A formalism by which the symmetry of the orbitals involved in this rearrangement can be deduced is as follows (Scheme 21): the σ-bond which is breaking is located (A) and the two radical species obtained by a (hypothetical) homolytic cleavage of this bond are identified (B). Consideration of the highest occupied molecular orbitals (HOMOs) of these two radical species (C) shows that the lobes required to overlap to form the new σ-bond have the same phase (the lobes of the orbitals produced from the breaking bond *must* have the same phase) and thus *concerted* rearrangement is allowed, giving the product (D).

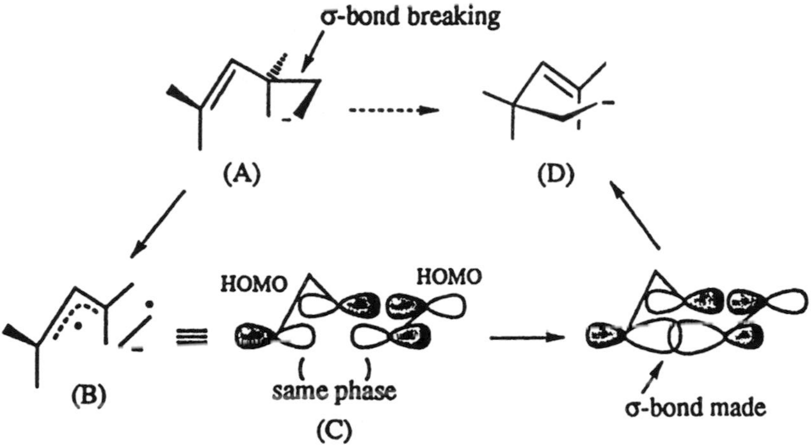

Scheme 21

The 'envelope' conformation shown for the transition state geometry in Scheme 21 is not necessarily that through which all [2,3] sigmatropic rearrangements pass. Nakai and Mikami[18] have suggested that the stereochemistry of some (Type II) [2,3] Wittig rearrangements is better accommodated using an alternative 'envelope' represented by (14) (Scheme 22).

Using the [2,3] Wittig rearrangement by way of illustration Scheme 23, it can be seen that there is a *syn* relationship between the C—C bond made and the C—O bond broken. Alternatively, one can say that the CH$_2$O unit is

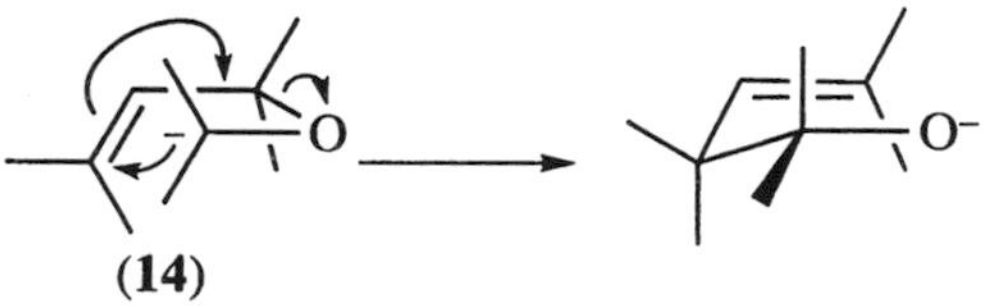

Scheme 22

transferred across one face of the allyl system (suprafacial migration). Like the [1,2] rearrangements discussed in Chapter 4, this rearrangement results in the conversion of a more readily available C—O bond-containing compound into a less readily available C—C bond-containing compound.

Even so, in an acyclic system there are always two (non-Nakai) conformations, (15) and (16) (Scheme 23), through which the rearrangement can proceed when a chiral centre is present in the starting material. Note that (15) and (16) have the *same* configuration at the chiral centre; they differ in that different diastereofaces of the double bond are being attacked in the two cases.

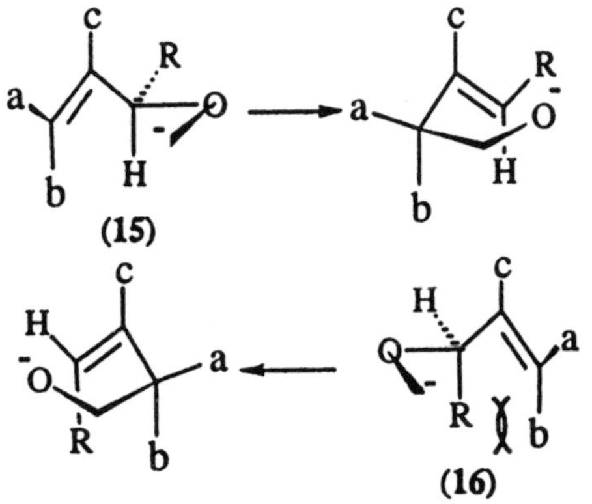

Scheme 23

The transition state geometry for the [2,3] sigmatropic rearrangement is reminiscent of that for the S_N2' reaction discussed previously (Scheme 5), which, in an acyclic case, can similarly proceed via two conformations involving *syn*-attack. (The S_N2' reaction can also proceed via two conformations involving *anti*-attack). The two possible products in Scheme 23, like those in Scheme 5, have opposite configurations for both their double bonds *and* their chiral centres, and this complementary relationship always holds.

Complete stereoselectivity in this chirality transfer will require reaction to proceed entirely through (15) or (16) in Scheme 23. As indicated (ʃ) there are steric interactions in both (15) and (16) of a magnitude that will depend on the nature of R, b and c.

A particularly valuable experimental finding by Still and Mitra[19] was that allyoxymethyl anions required for the [2,3] Wittig rearrangement can be generated by transmetallation of the corresponding trialkystannylmethyl ethers. Thus treatment of the ether (17), prepared as indicated in Scheme 24(a), with butyllithium, generates the corresponding alkyllithium (18) [Scheme 24(b)]. Spontaneous [2,3] rearrangement of (18) gives, after protonation, (19) with the double bond exclusively E.[20]

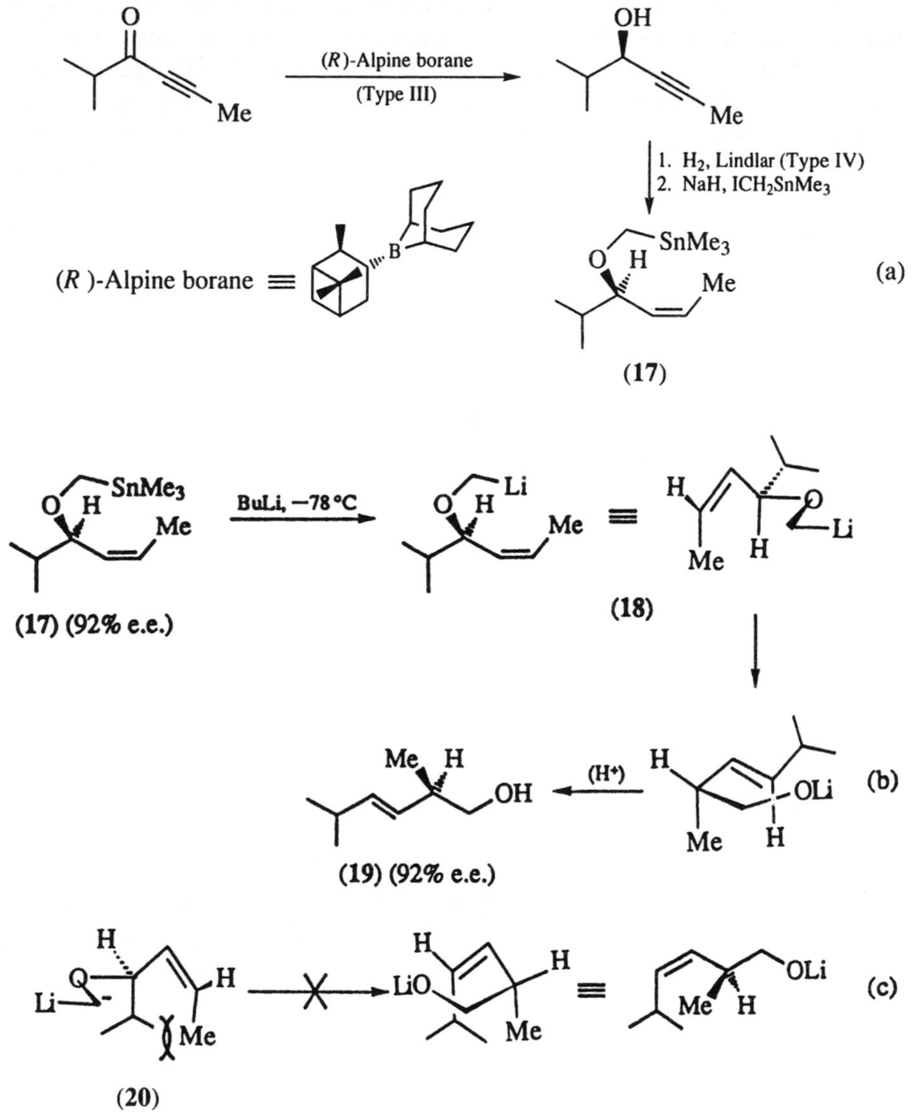

(a)

(b)

(c)

Scheme 24

The enantiomeric excess in the starting material (**17**) corresponds to that in the product (**19**) and the absolute configurations of (**17**) and (**19**) are in agreement with every molecule faithfully rearranging via a transition state resembling (**18**). No product arising from rearrangement via the alternative conformation (**20**) [Scheme 24(c)] is observed, presumably since the transition state in that case would be destabilised by the methyl–isopropyl interaction.

The preference in Scheme 23 for rearrangement via (15) rather than (16) is strong when the R group is large and when b is larger than hydrogen [cf. Scheme 24(b)]. However, this selectivity would be expected to decline when the allylic position bearing R is also substituted with a group R^1 of a similar bulk. Preferential reaction via (16) in Scheme 23 can be accomplished if c is a bulky group (methyl or larger) and b is a hydrogen as in the example in Scheme 25 (a Type IV reaction). Here the interaction between the butyl and methyl groups indicated in (22) is sufficient to direct all of the rearrangement via conformer (21).

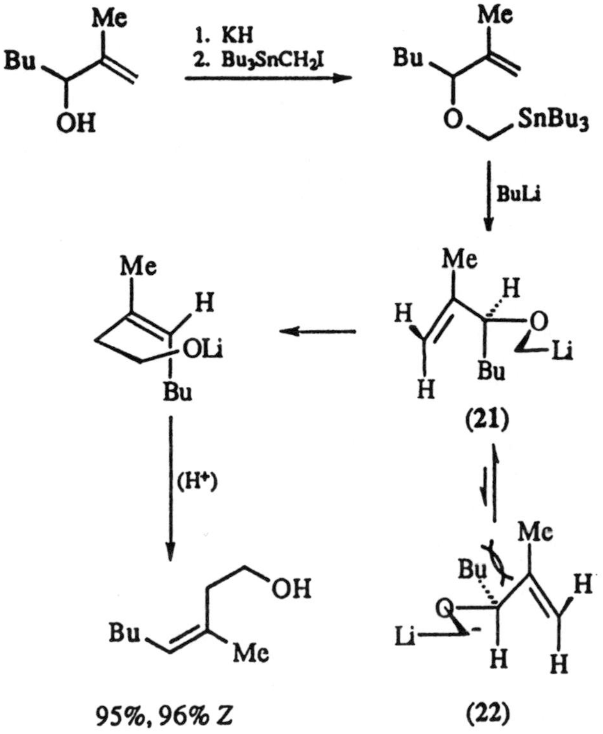

Scheme 25

Allyl sulphoxide–allyl sulphenate ester interconversion

The alternative location of the chiral centre in a [2,3] rearrangement is at the homoallylic position. This can be the case with sulphoxides since the sulphur is pyramidal and normally configurationally stable, i.e. the lone pair of electrons on sulphur does not invert at a measurable rate at room temperature.

However, enantiopure allylic sulphoxides, e.g. (23), racemise (Scheme 26)

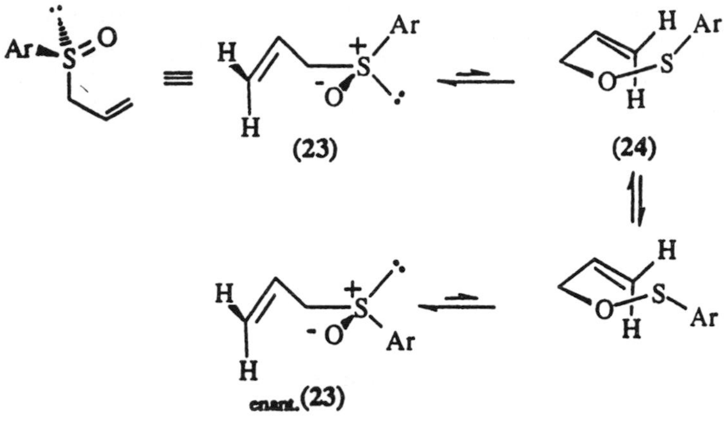

(23)   (24)

enant. (23)

Scheme 26

much more rapidly than those lacking an allyl group, by a mechanism which involves reversible formation of the achiral sulphenate ester (24).

[2,3] Sigmatropic rearrangement provides a route for the inversion configuration of the allyl double bond in allyl sulphoxides according to the route in Scheme 27. Thus with Ar = p-tolyl, rearrangement of (25) to an equilibrium mixture of (25) (77%) and (26) (23%) took place over a period of days at room temperature [but at a slower rate than racemisation of the enantiopure sulphoxide (25)].[21]

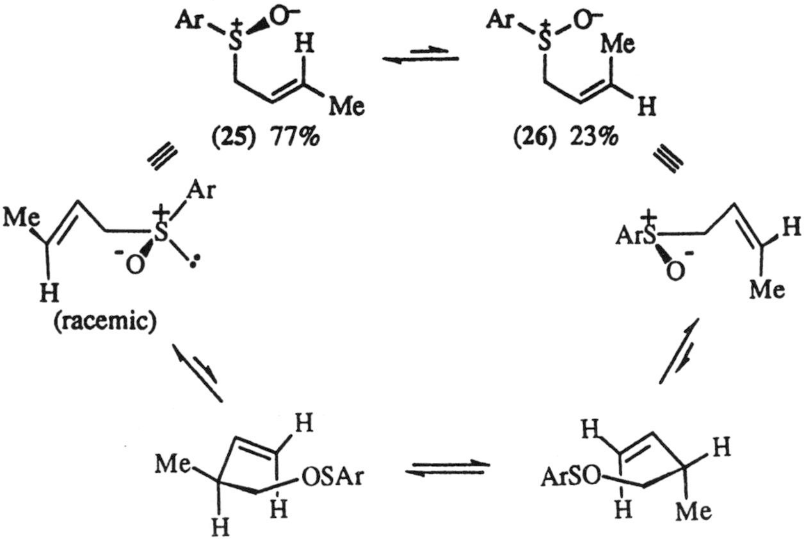

(25) 77%   (26) 23%

(racemic)

Scheme 27

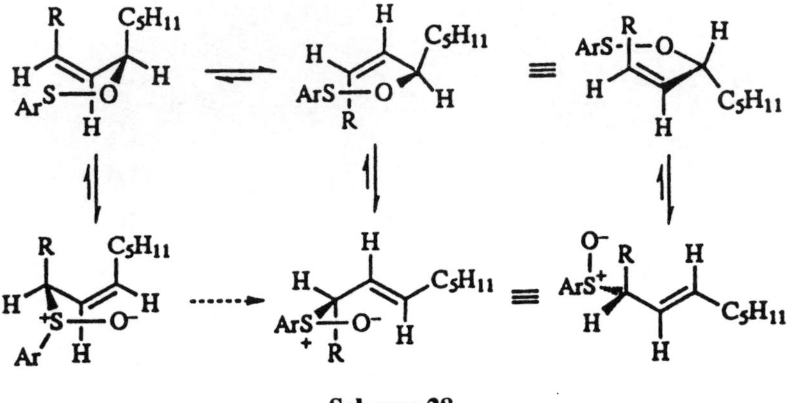

Scheme 28

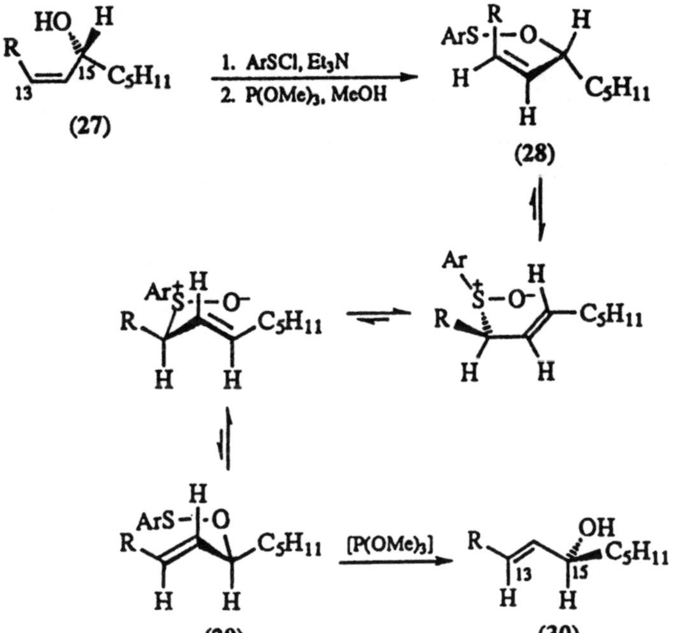

Ar = *p*-MeC$_6$H$_4$

R =

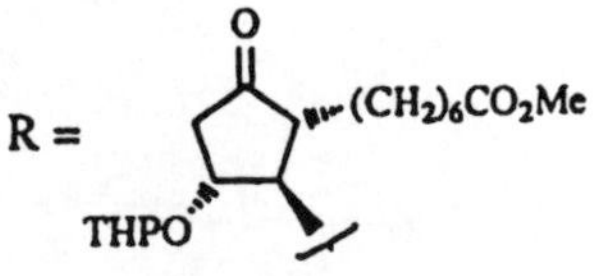

Scheme 29

If an allylic sulphoxide contains a configured double bond *and* a chiral centre at the allylic position, equilibration via sulphenate inverts the configuration of the double bond *and* the chiral centre according to Scheme 28, in which the chirality at sulphur is not specified. Note that in this equilibration there are two ways in which the sulphenate ester rearrangement can be depicted.

If a sulphenate ester is prepared from a chiral allylic alcohol containing a configured double bond, the [2,3] sigmatropic rearrangement sets up an equilibrium between two sulphenates (28) and (29) (and two sulphoxides, one of which is not shown) (Scheme 29). Trialkyl phosphites react with sulphenate esters, e.g. (28) and (29), by attack at the sulphur by phosphorus to produce the corresponding allylic alcohols.[22] In Scheme 29, the equilibrium which is set up between sulphenate esters (28) and (29) favours the latter. Consequently, attack of trimethyl phosphite produces allylic alcohol (30).

Overall, both the double bond and the chiral centre in allylic alcohol (27) have undergone inversion of configuration in the conversion to (30). This interconversion has been exploited by Stork and co-workers,[23] who converted the more easily prepared unnatural 13-*cis*-15β- into the natural 13-*trans*-15α-prostaglandin derivative (Scheme 29, R as shown) using this phosphite trapping device. The driving force in this rearrangement is presumably the conversion of a *cis*- to a *trans*-double bond.

IN CYCLIC SYSTEMS

It will become clear in succeeding chapters (if it is not apparent already) that stereoselection is more difficult to accomplish in reactions in which the substrate is acyclic. Reduction of the conformational freedom of a system undergoing [2,3] rearrangement by its incorporation into a cyclic system will usually lead to increased stereoselectivity. Thus, when the [2,3] Wittig rearrangement is constrained to take place in the ring D of a steroid (Scheme 30), the 16α-chirality is transmitted to C-20 in a wholly predictable way.[24] (A geometrical imperative is present in the reaction; see Chapter 11.) This is because only one face of the double bond is available for reaction, in contrast to the acyclic case, where both faces are accessible.

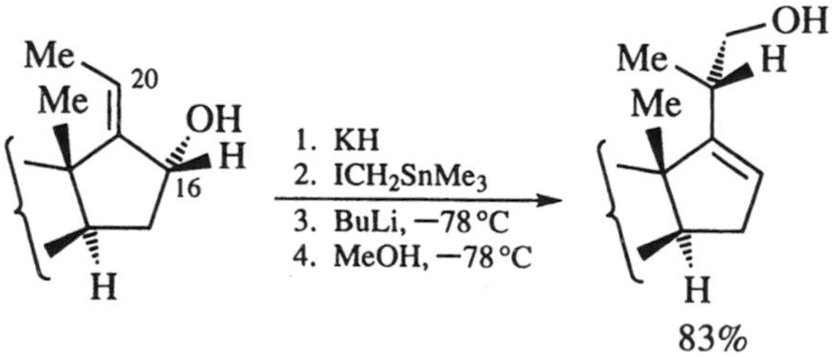

Scheme 30

Whereas the sulphoxide–sulphenate ester equilibrium usually lies on the side of the former, the selenoxide–selenate equilibrium favours the latter. The resulting selenate ester is readily hydrolysed to the alcohol.[25] In the reaction in Scheme 31 there is also no possible ambiguity about the stereochemical outcome of the rearrangement.

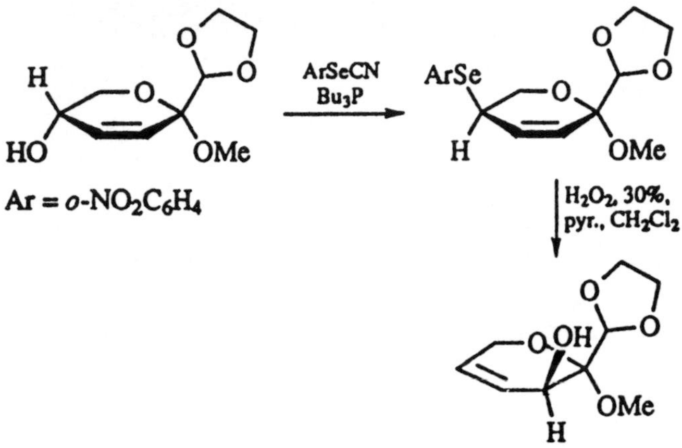

Ar = o-NO$_2$C$_6$H$_4$

Scheme 31

[2,3] Rearrangement of an allyl sulphonium ylide is exemplified by the conversion of (**31**) to (**32**) in Scheme 32. Since this rearrangement regenerates an allylic thioether, it may be carried out iteratively to provide larger ring compounds.[26]

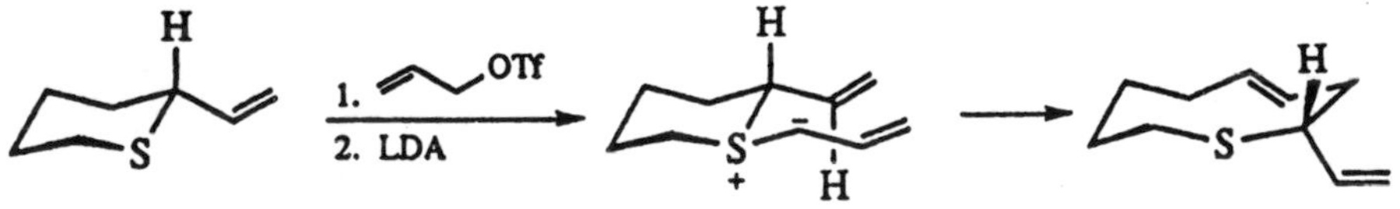

Scheme 32

Simple chirality transfer via [3,3] sigmatropic rearrangements

The transition-state geometry for [3,3] sigmatropic rearrangement can be either chair- or boat-shaped [Scheme 33(a)].

Like the [2,3] rearrangement, the concertedness of the [3,3] rearrangement is allowed since the symmetry of the orbitals of the two allyl radicals produced by (hypothetical) homolytic cleavage of the σ-bond allows for in-phase overlap of the terminal lobes to form a stable new σ-bond [Scheme 33(b)] (*cf.* Scheme 21).

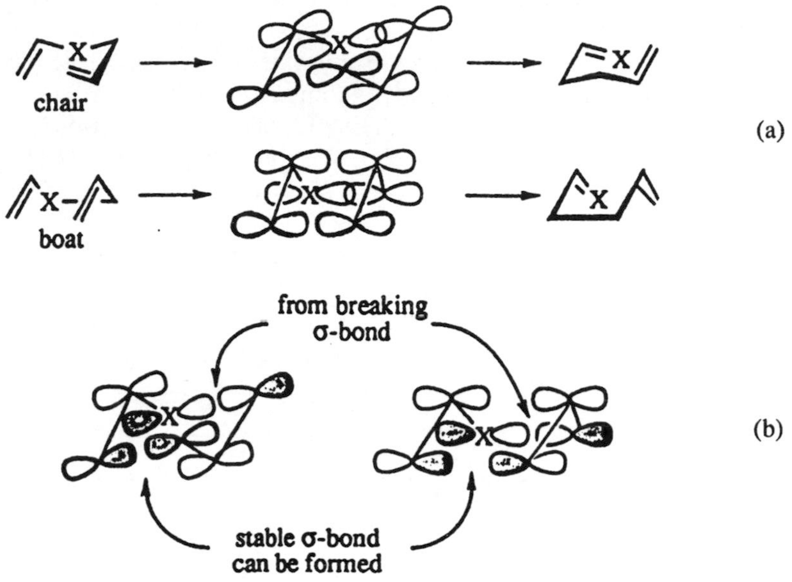

(a)

(b)

Scheme 33

The all-carbon [3,3] rearrangement (Scheme 33; X – CH$_2$)—the Cope rearrangement[27]—is useful in stereoselective synthesis but we shall illustrate chirality transfer by particular reference to the Claisen rearrangement (Scheme 33; X = O) since this reaction has been more widely studied and used in stereoselective synthesis.[28] The Claisen rearrangement is closely related to the [2,3] Wittig rearrangement and, as in the latter, a C—C bond is made at the expense of the C—O bond which is broken on a single face of the allyl system.

ACYCLIC CASES

For chirality transfer using [3,3] sigmatropic rearrangements, there must be at least one chiral centre present in the pericyclic array and one configured double bond, and since in the Claisen rearrangement there is only one sp^3-hybridised carbon, this must be the chiral centre. The chirality transfer which results from the Claisen rearrangement is 1,4 or 1,3 depending on whether

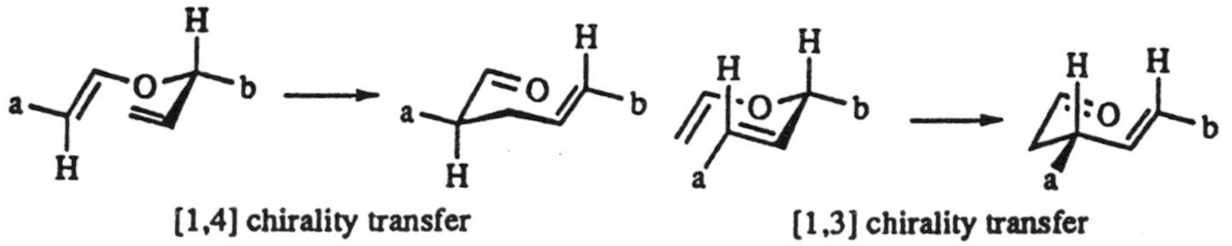

[1,4] chirality transfer   [1,3] chirality transfer

Scheme 34

the configured double bond is present in the vinyl or allyl ether moiety (Scheme 34).

In Scheme 35, using the single stereoisomer of the vinyl ether shown, there are four possible transition-state geometries (TSGs) through which 1,4-chirality transfer can be effected. The products from the two chair TSGs are diastereoisomers as are those from the two boat TSGs; the two diastereoisomers formed by the chair TSGs are enantiomers of those formed by the boat TSGs.

The great use of the Claisen rearrangement in synthesis derives from a number of factors including (a) the ease of preparation of substituted allyl vinyl ethers, (b) the equilibrium position which favours the carbonyl compound and (c) the usefulness of the products of the rearrangement for

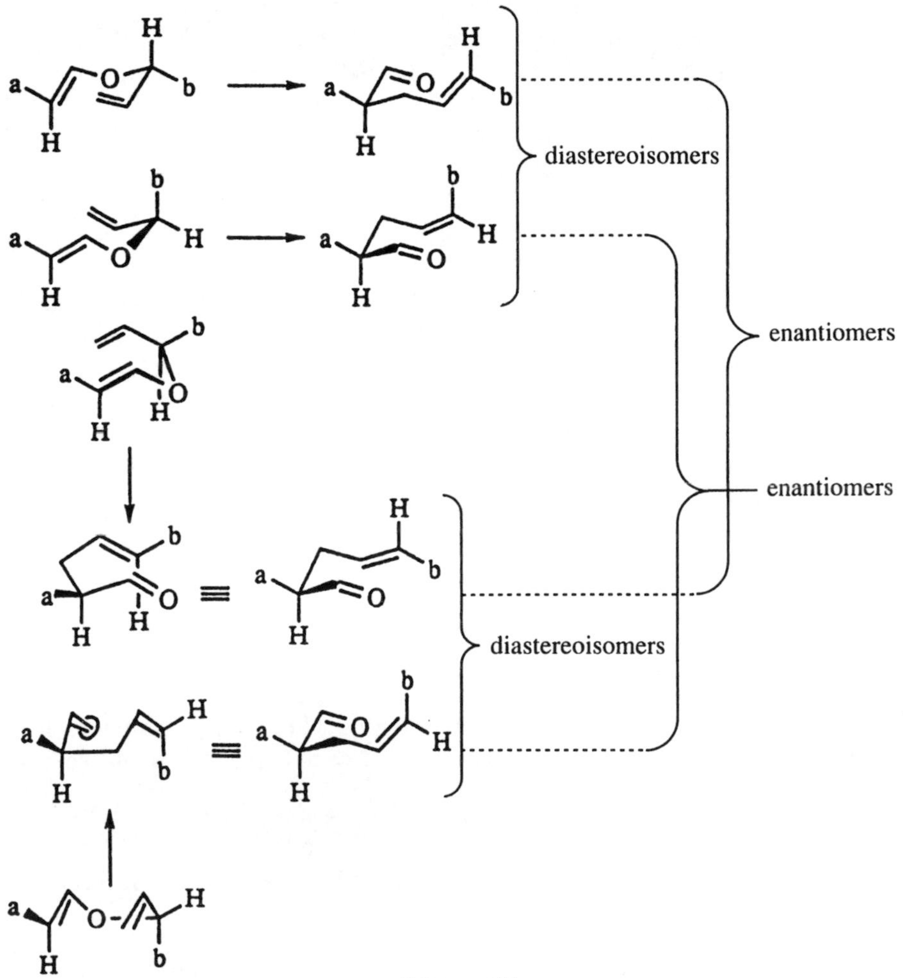

Scheme 35

further manipulation. Its value in *stereoselective* synthesis is the result not only of a clear preference either for the chair transition state (usual for acyclic substrates) or for the boat (common for cyclic substrates), but also for that chair TSG in which b is equatorial (**33**) rather than axial (**34**) (Scheme 36).

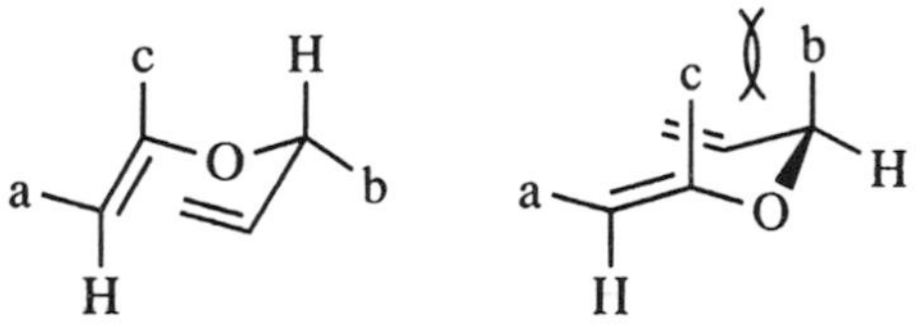

(**33**)   (**34**)

Scheme 36

There are two important consequences of this preference for TSG (**33**). The first is that there will be increasingly selective formation of an *E*-double bond as the 1,3-diaxial-type interaction in TSG (**34**) becomes more serious. This is revealed from comparison of the rearrangements in Scheme 37.[29]

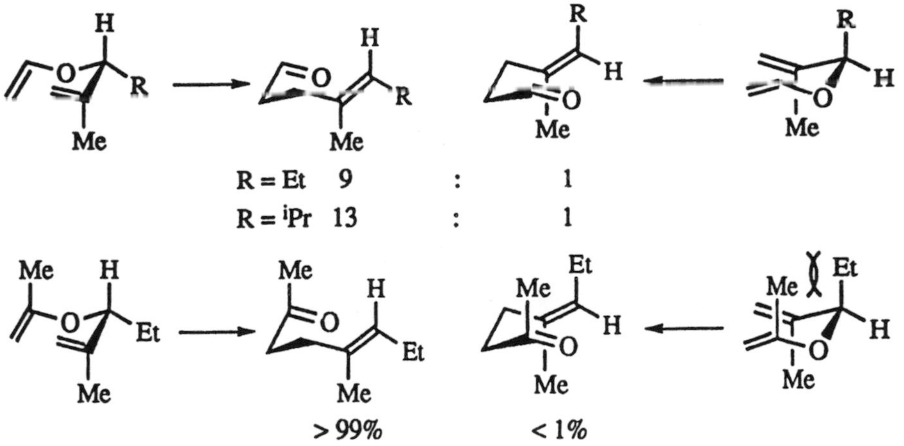

R = Et   9   :   1
R = iPr   13   :   1

> 99%   < 1%

Scheme 37

Relevant to the nature of c in Scheme 36 is the reaction by which the vinyl ether in the starting material is formed. Two valuable modifications of the Claisen rearrangement both generate a substituted vinyl ether and carry out the [3,3] rearrangement *in situ* and in both, the bulk of the c group (Scheme 36) leads to enhanced diastereoselectivity. In the Eschenmoser variant, the allyl alcohol is heated in the presence of *N,N*-dimethylacetamide dimethylacetal (**35**) (Scheme 38). The substituted vinyl ether (**36**), formed as shown, rearranges to the amide (**37**).[30]

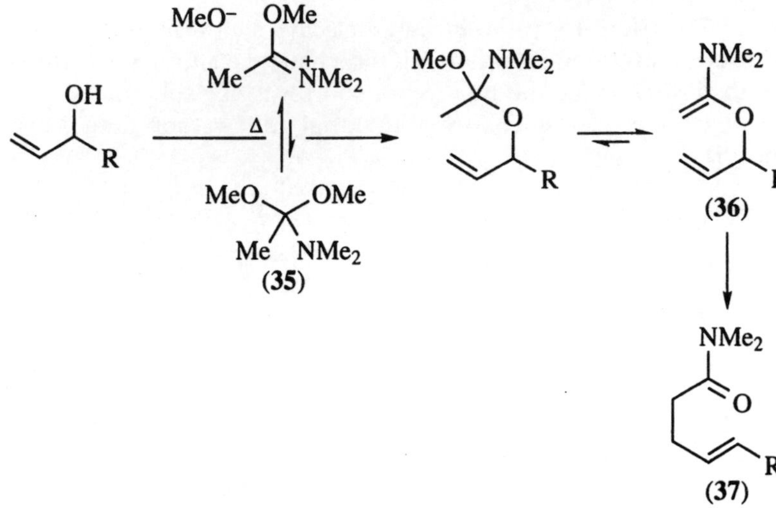

Scheme 38

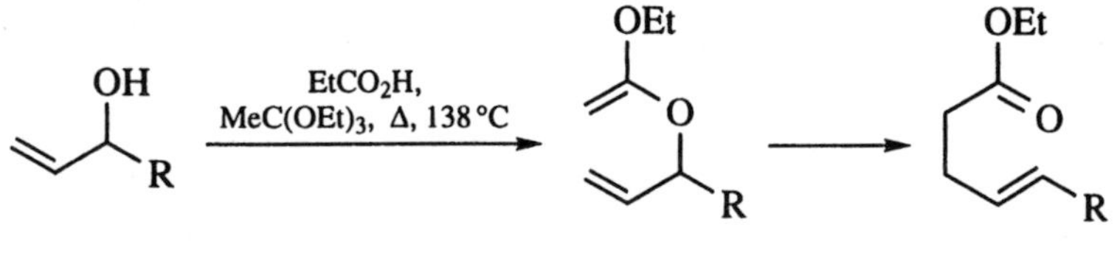

Scheme 39

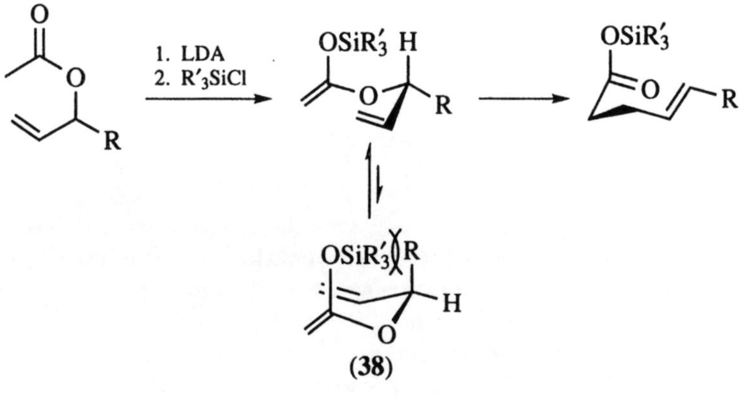

Scheme 40

A closely related procedure uses an *ortho*ester as the source of the vinyl ether carbons (Scheme 39) and the product is an ester; a weak acid is usually present here as a catalyst.[31]

In the Ireland–Claisen modification, a trialkylsilyloxyvinyl ether (a silyl ketene acetal) is produced from an allyl ester by the action of lithium diisopropylamide (LDA) followed by silylation (Scheme 40).[32] Not only does the trialkylsilyloxy group improve the stereoselectivity [large diaxial repulsion between R and $OSiR^1_3$ in the alternative chair TSG (**38**)] but it also facilitates the rearrangement such that it proceeds at or close to ambient temperature.

The second consequence of the preference for TSG (**33**) (Scheme 36) is that use of an enantiopure allyl vinyl ether will result in a product of predictable configuration at its sp^3-chiral centre. Thus reaction of (*E*)-pent-3-en-2-ol (**39**) with (**35**) in boiling xylene (Scheme 41) gave amide (**40**) with 1,3-chirality transfer.[33]

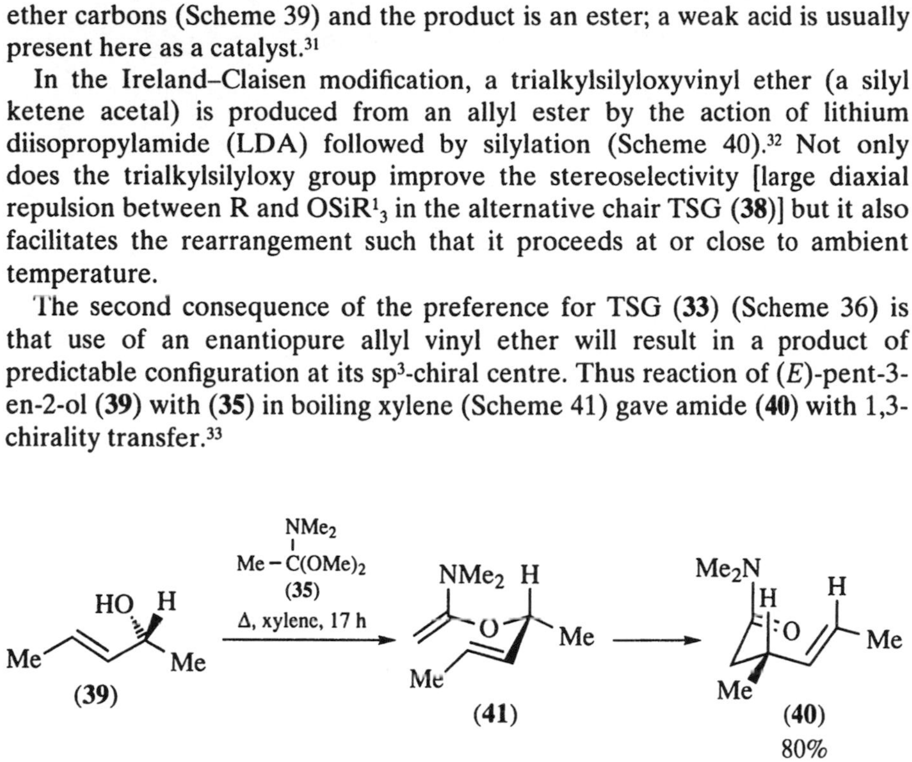

Scheme 41

Based on the enantiopurity of the starting material (**39**) and the product (**40**), the reaction goes with ≥90% enantioselectivity with the sense of induction consistent with the chair transition state (**41**) shown.

Enantiopurified secondary allyl alcohols required for the enantioselective Claisen rearrangement such as that in Scheme 41 are accessible by enantioselective reduction of α,β-acetylenic ketones with Alpine borane [cf. Scheme 24(a)] or by kinetic resolution using the enantioselective Sharpless epoxidation (see Chapter 14).

A clever way in which the preference for formation of (*E*)-alkenes referred to above can be overridden is by the use of sterically hindered aluminium catalysts (Scheme 42).[34]

As shown in Scheme 42, the bulky aluminium catalyst, by complexing to the ether oxygen, forces the isobutyl group to occupy the pseudo-axial position, particularly if the oxygen is more sp^2- than sp^3-hybridised. The result is a preference for formation of (**43**) with a *Z*-configured double bond.

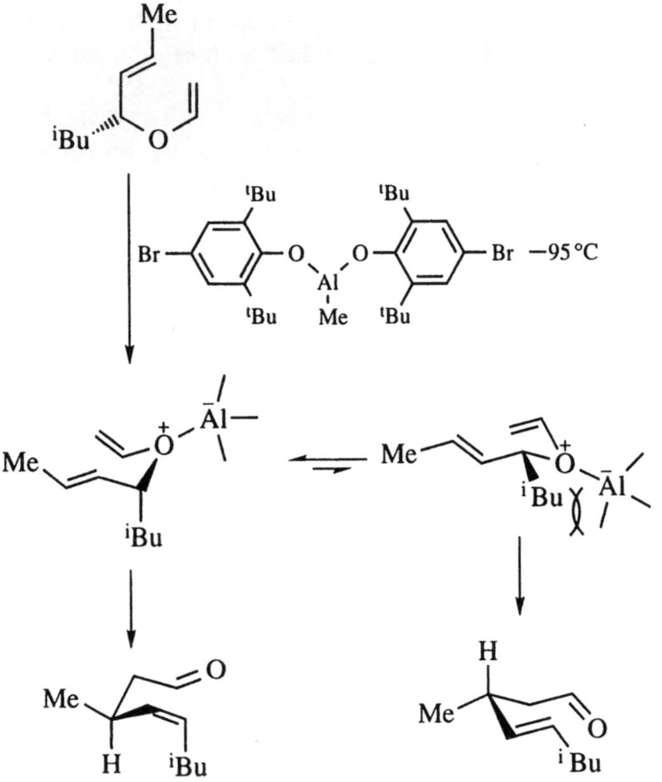

(43) 78% e.e.   **(42)**

Yield 74%

ratio **(42)** : **(43)** = 16 : 84

Scheme 42

A difficulty in practice in bringing about 1,4-chirality transfer in the Claisen rearrangement is the generation of a configurationally homogeneous vinyl ether double bond in the vinyl allyl ether undergoing rearrangement. With the Ireland–Claisen modification this calls for a method of conversion of the ester **(44)** into the silyl ketene acetal **(45)** or **(46)** as required (Scheme 43). Whilst a complete solution to this problem is not yet available, high selectivities (9:1) can often be achieved by the use of methods that generate one or other of the ester enolate intermediates **(47)** or **(48)** (see Chapter 8).

If the substituent a in **(33)** (Scheme 36) contains a chelating atom (O, S or N), the configurational purity of the silyl ketene acetal is often assured (Scheme 44) because the chelation in the enolate demands that the double bond be contained in a ring.[35]

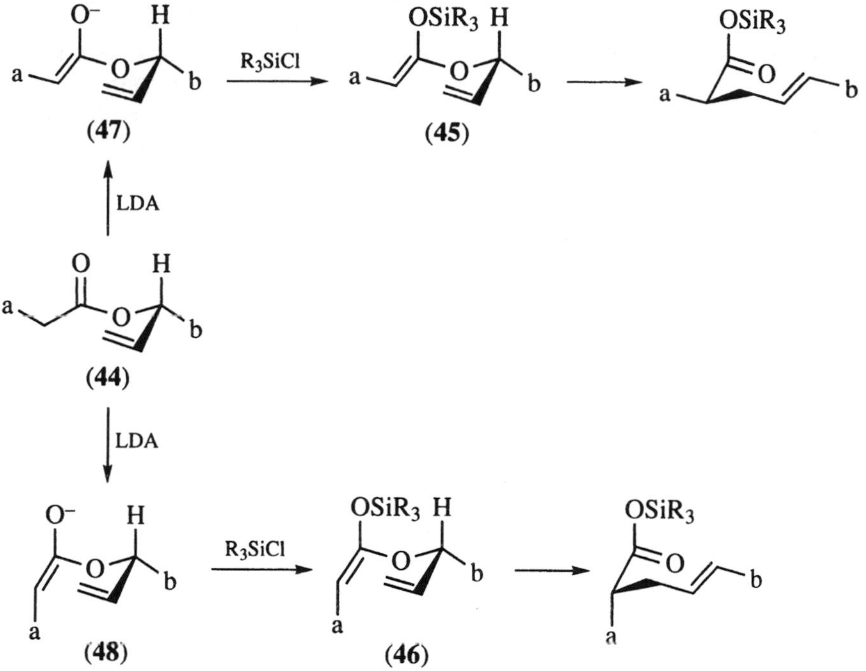

Scheme 43

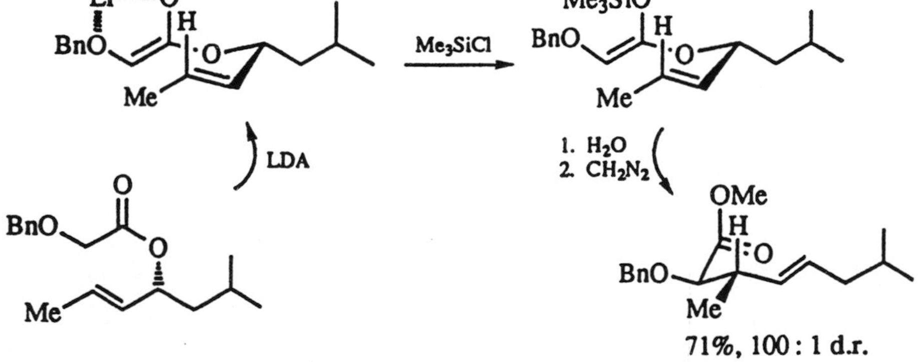

71%, 100 : 1 d.r.

Scheme 44

[3,3] SIGMATROPIC REARRANGEMENTS INVOLVING RINGS

If the allyl unit in the Claisen rearrangement is confined to a ring with eight or less members, e.g. Scheme 45, the four TSGs available in an acyclic case (Scheme 35) are reduced to two. Only one face of the allyl double bond can enter into the rearrangement and the configuration of the new chiral centre on the ring is predictable irrespective of whether a chair or boat TSG is involved.

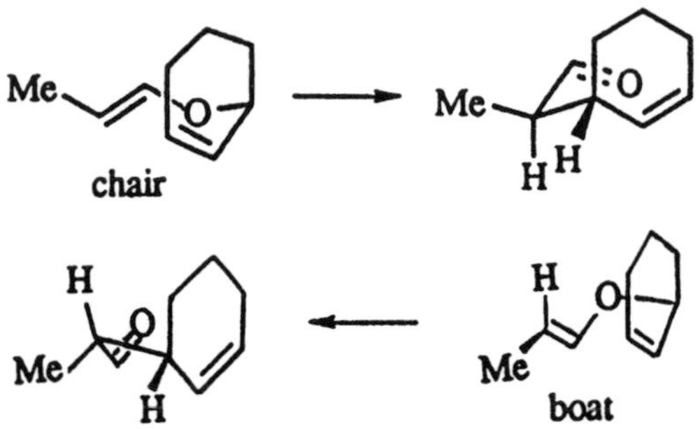

Scheme 45

A substituent R^2 on the vinyl ether double bond (Scheme 46) and *cis* substituents R^3, R^4 on the ring, as in (**49**), are likely to destabilise the chair more than the boat form, and it is in this case that the boat-derived stereostructure is most likely to be found for the product.

When the vinyl ether moiety is confined to a ring of eight or less atoms, the only possible TSG is the boat. As the example in Scheme 47 illustrates, the Ireland–Claisen rearrangement is capable of bringing about the formation of a quaternary chiral centre (directly bonded in this case to two tertiary chiral centres).[36]

Both the Claisen and Cope rearrangements may lead to products in which one of the double bonds in the first-formed product is part of an enolate which is converted into a ketone in the work-up (Scheme 48). In the retrosynthetic analysis of (**52**) as a target molecule, vinylcarbinol (**50**) does not immediately suggest itself as a precursor—the synthetic potential

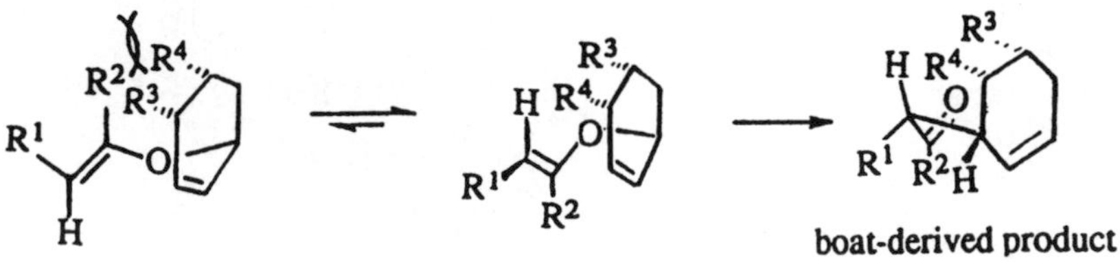

Scheme 46

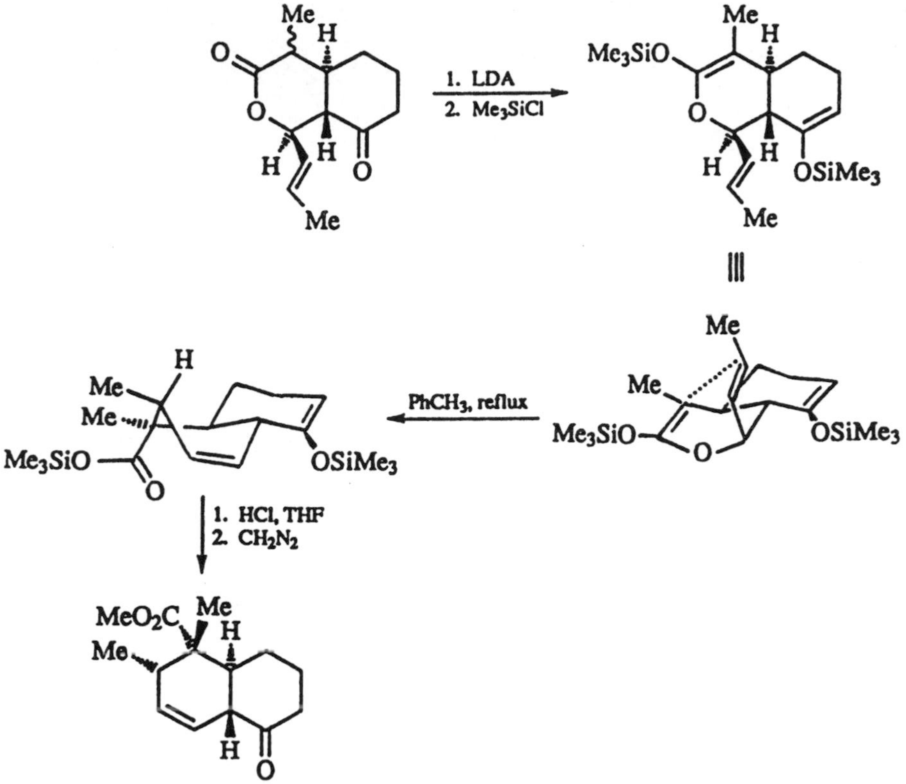

Scheme 47

of the [3.3] rearrangement is disguised. This disguise results from the need to draw the enolised form of (52) in an unfavourable boat conformation (51) for the possibility of the (reverse) sigmatropic rearrangement to become evident.

Nevertheless, blowing the cover of a disguised (reverse) [3,3] sigmatropic rearrangement may be rewarding, particularly since the anionic 'oxy-Cope' rearrangement can be carried out under very mild conditions. Evans and Golob[37] showed that conversion of alcohol (50) into its alkoxide brought about an increase of up to 10^{17}-fold in the rate of the Cope rearrangement. Thus, whereas thermal rearrangement of (50) was previously carried out at ~300 °C,[38] the potassium alkoxide, in the presence of a crown ether, rearranged below room temperature. Much use has been made of the mild conditions required for this anionic oxy-Cope rearrangement, particularly since the anionic oxygen has been found to have a preference for the 'equatorial' position (Scheme 49) which is greater than that of a hydroxyl group.[39]

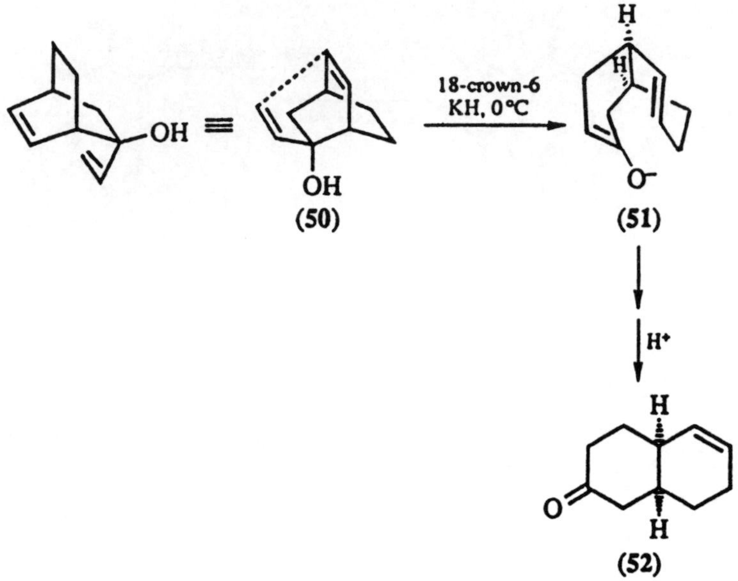

Scheme 48

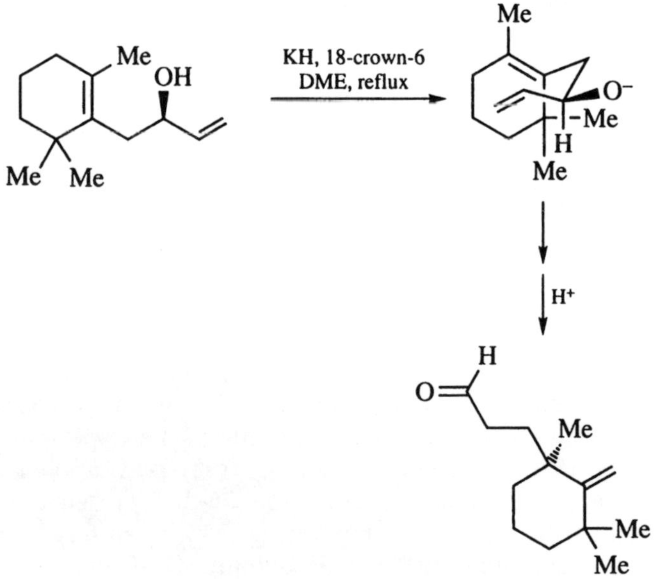

Scheme 49

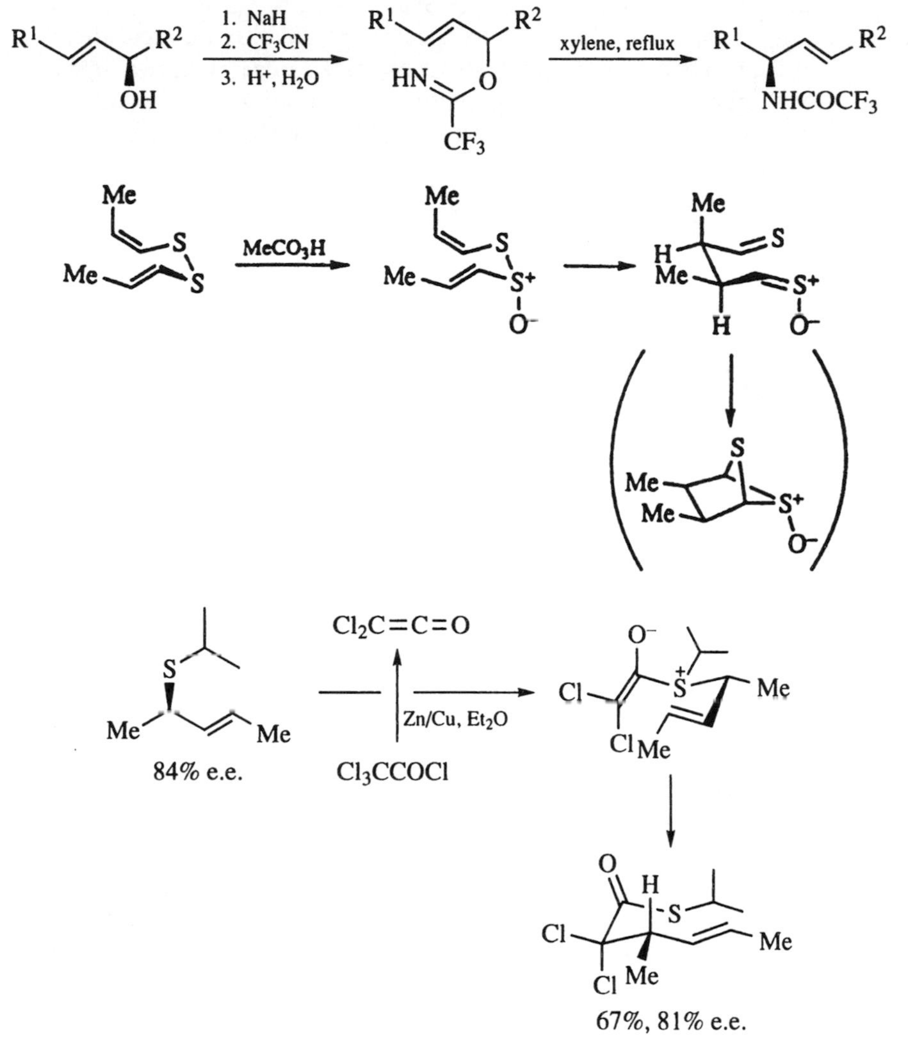

Scheme 50

Some examples of chirality transfer in [3,3] sigmatropic rearrangements other than the Claisen rearrangement are given in Scheme 50.[40]

[1,5] Sigmatropic rearrangements

Thermal [1,5] sigmatropic rearrangement of the C—R bond in Scheme 51 takes place suprafacially with retention of configuration in the migrating group R. The (hypothetical) homolytic cleavage of the C—R bond results in a pentadienyl radical whose terminal lobes have correct phase for this suprafacial rearrangement to occur (*cf.* Scheme 21).

Thus, given the configurations of the terminal double bond and the chiral centre in the starting material, the transition state geometry shown in Scheme 51 allows one to predict confidently the relationship between the configuration of the new chiral centre and that of the double bond in the product, although there remains the question of whether path (a) will be favoured over (b) or vice versa.

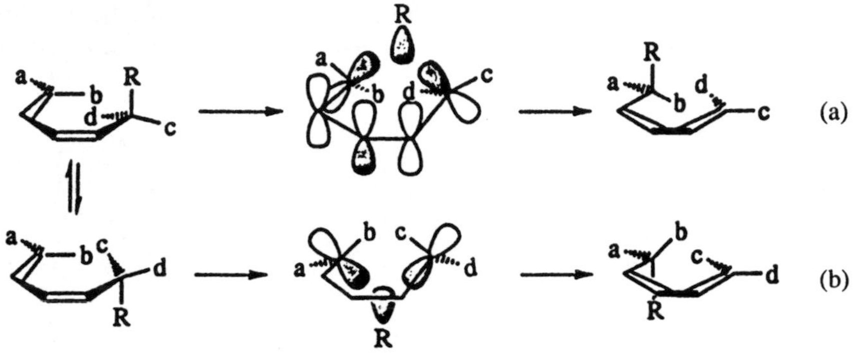

Scheme 51

In practice, the majority of [1,5] shifts have been those of hydrogen although other groups (CHO, $PhCH_2$, $SiMe_3$ and CH_3CO) have been shown to have higher migratory aptitudes than hydrogen.[41] Because of the difficulty in attaining the *s-cis*-diene conformation depicted in Scheme 51 in an acyclic substrate, [1,5] shifts are most commonly found in dienyl systems which are contained in a ring and especially in cyclopentadienyl, cyclohexadienyl, cycloheptadienyl or cycloheptatrienyl systems.[42] The ambiguity as to the stereochemical outcome which exists in the acyclic case (Scheme 51) is also removed since suprafacial migration is possible only to one face in these cyclic systems.

Unfortunately, it is often difficult in practice in these cases to limit the [1,5] shift to a single migration and serial [1,5] H rearrangements often result. In cyclopentadienyl systems, facile [1,5] sigmatropic rearrangement usually means that isomerically pure mono- or disubstituted derivatives are unobtainable (Scheme 52). From a synthetic point of view this is most

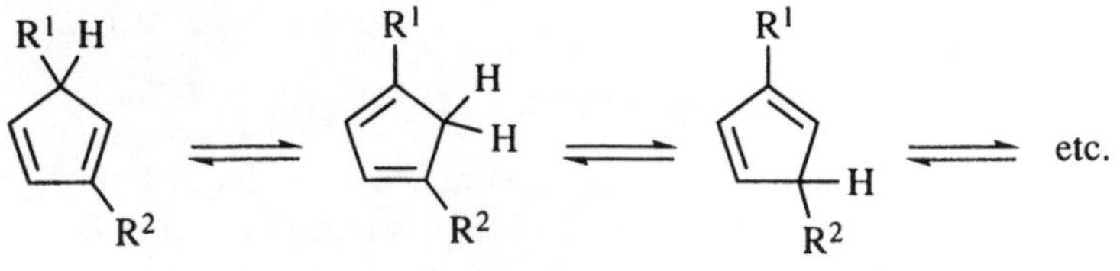

Scheme 52

unfortunate because it limits the use of substituted cyclopentadienes as dienes in Diels–Alder cycloadditions: a mixture of interconverting dienes will usually result in an intractable mixture of isomeric adducts.

Possibly the most useful [1,5] sigmatropic rearrangements are those used in tandem reactions: the diene from one or more [1,5] shifts is trapped, often by an intramolecular pericyclic reaction. Scheme 53 shows a synthesis of a functionalised spiro[4,5]decane (54), a skeleton found in many natural products; only one of the three possible isomers in the 1,5-sigmatropically rearranging mixture (53) is trapped by intramolecular (Type II) Diels–Alder cycloaddition.[43]

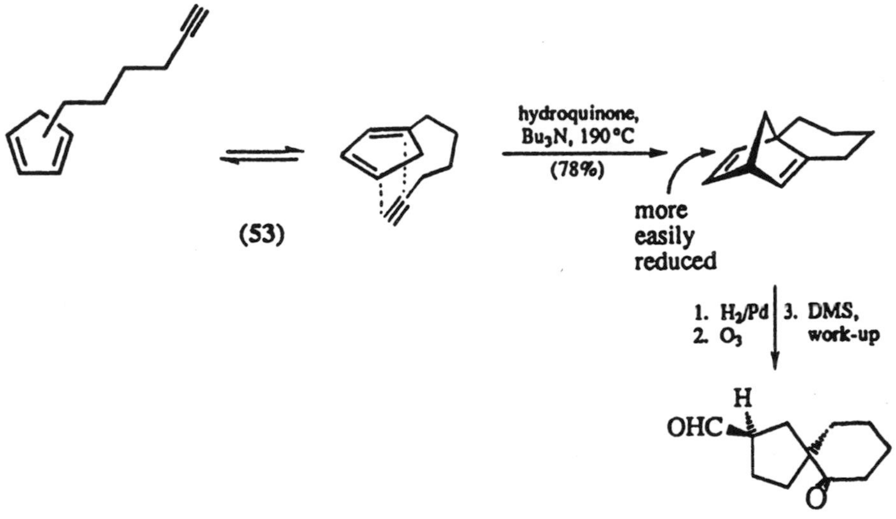

Scheme 53

More amenable to control are homo-[1,5] sigmatropic shifts in which a cyclopropane or other three-membered ring takes the place of one of the double bonds in Scheme 51 (Scheme 54);[44] the product is now non-conjugated and therefore cannot undergo a further [1,5] rearrangement. The

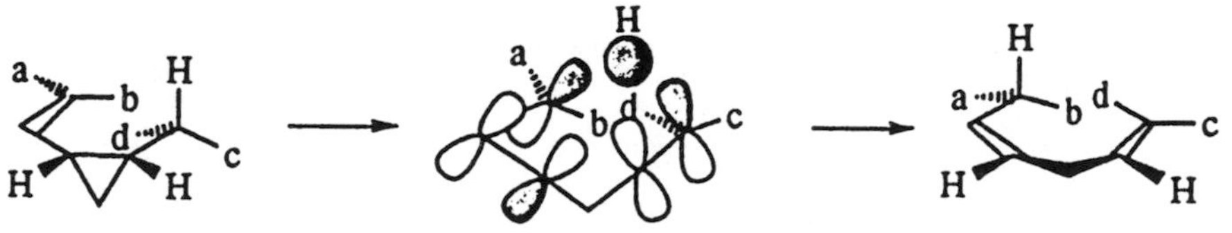

Scheme 54

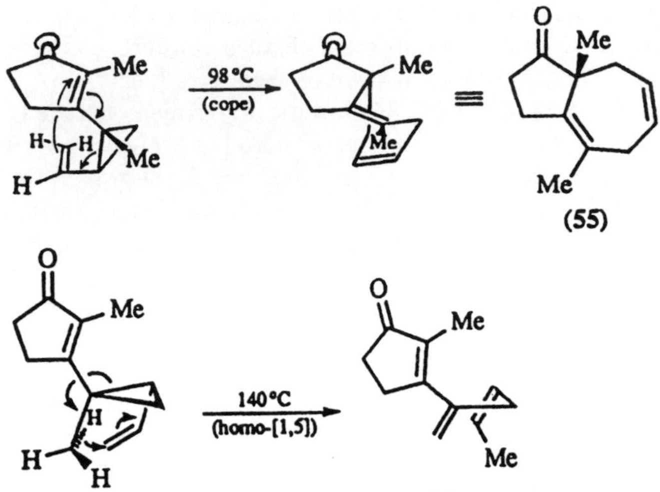

(55)

(56)

Scheme 55

migrating group here is again usually a hydrogen and the energetically favourable cleavage of the strained three-membered ring ensures that the reaction is irreversible. For example, when *cis*- and *trans*-cyclopropane diastereoisomers in Scheme 55 were heated, the *cis* isomer gave rise to the product (55) from a Cope rearrangement whereas the *trans* isomer gave (56) from the homo-[1,5] sigmatropic rearrangement.[45]

Simple chirality transfer in the ene (retroene) reaction[46]

The simplest all-carbon ene reaction is that shown in Scheme 56. This reaction has features in common with both the Diels–Alder reaction and the [1,5] sigmatropic rearrangement of hydrogen, but the transition-state

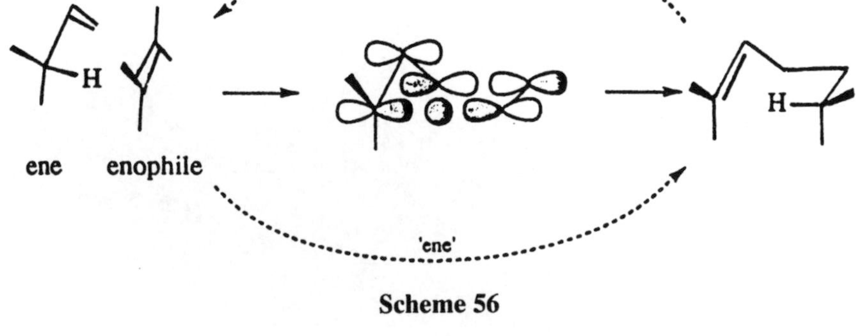

ene   enophile

'retro-ene'

'ene'

Scheme 56

geometry (TSG) which results from overlap of orbitals as shown in Scheme 56 is more like that of the [2,3] sigmatropic rearrangement.

The ene reaction and its reverse, the retroene reaction, are pericyclic reactions: bonds are made and broken concertedly in a cyclic transition state involving six electrons and the symmetry of the orbitals involved is appropriate for this to occur.

Since there is a single sp³-centre available in the ene reaction ensemble, simple chirality transfer requires this to be the chiral centre and one of two double bonds to be configured.[47] Scheme 57 shows that, like the 2,3-sigmatropic rearrangement, there are two transition states available which lead to diastereoisomeric products. Both diastereoisomeric products are formed in the reaction in Scheme 58 in which the S/R ratio at the new chiral centre in the product is the same as the D/H ratio at the vinyl position. This result was interpreted by Stephenson and Mattern[48] as support for the concertedness of the ene reaction.

Some of the most useful stereoselective ene/retroene reactions have one or

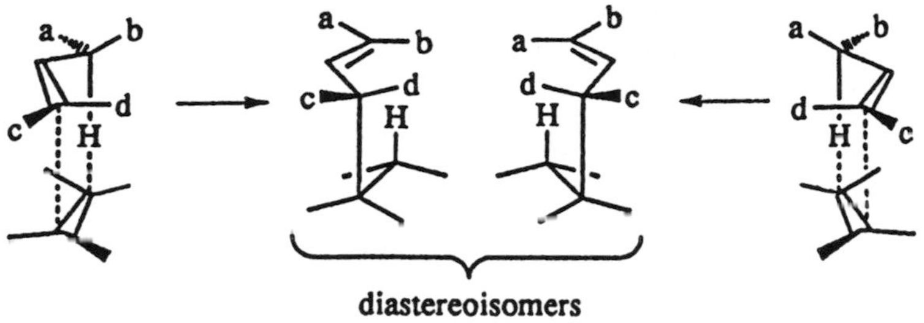

diastereoisomers

Scheme 57

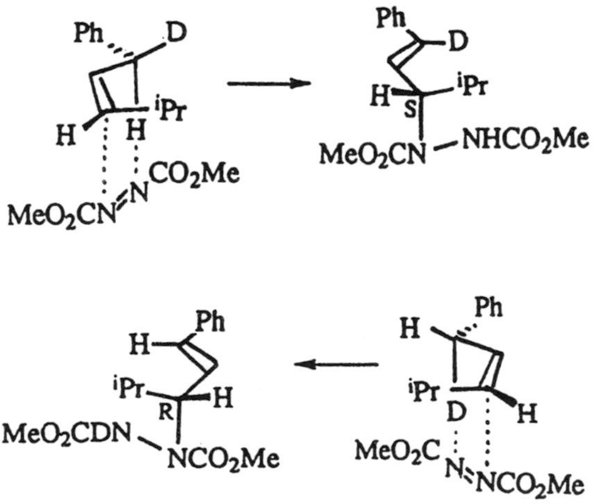

Scheme 58

more carbon atoms in the ene or enophile components replaced by heteroatoms (e.g. Scheme 58). Thus the role of the enophile is often taken on by N=N, C=O, O=O, etc., instead of C=C or C≡C. With a carbonyl group as the enophile, the ene reaction can be catalysed by Lewis acids, e.g. $SnCl_4$ or R_3Al, which coordinate with the oxygen of the carbonyl group and greatly reduce the temperature at which the reaction will take place.[49] In Scheme 59, the catalysed ene reaction takes place at −78 °C. Enantioselectivity in the sense obtained is thought to be assisted by an attraction in (57) between the negatively charged oxygen (of OTBS) and the positively charged sulphur.[50]

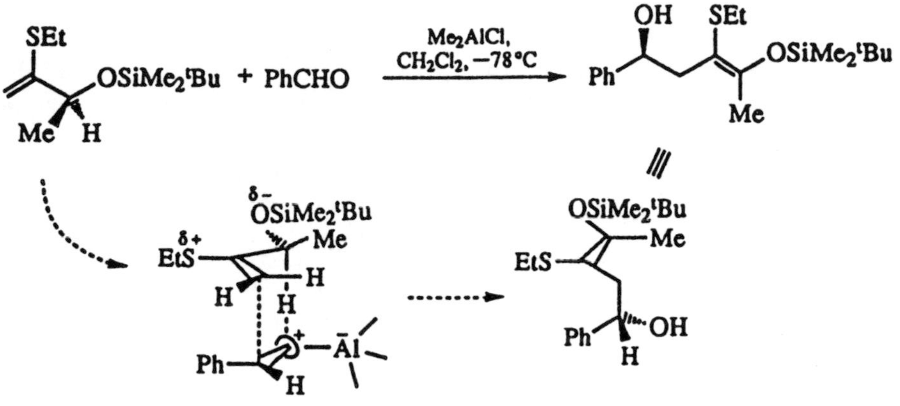

Scheme 59

Summary

Simple chirality transfer reactions are those in which a chiral centre in the starting material is lost and a new chiral centre is created elsewhere in the molecule; this transformation is mediated by the shift of one or more double bonds, at least one of which is configured, to a new position in the product.

Diastereoselectivity in these reactions is more likely when all or part of the system undergoing the chirality transfer is incorporated into a cyclic system, although the substitution present in the starting material can sometimes give rise to high diastereoselectivity even in acyclic cases.

The reversibility of the allylic sulphoxide–sulphenate ester [2,3] sigmatropic rearrangement allows the interconversion of diastereoisomeric sulphenate esters and hence provides a means of inverting the configuration in an enantiopure allylic alcohol at both the double bond and the chiral centre simultaneously.

The [3,3] sigmatropic rearrangement is a most valuable simple chirality transfer reaction because of the usual preference in an acyclic system for a chair-shaped transition state with the bulkier substituent located in an equatorial position. Using the Ireland modification of the Claisen rearrangement, this [3,3] sigmatropic rearrangement can be carried out under mild conditions. 1,4-Chirality transfer in the Ireland–Claisen rearrangement requires methods for diastereoselective enolate generation. The oxy-Cope [3,3] sigmatropic rearrangement can proceed at even lower temperatures than the Ireland–Claisen rearrangement.

For the [1,5] sigmatropic rearrangement to be useful in synthesis, ways must be found of circumventing the serial [1,5] sigmatropic rearrangements which otherwise tend to occur in cyclic dienes.

References

1. I. Fleming, *Org. React.*, 1989, **37**, 57; I. Fleming and N.K. Terrett, *Pure Appl. Chem.*, 1983, **55**, 1707.
2. T. Hayashi, M. Konishi, H. Ito and M. Kumada, *J. Am. Chem. Soc.*, 1982, **104**, 4962.
3. G. Stork and A.F. Kreft, *J. Am. Chem. Soc.*, 1977, **99**, 3850, 3850; G. Stork and A.R. Schoofs, *J. Am. Chem. Soc.*, 1979, **101**, 5081.
4. J.A. Marshall, *Chem. Rev.*, 1989, **89**, 1503.
5. J.A. Marshall, J.D. Trometer, B.E. Blough and T.D. Crute, *Tetrahedron Lett.*, 1988, **29**, 913.
6. T. Ibuka, M. Tanaka, S. Nishii and Y. Yamamoto, *J. Am. Chem. Soc.*, 1989, **111**, 4864.
7. H.L. Goering and S.S. Kantner, *J. Org. Chem.*, 1984, **49**, 422.
8. E.J. Corey and N.W. Boaz, *Tetrahedron Lett.*, 1984, **25**, 3063.
9. H.L. Goering, S.S. Kantner and C.C. Tseng, *J. Org. Chem.*, 1983, **48**, 715.
10. B.M. Trost, *Acc. Chem. Res.*, 1980, **13**, 385.
11. G. Stork and J.M. Poirier, *J. Am. Chem. Soc.*, 1983, **105**, 1073.
12. B.M. Trost and T.P. Klun, *J. Am. Chem. Soc.*, 1979, **101**, 6756.
13. B.M. Trost, G.-H. Kuo and T. Benneche, *J. Am. Chem. Soc.*, 1988, **110**, 621.
14. B.M. Trost and A.G. Romero, *J. Org. Chem.*, 1986, **51**, 2332.
15. H. Matsushita and E. Negishi, *Chem. Commun.*, 1982, 160; see also I. Stary and P. Kocovsky, *J. Am. Chem. Soc.*, 1989, **111**, 4981.
16. R.B. Woodward and R. Hoffmann, *The Conservation of Orbital Symmetry*, Verlag Chemie, Weinheim, 1970.
17. K. Fukui, *Acc. Chem. Res.*, 1971, **4**, 57; I. Fleming, *Frontier Orbitals and Organic Chemical Reactions*, Wiley, Chichester, 1976.
18. R.W. Hoffman, *Angew. Chem. Int. Ed. Engl.*, 1979, **18**, 563; T. Nakai and K. Mikami, *Chem. Rev.*, 1986, **86**, 885; E. Vedejs, *Acc. Chem. Res.*, 1984, **17**, 358.
19. W.C. Still and A. Mitra, *J. Am. Chem. Soc.*, 1978, **100**, 1927.
20. D. J.-S. Tsai and M.M. Midland, *J. Am. Chem. Soc.*, 1985, **107**, 3915; M.M. Midland and Y.C. Kwon, *Tetrahedron Lett.*, 1985, **26**, 5013.
21. P. Bickart, F.W. Carson, J. Jacobus, E.G. Miller and K. Mislow, *J. Am. Chem. Soc.*, 1968, **90**, 4869.
22. D.A. Evans and G.C. Andrews, *Acc. Chem. Res.*, 1974, **7**, 147.
23. J.G. Miller, W. Kurz, K.G. Untch and G. Stork, *J. Am. Chem. Soc.*, 1974, **96**, 6774.
24. L. Castedo, J.R. Granja and A. Mourino, *Tetrahedron Lett.*, 1985, **26**, 4959.
25. B. Szechner, *Tetrahedron Lett.*, 1989, **30**, 3829; P.A. Zoretic, R.J. Chambers, G.D. Marbury and A.A. Riebiro, *J. Org. Chem.*, 1985, **50**, 2981.

26. E. Vedejs, M.J. Arco, D.W. Powell, J.M. Renga and S.P. Singer, *J. Org. Chem.*, 1978, **43**, 4831.
27. R.K. Hill, in *Asymmetric Synthesis*, ed. J. Morrison, Academic Press, Orlando, 1984, Vol. 3, Chapt. 8; S.J. Rhoads and N.R. Rawlins, *Org. React.*, 1975, **22**, 1.
28. F.E. Ziegler, *Chem. Rev.*, 1988, **88**, 1423; F.E. Ziegler, *Acc. Chem. Res.*, 1977, **10**, 227.
29. D.J. Faulkner and M.R. Petersen, *Tetrahedron Lett.*, 1969, 3243.
30. D. Felix, K. Gschwend-Steen, A.E. Wick and A. Eschenmoser, *Helv. Chim. Acta*, 1969, **52**, 1030.
31. W.S. Johnson, L. Werthemann, W.R. Bartlett, T.J. Brocksom, T. Li, D.J. Faulkner and M.R. Petersen, *J. Am. Chem. Soc.*, 1970, **92**, 741.
32. R.E. Ireland and R.H. Mueller, *J. Am. Chem. Soc.*, 1972, **94**, 5897; R.E. Ireland, R.H. Mueller and A.K. Willard, *J. Am. Chem. Soc.*, 1976, **98**, 2868.
33. R.K. Hill, R. Soman and S. Sawada, *J. Org. Chem.*, 1972, **37**, 3737.
34. K. Nonoshita, H. Banno, K. Maruoka and H. Yamamoto, *J. Am. Chem. Soc.*, 1990, **112**, 316.
35. T.J. Gould, M. Balestra, M.D. Wittman, J.A. Gary, L.T. Rossano and J. Kallmerton, *J. Org. Chem.*, 1987, **52**, 3889.
36. S.J. Danishefsky and J.E. Audia, *Tetrahedron Lett.*, 1988, **29**, 1371.
37. D.A. Evans and A.M. Golob, *J. Am. Chem. Soc.*, 1975, **97**, 4765.
38. J.A. Berson and M. Jones, *J. Am. Chem. Soc.*, 1964, **86**, 5019.
39. E. Lee, I.-J. Shin and T.-S. Kim, *J. Am. Chem. Soc.*, 1990, **112**, 260.
40. I. Savage and E.J. Thomas, *Chem. Commun.*, 1989, 717; L.E. Overman, *J. Am. Chem. Soc.*, 1974, **96**, 597; E.W. Baxter, D. Labaree, S. Chao and P.S. Mariano, *J. Org. Chem.*, 1989, **54**, 2893; R. Oehrlein, R. Jeschke, B. Ernst and D. Bellus, *Tetrahedron Lett.*, 1989, **30**, 3517; M. Sworin and K.-C. Lin, *J. Am. Chem. Soc.*, 1989, **111**, 1815; E. Vedejs, M.J. Arco, D.W. Powell, J.M. Renga and S.P. Singer, *J. Org. Chem.*, 1978, **43**, 4831.
41. D.J. Field, D.W. Jones and G. Kneen, *J. Chem. Soc., Perkin Trans. 1*, 1978, 1050; M.J. Collett, D.W. Jones and S.J. Renyard, *J. Chem. Soc., Perkin Trans. 1*, 1986, 1471.
42. C.W. Spangler, *Chem. Rev.*, 1976, **76**, 187.
43. J.-E. Nyström, T.D. McCanna, P. Helquist and R.S. Iyer, *Tetrahedron Lett.*, 1985, **26**, 5393.
44. J.J. Gajewski, *Hydrocarbon Thermal Isomerisations*, Academic Press, New York, 1981, p.186.
45. P.A. Wender, M.A. Eissenstat and M.P. Filosa, *J. Am. Chem. Soc.*, 1979, **101**, 2196.
46. W. Oppolzer and V. Snieckus, *Angew. Chem. Int. Ed. Engl.*, 1978, **17**, 476; H.M.R. Hoffmann, *Angew. Chem. Int. Ed. Engl.*, 1969, **8**, 556; W. Oppolzer, *Angew. Chem. Int. Ed. Engl.*, 1989, **28**, 38; D.L. Taber, *Intramolecular Diels–Alder and Alder–Ene Reactions*, Springer Verlag, Berlin, 1984.
47. R.K. Hill and M. Rabinovitz, *J. Am. Chem. Soc.*, 1964, **86**, 965.
48. L.M. Stephenson and D.C. Mattern, *J. Org. Chem.*, 1976, **41**, 3614.
49. B.B. Snider, *Acc. Chem. Res.*, 1980, **13**, 426; G.B. Gill, K. Morrison, S.J. Parrott and B. Wallace, *Tetrahedron Lett.*, 1979, 4867, and references cited therein.
50. K. Tanino, H. Shoda, T. Nakamura and I. Kuwajima, *Tetrahedron Lett.*, 1992, **33**, 1337.

6 TYPE II REACTIONS: SIMPLE DIASTEREOSELECTIVITY IN 1,2-ADDITIONS TO ALKENES, DIELS–ALDER AND 1,3-DIPOLAR CYCLOADDITIONS AND THE ENE REACTION

General

In Type II reactions, two or more chiral centres are formed simultaneously from one or more precursors, at least one of which is prochiral but neither of which is chiral; if a reagent is used, this is achiral.

The largest number of highly diastereoselective Type II reactions are those in which the prochiral precursors are double bonds, usually C=C, C=O, C=N, or C=C—C=C. These reactions include (a) those which result in the addition of X, X—X or X—Y across a *configured* double bond and (b) many pericyclic reactions including electrocyclic reactions, cycloadditions, sigmatropic rearrangements and cheletropic additions. Almost all of the products from these Type II reactions are either cyclic or are formed via cyclic transition states and the diastereoselectivity which arises is referred to as *simple diastereoselectivity*.

$$\text{(a)}$$

$$\text{(b)}$$

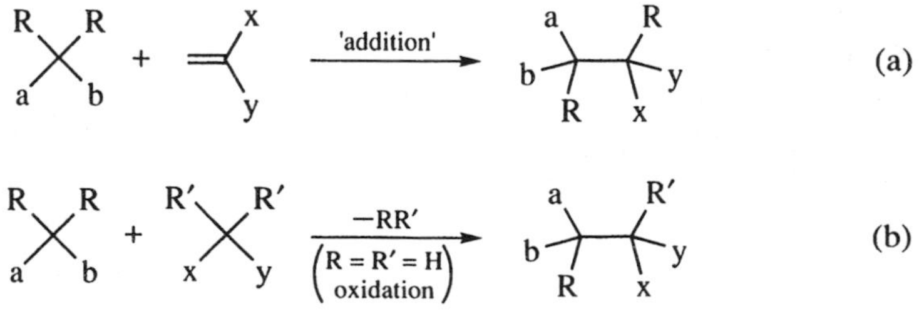

Scheme 1

Type II reactions also include those in Scheme 1 in which two chiral centres are formed (a) from addition of a precursor having a prochiral *centre* to a prochiral double bond or (b) from the combination of two precursors each having prochiral centres. However, diastereoselective versions of these reactions are as yet uncommon and so this chapter will be concerned with reactions of the first kind above.

Relative configuration in Type II reactions

The absence of chirality in the precursor(s) in Type II reactions has two important consequences: (a) the products are always racemic and therefore any stereoselectivity in Type II reactions will be diastereoselectivity; and (b) there can be no asymmetric induction. It is this latter consequence which distinguishes Type II from Type III reactions: Type II reactions become Type III (or Type II/III) when one of the components contains a chiral centre (see Chapter 13).

A characteristic of many Type II reactions which proceed via cyclic transition states is that they show inherent diastereoselectivity (stereospecificity). This arises from the single well defined transition state geometries (TSGs) through which the reactions proceed. The relative configuration at the two more chiral centres that are formed derives from the TSG and from the configuration(s) of the prochiral precursor(s).

Other Type II reactions (e.g. the aldol reaction) lack this inherent diastereoselectivity. Nevertheless, such reactions can occasionally exhibit high diastereoselectivity because of the particular combination of substituents on their reacting components. The *occasional* diastereoselectivity which results in Type II versions of these reactions will clearly be less reliable than inherent diastereoselectivity.

For some Type II reactions proceeding via cyclic transition states (e.g. Diels–Alder reactions), the diastereoselectivity may be inherent or occasional or both, depending on the substitution of the reacting components.

EPOXIDATION, AZIRIDINATION AND CYCLOPROPANATION

In the Bartlett mechanism for the epoxidation of alkenes using peroxyacids, shown in Scheme 2, both bonds to the peroxy oxygen are formed from the same face of the (configured) double bond in a single step. Of course, both faces of the alkene are equally likely to react in this way leading to a racemic product. The *syn* addition of oxygen to the double bond results in the inherent diastereoselectivity of the reaction and is a feature of all peroxyacid epoxidations of configured double bonds.[1] The same *syn* addition is found in the formation of aziridines from alkenes using the aziridinating agent in Scheme 3(a)[2] (a nitrogen analogue of a peroxyacid) and in cyclopropanation of alkenes using a carbenoid (the Simmons–Smith reaction)[3] (Scheme 3(b)).

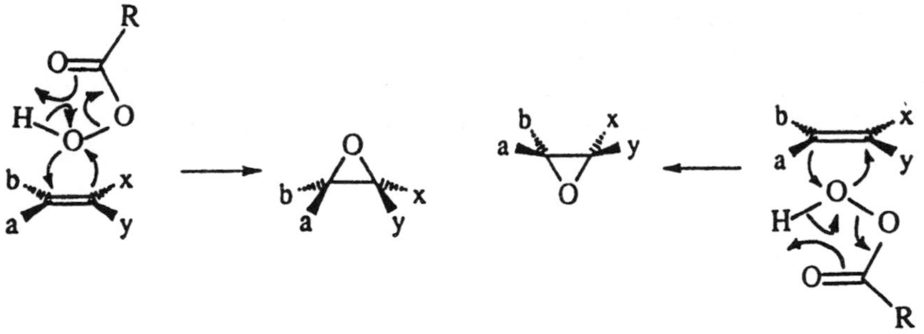

Scheme 2

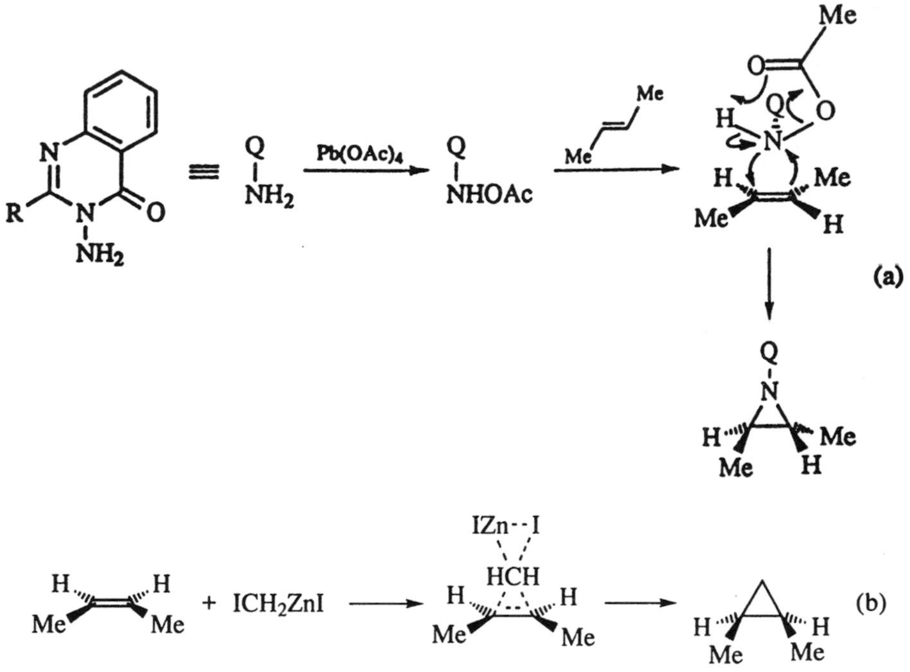

(a)

(b)

Scheme 3

Type II aziridination and cyclopropanation of alkenes can also be brought about by cheletropic addition of singlet nitrenes and carbenes, respectively (see later).

BROMINATION AND SIMILAR ADDITIONS TO ALKENES

Bromination of a configured double bond is similar to epoxidation in that a cyclic bromonium ion is formed by *syn* addition to each face of the double bond. However, the bromonium ion undergoes spontaneous ring opening by

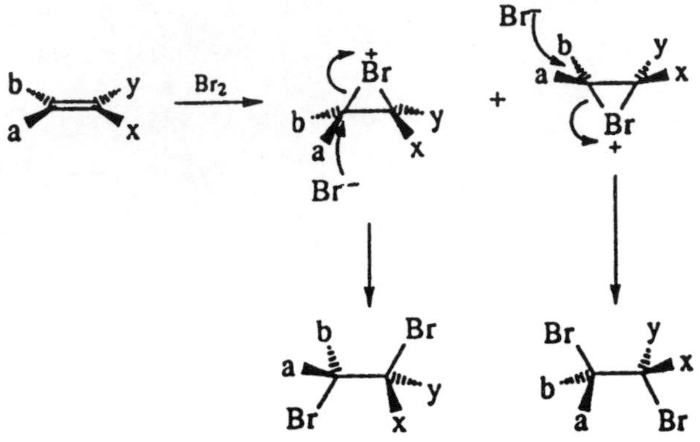

Scheme 4

bromide anion in an S_N2 fashion (Type I) leading to overall *anti* addition of bromine (Scheme 4).

In Scheme 4 we are assuming that ring opening of the bromonium ion is regiospecific with attack of bromide at the a,b-substituted carbon as shown, although in fact the same racemic dibromide would have been obtained had the bromide attacked in the opposite regio-sense, at the x,y-substituted carbon (in the same S_N2 fashion). However, ring opening of the bromonium ion may be brought about by nucleophiles other than bromide if these are present (including the solvent itself) and in these cases there will be the possibility of regioisomer formation (Scheme 5).

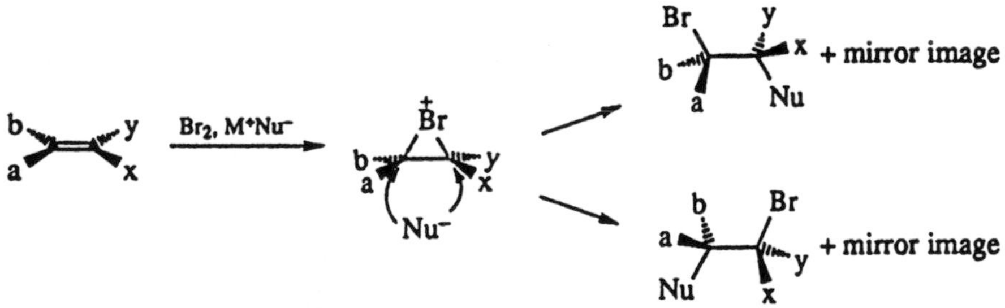

Scheme 5

Bromonium ion formation and ring opening in a stereo-defined way, if not always in a regio-defined way, is analogous to a number of other *anti* additions to alkenes which proceed via ring opening of -onium ions including episulphonium (**1**) and selenenium (**2**) as well as other halonium ions of which the iodonium ion is the most important.

In this category also we can include the oxonium ion (**3**) formed by reaction of an epoxide with an electrophile—usually a proton or Lewis acid; this, like the bromonium ion in Scheme 5, then undergoes rapid S_N2 ring opening by nucleophiles.

These overall *anti* additions are the more valuable because most other highly diastereoselective Type II additions to double bonds take place in a *syn* fashion (see below).

HYDROBORATION–OXIDATION

Although the product from hydroboration of an alkene is not cyclic, the transition state is regarded as being cyclic in character (Scheme 6).

Note that the overall addition of water to the double bond in Scheme 6 is *syn* since replacement of the boron–carbon bond by oxygen–carbon proceeds with complete retention of configuration at carbon (Type I; see Chapter 4).

R R E
| | |
$^+$S $^+$Se $^+$O

(1)   (2)   (3)

$$\text{b} \diagdown \diagup \text{y} \quad \xrightarrow[\text{(Type II)}]{R_2BH} \quad \text{H--BR}_2 \quad \longrightarrow \quad \text{H} \quad \text{BR}_2$$

$$\downarrow \begin{array}{l} H_2O_2,\ NaOH \\ \text{(Type I)} \end{array}$$

$$\text{H} \quad \text{OH}$$

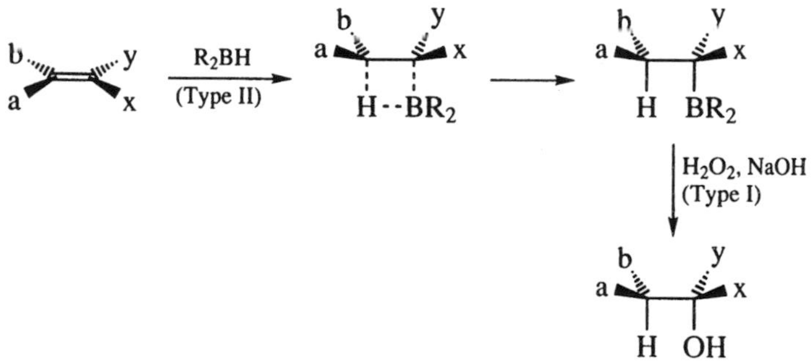

(+ mirror image)

Scheme 6

In Scheme 6, complete regioselectivity has been assumed with the boron adding to the x,y-substituted alkene carbon. This regioselectivity is normally determined by the bulk of the substituents a,b or x,y on the alkene, with the boron adding to the less substituted carbon.

HYDROXYLATION

Epoxidation of a configured double bond followed by treatment of the epoxide with aqueous acid brings about *anti* hydroxylation (also referred to as dihydroxylation) (Scheme 7).

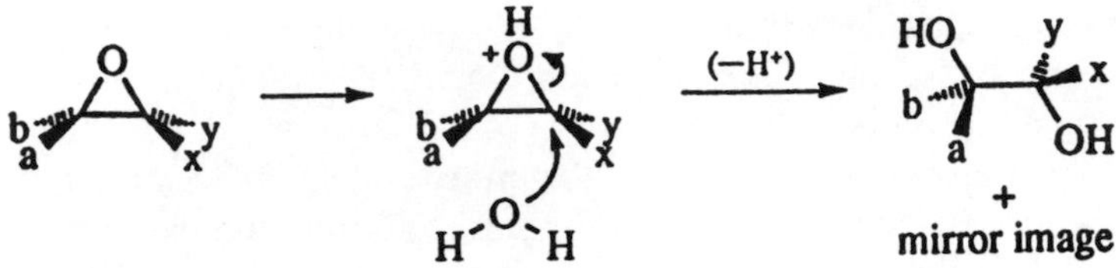

Scheme 7

A stereochemically complementary method for hydroxylation of an alkene giving the product of *syn* addition uses osmium tetroxide followed by cleavage of the osmate ester (Scheme 8). It is possible that the osmate ester is formed by rearrangement of the four-membered ring species (**4**) (see Chapter 14). The same *syn*-hydroxylation can be accomplished (Scheme 8) using potassium permanganate; in this case a cyclic manganate ester is an intermediate.

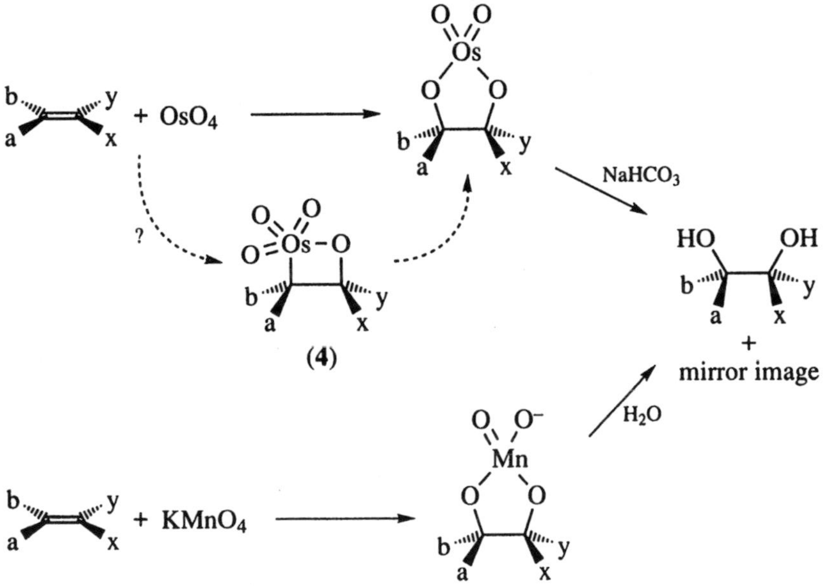

Scheme 8

Type II pericyclic reactions[4]

ELECTROCYCLIC REACTIONS

In Type II $4\pi \rightarrow 2\pi$ thermal electrocyclic ring closure, two configured double bonds are converted into two adjacent chiral centres. Thus thermal electrocyclic ring closure of the substituted butadiene in Scheme 9 gives a

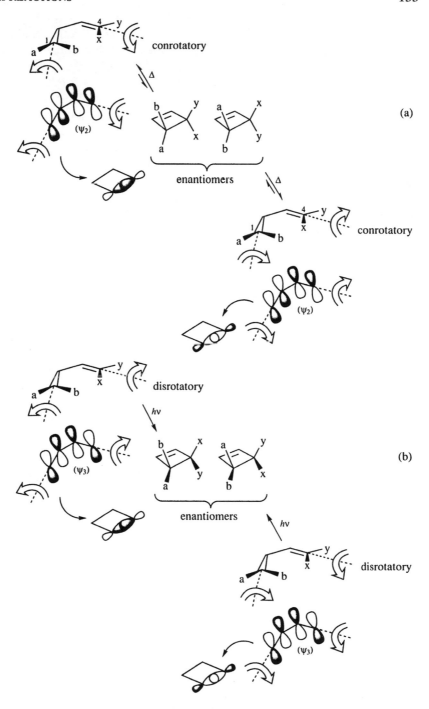

Scheme 9

single racemic cyclobutene (although in practice the equilibrium in this reaction more often lies on the side of the diene). The relative configuration shown for the two chiral centres is a consequence of the conrotatory mode of ring closure. This conrotatory ring closure is dictated by the symmetry of the highest occupied molecular orbital (HOMO), ψ_2; only by rotating around the axes shown [Scheme 9(a)] in the same sense does in-phase overlap of the p-orbitals on C-1 and C-4 occur (Frontier Molecular Orbital Theory).

Ring closure of the same butadiene by photochemical means via the HOMO of first excited single state (ψ_3) gives a cyclobutene having the alternative relative configuration since the orbital symmetry-directed cyclisation in this case is disrotatory [Scheme 9(b)].

CYCLOADDITIONS—INHERENT AND OCCASIONAL DIASTEREOSELECTIVITY

In the Type II reactions we have considered so far, two adjacent chiral centres are formed in the product. In cycloadditions, more than two chiral centres may be formed and, when only two are formed, they are not necessarily on adjacent (carbon) atoms.

For example, the singlet carbene in Scheme 10(a) adds in completely *syn* fashion to the configured alkene, i.e. the *trans* relationship of x and y in the alkene is retained in the cyclopropane product.[5] Since addition to the corresponding *cis*-alkene in Scheme 10(b) gives only the product in which x and y are *cis*, this cyclopropanation by singlet carbenes is stereospecific (see Chapter 1). This complete *syn* diastereoselectivity is inherent in the carbene cycloaddition mechanism and, in fact, is used to determine the spin state of the carbene since

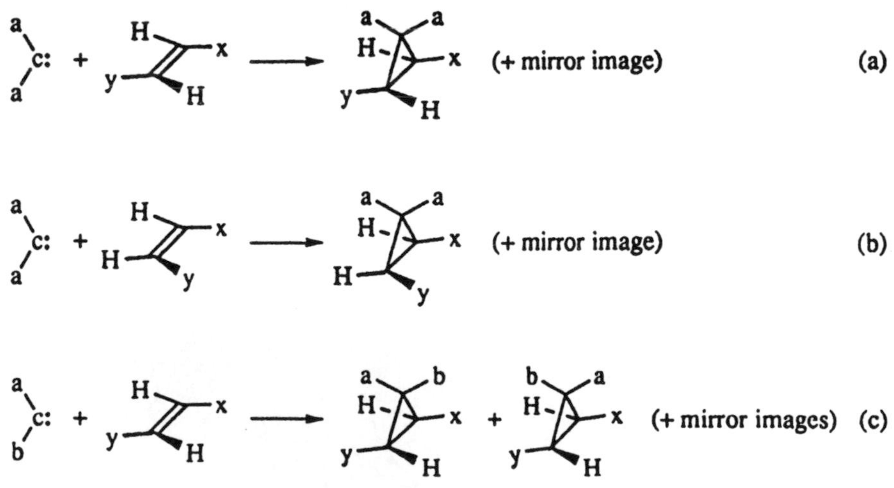

Scheme 10

triplet carbenes do not normally show such complete diastereoselectivity.

If, however, the carbene bears two different substituents, then a third chiral centre is formed in the product [Scheme 10(c)].

Although both cyclopropanes in Scheme 10(c) have x and y *trans* (inherent diastereoselectivity), prediction of the relative configuration of the third (a,b-substituted) chiral centre is far from safe and both diastereoisomers will in general be formed, albeit in unequal amounts.

The inherently diastereoselective part of the carbene addition in Scheme 10(a) is the result of concerted addition of the electron-deficient carbene to either *enantioface* of the alkene, a process analogous to epoxidation in Scheme 2 and giving rise to racemic cyclopropane.

Using a carbene bearing two different substituents, however, the two enantiofaces of the alkenes become *diastereofaces* when associated with the carbene in the transition state as shown in (5) and (6) (Scheme 11). As a

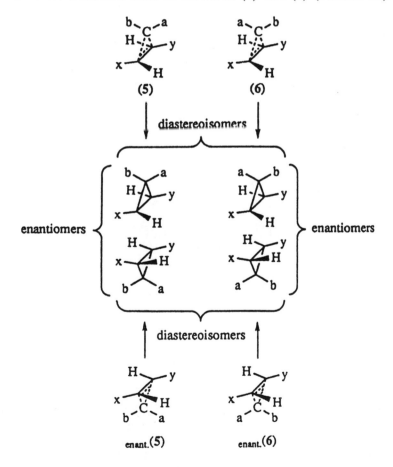

Scheme 11

result, the two transition states are diastereoisomeric and give rise to two racemic diastereoisomeric products as indicated in Scheme 11; in each case the inherent diastereoselectivity referred to above is still manifested.

Whether the creation of the third chiral centre at the erstwhile carbene carbon results in a single diastereoisomer in the cyclopropanation (*occasional* diastereoselectivity) depends on the relative energies of (**5**) and (**6**), which in turn will depend on the steric and/or electronic interactions between a and x (or y) and b and y (or x) in these transition states. These interactions *can* lead to a single (relative) configuration at the third chiral centre (most often when metal-coordinated carbenes or carbenoids are used). However, complete diastereoselectivity in this sense is not inherent to the cyclopropanation.

As previously mentioned, the division of the diastereoselectivity obtaining in cyclopropanation into inherent and occasional is typical of many of the Type II cycloadditions which follow.

Diels–Alder reaction[6]

The transition-state geometry for the Diels–Alder reaction is that shown in Figure 1, where both components add suprafacially. This is the best way in which the frontier molecular orbitals of the diene and dienophile can overlap in-phase.

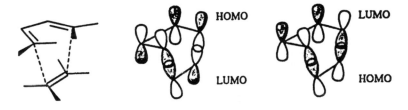

Figure 1

In the Type II Diels–Alder reaction, both the diene and the dienophile are achiral and if a catalyst is used, this is also achiral. Diels–Alder reactions in which chiral centres are present in the substituents on the diene, dienophile or catalyst are Type III or Type II/III and will be considered in Chapters 9, 10 and 13.

As the transition-state geometry in Figure 1 shows, *syn* addition takes place on one face of the dienophile by the diene and reciprocally on one face of the diene by the dienophile. This mutual (stereospecific) *syn* addition of diene and dienophile is an inherent stereochemical characteristic of the Diels–Alder reaction (see above). It is reflected in the stereostructure of the product cyclohexene (**7**) (Scheme 12), in which four chiral centres have been created in the reaction of diene (**8**) and dienophile (**9**). Thus in (**7**), the *trans* disposition of x and y in the dienophile (**9**), and of the substituents a (relative

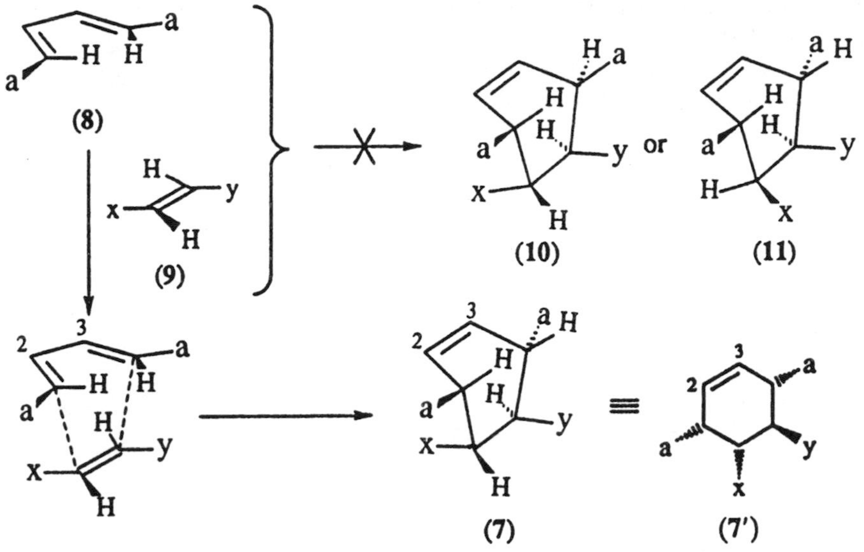

Scheme 12

to the C$_2$—C$_3$ bond) in the diene (8), has been retained.

There is now little doubt that the Diels–Alder reaction can proceed via a continuum of transition states in terms of the degree to which the new σ-bonds between the diene and dienophile are made. However, complete *syn* diastereoselectivity in additions to diene and dienophile is a stereochemical hallmark of all Diels–Alder reactions and means, for example, that for the reaction in Scheme 12 the cyclohexene diastereoisomers (10) and (11) are *not* obtained.

Scheme 12 depicts for convenience the formation of a single enantiomer of the product cyclohexene. Since this is a Type II reaction, however, the product must in reality be racemic. In Scheme 13 the enantiomeric product, enant.(7), and the transition state which leads to it are drawn as the mirror images of those shown in Scheme 12. Reacting opposite faces of both diene and dienophile to those used in Scheme 12 will also lead to enant.(7) (Scheme 13). In the remainder of this chapter, the products from Type II reactions will usually be drawn as single enantiomers but will in fact always have been formed as racemates.

In the Diels–Alder reaction in Scheme 12, cyclohexene (7) (and its enantiomer) is not the only diastereoisomer which can be formed; this is because the diene and dienophile can react together as shown in Scheme 14.

Here the transition state (12) has y located on the same side as the diene (the *endo* position) and x on the opposite side (the *exo* position) and the product is therefore (13): in the transition state in Scheme 12, the location of

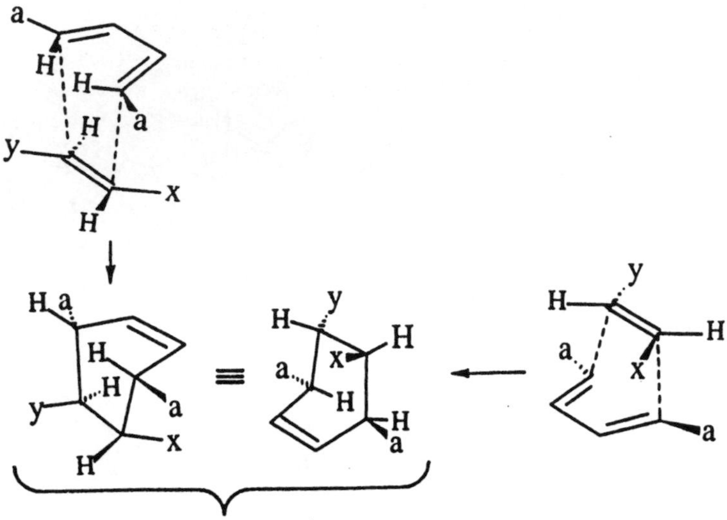

Scheme 13

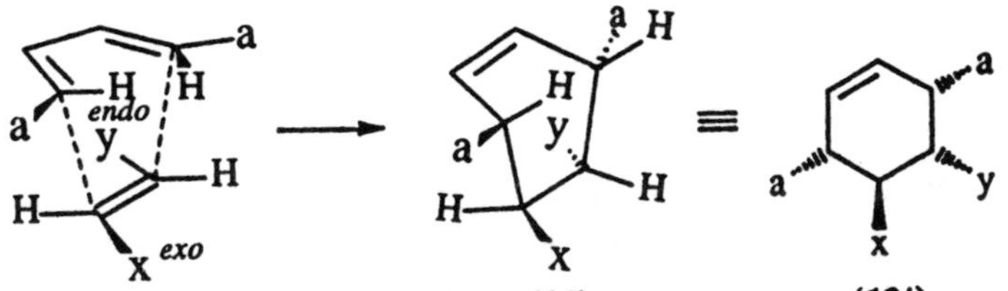

(12)   (13)   (13')

Scheme 14

x and y is reversed, x being *endo* and y *exo*, so that the product is (7). Notice that in Scheme 14, the inherent *syn* addition of diene to dienophile and dienophile to diene still obtains.

Comparison of the two cyclohexene products (7') (Scheme 12) and (13') (Scheme 14) reveals that these are indeed diastereoisomers and not enantiomers or regioisomers. Whether one or other or both of these diastereoisomers is produced depends on the preference of x (or y) for the *endo* (or *exo*) position. This *exo/endo* diastereoselectivity of the Diels–Alder reaction is *occasional*, although in either case the inherent *syn* addition to the dienophile and diene still prevails.

The situation at this point is very similar to that described earlier for addition of an unsymmetrical carbene to a configured alkene: diastereoselectivity is in part inherent and in part occasional. However, there is an additional complication in the Diels–Alder reaction when both the diene and the dienophile are unsymmetrically substituted. This is a problem not of diastereoselectivity but of regioselectivity. Thus in the reaction of the unsymmetrical diene and dienophile in Scheme 15 there are four possible products (each a racemate; only one enantiomer of each is shown).

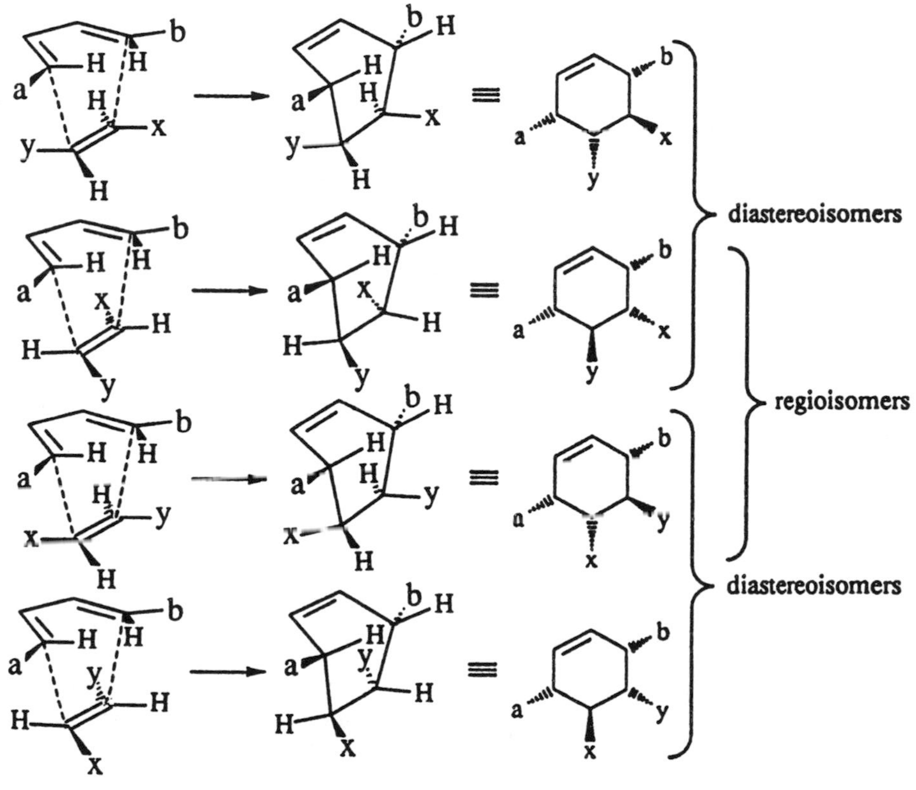

Scheme 15

Consequently, for the formation of a single product, the Type II Diels–Alder must be completely diastereoselective in the *exo/endo* sense and also completely regioselective; *syn* addition to the diene and the dienophile can be taken for granted.

Regioselectivity in the Diels–Alder reaction[7]

Frontier molecular orbital theory can be used to predict the regioselectivity of the Diels–Alder reaction by
(a) identifying the HOMO and LUMO of diene and dienophile which are closer in energy;

(b) calculating the orbital coefficients on the terminal atoms of the diene and dienophile in this HOMO–LUMO combination;
(c) matching the larger coefficient at the 1 and 4 positions of the diene with the larger coefficient at the 1 and 2 positions of the dienophile.

Thus in Scheme 16, where the size of the circles represents the magnitude of the coefficients on the interacting lobes, formation of the '*ortho*' isomer is favoured.

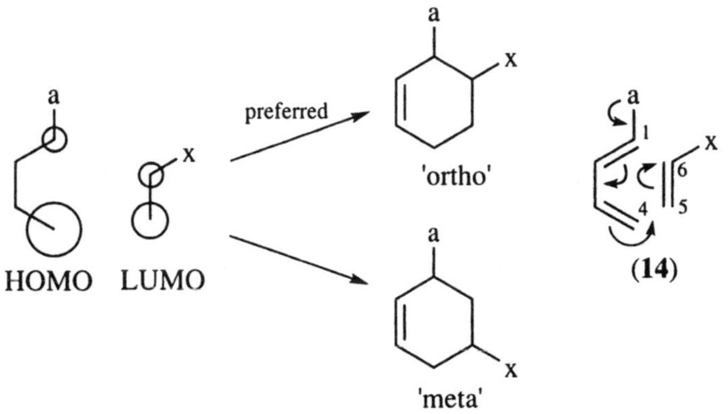

Scheme 16

In practice, formation of a single regioisomer in the sense shown in Scheme 16 is often the case when a is a strongly electron-donating substituent and x a strongly electron-withdrawing substituent. This regioselectivity is that which would result if, in the transition state (**14**) (Scheme 16), formation of the 4,5-bond has run slightly ahead of that of the 1,6-bond leading to stabilisation by a and x of the fractional charges thereby generated.

The regioselectivity of the Diels–Alder reaction may be affected by a number of factors including catalysis, pressure and solvent, but for the many reactions which are not very regioselective, one reliable solution is to tether the diene and dienophile together in such a way that the reaction—now intramolecular—can form only one regioisomer (see below).

exo/endo-*Diastereoselectivity (occasional diastereoselectivity) in the Diels–Alder reaction*

From our previous discussion, *exo/endo* diastereoselectivity is determined by the preference of the dienophile substituents for the *exo* or *endo* positions in the cycloaddition transition state (Schemes 12 and 14).

There are a few Diels–Alder reactions which show complete *exo* diastereoselectivity (e.g. Scheme 17),[8] usually as a result of steric repulsion in the *endo* transition state.

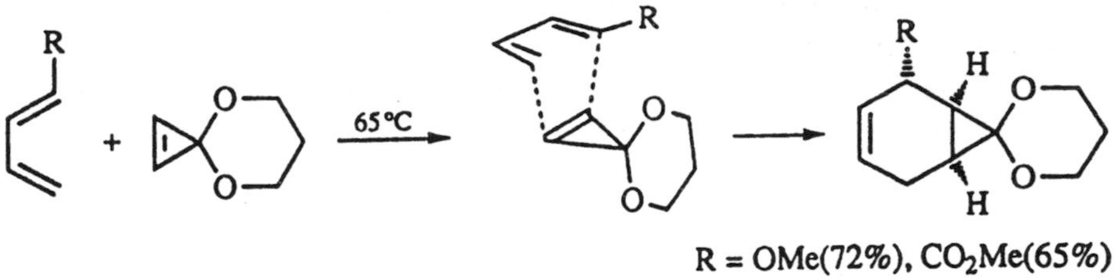

R = OMe(72%), CO$_2$Me(65%)

Scheme 17

endo Selectivity is found when the dienophile bears a single carbonyl group (Scheme 18). This is thought to result from an attractive secondary orbital interaction in the *endo* transition state between orbitals on the carbonyl group and the diene although other explanations have been suggested. In frontier orbital terms,[9] this attractive secondary orbital interaction can be ascribed to overlap of orbitals of matching symmetry on the carbonyl carbon and the diene C-2 in the dominant HOMO–LUMO pair (15) (Scheme 18).[7] The dienophile reacts as a 2π system but is, in fact, part of a 4π system. The LUMO of this 4π system is that shown in (15), and it has the appropriate symmetry for both primary and secondary orbital overlap with the HOMO of the 4π system of the diene.

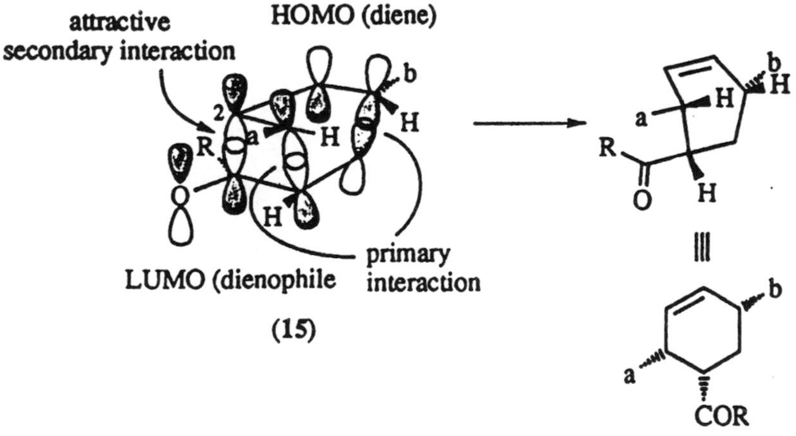

Scheme 18

Attractive secondary interactions of this type are the basis of the Alder *endo* rule which is more usually applied to bicyclic systems, e.g. the formation of the *endo* isomer (16) from reaction of cyclopentadiene and maleic anhydride [Scheme 19(a)].[10]

The same *endo* selectivity is responsible for the stereostructure of the product from reaction of the acyclic diene (17) (Danishefsky's diene) and maleic anhydride [Scheme 19(b)].[11]

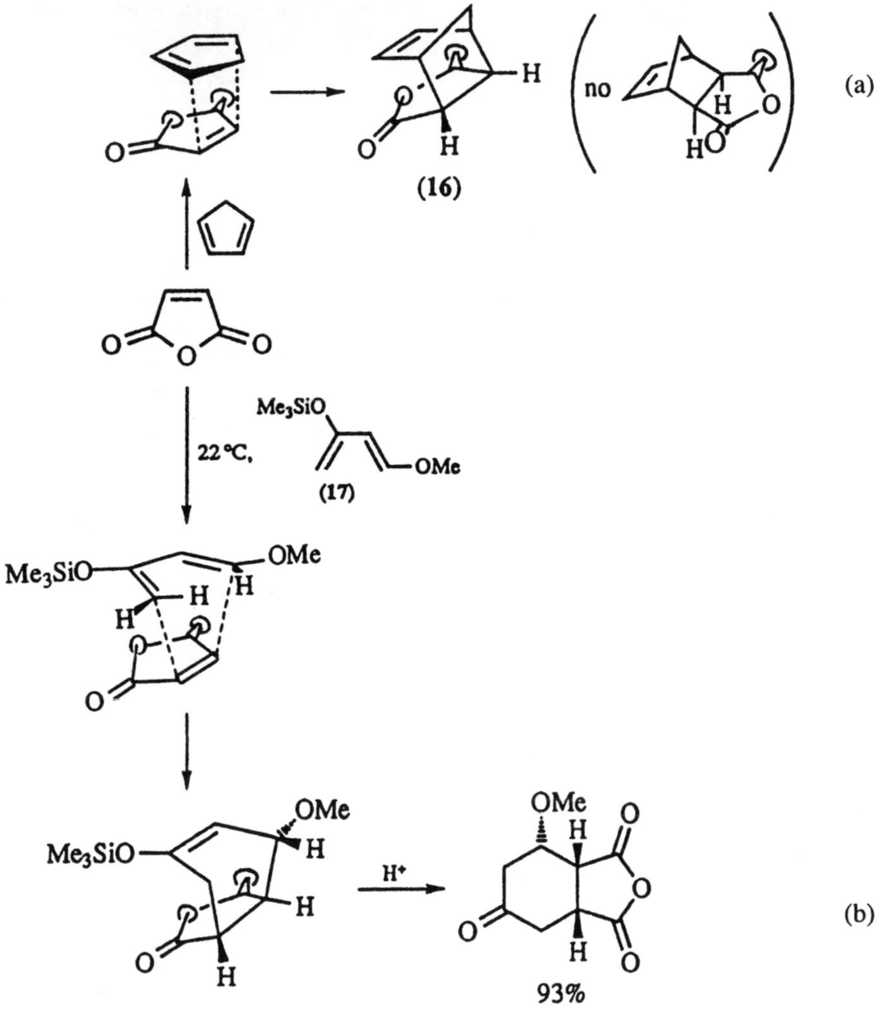

Scheme 19

Complete *endo* selectivity in the Diels–Alder reaction is the exception rather than the rule.[12] Thus the reactions of cyclopentadiene with various carbonyl-substituted dienophiles in Scheme 20 show that the *endo*(carbonyl)-substituted diastereoisomer is not the exclusive product nor even sometimes

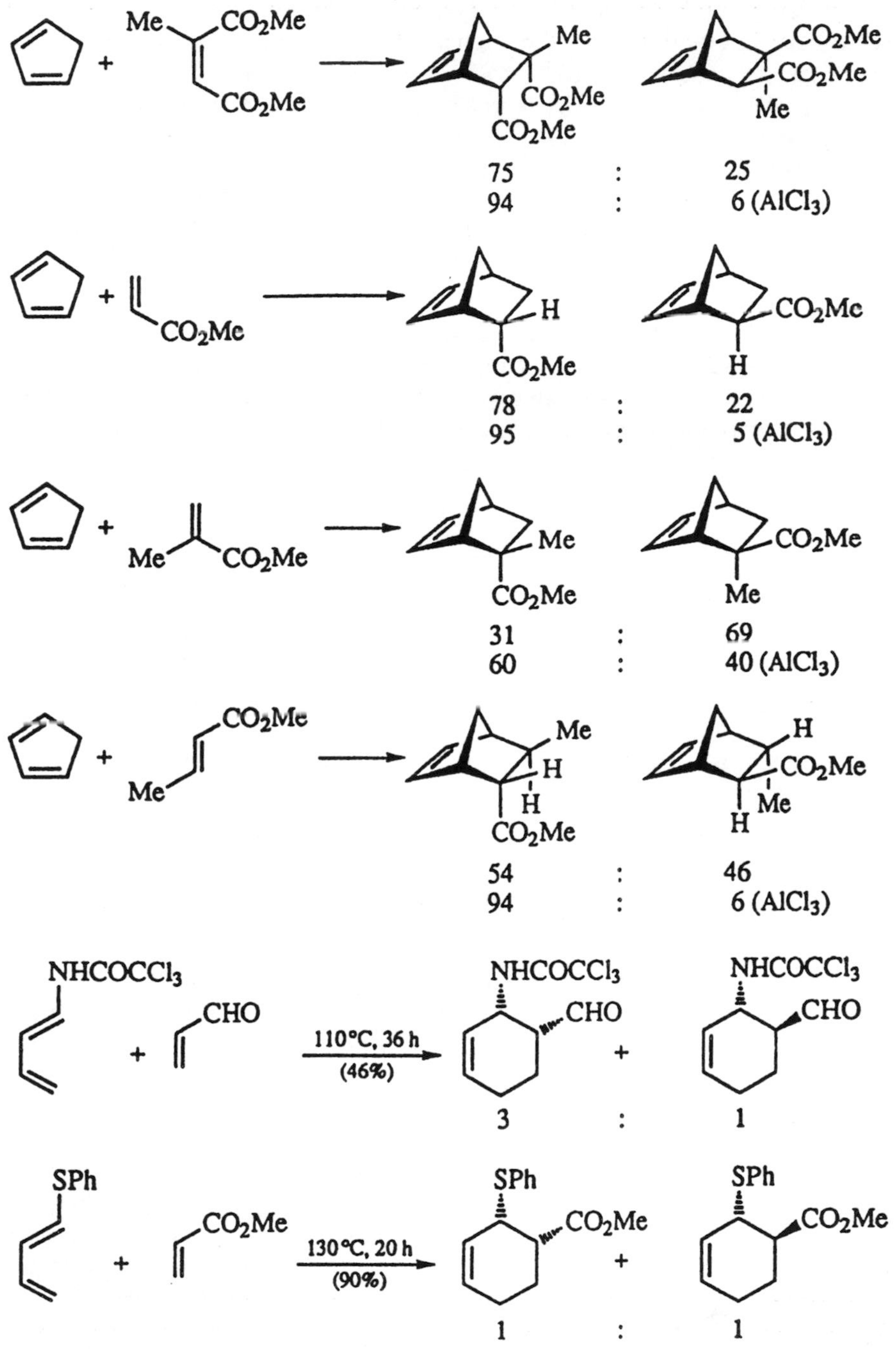
Me
CO2Me
CO2Me
+
Me
CO2Me
CO2Me
CO2Me
CO2Me
Me
75 : 25
94 : 6 (AlCl3)
+
CO2Me
H
CO2Me
CO2Me
H
78 : 22
95 : 5 (AlCl3)
+
Me
CO2Me
Me
CO2Me
CO2Me
Me
31 : 69
60 : 40 (AlCl3)
+
CO2Me
Me
Me
H
H
CO2Me
H
CO2Me
Me
H
54 : 46
94 : 6 (AlCl3)
NHCOCCl3
CHO
+
110°C, 36 h
(46%)
NHCOCCl3
CHO
+
NHCOCCl3
CHO
3 : 1
SPh
CO2Me
+
130°C, 20 h
(90%)
SPh
CO2Me
+
SPh
CO2Me
1 : 1

Scheme 20

the major product. It appears that a methyl group has its own attractive secondary interaction with the diene and can compete with the carbonyl group for the *endo* position.

Incomplete *endo* selectivity is not restricted to Diels–Alder reactions with cyclopentadiene as the diene, as the examples using acyclic dienes in Scheme 20 show;[13,14] only the *cis*-substituted products are *endo* derived. Not surprisingly, *endo* selectivity increases when dienophiles such as maleic anhydride are used in which both alkene carbons are substituted with carbonyl groups, and these are *cis* [Scheme 19].

As Scheme 20 also shows, *endo*(carbonyl) diastereoselectivity is much improved by catalysis with Lewis acids (AlCl₃) although, in the case of methyl methacrylate, this only changes the *endo/exo* ratio from 31:69 to 60:40. Increased diastereoselectivity may also be accompanied by increased regioselectivity and both can be accounted for in frontier orbital terms by an augmented secondary interaction (Scheme 21).[7]

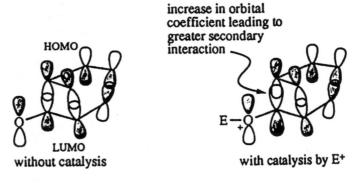

Scheme 21

endo Diastereoselectivity for many Diels–Alder reactions (along with regioselectivity and rate constants) can be markedly increased by changes in other experimental conditions and particularly by using water as the solvent, by addition of metal salts to organic solvents (e.g. LiClO₄ in diethyl ether), and by an increase in pressure.

Thus, *endo–exo* selectivity in the reaction of cyclopentadiene and methyl vinyl ketone is increased in water and is increased further using water containing dissolved lithium chloride (Scheme 22). It is thought that, as a result of hydrophobic reactions with the water, the reactants are aggregated and subjected to an internal pressure in the solvent cavities that they occupy. This effect is magnified in the presence of lithium chloride ('salting in').[15]

Lithium perchlorate dissolves in diethyl ether; in the reaction of cyclopentadiene and methyl acrylate in this solvent, the *exo/endo* ratio is increased over that obtained using water (Scheme 23).[16] The origin of this

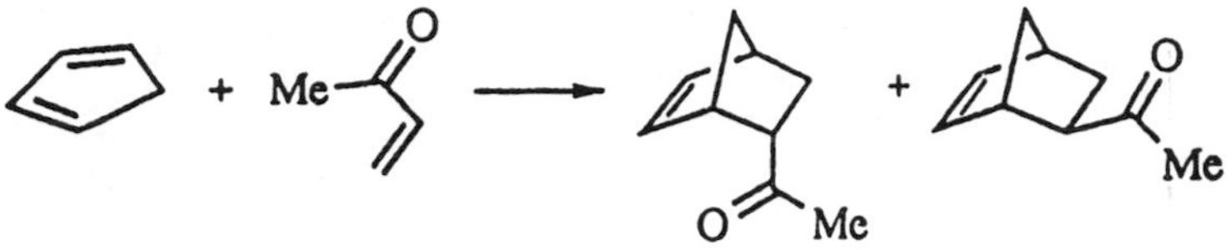

in cyclopentadiene	3.9	:	1
in H₂O	21.4	:	1
in H₂O, LiCl	28	:	1

Scheme 22

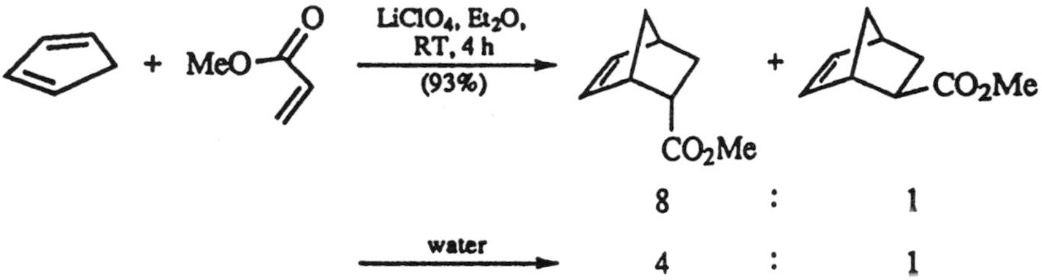

8	:	1
4	:	1

Scheme 23

increased selectivity is thought to be the result of the lithium functioning as a Lewis acid.

Diels–Alder reactions using furans as dienes are potentially useful because the furans are often readily available and the oxygen bridge in the bicyclo[2.2.1] adducts is easily broken (Type 0: see Chapter 3, Scheme 23). However, the reaction of simple furans in this reaction is often reversible and the equilibrium may lie on the side of the furan and dienophile rather than the adduct.[17]

A further consequence of this reversibility is that, when the equilibrium does favour the adduct, a thermodynamically formed mixture of *endo* and *exo* products may be obtained. Occasional diastereoselectivity under these conditions is, in general, less likely than in kinetically controlled irreversible Diels–Alder reactions.

Characteristic of Diels–Alder reactions is a high negative volume of activation (from -25 to $-45 \, cm^3 \, mol^{-1}$ for intermolecular reactions). The application of high pressures to these cycloadditions (8–20 kbar), therefore, increases the rate of reaction. Using simple furans, the application of high pressure often brings about Diels–Alder reactions in greater yields than can

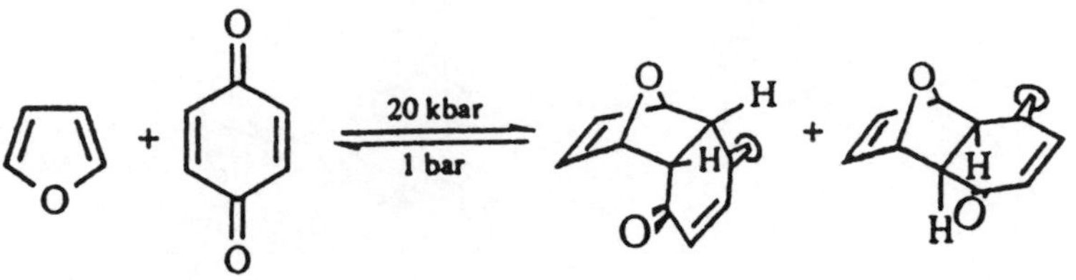

Scheme 24

be achieved at atmospheric pressure and the products are kinetically formed. Formation of the *endo* diastereoisomer is usually favoured but high diastereoselectivity is uncommon.[18]

As Scheme 24 indicates, the Diels–Alder adducts, formed at high pressure, may start to dissociate at atmospheric pressure. If these adducts are required for synthetic purposes, therefore, they must be reacted directly in a way which eliminates this possibility of cycloreversion.

Type II intramolecular Diels–Alder (IMDA) reaction[19]

The problems of both incomplete regioselectivity and *endo/exo* diastereoselectivity in the intermolecular Diels–Alder reaction arise because of the ability of diene and dienophile to react via different transition states of comparable energies.

Complete control of the regioselectivity in the Diels–Alder reaction can be exercised by linking the diene and dienophile together using a tether of limited length, i.e. comprised of a limited number of atoms. In practice, the tethers most often used contain three or four atoms with the diene linked at C-1 or C-4. These intramolecular Diels–Alder (IMDA) reactions lead, in the all-carbon cases, to hydroindene or hydronaphthalene systems (Scheme 25).

Scheme 25

It is important to be able to represent the transition states for these IMDA reactions adequately so that interactions of substituents on the diene, dienophile and tether can be appreciated. For the hydroindene system, the approach of diene and dienophile can be depicted as in the intermolecular versions (Scheme 26).

For the hydronaphthalene system it is often more appropriate to draw the corresponding transition states such that the chair-like perspective of the tether becomes apparent (Scheme 27).

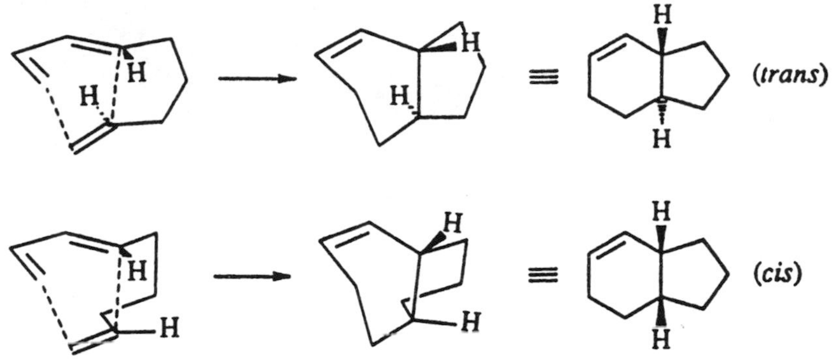

Scheme 26

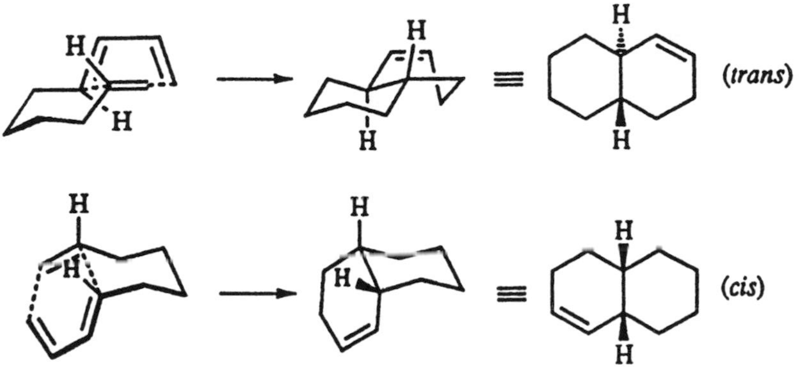

Scheme 27

A single regioisomer is formed in these IMDA reactions (when $n \leqslant 4$) because the alternative bicyclo[n.3.1] products (Scheme 28) would be considerably more strained and the transition states leading to them would be correspondingly raised in energy. Although suitable tethering of the diene and dienophile will dictate the regioselectivity of the Diels–Alder reaction, it does not necessarily dictate the *exo/endo* diastereoselectivity; as indicated in Schemes 26 and 27, either *cis*- or *trans*-fused bicyclic compounds can be formed. However, when a cycloaddition such as the Diels–Alder reaction is carried out intramolecularly, particularly with a tether of limited length, new steric and electronic interactions may come into play by comparison with the intermolecular version of the reaction. In general the difference between *exo* and *endo* transition-state energies will be augmented by such interactions and higher occasional diastereoselectivity will result.

An additional benefit of intramolecularity is entropy derived: the probability of the diene and dienophile coming together is increased if they

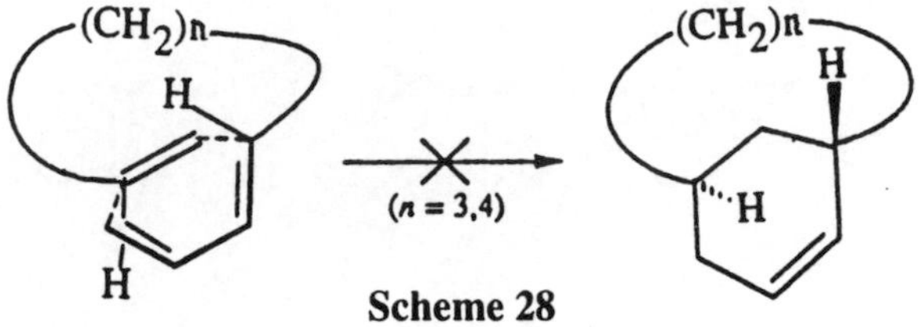

Scheme 28

are appropriately linked. This will manifest itself in reaction at a lower temperature than is needed for analogous intermolecular reactions and, consequently, by and large, to greater *exo/endo* selectivity (see Chapter 1).

We have seen that secondary orbital interactions can favour *endo* diastereoselectivity with a carbonyl-bonded substituent on a dienophile (Scheme 18), and that this often feeble interaction can be augmented by complexation of the carbonyl group with a Lewis acid (Scheme 21). This same increase in *endo* diastereoselectivity can be brought about in an acid-catalysed IMDA reaction.

Thus, the triene (18) (Scheme 29) is cyclised thermally to *cis-* and *trans-*hydroindenes (19) and (20) in a 28:72 ratio, whereas in the presence of ethylaluminium dichloride as a catalyst, only the *trans*-fused product (20) is obtained, in 80% yield.[20]

The effect of a catalyst cannot, however, always be relied upon; the corresponding triene containing the *cis*-α,β-unsaturated ester (21) (Scheme 30) likewise gives a mixture of diastereoisomers but their ratio is almost unaffected by carrying out the reaction in the presence of Lewis acids such as ethylaluminium dichloride.[20]

Higher *endo* diastereoselectivity appears to be more common in the formation of *cis*-hydronaphthalenes than *cis*-hydroindenes; possibly the four atom tether gives rise to a marginally looser transition state which allows more efficient secondary orbital interaction [Scheme 31(a)].[21] The Lewis acid-catalysed reaction in Scheme 31(b) also gives the *endo* (carbonyl) adduct.[22]

The *exo/endo* ratio in these IMDA reactions can be influenced by steric interactions between substituents in the diene and the tether (including 'axial' hydrogens) (Scheme 32). The steric interaction between an alkyl group R and the axial hydrogen shown in (23) is greater than that between R and the axial hydrogen shown in (22),[23] so *exo* becomes the preferred mode of addition.

The presence of a heteroatom in the tether can also affect the *endo/exo* ratio. In Scheme 33, complete diastereoselectivity arises from the preference for *endo* transition state (24) in which overlap of the urethane and diene systems is largely conserved and the $A_{1,3}$-strain (see Chapter 9) present in (25) is absent.[24]

Heating the triene (26) (Scheme 34), whose diene component is linked to the dienophile via a non-terminal carbon atom, gave the

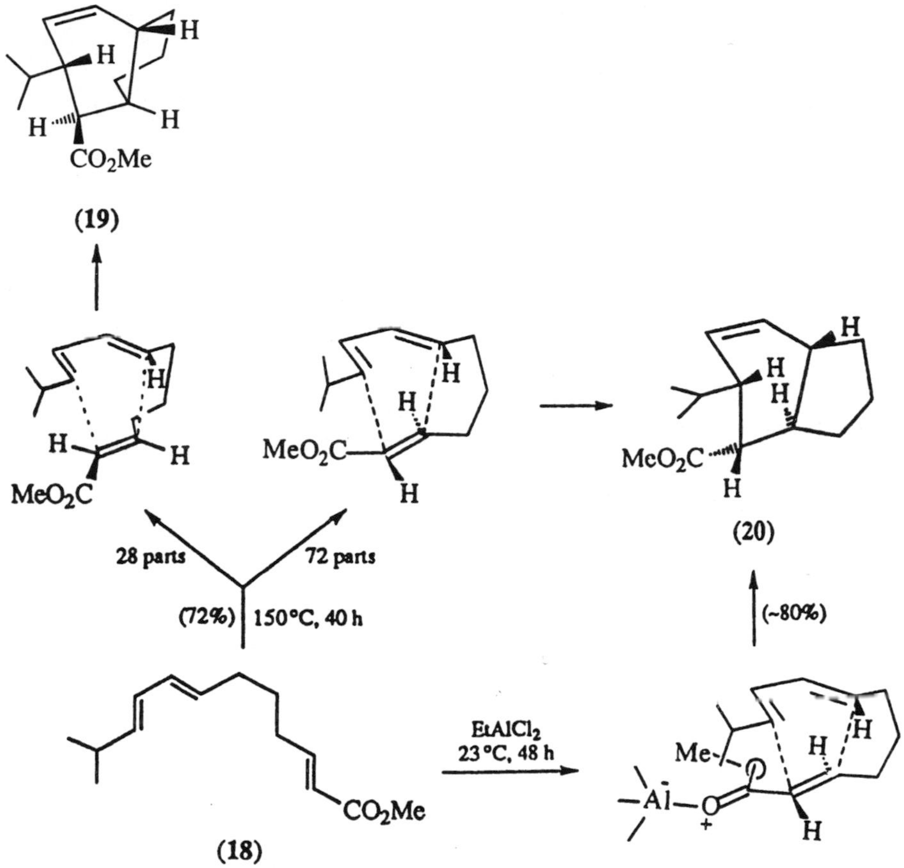

(19)

(20)

28 parts   72 parts

(~80%)

(72%) | 150°C, 40 h

EtAlCl$_2$
23°C, 48 h

(18)

Scheme 29

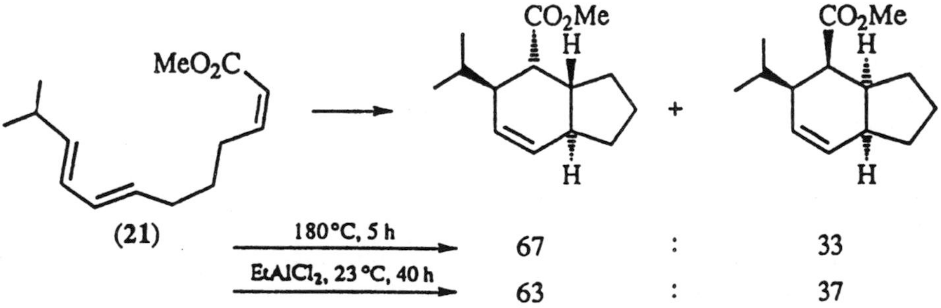

180°C, 5 h

EtAlCl$_2$, 23°C, 40 h

(21)

CO$_2$Me   CO$_2$Me

67   :   33

63   :   37

Scheme 30

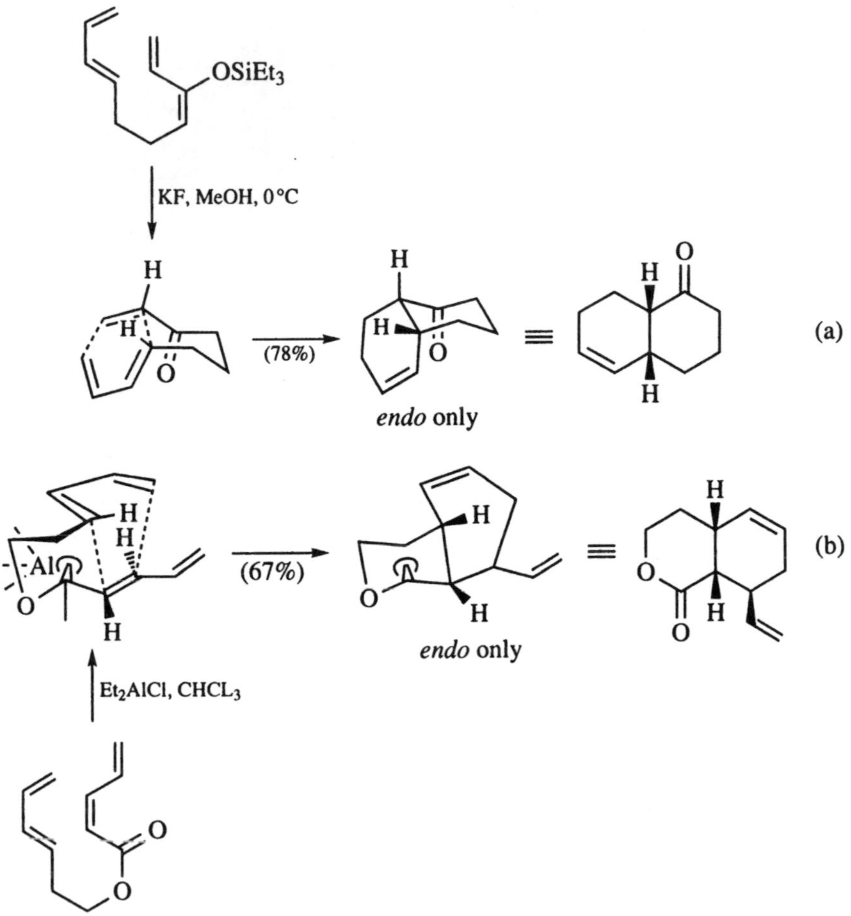

Scheme 31

oxabicyclo[4.3.1]decane **(27)**, which contains a bridgehead double bond.[25] The tightness of the transition state here ensures that the reaction is completely diastereoselective (*exo*-ethoxycarbonyl) and, of course, completely regioselective. Reduction of the double bond in **(27)** takes place exclusively from the *exo* face and ring opening gives **(28)**. To appreciate the selectivity inherent in this synthesis of **(28)** one need only consider the problems that would arise in its attempted synthesis starting with the *intermolecular* Diels–Alder shown in Scheme 34.

Linking together the two partners in a cycloaddition reaction by a tether designed to be easily broken (cf. Scheme 34; tether an ester) is valuable for the stereoselective synthesis of both cyclic and acyclic target molecules (see Chapter 3).

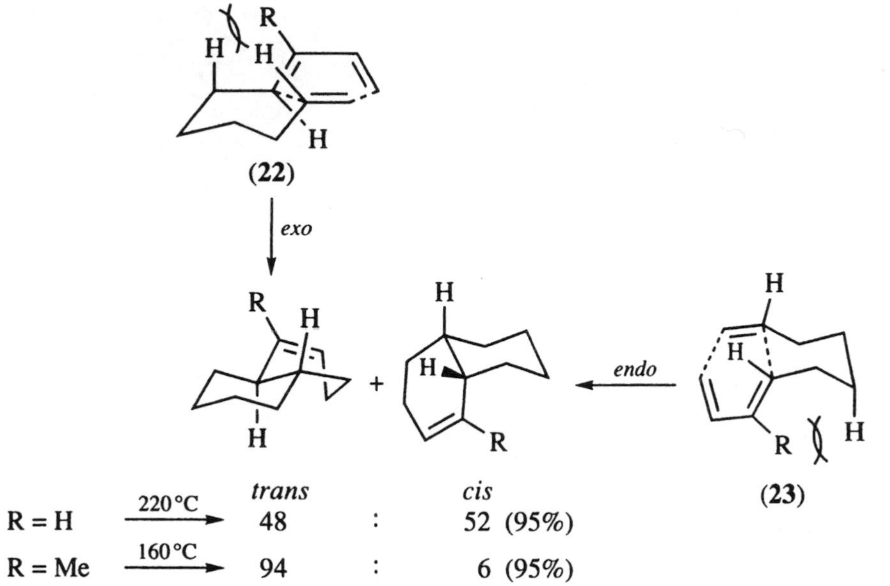

(22)

exo

endo

(23)

		trans		*cis*	
R = H	$\xrightarrow{220\,°C}$	48	:	52	(95%)
R = Me	$\xrightarrow{160\,°C}$	94	:	6	(95%)

Scheme 32

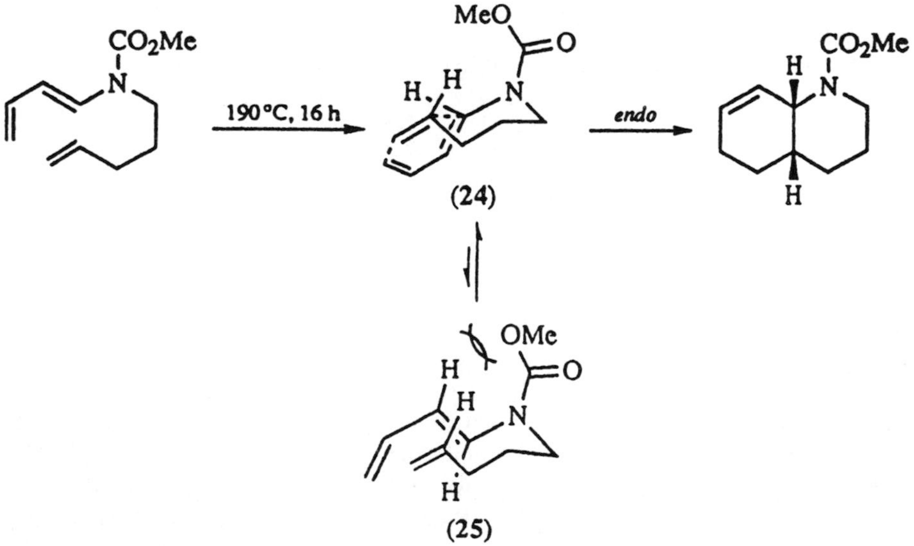

$190\,°C, 16\,h$

(24)

endo

(25)

Scheme 33

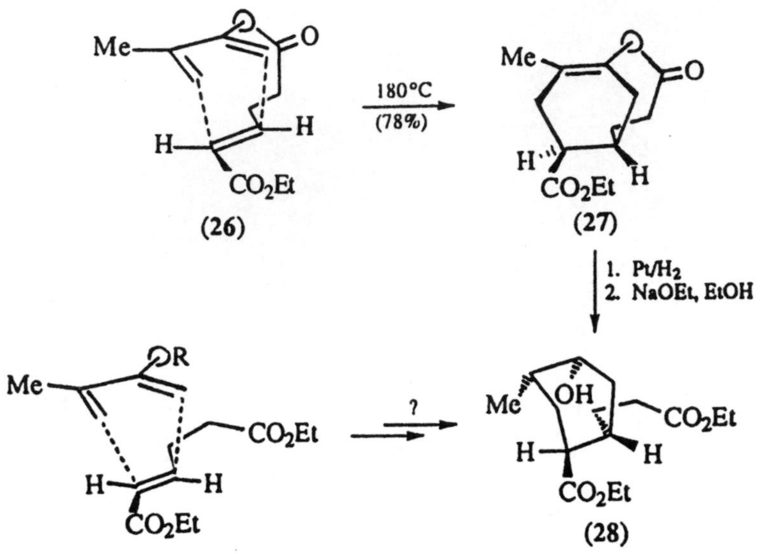

Scheme 34

Heterocycles via Diels–Alder reactions

Replacement of one or more atoms in the all-carbon Diels–Alder reaction by heteroatoms gives rise to six-membered-ring heterocycles. The double bonds of heteroatom-containing dienophiles[26,27] include $C=S$, $C=N$, $C=O$, $N=O$, $O=O$, $N=S=O$ and $N=N$ and heteroatom-containing dienes include (29)–(32).[6]

Of particular interest in stereoselective synthesis are the reactions of

electron-rich dienes with carbonyl groups (usually aldehydes) as the dienophile. These reactions are catalysed by Lewis acids, e.g. zinc chloride or europium salts, and are highly *endo* diastereoselective (Scheme 35).[28]

In Scheme 35, the occasional diastereoselectivity is dominated by the large bulk of the europium with its associated ligands; its coordination to the carbonyl group *anti* to the methyl and its *exo* placement in the transition state (33) to minimise steric effects lead to an *endo* placement of the aldehyde methyl group. The 2-methoxy substituent in the adduct can be retained or eliminated, depending on the work-up, and the products are related to natural sugars.

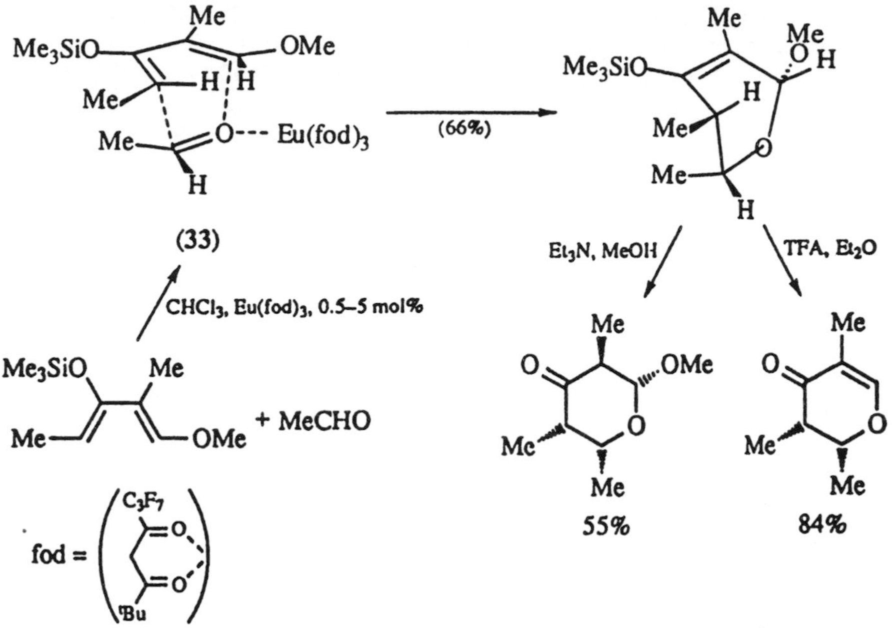

Me₃SiO—Me—OMe
—H H
Me
Me——O--- Eu(fod)₃   (66%)
H

(33)

/ CHCl₃, Eu(fod)₃, 0.5–5 mol%

Me₃SiO Me
Me——OMe + MeCHO

fod = (C₃F₇ =O, =O', ᵗBu)

Me₃SiO—Me—Me O H
H
Me
Me——O
H

Et₃N, MeOH /   \ TFA, Et₂O

Me
O—OMe
Me——O
Me
55%

Me
O
Me——O
Me
84%

Scheme 35

The ability of other metal ions (Ti⁴⁺, Mg²⁺) to chelate α- and β-alkoxyaldehydes allows diastereoselective cycloaddition of aldehydes with chiral centres α to the carbonyl group (Type III reactions; see Chapter 10).

Type II 1,3-dipolar cycloadditions[29]

Like Diels–Alder reactions, 1,3-dipolar cycloadditions involve $4\pi + 2\pi$ concerted reaction of a 1,3-dipolar species (the 4π component) and a dipolarophile (the 2π component).

1,3-Dipoles can be divided into two classes: those in which the three atoms comprising the 1,3-dipole are linear and those in which they are not. Examples are given in Table 1.

Since the products of these cycloadditions are five-membered rings in which the 1,3-dipolar residue must be bent, one might expect the linear type to be less reactive than those which are already bent. This, however, does not seem to be important; the energy required to bend a linear dipole in the transition state is apparently relatively small.

As indicated in Scheme 36, with an unsymmetrical 1,3-dipole and dipolarophile, the possibility of regioisomers arises. The major regioisomer formed in this case can be predicted using the same procedure applied to the Diels–Alder reaction—the HOMO/LUMO pair closer in energy is identified and, within this pair, those terminal atoms on the dipole and dipolarophile bearing the largest orbital coefficients became bonded[7] Steric effects may control the sense of regioselectivity; nitrile oxides give adducts in which the oxygen atom becomes bonded to the more hindered end of the dipolarophile (Scheme 37) in the major regioisomer.[30]

Table 1. Examples of 1,3-dipoles

Linear		Bent	
$-\overset{+}{C}=N-\overset{-}{O}$	nitrile oxide	$\overset{+}{O}\overset{O}{\diagup}\overset{=}{O}$	ozone
$-\overset{+}{C}=N-\overset{/}{C}_{\backslash}$	nitrile ylide	$\overset{+}{C}\overset{O}{\diagup}\overset{=}{C}$	carbonyl ylide
$\overset{+}{N}=N-\overset{/}{N}_{\backslash}$	azide	$\overset{+}{C}\overset{N}{\diagup}\overset{-}{O}$	nitrone
$\overset{+}{N}=N-\overset{/}{C}_{\backslash}$	diazoalkane	$\overset{+}{C}\overset{N}{\diagup}\overset{-}{C}$	azomethine ylide

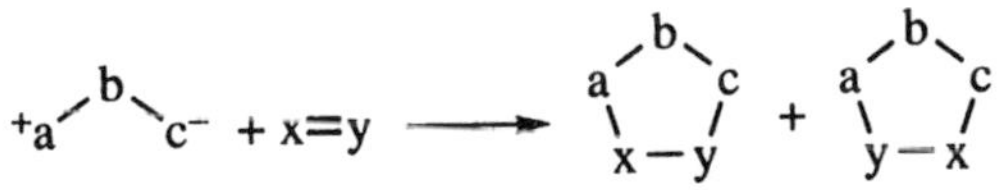

Scheme 36

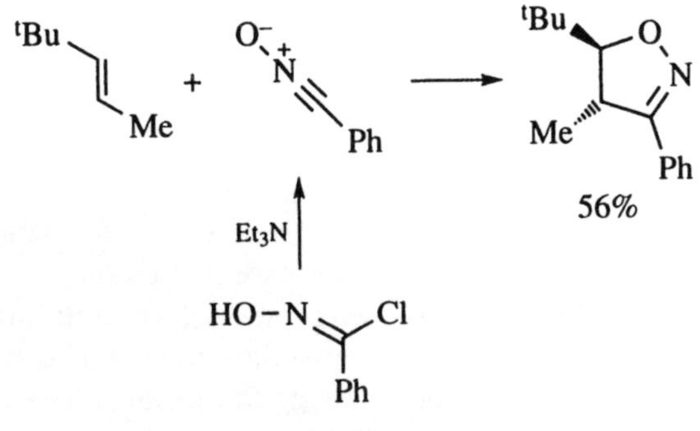

Scheme 37

Stereochemistry of 1,3-dipolar cycloadditions

Some linear 1,3-dipoles, e.g. nitrile oxides, are not prochiral in these cycloadditions and so the only chiral centres in the product are those derived from *syn* addition to the dipolarophile (Scheme 37).

Nitrile oxides are unstable and must be generated *in situ*. In the reaction in Scheme 38 this carried out by thermolysis of the dimer (34), which is in equilibrium with a small concentration of the monomer (35). The relatively high temperature allows cycloaddition of (35) not only to 1,2-disubstituted alkenes but also to trisubstituted alkenes, which are normally less reactive.[31]

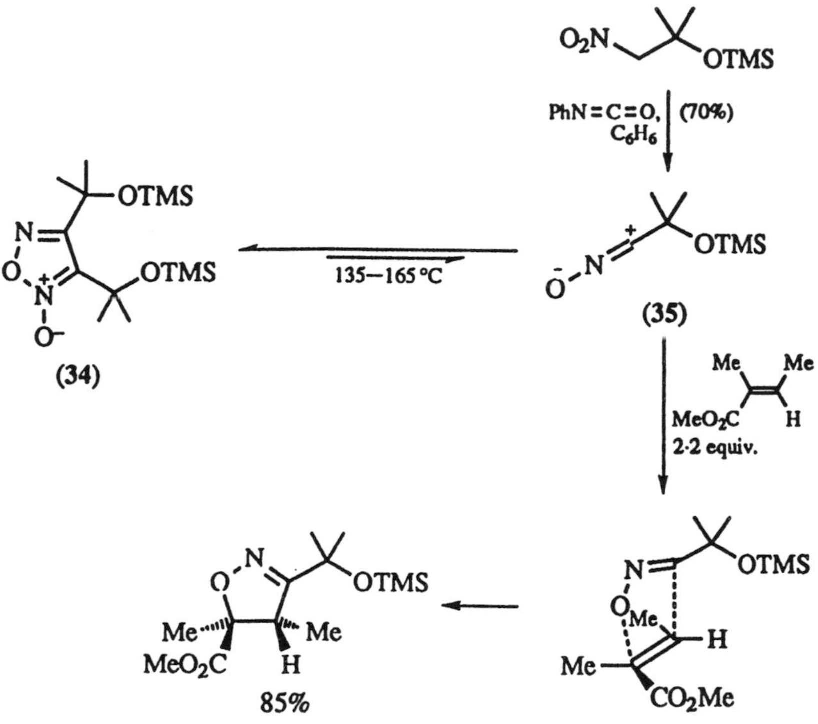

Scheme 38

For prochiral 1,3-dipoles, there can be reciprocal *syn* addition to 1,3-dipole and dipolarophile, as in the Diels–Alder reaction (stereospecificity; inherent diastereoselectivity). However, unlike the Diels–Alder reaction in which the diene is invariably configurationally stable, the barrier to configurational interconversion (stereomutation) in bent 1,3-dipoles may be low as in, for example, azomethine ylides (Scheme 39).

Consequently, a mixture of diastereoisomeric cycloadducts may be

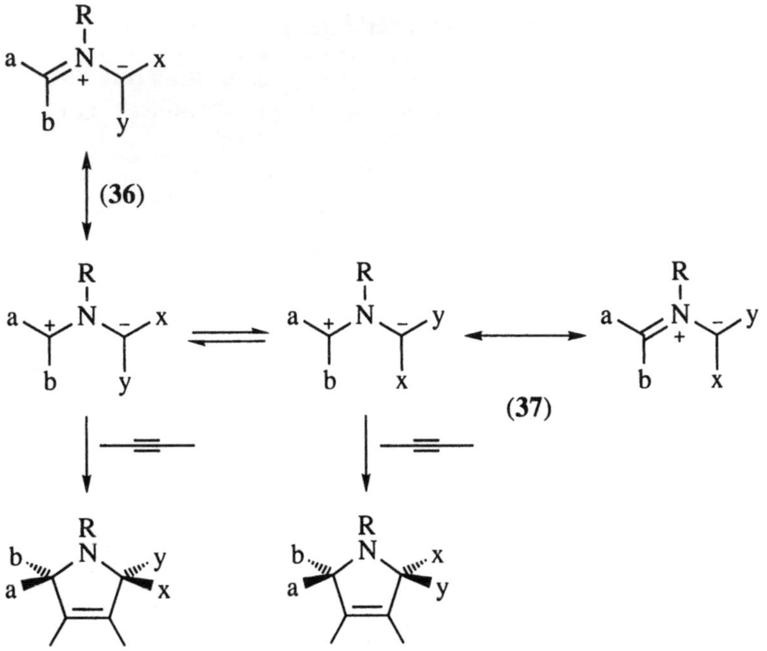

Scheme 39

obtained (with each of the stereoisomeric 1,3-dipolar species **36** and **37** in Scheme 39 reacting with complete *syn* selectivity). Using unreactive dipolarophiles, this may be the result even if the 1,3-dipolar species was initially generated as a single diastereoisomer [(**36**) or (**37**) in Scheme 39]. However, for cycloaddition of prochiral 1,3-dipoles which react via a single configuration with a prochiral dipolarophile the consequences are analogous to those in the Diels–Alder reaction (cf. Scheme 15); four (racemic) products may be formed when regioisomers are also considered.

exo/endo (occasional) Diastereoselectivity in 1,3-dipolar cycloadditions is, as in the Diels–Alder reaction, determined by the preference of the dipolarophile substituents for the *endo* or *exo* position. However, it appears that secondary orbital interactions which usually favour placement of, e.g., carbonyl substituents in the *endo* position in the Diels–Alder reaction (Scheme 18), are much weaker in the corresponding 1,3-dipolar cycloaddition, if they are present at all. That being so, diastereoselectivity, which may still favour the formation of the *endo* product, may be more the result of steric or other electronic effects.

Like nitrile oxides, azomethine ylides must also be generated *in situ* and

Scheme 40 shows some of the different ways in which this can be done.

Scheme 41 shows examples of diastereoselective nitrone and carbonyl ylide 1,3-dipolar cycloadditions.

(a) aziridine ring opening

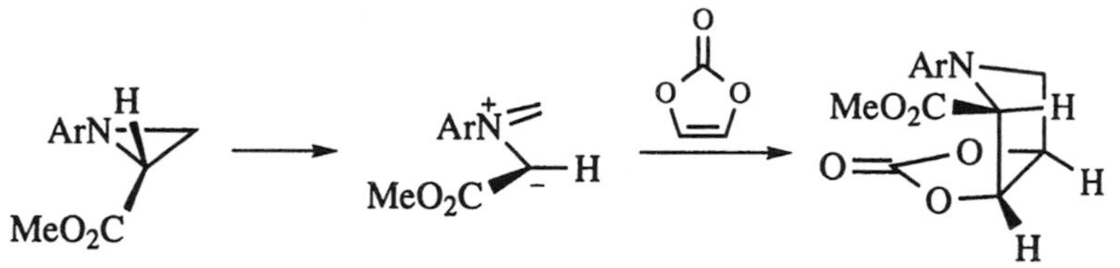

(ref. 32)

(b) α-amino acid decarboxylation

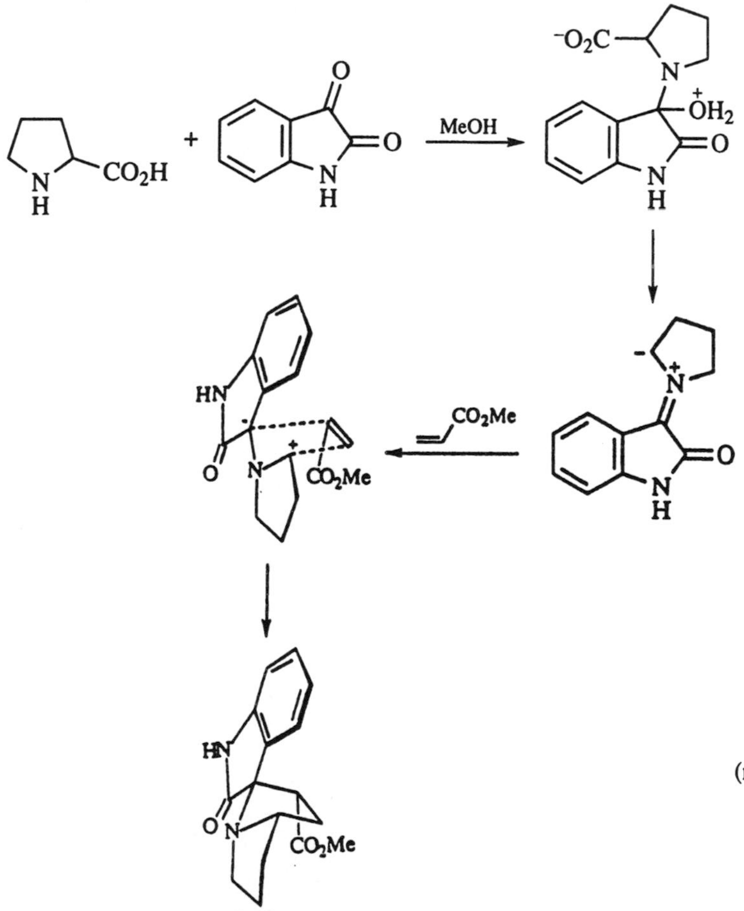

(ref. 33)

(73%)

Scheme 40   *(continued)*

Scheme 40 *(continued)*

(c) 1,2-prototropy

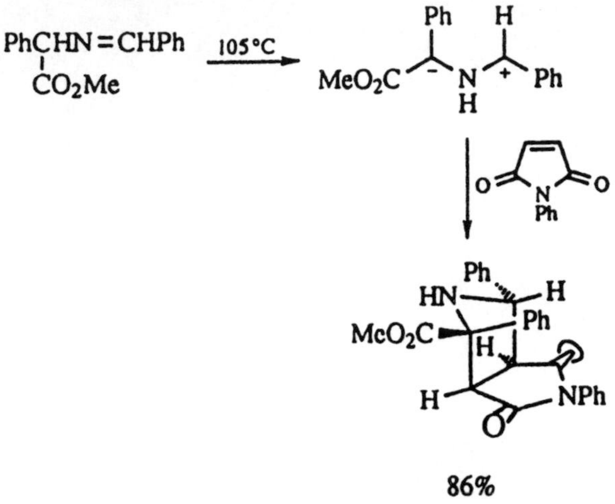

(ref. 34)

86%

(d) fluorodesilylation

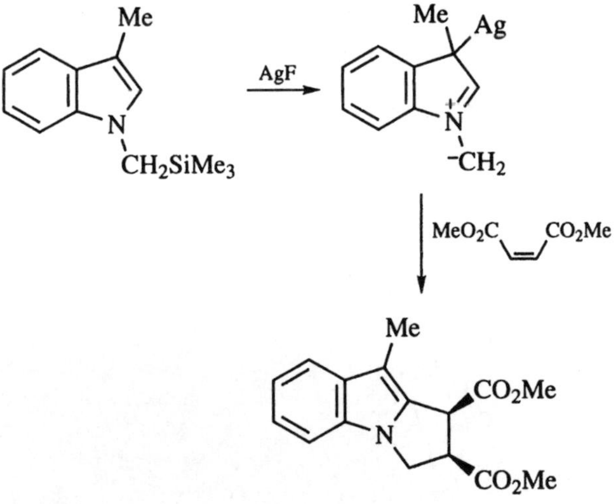

(ref. 35)

Scheme 40

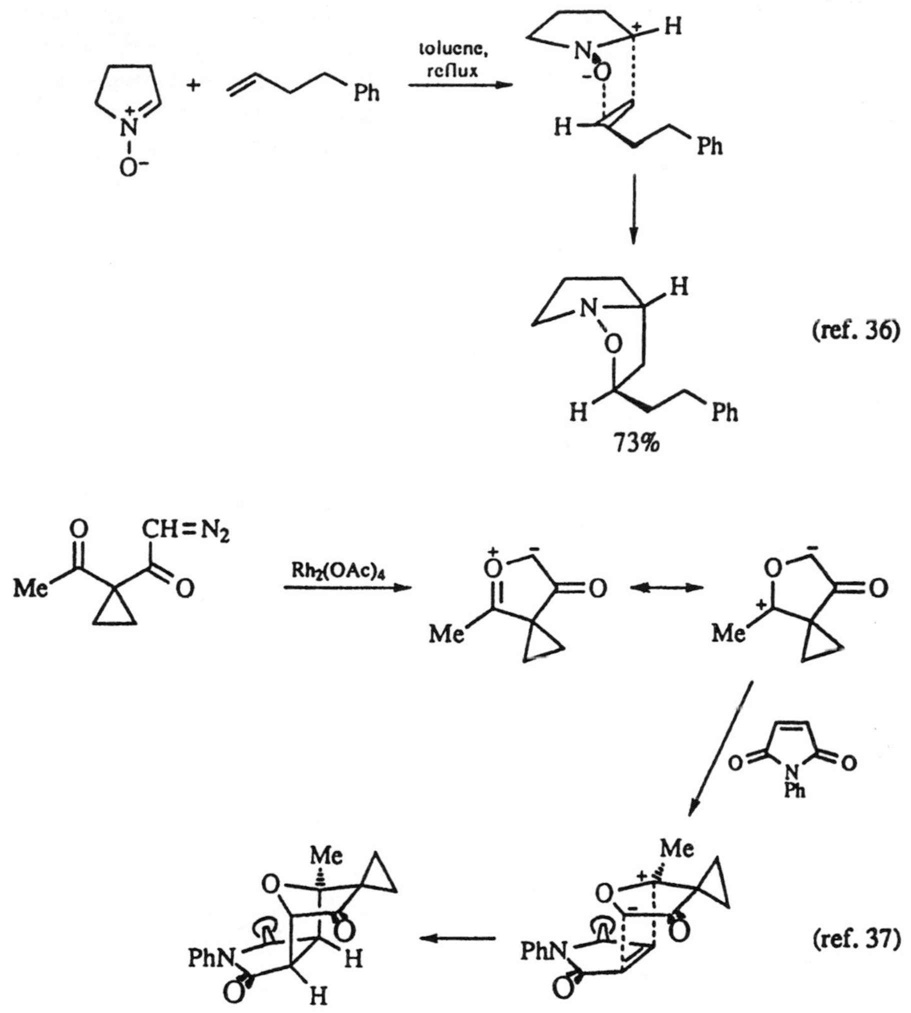

(ref. 36)

73%

(ref. 37)

Scheme 41

Intramolecular 1,3-dipolar cycloaddition[29,38]

The weakened *endo*-directing secondary orbital interaction in the transition state for 1,3-dipolar cycloaddition means that selectivity in the *endo* sense is reduced and may easily be overridden by other effects leading to *exo* diastereoselectivity. High regioselectivity also cannot be relied upon and in any case, where it obtains, the alternative regioisomer may be the one that is actually desired.

Intramolecularity in 1,3-dipolar cycloadditions is an especially valuable device for directing the regiochemical course of the reaction, and often results in enhanced occasional (*exo/endo*) selectivity also. The latter is usually the result of the limited length of the tether, coupled with non-bonded interactions of substituents on the tether and the 1,3-dipole or dipolarophile. Electronic effects resulting from enforced orientations of substituents may also be introduced or magnified in these intramolecular 1,3-dipolar cycloadditions.

Some examples of completely diastereoselective and regiospecific intramolecular 1,3-dipolar cycloadditions are illustrated in Scheme 42. In these reactions a five-membered ring is bound to be present in the bicyclic product. When the second ring is also five-membered and fused to the first, a bicyclo[3.3.0] system results and the ring fusion is invariably *cis*, as the examples in Scheme 42 illustrate.

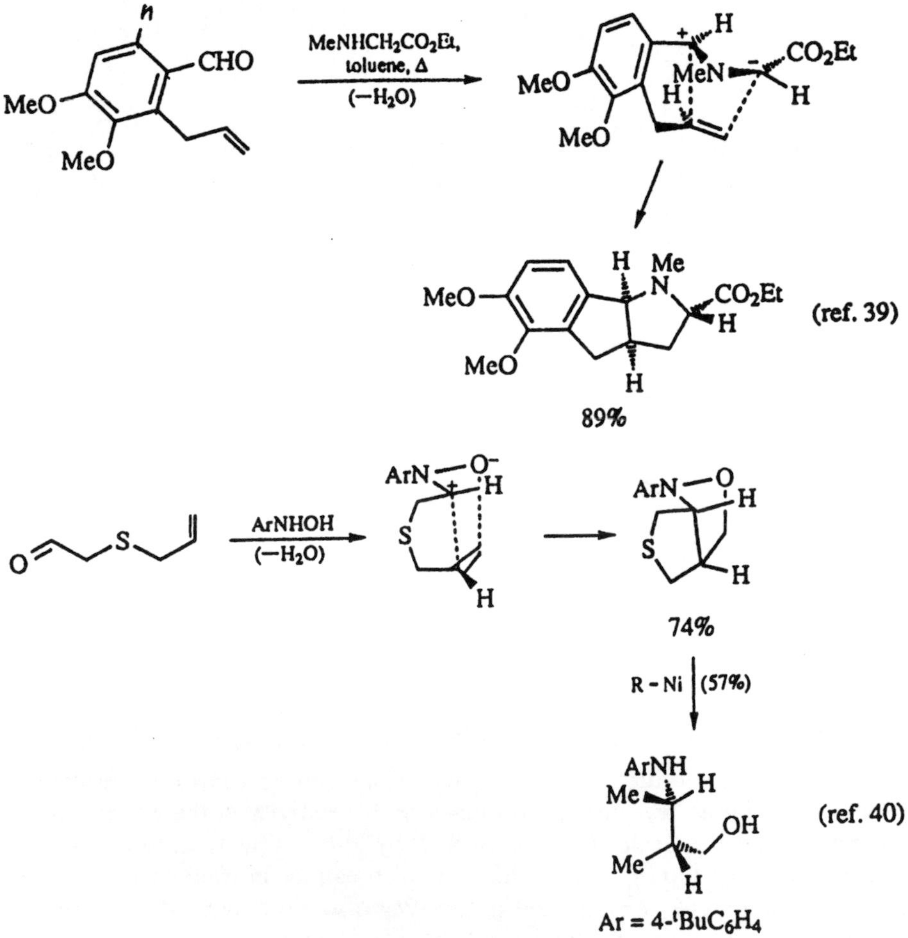

Scheme 42

THE ENE REACTION

The ene and the Diels–Alder reactions are related pericyclic reactions which frequently co-occur in the reaction of a diene bearing an allylic C—H bond (an ene component) with dienophile (enophile) (Scheme 43).

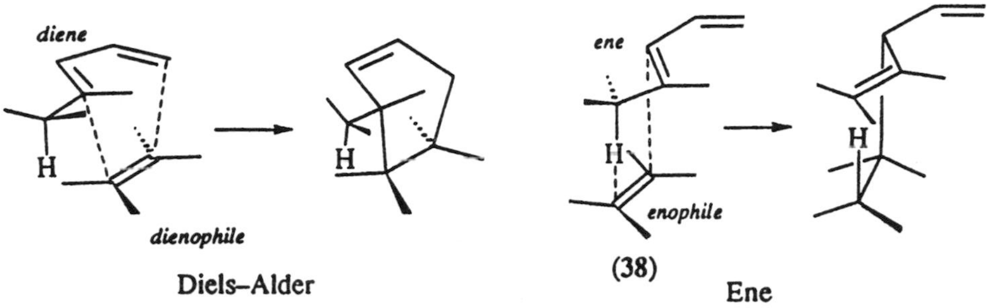

diene

dienophile

Diels–Alder

ene

enophile

(38)

Ene

Scheme 43

The inherent diastereoselectivity (stereospecificity) in the ene reaction is evident from inspection of (**38**): both bonds are made *syn* to the enophile (the configuration of the enophile is retained in the product) and the bond made to the double bond of the ene component is *syn* to the hydrogen which is transferred.

The role of the enophile can be assumed by the same double bonds that can function as dienophiles in the Diels–Alder reaction, e.g. C=O, C=N, C=S, N=N, N=O as well as C=C. Preparatively, aldehydes are particularly useful as enophiles because their reactivity is greatly increased by complexation with Lewis acids, allowing them to be used at low temperatures. The products using aldehydes as enophiles are homoallylic alcohols, i.e. the ene reaction brings about the same transformation as is accomplished by using allylmetal addition to the aldehyde (Scheme 44).

The transition-state geometries of lowest energy for the ene reactions of propene with ethylene and of propene with formaldehyde have been calculated by Loncharich and Houk[42] and found to be very similar. For the carbonyl–ene reaction, the substituents on carbons 1 and 1′ are staggered, as illustrated in the Newman projection (**39**) along the (forming) 1—1′ carbon–carbon bond (Figure 2).

We shall illustrate the most important stereochemical features of the Type II ene reaction by reference to the Lewis acid-catalysed reactions of aldehydes with allyl systems.[43] With some combinations of substituted ene/enophile the mechanism of the catalysed reaction may involve ion-pair

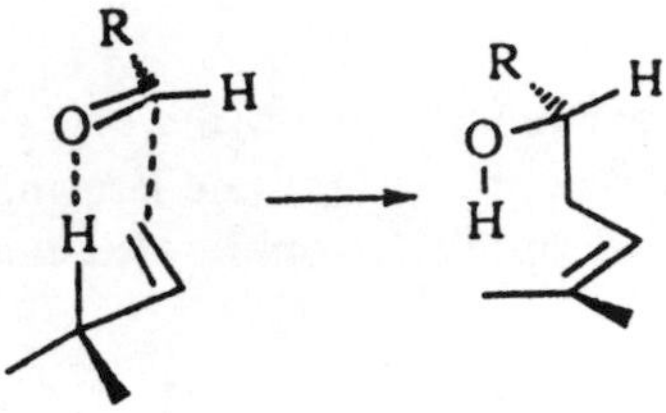

ene using aldehyde

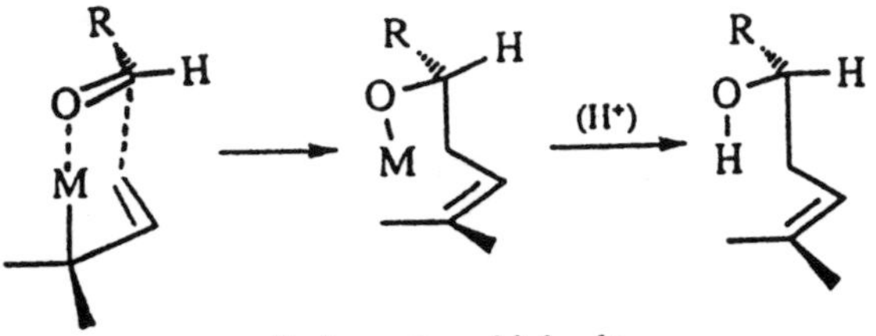

allylmetal + aldehyde

Scheme 44

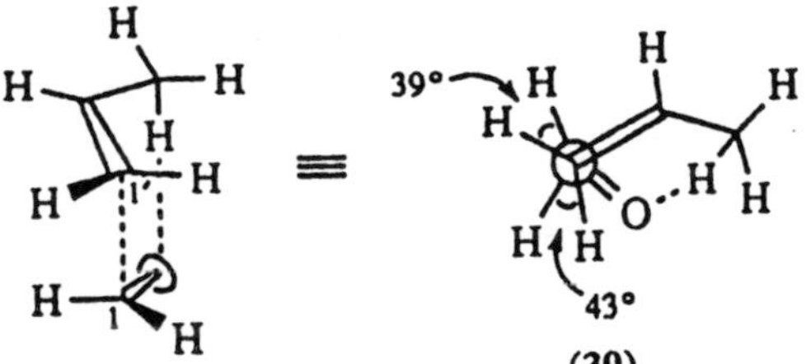

(39)

Figure 2

intermediates, but this does not appear to be important for the reactions discussed below.

In the reaction of a 1,1-disubstituted propene with an aldehyde, two transition states, *exo* and *endo*, are possible (Scheme 45). In these transition states, the carbonyl oxygen is assumed to be sp²-hybridised with the Lewis acid AL_n complexed to the lone pair *syn* to the aldehyde hydrogen to minimise steric effects.

The major factors which control the occasional (*exo/endo*) diastereoselectivity in these ene reactions are (i) steric interactions between R and a in (**40**) and between R and b in (**41**) and (ii) steric interactions between the Lewis acid AL_n and substituents on the ene component.

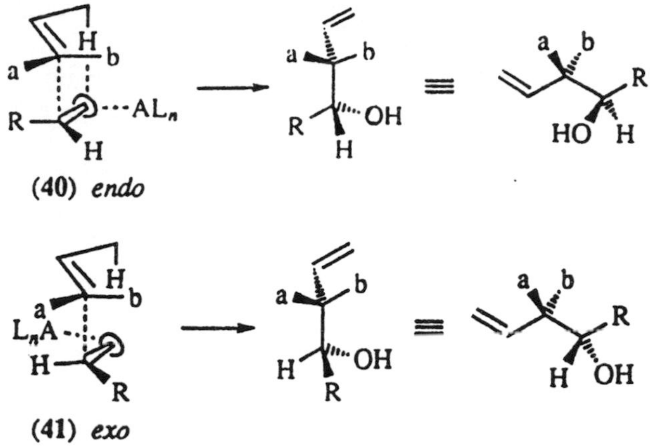

(40) *endo*

(41) *exo*

Scheme 45

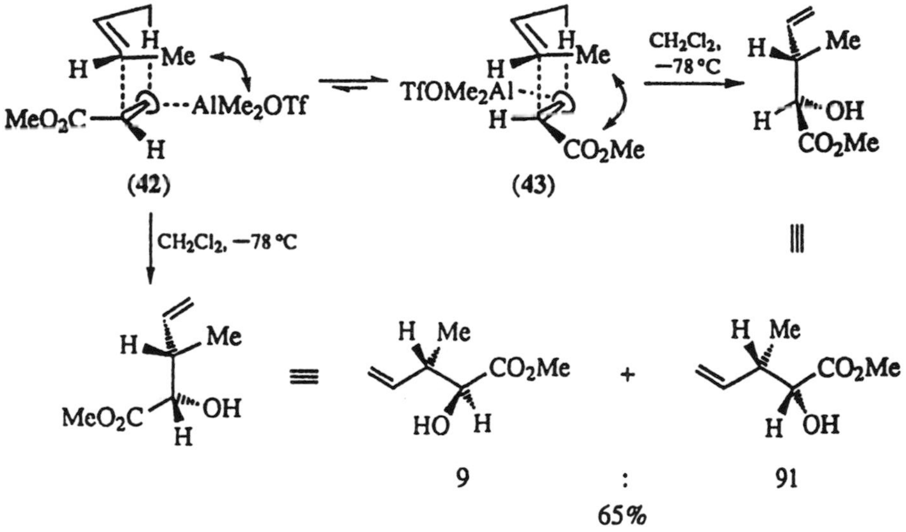

(42)

(43)

9 : 91

65%

Scheme 46

In Scheme 46, methyl glyoxylate reacts with an excess of (Z)-but-2-ene in the presence of dimethylaluminium triflate to give a 9 : 91 ratio of diastereoisomers resulting from *endo* and *exo* addition, respectively. This is rationalised on the basis of greater steric interaction between substituents on the coordinated aluminium and the ene methyl group in (42) than between the methoxycarbonyl group and the same methyl group in (43).[44]

Catalysis of the ene reaction of alkyl glyoxylates by tin(IV) chloride is different (Scheme 47) because the tin atom is able to complex with the

oxygens of both carbonyl groups. Reaction with (E)-but-2-ene gives mainly
trans-diastereoisomer (44) because of greater repulsion between the Lewis
acid and the methyl group in (45). Cis and trans in Scheme 47 refer to the
relative positions of the substituents (Me, OH) with the longest chain drawn
in the zig-zag conformation; the hydrogens at these chiral centres are omitted
for clarity. (These diastereoisomers are often referred to as syn and anti
respectively; in this book cis and trans refer to the spatial relationship
between two substituents in a starting material or product and syn and anti to
this relationship between two (interacting) substituents in a reaction).

The same trans isomer is the major diastereoisomer obtained when (Z)-
but-2-ene is used, although in this case the diastereoselectivity is not high

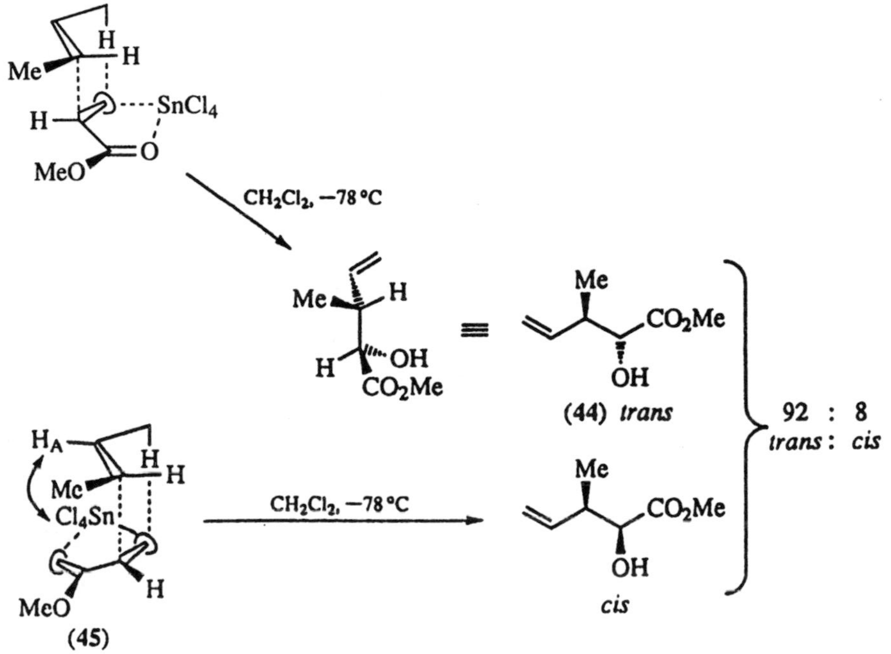

Scheme 47

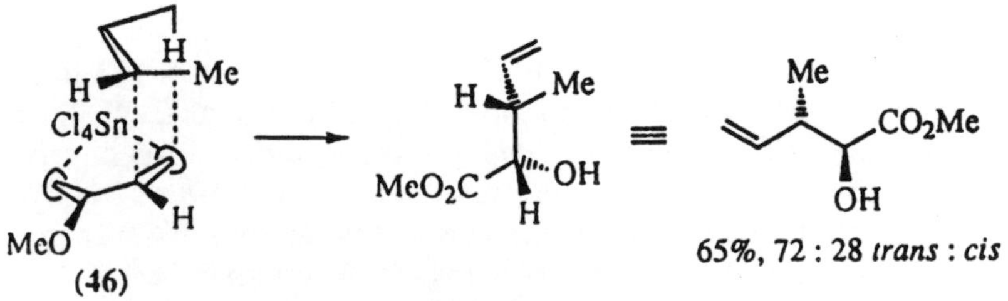

Scheme 48

(Scheme 48); probably there is some interaction between the olefinic H_A proton and the substituents on the tin (see below).

The presence of an additional substituent on the ene component can further increase the diastereoselectivity[42] (Scheme 49) (cf. Scheme 47). On the other hand, this 2-trimethylsilyl substituent brings about a reversal of the sense of diastereoselectivity using the 1,2-*cis*-dimethyl isomer (Scheme 50) (cf. Scheme 48).[45]

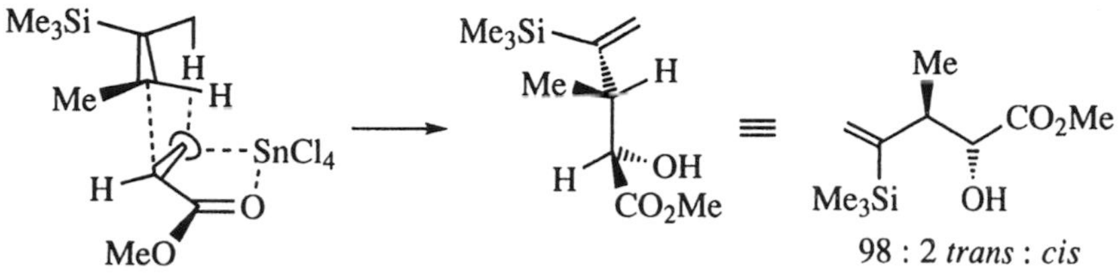

Scheme 49

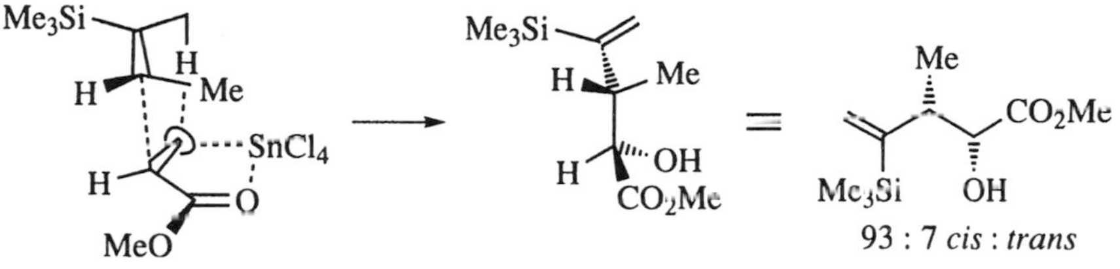

Scheme 50

It appears that in this case, steric repulsion between the trimethylsilyl group and the substituents on the coordinated Lewis acid in a transition state resembling (**46**) outweighs that between the methyl group and the co-ordinated Lewis acid in (**47**) (Scheme 50).

Regioselectivity in the ene reaction

Two major drawbacks to the use of the intermolecular ene reaction in synthesis have been the high temperatures previously needed for reaction and the lack of regioselectivity generally exhibited in such cases. The discovery that Lewis acid catalysis enables the intermolecular carbonyl ene reaction to be accomplished at temperatures as low as $-78\,°C$ with beneficial effects on the *diastereoselectivity* (see above) has stimulated efforts to bring about *regioselectivity* in the reaction. As usual, this can be accomplished by carrying out the reaction intramolecularly (see below), but work by Nakai and co-workers[46] has shown that, in an intermolecular reaction, an allylic oxygen or even a homoallylic oxygen can retard hydrogen transfer from this allylic

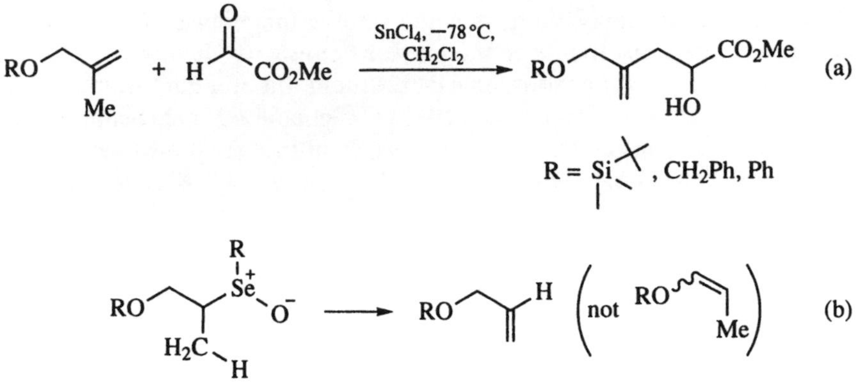

Scheme 51

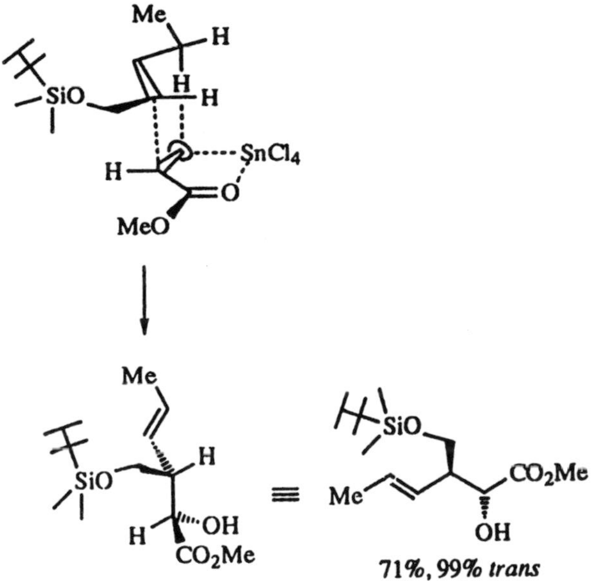

Scheme 52

position [Scheme 51(a)], thereby favouring transfer from an alternative allylic position. This effect seems to be partly steric and partly electronic in origin and is reminiscent of the complete regioselectivity which obtains in the *syn* elimination reactions of β-alkoxyalkylselenoxides [Scheme 51(b)].

Thus a type II ene reaction using the 1-silyloxy-substituted (*E*)-but-2-ene with methyl glyoxylate and tin(IV) chloride is both highly regio- and diastereoselective (Scheme 52).[46]

Scheme 52 also shows that when the hydrogen which is transferred in the ene reaction is derived from a secondary centre (as opposed to a methyl group in the schemes above), a configured double bond is formed, which, in the absence of a substituent at the 2-position on the ene component, is usually *E*. This is the case irrespective of the double band configuration in the parent ene component.

Intramolecularity provides another solution to the problem of regiocontrol in the ene reaction.[19a] In most reactions in this category, there is an additional chiral centre present in the starting material and these are, therefore, Type II/III reactions (see Chapter 13).

Summary

The Type II reactions considered in this chapter are those which are in part stereospecific (*inherently* diastereoselective). Thus, *syn* additions of achiral reagents XY to configured double bonds give single diastereoisomers as a result of the mechanism of the reaction in which both chiral centres are created more or less simultaneously: overall *anti* additions result from Type I reactions on these *syn* addition products.

Likewise, cycloaddition reactions, e.g. the Diels–Alder reaction, also contain an inherently diastereoselective element arising from the reciprocal *syn* (suprafacial) addition of the components. However, these cycloaddition reactions may also be diastereoselective in another sense (*exo/endo* for the Diels–Alder reaction) which is *occasional*, i.e. depends on the substitution of the interacting components and, unlike the inherent diastereoselectivity, will not be complete in most cases.

References

1. G. Berti, *Top. Stereochem.*, 1973, **7**, 93.
2. R.S. Atkinson, M.J. Grimshire and B.J. Kelly, *Tetrahedron*, 1989, **45**, 2875.
3. H.E. Simmons, T.L. Cairns, S.A. Vladuchick and C.M. Hoiness, *Org. React.*, 1972, **20**, 1.
4. T. Gilchrist and R.C. Storr, *Organic Reactions and Orbital Symmetry*, Cambridge University Press, 2nd edn, 1979; G.B. Gill and M.R. Willis, *Pericyclic Reactions*, Chapman and Hall, London, 1974; R.B. Woodward and R. Hoffman, *The Conservation of Orbital Symmetry*, Academic Press, New York, 1970.
5. R.A. Moss, *Acc. Chem. Res.*, 1980, **13**, 58, and references cited therein.
6. W. Carruthers, *Cycloaddition Reactions in Organic Synthesis*, Pergamon Press, Oxford, 1990; F. Fringuelli and A. Taticchi, *Dienes in the Diels–Alder Reaction*, Wiley, New York and Chichester, 1990.
7. I. Fleming, *Frontier Orbitals and Organic Chemical Reactions*, Wiley, Chichester, 1976.
8. D.L. Boger and C.E. Brotherton, *Tetrahedron*, 1986, **42**, 2777.
9. M. Burdisso, R. Gandolfi, P. Grünager and A. Rastelli, *J. Org. Chem.*, 1990, **55**, 3427, and references cited therein.

10. K. Alder and G. Stein, *Justus Liebigs Ann. Chem.*, 1933, **524**, 222.
11. S. Danishefsky, T. Kitahara, C.F. Yan and J. Morris, *J. Am. Chem. Soc.*, 1979, **101**, 6996.
12. J.A. Berson, Z. Hamlet and W.A. Mueller, *J. Am. Chem. Soc.*, 1962, **84**, 297; T. Inukai and T. Kojima, *J. Org. Chem.*, 1966, **31**, 2032.
13. L.E. Overman and L.A. Clizbe, *J. Am. Chem. Soc.*, 1976, **98**, 2352.
14. T. Cohen, A.J. Mura, D.W. Shull, E.R. Fogel, R.J. Ruffner and J.R. Falck, *J. Org. Chem.*, 1976, **41**, 3218; see also P.V. Alston, R.M. Ottenbrite and T. Cohen, *J. Org. Chem.*, 1978, **43**, 1864.
15. R. Breslow, U. Maitra and D. Rideout, *Tetrahedron Lett.*, 1983, **24**, 1901; R. Breslow, *Acc. Chem. Res.*, 1991, **24**, 159; P.A. Grieco, P. Galatsis and R.F. Spohn, *Tetrahedron*, 1986, **42**, 2847, and references cited therein.
16. P. Grieco, J.J. Nunes and M.D. Gaul, *J. Am. Chem. Soc.*, 1990, **112**, 4595; M.A. Foreman and W.P. Dailey, *J. Am. Chem. Soc.*, 1991, **113**, 2761.
17. J. Staunton, in *Comprehensive Organic Chemistry*, Vol. 4, ed. P.G. Sammes, Pergamon Press, Oxford, 1979.
18. K. Matsumoto and A. Sera, *Synthesis*, 1985, 999; R. Van Eldick, T. Asano and W.J. Le Noble, *Chem. Rev.*, 1989, **89**, 549.
19. (a) D.L. Taber, *Intramolecular Diels–Alder Reactions and Alder-Ene Reactions*, Springer, New York, 1984; (b) D. Craig, *Chem. Soc. Rev.*, 1987, **16**, 187.
20. W.R. Roush and H.R. Gillis, *J. Org. Chem.*, 1982, **47**, 4825; W.R. Roush, H.R. Gillis and A.I. Ko, *J. Am. Chem. Soc.*, 1982, **104**, 2269.
21. W. Oppolzer, R.L. Snowden and D.P. Simmons, *Helv. Chim. Acta*, 1981, **64**, 2002.
22. C. Chen and D.J. Hart, *J. Org. Chem.*, 1990, **55**, 6236.
23. Y.-T. Lin and K.N. Houk, *Tetrahedron Lett.*, 1985, **26**, 2269; S.R. Wilson and D.T. Mao, *J. Am. Chem. Soc.*, 1978, **100**, 6289.
24. W. Oppolzer and W. Fröstl, *Helv. Chim. Acta*, 1975, **58**, 587, 590.
25. K.J. Shea and E. Wada, *J. Am. Chem. Soc.*, 1982, **104**, 5715.
26. D. Boger and S.M. Weinreb, *Hetero Diels–Alder Methodology in Organic Synthesis*, Academic Press, San Diego, 1987.
27. S.M. Weinreb and P.M. Scola, *Chem. Rev.*, 1989, **89**, 1525; S.M. Weinreb, *Acc. Chem. Res.*, 1988, **21**, 313; S.M. Weinreb and R.R. Staib, *Tetrahedron*, 1982, **38**, 3087.
28. M. Bednarski and S. Danishefsky, *J. Am. Chem. Soc.*, 1983, **105**, 3716.
29. A. Padwa, *1,3-Dipolar Cycloaddition Chemistry*, Wiley, New York, 1984; 'Synthetic applications of 1,3-dipolar cycloaddition reactions,' Symposia in Print, *Tetrahedron*, 1985, **41**, 3447, *et seq.*
30. S.F. Martin and B. Dupré, *Tetrahedron Lett.*, 1983, **24**, 1337.
31. D.P. Curran and C.J. Fenk, *J. Am. Chem. Soc.*, 1985, **107**, 6023.
32. P. DeShong, D.A. Kell and D.R. Sidler, *J. Org. Chem.*, 1985, **50**, 2309.
33. R. Grigg, *Chem. Soc. Rev.*, 1987, **16**, 89.
34. R. Grigg, H.Q. Nimal Gunaratne and J. Kemp, *J. Chem. Soc., Perkin Trans. 1*, 1984, 41.
35. A. Padwa, G.E. Fryxell, J.R. Gasdaska, M.K. Venkatramanan and G.S.K. Wong, *J. Org. Chem.*, 1989, **54**, 644; see also E. Vedejs and G.R. Martinez, *J. Am. Chem. Soc.*, 1980, **102**, 7993.
36. J.J. Tufariello and J.M. Puglis, *Tetrahedron Lett.*, 1986, **27**, 1265.
37. A. Padwa, R.L. Chinn, S.F. Hornbuckle and L. Zhi, *Tetrahedron Lett.*, 1989, **30**, 301.
38. A. Hassner, R. Maurya, A. Padwa and W.H. Bullock, *J. Org. Chem.*, 1991, **56**, 2775, and references cited therein.
39. P.N. Confalone and E.M. Huie, *J. Am. Chem. Soc.*, 1984, **106**, 7175.

40. H.G. Aurich and K.-D. Möbus, *Tetrahedron Lett.*, 1988, **29**, 5755.
41. H.M.R. Hoffmann, *Angew. Chem., Int. Ed. Engl.*, 1969, **8**, 556; B.B. Snider, *Acc. Chem. Res.*, 1980, **13**, 426.
42. R.J. Loncharich and K.N. Houk, *J. Am. Chem. Soc.*, 1987, **109**, 6947.
43. J.P. Benner, G.B. Gill, S.J. Parrott, B. Wallace and M.J. Begley, *J. Chem. Soc., Perkins Trans. 1*, 1984, 315.
44. K. Mikami, T.-P. Loh and T. Nakai, *Tetrahedron Lett.*, 1988, **29**, 6305.
45. K. Mikami, T.-P. Loh and T. Nakai, *J. Am. Chem. Soc.*, 1990, **112**, 6737.
46. K. Mikami, M. Shimizu and T. Nakai, *J. Org. Chem.*, 1991, **56**, 2952.

7 TYPE II REACTIONS: OCCASIONAL DIASTEREOSELECTIVITY

In this chapter, Type II reactions will be considered whose diastereoselectivity is in the main brought about by the particular substitution on the interacting components, i.e. it is occasional rather than inherent in the mechanism of the reaction.

Type II $2\pi + 2\pi$ cycloadditions

PHOTOCHEMICAL[1]

From orbital symmetry considerations, concerted cycloaddition of two alkenes to give a cyclobutane is allowed suprafacially–suprafacially, i.e. *syn* with respect to both alkenes, provided that one reacts via its singlet excited state. Unfortunately, in practice there are number of factors which conspire to reduce the theoretical diastereoselective promise of this reaction, at least with acyclic alkenes. One factor is the tendency for alkenes to undergo photochemical *trans/cis* (*E/Z*) diastereoisomerism. Another factor is that in many intermolecular reactions, the photoexcited state involved is a triplet from which concerted cycloaddition is precluded because of the necessity for spin inversion.

In general, therefore, photochemical $2\pi + 2\pi$ cycloadditions do not have an inherent diastereoselectivity of the kind that is present in, e.g., Diels–Alder additions. However, if appropriate constraints are applied, this cycloaddition can display high *occasional* diastereoselectivity. Thus, photochemical *E/Z* diastereoisomerism can be eliminated by incorporation of the alkene into a ring of sufficiently small size (five-membered or less). By carrying out the reaction *intramolecularly*, there is a greater chance of reaction occurring via a singlet state and regioselectivity is more likely to be obtained using a tether of an appropriate length.

When both alkenes are contained in rings ($n \leqslant 5$) the cycloaddition is invariably *syn* on both of them although *exo* and *endo* products may still be formed as in the photochemical dimerisation of maleic anhydride (Scheme 1).[2] With unsymmetrical cyclic alkenes, head-to-head or head-to-tail

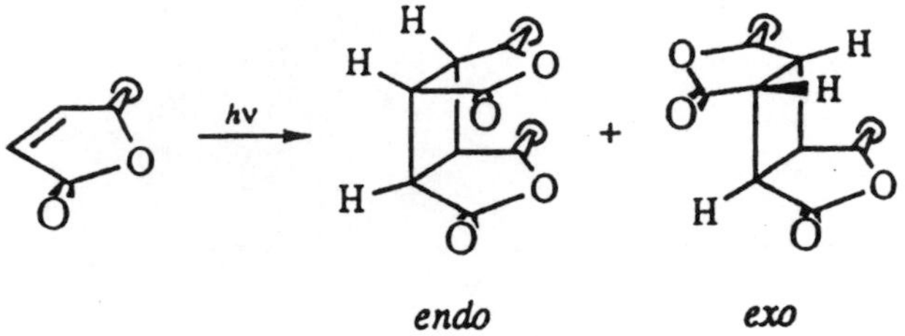

endo   exo

Scheme 1

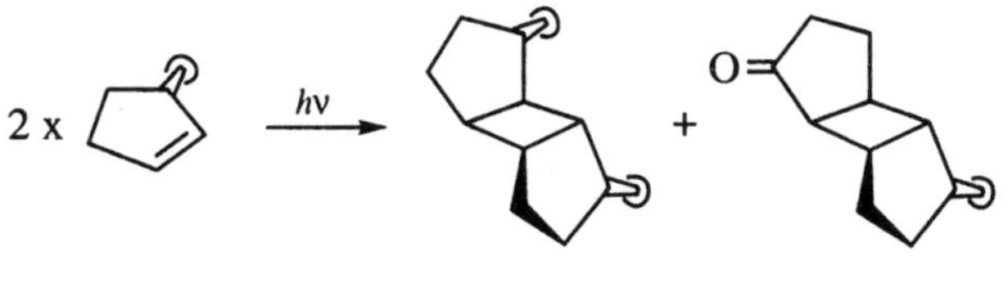

Scheme 2

dimerisation (regioisomerism) is also possible (Scheme 2)[3] (in addition to *exo/endo* addition).

Most stereoselective syntheses which make use of photochemical $2\pi + 2\pi$ reactions take advantage of the E/Z configurational stability of cyclic alkenes (Scheme 3). The use of α,β-unsaturated carbonyl compounds with their longer wavelength absorption allows their selective excitation in the presence of other alkenes [Scheme 3(b) and (c)].[4-6]

Ring opening of the strained cyclobutane intermediate or product is also a pervasive feature of syntheses involving these $2\pi + 2\pi$ photochemical cycloadditions as in the de Mayo reaction [Scheme 3(c)]. The direct creation of two adjacent quaternary chiral centres diastereoselectively, by *syn* addition to the alkene C=C bond with the formation of two C—C bonds as in Scheme 3(c), is a conversion for which there are few other methods available. Any lack of diastereoselectivity in addition to the enol moiety is immaterial since retroaldolisation destroys the chirality at these two centres.

Although photostimulated $2\pi + 2\pi$ cycloadditions involving *cyclohexenones* are not always reliably *syn* (Scheme 4), the strain present in the *trans*-fused diastereoisomer (**1**) allows its easy conversion into the *cis*-fused form (**2**) by base-catalysed equilibration.[7]

Scheme 5 uses a reductive cleavage of the photochemically-derived cyclobutane (**3**) as a model for a synthesis of ($\pm$)-10-epijuneol.[8]

Lineatin (**8**) (Scheme 6) is an aggregation pheromone of the female ambrosia beetle *Trypodendron lineatum*, a pest causing damage to sawn

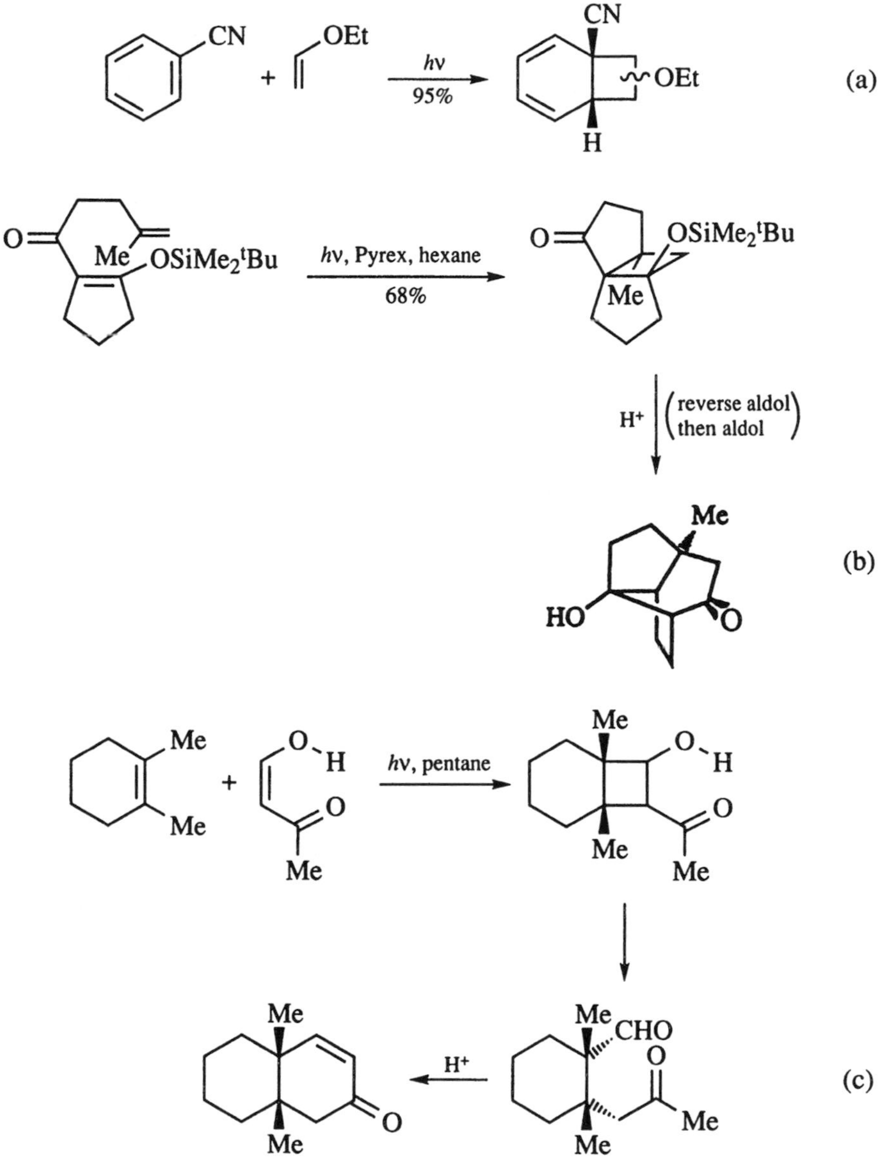

Scheme 3

timber. The obvious disconnection of this target molecule is to the retro-acetalisation product (**9**). However, White *et al.*[9] used the less obvious retrosynthetic step to give (**7**) as the target. Synthesis of (**7**) as in Scheme 6 uses $2\pi + 2\pi$ photoaddition of ethyne to (**4**) in a key step and also makes

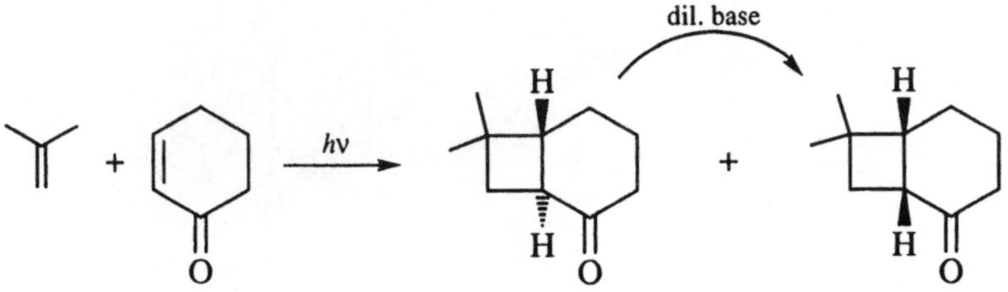

(1) 4 parts   **(2)** 1 part

Scheme 4

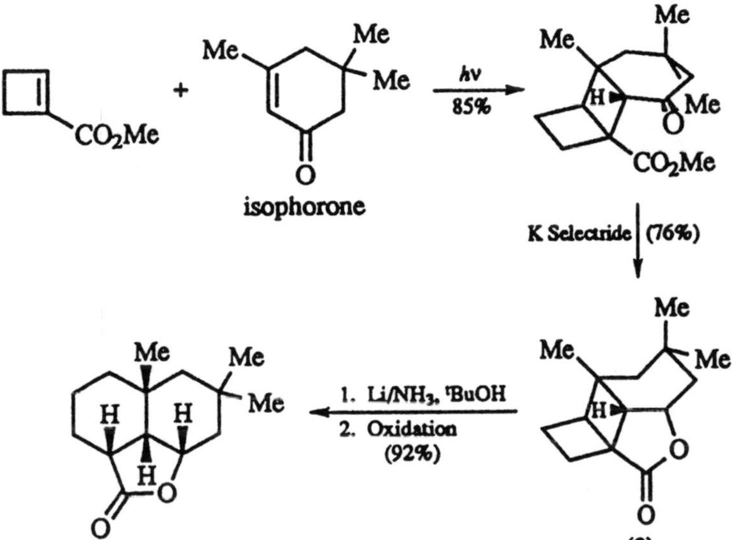

isophorone

K Selectride (76%)

1. Li/NH₃, 'BuOH
2. Oxidation
(92%)

(3)

$$\text{K-Selectride} = \text{K}\left(\text{C}_2\text{H}_5\overset{\overset{\text{CH}_3}{|}}{\text{CH}}\right)_3\text{BH}$$

Scheme 5

good use of the shape of the unsaturated [4.2.0] skeleton in directing the hydroboration of (**5**) and in DIBAL reduction of (**6**) (after tosylation). Intramolecular S_N2 displacement by the hemiacetal hydroxyl group gives ($\pm$) lineatin.[9]

Photochemical $2\pi + 2\pi$ cycloadditions of (cyclic) alkenes with carbonyl

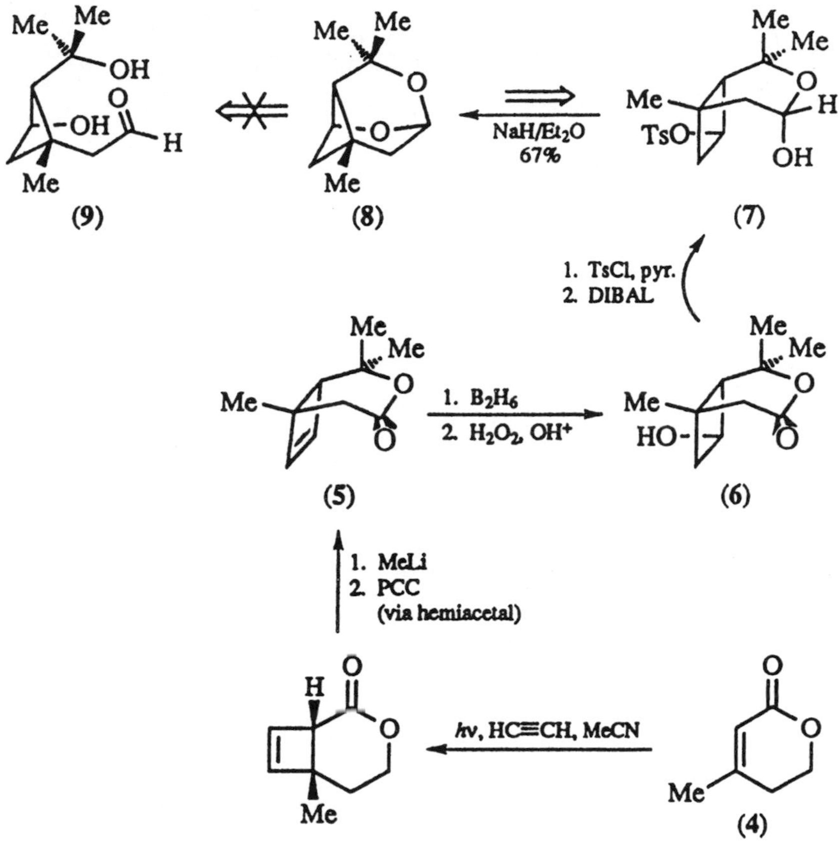

Scheme 6

groups (the Paterno–Büchi reaction) have been widely studied, Schreiber and Hoveyda[10] used the reaction in Scheme 7(a), carried out on a 30 g scale, in a synthesis of avenaciolide, an anti-fungal mould metabolite. Using aliphatic aldehydes, singlet excited states are involved and higher diastereoselectivity results than with, for example, aromatic ketones.

Again, the strain in the four-membered (oxetane) ring of the product is used to advantage. In Scheme 7(b) (Type 0) hydrolysis of the oxetane–acetal unit in (10) leads to a single aldol diastereoisomer (11). The shape of the bicyclo[3.2.0] system allows completely diastereoselective hydrogenation of (10) (Type III) from the more accessible face; after hydrolysis, the product is (12).[11]

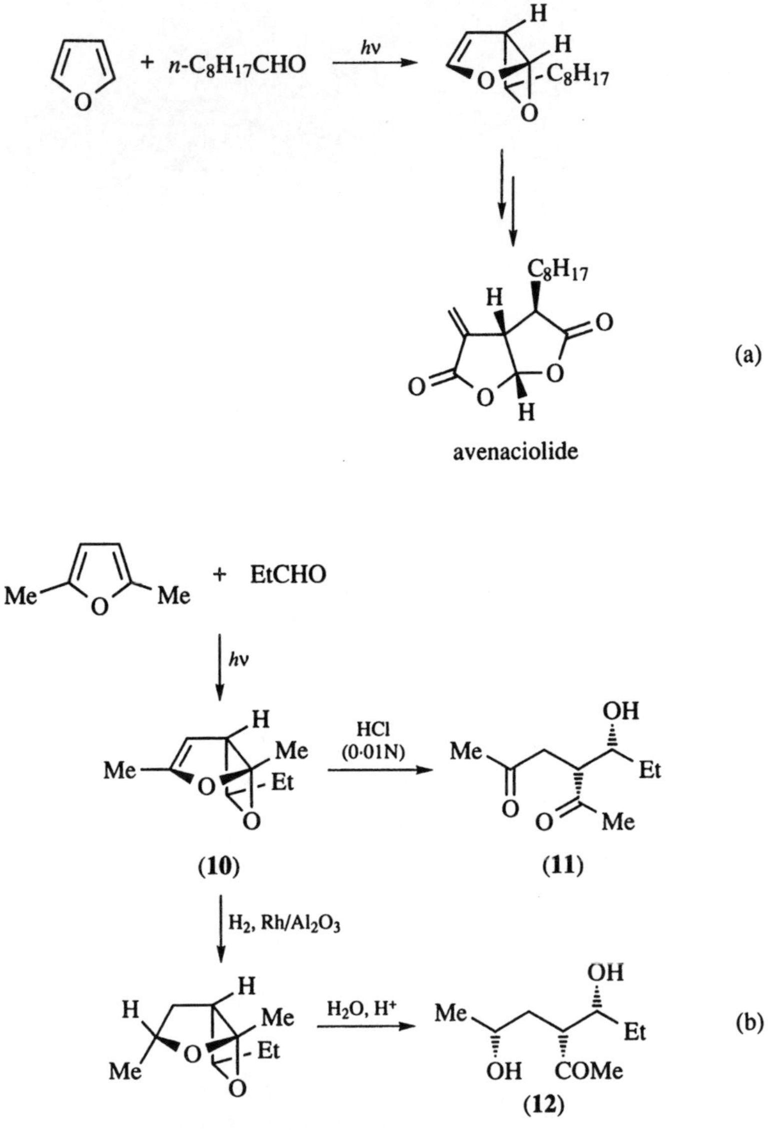

(a)

avenaciolide

(b)

Scheme 7

Type II photochemical 1,3-cycloaddition of arenes[12]

1,3-Cycloaddition of benzene and substituted benzenes to alkenes under the influence of light can be a reaction of high diastereoselectivity. In Scheme 8(a), *syn* addition to the alkene is accompanied by complete *endo* diastereoselectivity (methyl groups *cis* to the double bond). The reaction with substituted benzenes is also often highly regioselective [Scheme 8(b)].

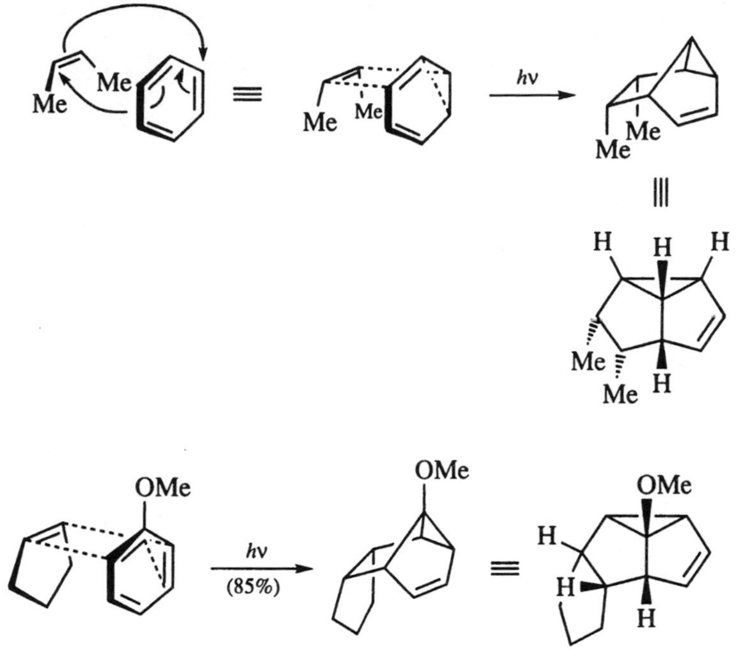

(a)

(b)

Scheme 8

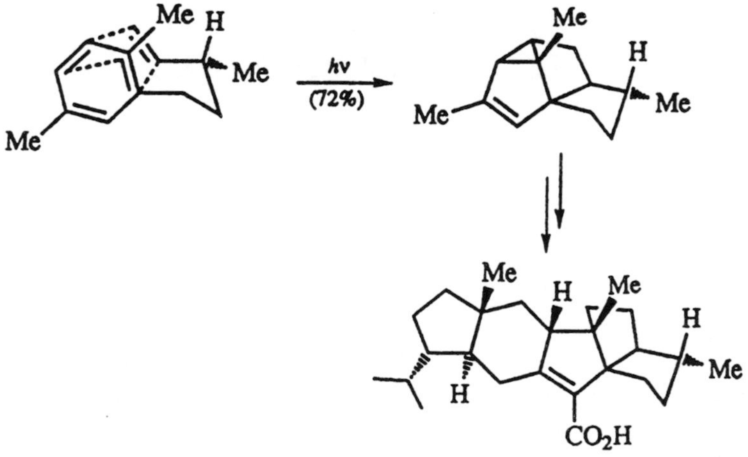

(–)-retigeranic acid

Scheme 9

This photochemical 1,3-cycloaddition, and particularly the intramolecular version of it, has been brilliantly exploited by Wender and his co-workers in syntheses of a number of natural products, e.g. Scheme 9, which is part of a synthesis of $(-)$-retigeranic acid.[13]

Photochemistry in crystals[14]

One of the advantages of photochemistry is that it can be carried out on compounds in the crystalline state as well as in solution.

In a crystal, molecules are stacked in a highly ordered way, usually in the same conformation, and there is little variation in the atomic distances separating nearest neighbours. For intermolecular reactions to occur, therefore, the separation between two molecules and the relative orientation of the reacting functional groups must be appropriate. When this is the case, completely diastereoselective reaction may ensue as in the photochemical dimerisation of hexadienoic acid and its derivatives (Scheme 10).[15] Photochemical reaction (sensitised) of these compounds in solution gives mixtures of diastereoisomers.

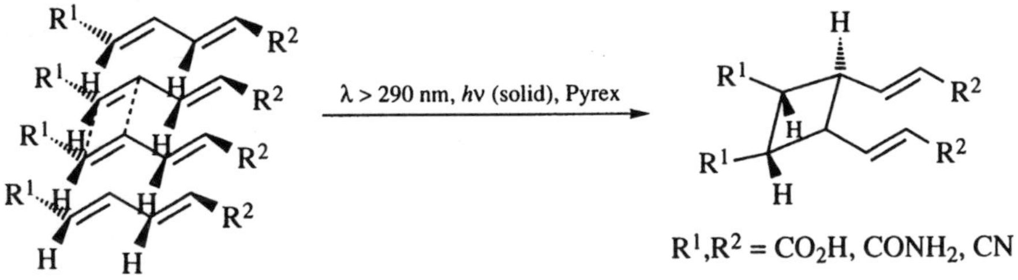

$\lambda > 290$ nm, $h\nu$ (solid), Pyrex

$R^1, R^2 = CO_2H, CONH_2, CN$

Scheme 10

Likewise, the photolysis of maleic anhydride (Scheme 1), when carried out in the crystalline phase, gives only the *endo* diastereoisomer.[3]

OTHER $2\pi + 2\pi$ CYCLOADDITIONS

Orbital symmetry considerations not only predict concerted suprafacial–suprafacial cycloaddition of two alkenes when one is in its excited state ($2\pi_s + 2\pi_s$) but also concerted cycloaddition of two ground state alkenes with one alkene acting antarafacially ($2\pi_s + 2\pi_a$).

In practice, alkenes fail to undergo concerted $2\pi_s + 2\pi_a$ cycloadditions, presumably because steric interactions between substituents on the alkene sp^2-carbons and other strain factors are too severe (Figure 1).

Reduction of this steric interaction would be expected if the substituents at one terminus of the alkene were absent which would be the case using a ketene (Figure 2). Ketenes do react with alkenes to give cyclobutanones but

Figure 1

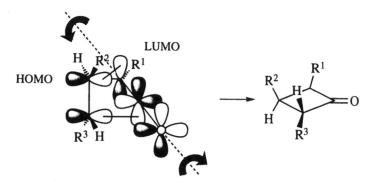

Figure 2

it is not clear that this cycloaddition is a simple $2\pi_s + 2\pi_a$ with the ketene acting antarafacially. This is because the orthogonal π^*-orbital of the carbonyl group in the ketene may be involved, making the reaction a $2\pi + 2\pi + 2\pi$ type.[16]

Although ketene-alkene cycloadditions are inherently *syn* diastereoselective (*syn* stereospecific) in addition to the alkene (Scheme 11) there is evidence to suggest that in the transition state, formation of the bond to the electron-deficient ketene carbonyl has run ahead of that between the other two carbons, i.e. the reaction has dipolar character.[17]

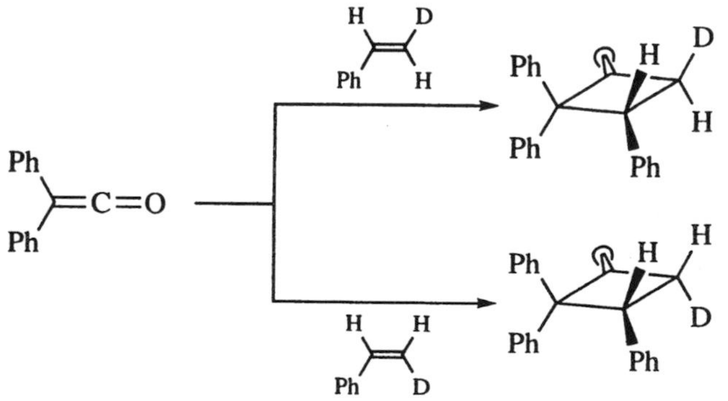

Scheme 11

Wang and Houk[18] have calculated that the transition state of lowest energy in the cycloaddition of ethene and ketene is that shown in Figure 3, which illustrates that whilst the C_1—C_3 bond is almost fully formed, the C_2—C_4 bond is far less developed.

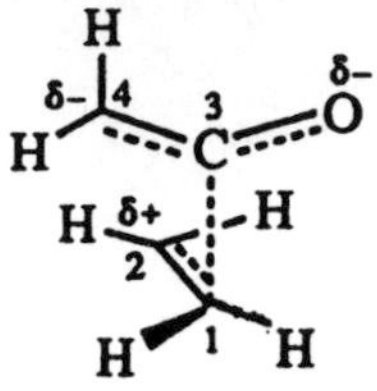

Figure 3

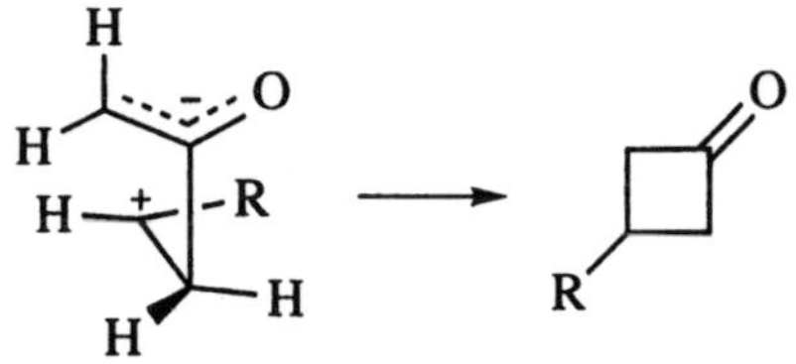

Scheme 12

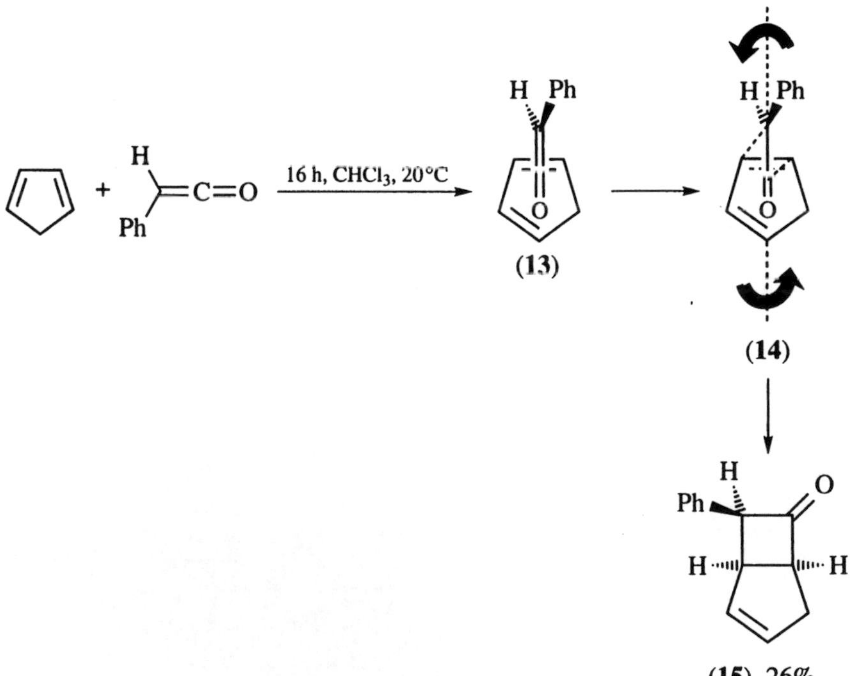

Scheme 13

Asynchronous bonding in the transition state helps to explain the regiochemistry of this reaction; the alkene carbon less able to sustain the partial positive charge becomes bonded to the ketene carbonyl carbon (Schemes 12 and 13).

Antarafacial addition of the ketene as in Figure 2 also explains a subtle stereochemical feature of these $2\pi_s + 2\pi_a$ cycloadditions which leads to the larger ketene substituent being located in the more sterically crowded environment in the product. Thus, addition of phenylketene to cyclopentadiene (Scheme 13) gives an excess of the bicyclo[3.2.0]heptenone diastereoisomer (15) in which the phenyl group is *endo*.[19]

This stereochemical outcome is a consequence of the preferred approach of the ketene in (13), in which the hydrogen of the ketene is directed towards the diene, coupled with the required direction of twisting around the ketene C=C bond as shown in (14). Like the Diels–Alder reaction, therefore, the diastereoselectivity of the $2\pi_s + 2\pi_a$ addition of an alkene and a ketene can have an inherent component (*syn* addition to the alkene) and an occasional

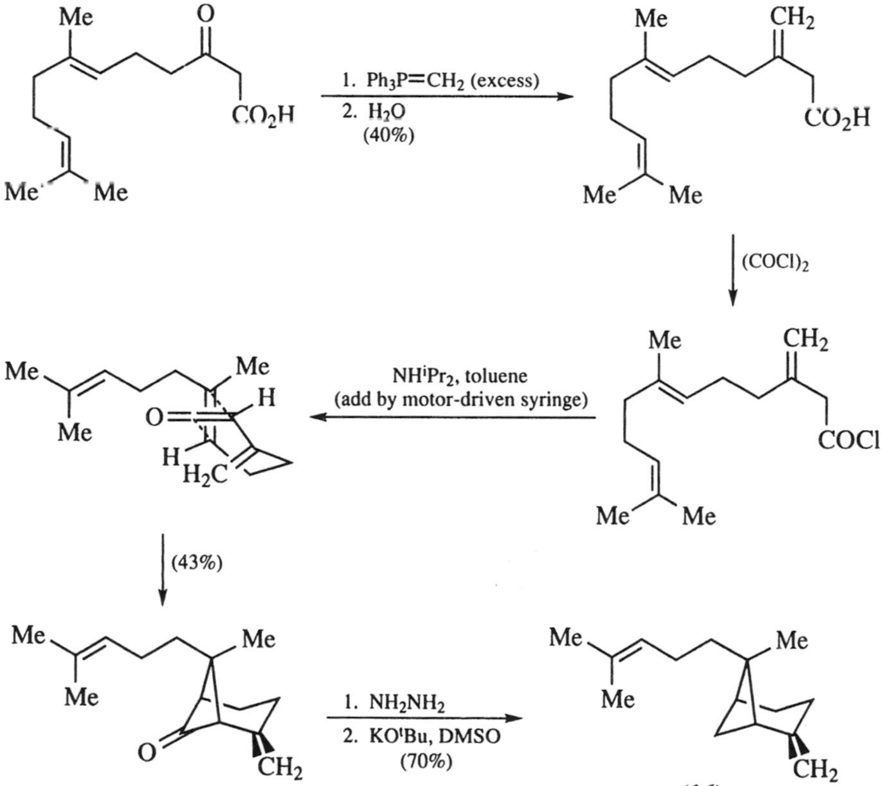

Scheme 14

component (*endo/exo* addition). The *endo* diastereoselectivity which is found
in Scheme 13 cannot be expected in all cases, particularly where the ketene is
disubstituted with groups of a similar bulk.

Intramolecular ketene–alkene cycloadditions[20] have been used to
synthesise four-membered ring-containing products, e.g. β-*trans*-
bergotamene (**16**) (Scheme 14).[21]

Keteniminium salts (**17**) can also play the role of the antarafacial
component in $2\pi_s + 2\pi_a$ cycloadditions (Scheme 15).[22]

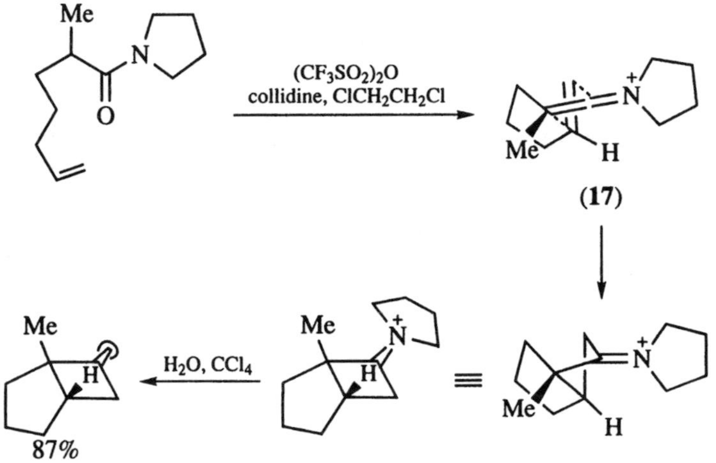

Scheme 15

Other cycloadditions

In recent years, considerable effort has been directed towards developing
cycloaddition reactions which would give carbocyclic five-membered rings
with the facility and stereoselectivity with which the Diels–Alder reaction
gives six-membered rings. This effort has been necessary because of the
dearth of available all-carbon 1,3-dipoles and their generally inefficient
reactions with alkenes to give five-membered rings. An all-carbon 1,3-dipole
which might be expected to have enhanced stability would be one in which a
methylene group was present at C-2 (**18**). However, the species (**20**)
generated by decomposition of the azo compound (**19**) is biradicaloid in its
behaviour rather than dipolar and, whilst addition to alkenes to give five-

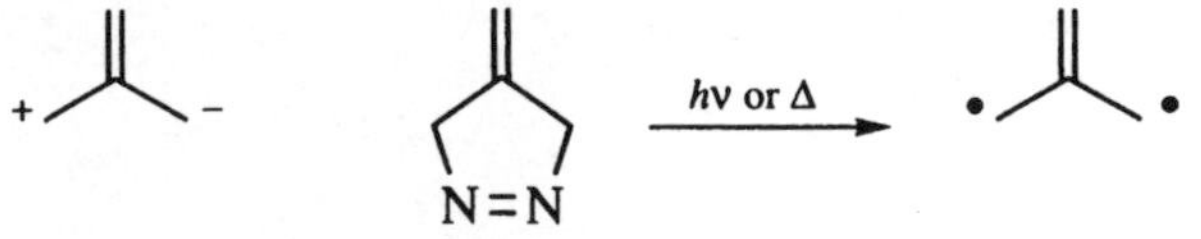

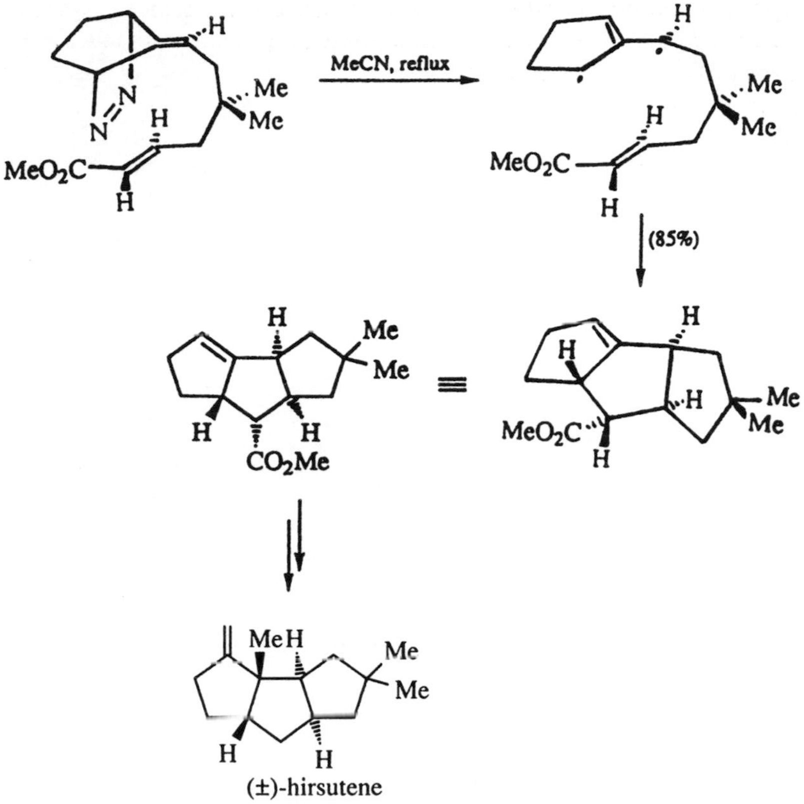

Scheme 16

membered rings occurs, diastereo- and regioselectivity are high only when the reaction is intramolecular as in Scheme 16, in which the tricyclic skeleton of hirsutene is formed.[23]

Trost[24] has successfully used palladium complexes of the 1,3-dipole (**18**) in cycloaddition reactions. These are generated *in situ*, e.g. by palladium-induced decomposition of 2-(trimethylsilylmethyl)allyl acetate (Scheme 17).

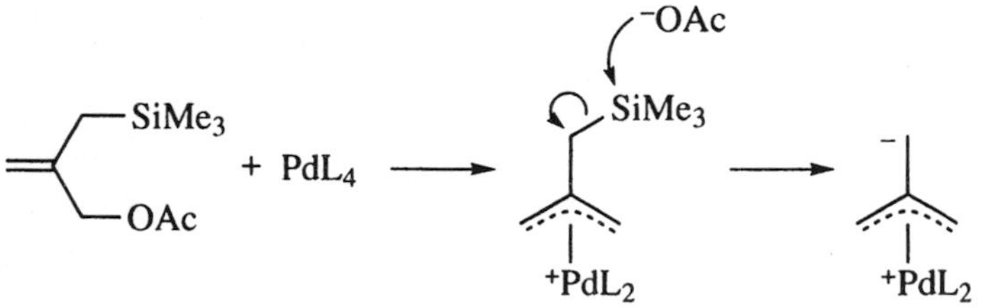

Scheme 17

These trimethylenemethane intermediates cycloadd efficiently to double bonds which are conjugated with at least one electron-withdrawing group (Scheme 18).

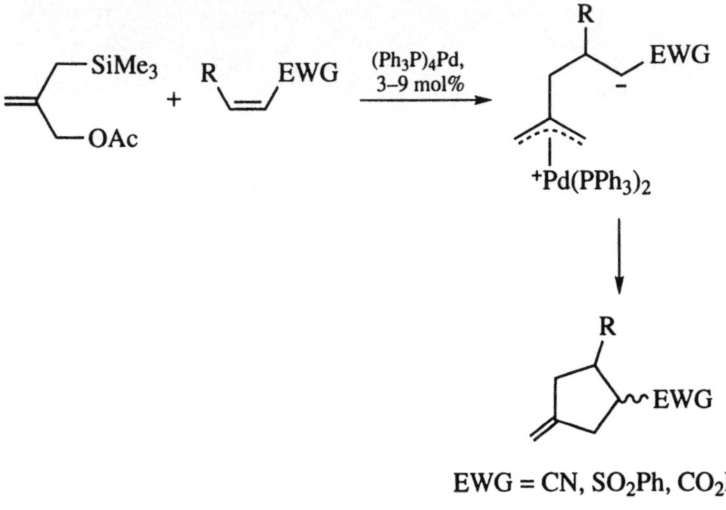

EWG = CN, SO$_2$Ph, CO$_2$R′

Scheme 18

Although high diastereoselectivity is obtained in addition to *trans*-alkenes, the configuration of *cis*-alkenes is incompletely retained in the product (Scheme 18). It appears, therefore, that the cycloaddition is not concerted and that formation of one σ-bond is complete before any formation of the second σ-bond has taken place allowing intervention of some C—C bond

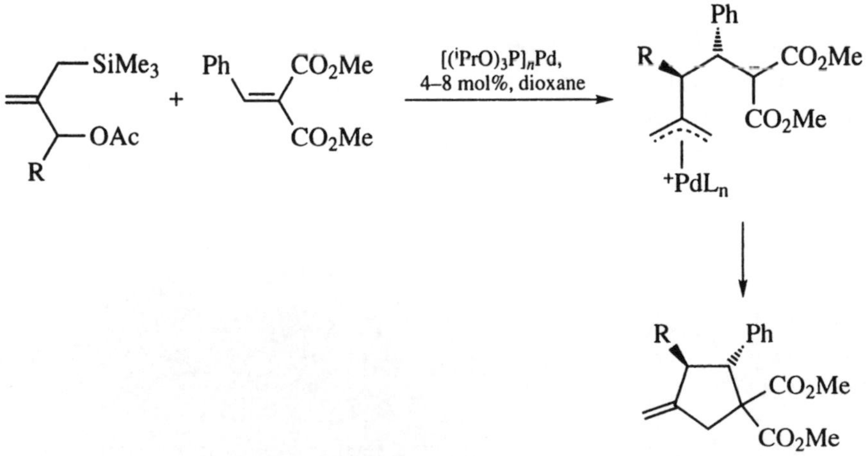

Scheme 19

rotation around the original double bond of the *cis*-alkene.

The diastereoselectivity of this cycloaddition, therefore, is occasional. Of course, configurational stability of *cis*-alkenes is assured by containing them in an eight-membered or smaller ring. In these cases, addition of the complexed 1,3-dipole is highly *syn* selective.

In Scheme 19, in which a substituted version of the palladium-complexed 1,3-dipole is generated, the cycloaddition is highly diastereoselective when R is electron-withdrawing (CN, COEt), giving a *trans* relationship between R and Ph; the diastereoselectivity is less when R is electron donating.[24] However, if the exocyclic methylene group in the product is removed by ozonolysis, thermodynamic equilibration then generates the *trans* diastereoisomer (**21**) as in Scheme 20, the first steps in a synthesis of rocaglamide.[25]

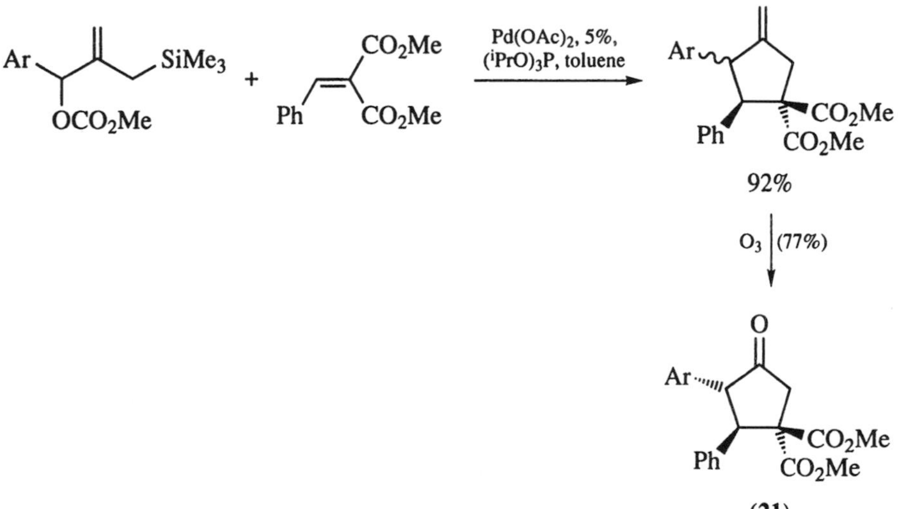

Scheme 20

1,4-CHELETROPIC ADDITIONS TO 1,3-DIENES

Addition of sulphur dioxide across a 1,3-diene is a cheletropic addition in which two new bonds to sulphur are made concertedly and the diene reacts suprafacially.[26]

In this reaction, therefore, using a diene having two prochiral double bonds, two chiral centres are created of defined relative configuration (inherent diastereoselectivity). Unfortunately, there is a dearth of molecules containing an atom (S in SO_2) having the filled and empty orbitals necessary for cheletropic addition (singlet carbenes do not, in general react with dienes by 1,4-addition: they add 1,2 to one of the double bonds). The reaction in Scheme 21 is, in practice, more often encountered in the reverse elimination mode, i.e. as a route to configurationally defined dienes (Type IV) from the corresponding dihydrothiophene dioxides (22) and (23), prepared by other means.[27]

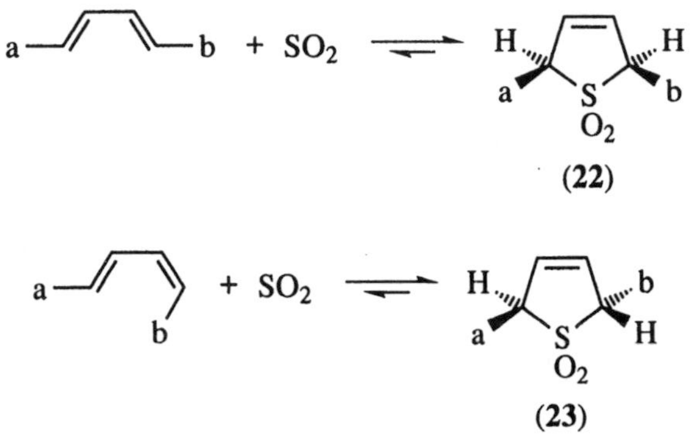

(22)

(23)

Scheme 21

OTHER 1,4-ADDITIONS TO 1,3-DIENES

Bäckvall et al.[28] have shown that cyclic and acyclic dienes can be converted completely diastereoselectively into 1,4-addition products using palladium chemistry. Thus cyclohexa-1,3-diene is converted into cis-1-acetoxy-4-chlorocyclohex-2-ene (24), cis-1,4-diacetoxycyclohex-2-ene (25) or trans-1,4-diacetoxycyclohex-2-ene (26), depending on the conditions (Scheme 22).

The mechanism suggested for cis-1,4-acetoxychlorination (Scheme 23) involves coordination of the diene to palladium followed by anti addition of acetate to give a π-allyl complex (27). Coordination by benzoquinone is followed by external anti attack of chloride to give overall syn addition. Similar mechanisms account for the formation of (25) and (26) with the acetoxy group being intramolecularly delivered from palladium in the latter case.

These products of 1,4-addition to conjugated dienes are versatile starting materials for nucleophilic substitution reactions with retention [using further Pd(0) chemistry; see Chapter 4] or inversion of configuration as exemplified in Scheme 24.

Scheme 25 illustrates an application of this tandem 1,4-addition–nucleophilic substitution in a synthesis of (28), a sex pheromone of the carpenter bee.[29]

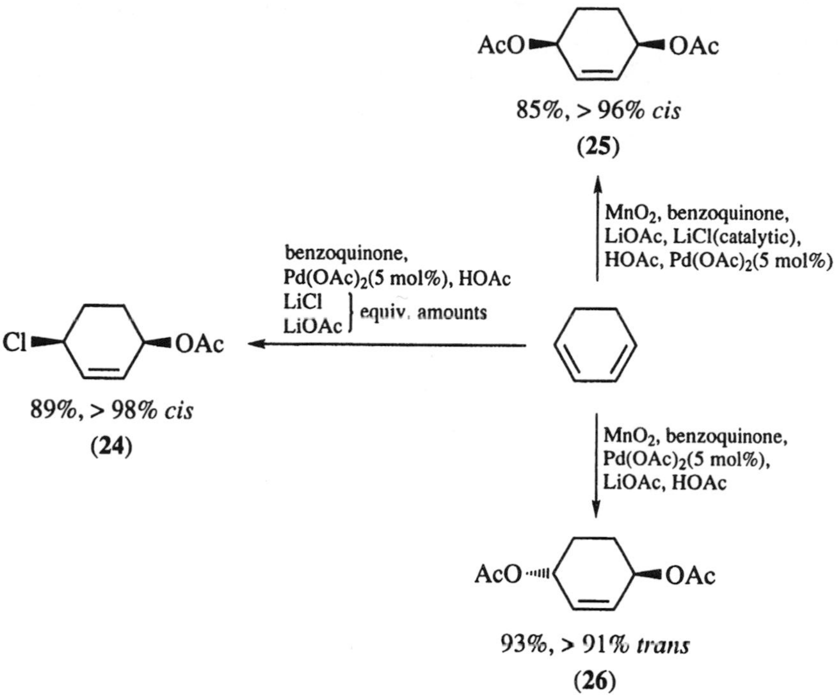

AcO——OAc

85%, > 96% *cis*
(**25**)

MnO$_2$, benzoquinone,
LiOAc, LiCl(catalytic),
HOAc, Pd(OAc)$_2$(5 mol%)

benzoquinone,
Pd(OAc)$_2$(5 mol%), HOAc
LiCl ⎱ equiv. amounts
LiOAc ⎰

Cl——OAc

89%, > 98% *cis*
(**24**)

MnO$_2$, benzoquinone,
Pd(OAc)$_2$(5 mol%),
LiOAc, HOAc

AcO····——OAc

93%, > 91% *trans*
(**26**)

Scheme 22

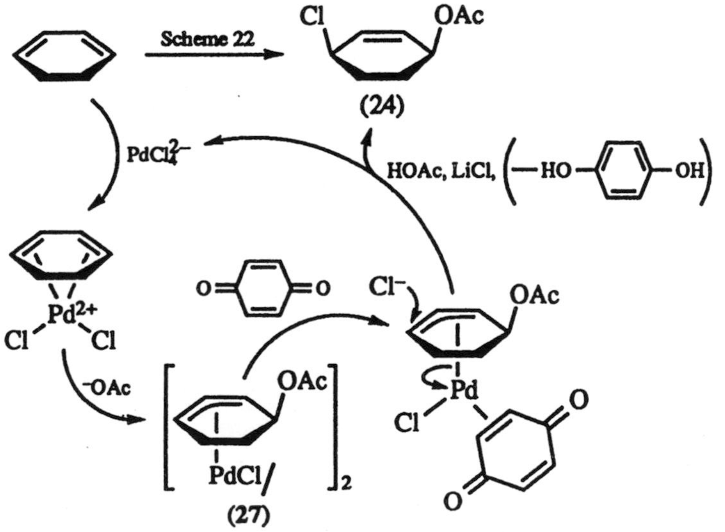

Cl   OAc

(**24**)

PdCl$_4^{2-}$

HOAc, LiCl, (—HO——OH)

Pd^{2+}
Cl   Cl

O——O   Cl$^-$ OAc

–OAc

OAc

PdCl /   ]$_2$
(**27**)

Pd
Cl

O

O

Scheme 23

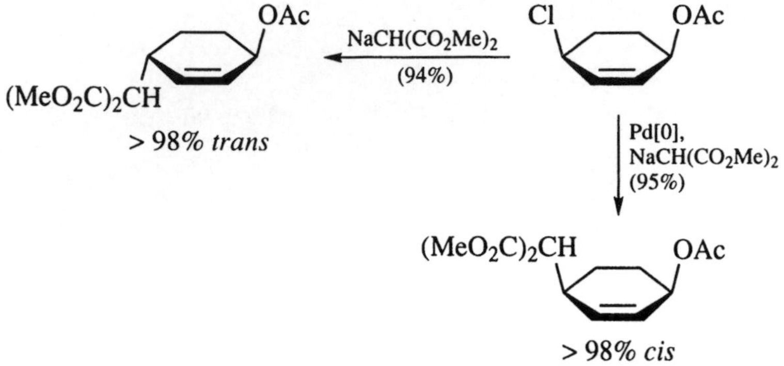

Scheme 24

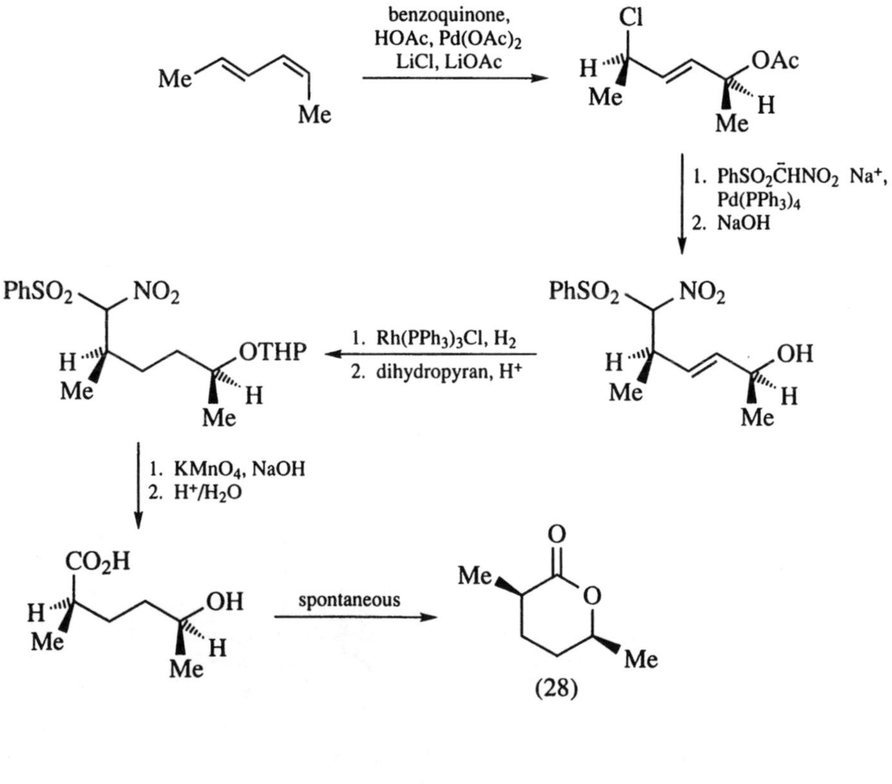

Scheme 25

[3,3] Sigmatropic rearrangements[30]

Simple diastereoselectivity in (Type II) [3,3] sigmatropic rearrangements results in the transformation of two configured double bonds, separated by two saturated atoms, into two adjacent chiral centres as illustrated for the Claisen rearrangement in Scheme 26.

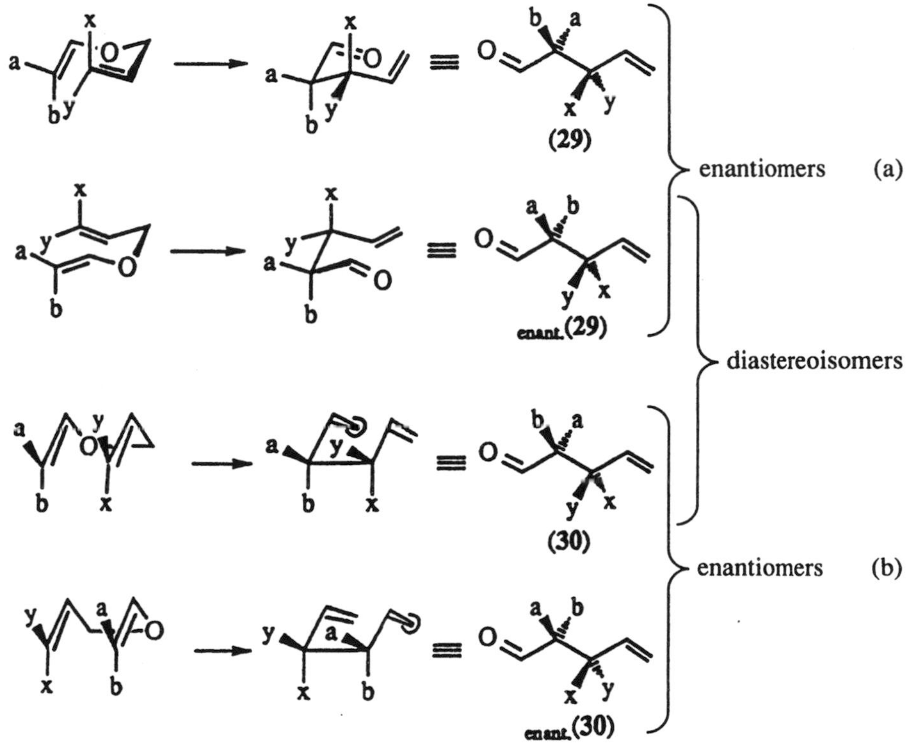

Scheme 26

The chair transition state is usually sufficiently lower in energy than the boat for formation of (29) [+ $_{enant.}$(29)] to be favoured over formation of (30) (+ $_{enant.}$(30)] by a factor of 20:1 or more.

A change in configuration at one double bond in Scheme 26(a) would lead, after rearrangement via a chair transition state, to (30) (+ $_{enant.}$(30)]. Accordingly, the use made of Type II [3,3] rearrangements to prepare two adjacent chiral centres will depend on the availability of 1,5-dienes, configurationally defined at both double bonds (Type IV reactions).

For the much-used Claisen rearrangement of allyl vinyl ethers, the more difficult bond to obtain configurationally defined is, in general, that of the vinyl ether because of the dearth of methods for their diastereoselective synthesis.

The Ireland modification of the Claisen rearrangement (Chapter 5) is a valuable variant, not least because of the much lower temperature required. Diastereoselectivity in this variant requires the generation and silylation of a configurationally defined enolate (Scheme 27) in a molecule already containing a configurationally pure allyl double bond.

In Scheme 27, the (E)-enolates are generated under kinetically controlled

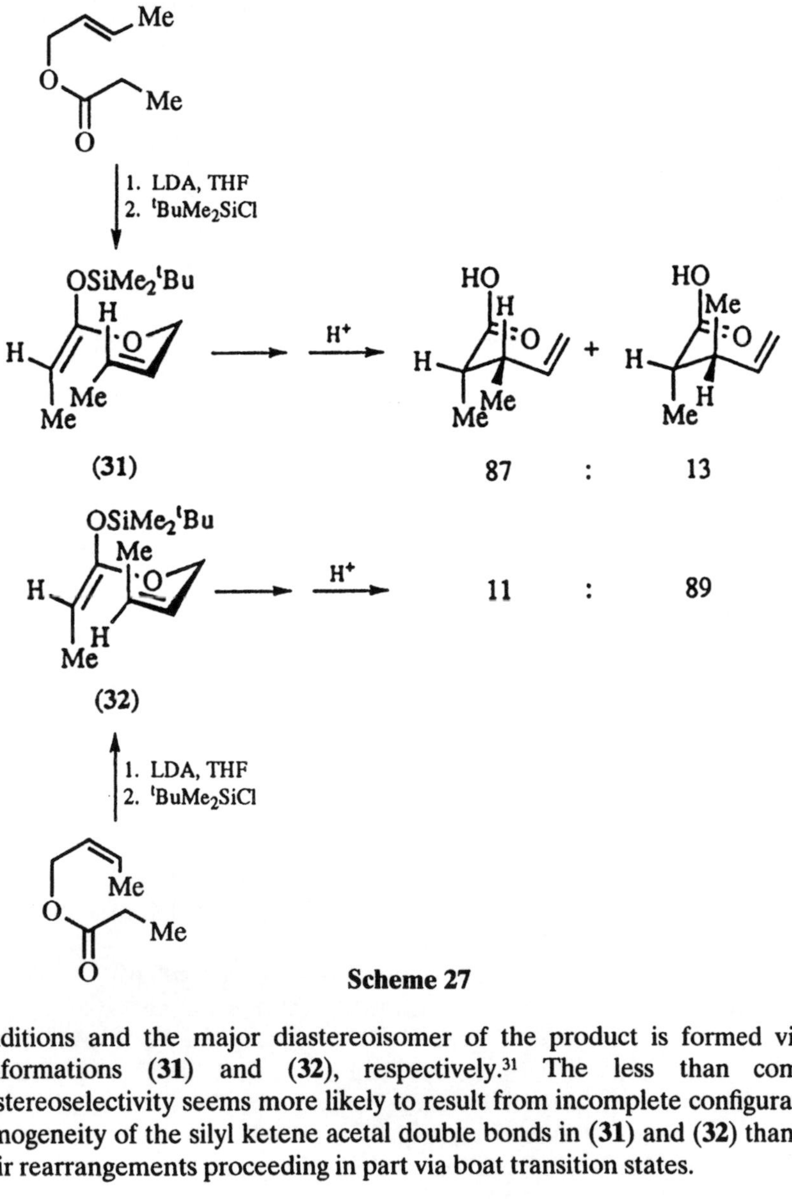

Scheme 27

conditions and the major diastereoisomer of the product is formed via the conformations (31) and (32), respectively.[31] The less than complete diastereoselectivity seems more likely to result from incomplete configurational homogeneity of the silyl ketene acetal double bonds in (31) and (32) than from their rearrangements proceeding in part via boat transition states.

The use of dipolar aprotic cosolvents (HMPA or DMPU) in Scheme 27 favours the formation of enolates having the alternative configurations to those shown and thus products with the opposite sense of diastereoselectivity. The problem of specific enolate generation is one to which we will return in Chapter 8.

[2,3] Sigmatropic rearrangements

Simple diastereoselectivity (Type II) in [2,3] rearrangements has been investigated in some detail by Nakai and Mikami.[32] It is suggested that in the Wittig rearrangement in Scheme 28 the diastereoselectivity is controlled by the balance between two factors. Thus, using the envelope conformation (33), there is an eclipsing interaction between R^2 and the hydrogen at C-2, whereas in conformation (33'), there is a *gauche* interaction between R^1 and R^2.

Using an *E*-configured alkene, these two interactions appear to be comparable and diastereoselectivity is not high, although the balance can be tipped by, for example, introducing a bulky substituent at C-2 [Scheme 29(a)]. In this case, reaction from conformation (34) is favoured.

For *Z*-configured alkenes, the stereoselectivity is higher and favours the formation of the *cis* diastereoisomer with, for example, a preference for reaction from conformation (35) in Scheme 29(b).

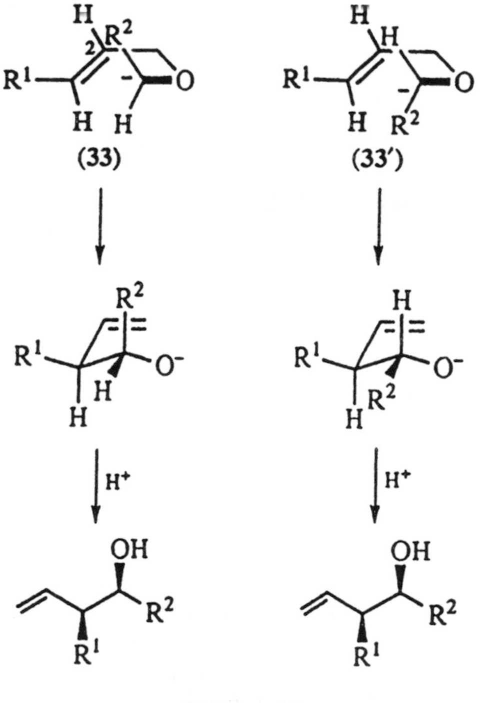

Scheme 28

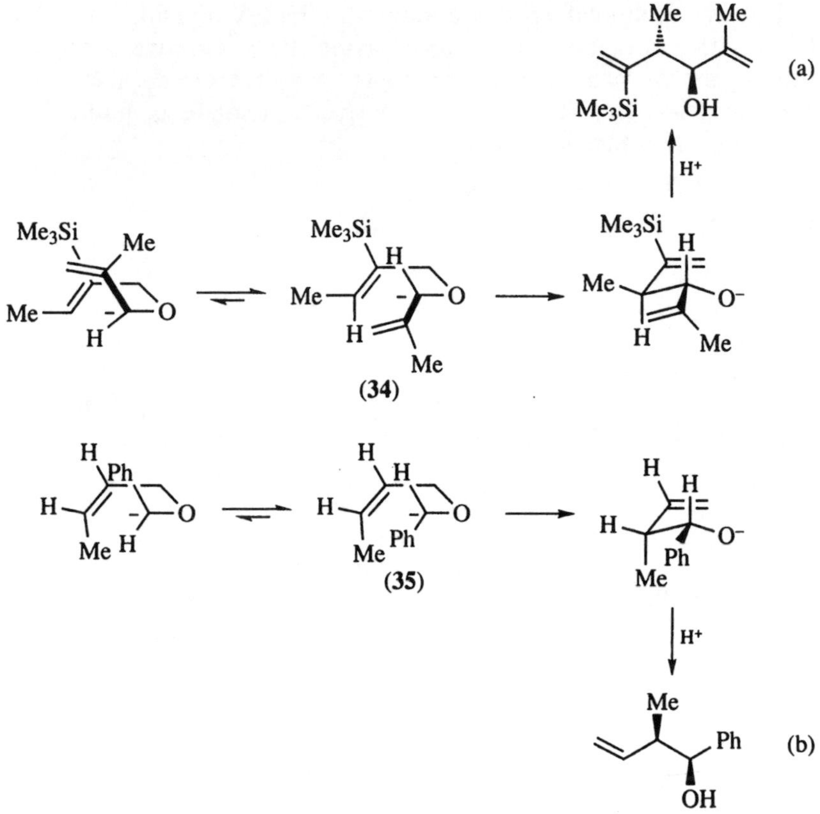

Scheme 29

Remarkably high diastereoselectivity with both (*E*)- and (*Z*)-alkenes is found using ethynyl groups as substituents on the oxymethylene group in the Wittig rearrangement in Scheme 30.

Calculations by Houk and co-workers[33] suggest that the preferred transition-state geometry for the Wittig rearrangement is the alternative envelope shown in (36), but this was only located when the lithium ion was included in the calculation. The preference of an ethynyl group for the *exo* position in this envelope does not arise, according to Houk's calculations, from avoidance of an eclipsing interaction between the *endo*-ethynyl group and the C_2-H.

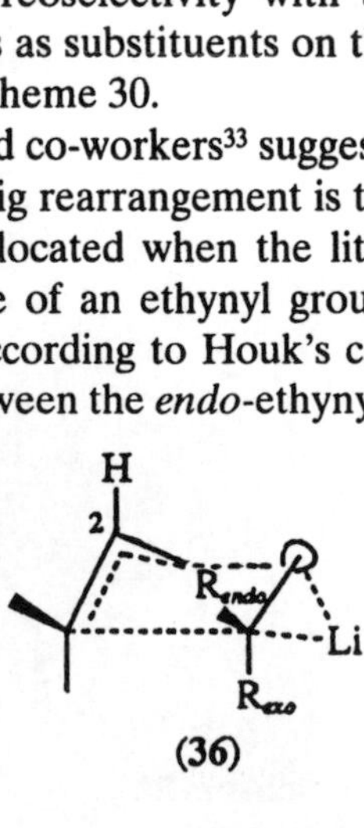

(36)

Whether the envelope conformation suggested by Nakai and Mikami or that in (**36**) (see also Chapter 5) is preferred for the transition state of the [2,3] sigmatropic rearrangement remains an open question.

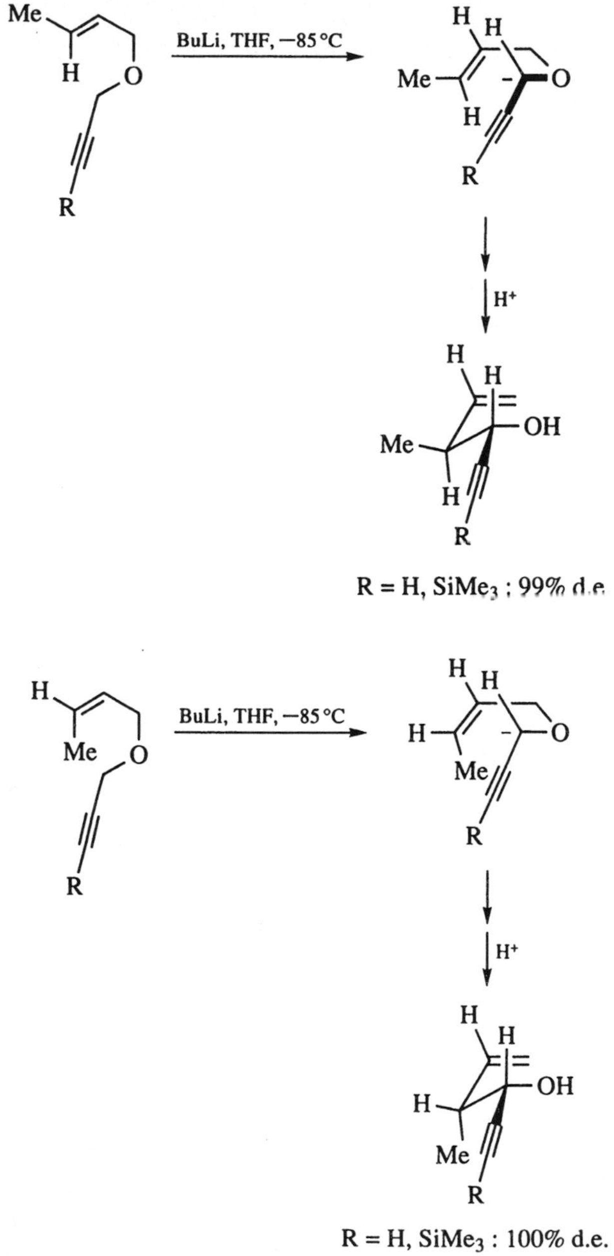
BuLi, THF, −85 °C
H⁺
R = H, SiMe3 ; 99% d.e.
BuLi, THF, −85 °C
H⁺
R = H, SiMe3 : 100% d.e.

Scheme 30

Summary

Type II photochemical $2\pi_s + 2\pi_s$ cycloadditions involving alkenes are normally reliably diastereoselective only when the photostimulated alkene is contained in a ring of less than six members.

Cyclobutanes and oxetanes resulting from such $2\pi + 2\pi$ cycloadditions are useful because of the ease with which their ring opening can be accomplished.

Type II thermal cycloaddition of ketenes and alkenes can be highly diastereoselective since the ketene can apparently function as the $2\pi_a$ component in a $2\pi_s + 2\pi_a$ cycloaddition.

Cycloaddition of (substituted) trimethylenemethane palladium complexes with alkenes gives five-membered rings, but the two carbon–carbon bonds are not formed simultaneously and the reaction is not inherently diastereoselective.

Some highly diastereoselective 1,4-additions to 1,3-dienes can be accomplished using palladium chemistry.

Type II [3,3] sigmatropic rearrangements are usually highly diastereoselective and proceed via chair transition states, but depend on the availability of 1,5-dienes with defined configurations for the two double bonds.

Some Type II [2,3] Wittig rearrangements are highly diastereoselective, particularly those with an ethynyl substituent on the oxymethylene group and an E- or Z-disubstituted double bond.

References

1. (a) A.C. Weedon, in *Synthetic Organic Photochemistry*, ed. W.M. Horspool, Plenum Press, New York, 1984; (b) J.D. Coyle (ed.), *Photochemistry in Organic Synthesis*, Special Publication No. 57, Royal Society of Chemistry, London; (c) I. Ninomiya and T. Naito, *Photochemical Synthesis*, Academic Press, London, 1989; (d) W. Oppolzer, *Acc. Chem. Res.*, 1982, **15**, 135; (e) M.T. Crimmins, *Chem. Rev.*, 1988, **88**, 1453.
2. J.M. Coxon and B. Halton, *Organic Photochemistry*, Cambridge University Press, Cambridge, 1974, p. 121.
3. P.E. Eaton, *J. Am. Chem. Soc.*, 1962, **84**, 2344; W.C. Agosta and S. Wolff, *Pure and Appl. Chem.*, 1982, **54**, 1579.
4. A. Gilbert and J. Baggott, *Essentials of Molecular Photochemistry*, Blackwell, Oxford, 1991, p. 373.
5. G. Pattenden and G.M. Robertson, *Tetrahedron Lett.*, 1986, **27**, 399.
6. S.W. Baldwin and R.E. Gawley, *Tetrahedron Lett.*, 1975, 3969.
7. E.J. Corey, R.B. Mitra and H. Uda, *J. Am. Chem. Soc.*, 1963, **85**, 362.
8. P.A. Wender and J.C. Lechleiter, *J. Am. Chem. Soc.*, 1978, **100**, 4321.
9. J.D. White, M.A. Avery and J.P. Carter, *J. Am. Chem. Soc.*, 1982, **104**, 5486.
10. S.L. Schreiber and A.H. Hoveyda, *J. Am. Chem. Soc.*, 1984, **106**, 7200.
11. S.L. Schreiber, A.H. Hoveyda and H.-J. Wu, *J. Am. Chem. Soc.*, 1983, **105**, 660.

12. P.A. Wender and T.W. Von Geldern, in ref. 1(b), p. 226; D. Bryce-Smith, *Pure Appl. Chem.*, 1968, **16**, 47.
13. P.A. Wender and S.K. Singh, *Tetrahedron Lett.*, 1990, **31**, 2517.
14. V. Ramamurthy and K. Venkatesan, *Chem. Rev.*, 1987, **87**, 433.
15. M. Lahav and G.M.J. Schmidt, *J. Chem. Soc. B*, 1967, 312; B.S. Green, M. Lahav and G.M.J. Schmidt, *J. Chem. Soc. B*, 1971, 1552.
16. E. Valenti, M.A. Pericas and A. Moyano, *J. Org. Chem.*, 1990, **55**, 3582.
17. J.E. Baldwin and J.A. Kapeck, *J. Am. Chem. Soc.*, 1970, **92**, 4874.
18. X. Wang and K.N. Houk, *J. Am. Chem. Soc.*, 1990, **112**, 1754.
19. M. Rey, S. Roberts, A. Dieffenbacher and A.S. Dreiding, *Helv. Chim. Acta*, 1970, **53**, 417; P.R. Brook, J.M. Harrison and A.J. Duke, *Chem. Commun.*, 1970, 589.
20. Review of intramolecular $2\pi_s + 2\pi_a$; B.B. Snider, *Chem. Rev.*, 1988, **88**, 793.
21. E.J. Corey and M.C. Desai, *Tetrahedron Lett.*, 1985, **26**, 3535; B.B. Snider, R.A.H.F. Hui and Y.S. Kulkarni, *J. Am. Chem. Soc.*, 1985, **107**, 2194.
22. I. Marko, B. Ronsmans, A.-M. Hesbain-Frisque, S. Dumas, L. Ghosez, B. Ernst and H. Greuter, *J. Am. Chem. Soc.*, 1985, **107**, 2192.
23. R.D. Little and G.W. Muller, *J. Am. Chem. Soc.*, 1981, **103**, 2744.
24. B.M. Trost, *Angew. Chem., Int. Ed. Engl.*, 1986, **25**, 1.
25. B.M. Trost, P.D. Greenspan, B.V. Yang and M.G. Saulnier, *J. Am. Chem. Soc.*, 1990, **112**, 9022.
26. W.L. Mock, *J. Am. Chem. Soc.*, 1970, **92**, 6918, and references cited therein.
27. See, for example, S.F. Martin, C. Tu, M. Kimura and S.H. Simonsen, *J. Org. Chem.*, 1982, **47**, 3634.
28. J.E. Bäckvall, J.O. Vagberg and K.L. Granberg, *Tetrahedron Lett.*, 1989, **30**, 617; J.E. Bäckvall, J.E. Nyström and R.E. Nordberg, *J. Am. Chem. Soc.*, 1985, **107**, 3676.
29. J.E. Bäckvall, S.E. Bystrom and J.E. Nyström, *Tetrahedron*, 1985, **41**, 5761.
30. F.E. Ziegler, *Chem. Rev.*, 1988, **88**, 1423.
31. R.E. Ireland, R.H. Mueller and A.K. Willard, *J. Am. Chem. Soc.*, 1976, **98**, 2868; R.E. Ireland, P. Wipf and J.D. Armstrong, *J. Org. Chem.*, 1991, **56**, 650; S. Pereira and M. Srebnik, *Aldrichim. Acta*, 1993, **26**, 17.
32. T. Nakai and K. Mikami, *Chem. Rev.*, 1986, **86**, 885.
33. Y.-D. Wu, K.N. Houk and J.A. Marshall, *J. Org. Chem.*, 1990, **55**, 1421.

8 TYPE II REACTIONS: SIMPLE DIASTEREOSELECTIVITY IN THE ALDOL AND RELATED REACTIONS

Two particularly important Type II reactions in stereoselective synthesis are the aldol and the related reactions of alkenylmetal or alkenylmetalloids with aldehydes (Scheme 1). In both of these reactions, the disastereoselectivity is wholly occasional (see Chapter 6), i.e. neither of these reactions is inherently diastereoselective.

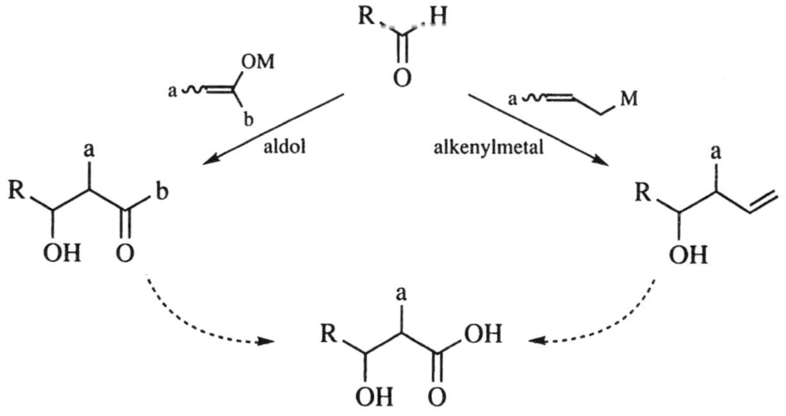

Scheme 1

Both reactions involve the union of two sp^2-hybridised carbons and when these carbons are contained in prochiral double bonds, two chiral centres are formed as in Scheme 1. The products from the two reactions can be interrelated by, e.g., conversion to the hydroxy acid (1), since reliable methods for the necessary functional group interconversions are available. The reactions have much in common stereochemically.

Type II aldol reactions

VIA CLOSED TRANSITION STATES

The aldol is the reaction of an enolate, usually derived from a ketone, ester or amide, with an aldehyde or ketone. In many cases it is believed to proceed preferentially by a (closed) chair transition state (Scheme 2) which is reminiscent of the preferred transition state for the [3,3] sigmatropic rearrangement (Chapter 5).[1]

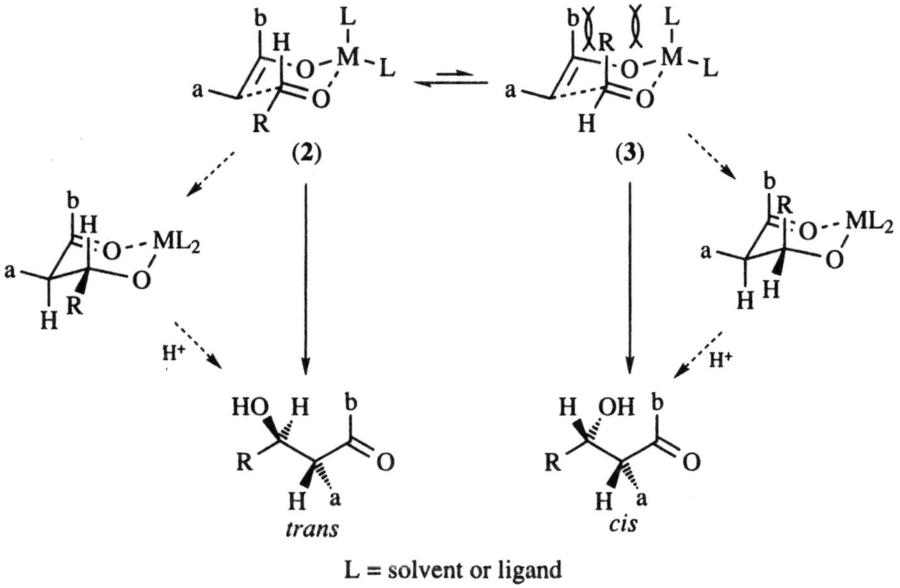

L = solvent or ligand

Scheme 2

As indicated in Scheme 2, unfavourable interactions of the R group of the aldehyde, when axial, with both b and L raise the energy of the chair (3) relative to the alternative (2) (R equatorial). The preferred product is, therefore, that having the *trans* configuration (*cis* and *trans* refer to the relative positions of OH and a when the backbone of the aldol is drawn in the zigzag form as shown).[2] It should be clear from Scheme 2 why ketones will, in general, be less diastereoselective than aldehydes in their reactions with enolates.

The role of M in these aldol reactions can be taken by a variety of elements including Li, Na, Zn, B, Al, Mg, Zr, Ti, Sn and Si (siliconates).

One particular factor which can be expected to influence the diastereoselectivity is the length of the M—O bond in Scheme 2, because as it decreases, the magnitude of the R–L interaction in (3) will increase.

It is the chelation of the metal M to both the carbonyl group and the oxyanion in the aldol product that stabilises it relative to the starting materials. Nevertheless, formation of the aldol can be reversible and the kinetically controlled aldol diastereoisomer ratio may be very different from the thermodynamically controlled one, as indicated in Scheme 3.[3] In this case the diastereoselectivity is much greater under thermodynamic control, but more usually, thermodynamic equilibration of aldols will result in a less disparate mixture of diastereoisomers.

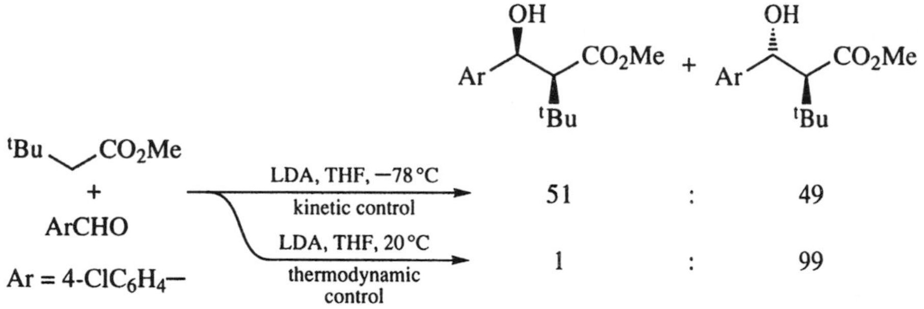

	kinetic control		
LDA, THF, −78 °C	51	:	49
LDA, THF, 20 °C thermodynamic control	1	:	99

$^tBu\diagup CO_2Me$ + ArCHO

Ar = 4-ClC_6H_4-

Scheme 3

Given the chair transition state through which the aldol reaction can proceed, the relative configuration of the two chiral centres in the product will depend on the configuration of the double bond in the enolate, as shown in Scheme 4. 'Directed' aldol formation, therefore, is dependent on the availability of enolates having the appropriate double bond geometry.[4]

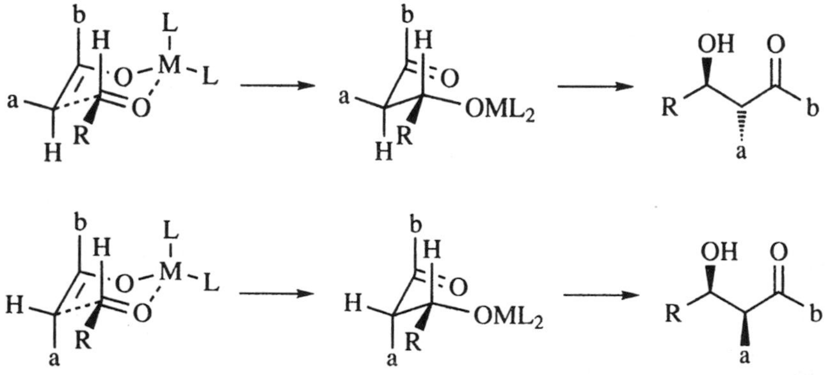

Scheme 4

The use of boron enolates has two advantages: thermodynamic equilibration of the aldol products is not a problem, as it may be using lithium enolates, and the ligands on boron can be chosen to maximise the preference of the group R in Scheme 4 for the equatorial position.[2] Completely stereoselective conversion of propiophenone into the (Z)-enolborinate (Scheme 5) can be achieved using BBN (9-borabicyclo[3.3.1]nonyl chloride) and a hindered base (see diastereoselective enolate formation below); reaction with benzaldehyde then gives only the *cis*-diastereoisomer.

For completely diastereoselective formation of the corresponding (E)-enolborinate it is necessary to use the bulky dicyclohexylboronyl chloride and base (Scheme 5); reaction with benzaldehyde is again completely diastereoselective, now giving the *trans*-aldol.[5]

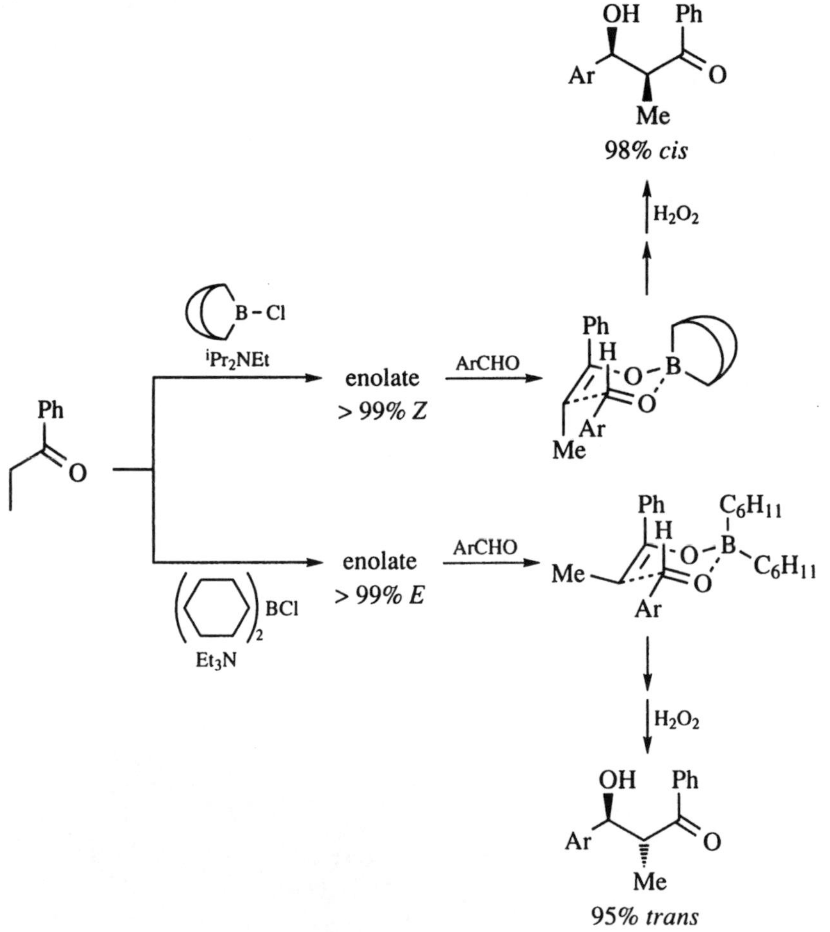

Scheme 5

Although methods are available for the generation *in situ* of lithium enolates of defined configuration (see below), their diastereoselectivity in aldol formation with aldehydes is less than that obtained using the corresponding boron enolate unless b (Scheme 2) is a large group. Apart from the thermodynamic equilibration referred to above, another factor is that whatever the ligands, they are not covalently bound to lithium as is the case with boron.[6] Consequently, the R–L interaction in (3) (Scheme 2) is less severe, giving a less pronounced difference in energy between (2) and (3). Also, the Li—O bond is longer than the B—O bond, and this will further reduce the difference in energy between (2) and (3).

Although the chair is thought to be the preferred transition-state geometry in many aldolisations, changes in substituents on the enolate or ligands on the metal can make the boat more favourable.[7] A reduction in the energy difference between these chair and boat transition states by comparison with chair and boat conformations of cyclohexane (Chapter 10) can be expected because of the reduced torsional strain present in the boat transition state.

The increased metal–oxygen bond length in the enolates of other metals, e.g. Zn, Sn and Zr, may also favour boat transition states or open transition states (see below) for their reactions with aldehydes. The higher diastereoselectivity found in aldolisations using persubstituted enolates of some metals [Scheme 6 (a)] is ascribed to destabilisation of the boat relative

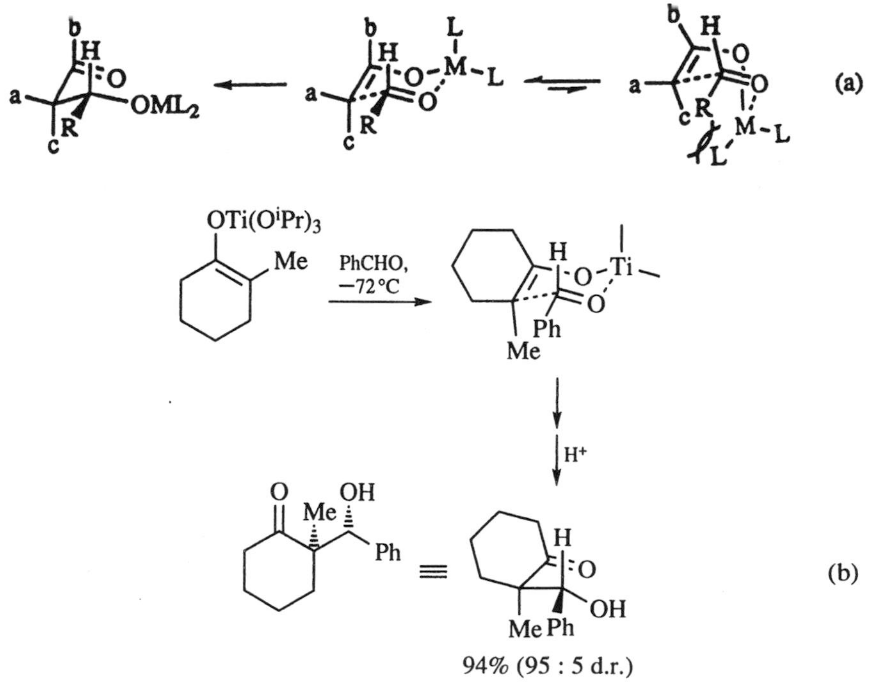

(a)

(b)

94% (95 : 5 d.r.)

Scheme 6

to the chair by interaction of c with one of the ligands on the metal.[8] In this reaction of a persubstituted enolate, a quaternary centre is created [Scheme 6 (b)].

Diastereoselective enolate formation (Type IV reactions)

In aldol condensations which proceed via (closed) chair transition states, the double bond configuration of the enolate determines which aldol diastereoisomer is formed (see above). Configurationally controlled double bond formation of enolates is therefore a prerequisite for stereochemical control of aldol reactions. Similarly, the stereochemistry of the Claisen rearrangement of enolate or enolate derivatives (Chapter 7) is controlled in part by the configuration of the double bond.

Since the stability of many metal enolates does not allow their isolation and purification, formation of a specific enolate or enol derivative from an unsymmetrical ketone requires regio- and stereocontrol, and aldolisation requires *in situ* reaction of the enolate with the aldehyde. The problem of regioisomers in the enolisation (Scheme 7) does not arise when the ketone is symmetrical, or has one substituent lacking enolisable protons or when enolisation of esters or amides is considered.

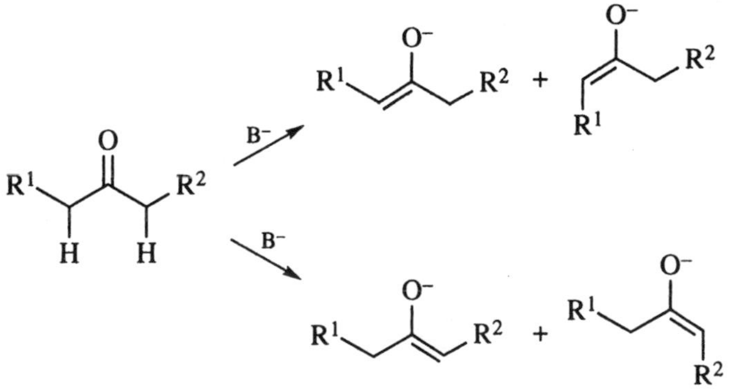

Scheme 7

Stereoselective formation of the enolate can, of course, be enforced by inclusion of the carbonyl group and the adjacent enolisable proton in a ring of eight or less members. In this case the enolate double bond is bound to be *cis*. The great use made of five- and six-membered lactones in synthesis derives from the fact that enolisation is perforce completely regio- and stereoselective.

In the stereoselective formation of acyclic enolates, an additional problem can arise which is not usually met in synthesis of other acyclic double bonds. This relates to the lack of configurational stability of the

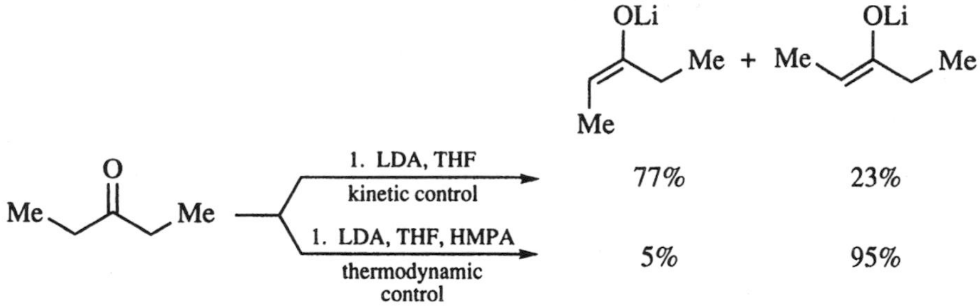

Scheme 8

enolate under the conditions of its formation. In some cases the consequent thermodynamic control of the double bond configuration may favour one enolate isomer over the other (Scheme 8),[9] but normally for highly stereoselective enolate formation in acyclic substrates one must make use of kinetic control. Removal by base of a proton from the carbon α to the carbonyl group requires the C_α—H σ-bond to be aligned parallel to the π-system of the carbonyl group; this is necessary for delocalisation of the electrons from the σ-bond into the carbonyl group in the transition state (Scheme 9).

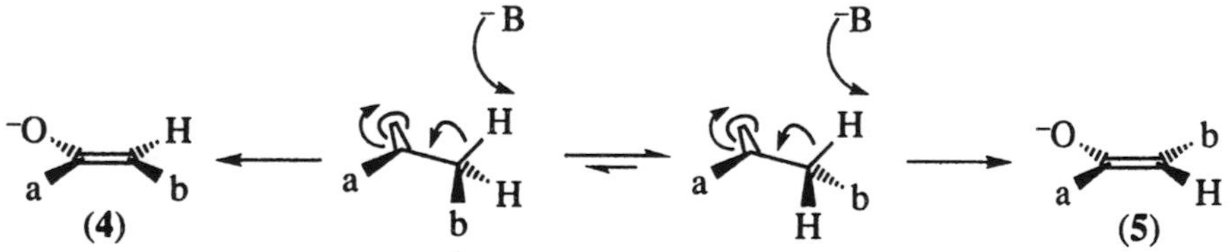

Scheme 9

For those carbonyl groups in which a or b (or both) is large, the transition state leading to (5) will be lower in energy than that leading to (4). This is the basis of the diastereoselective (Z)-enolate formation and aldolisation reactions of Heathcock et al.[10] [Scheme 10 (a)]. Note that the bulky group R in (7), having served its purpose as a stereocontrol element, can be removed to yield a more useful functional group [Scheme 10 (b)].[11]

Similar stereocontrol in boron enolate formation from ketones can be accomplished in the sense shown in Scheme 11 (cf. Scheme 5).[2] In this case, the orientation of the bulky dibutylboron on the carbonyl oxygen will be important in controlling the conformation (8) from which proton removal takes place; using a bulky base this will be from the methylene group distant from the dibutylboron.

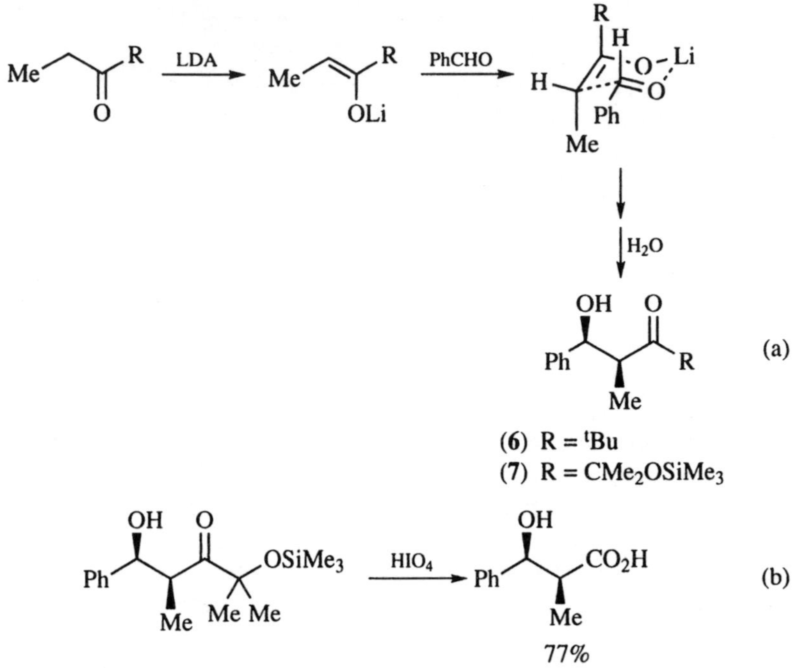

(a)

(6) R = ^{t}Bu
(7) R = CMe$_2$OSiMe$_3$

(b)

77%

Scheme 10

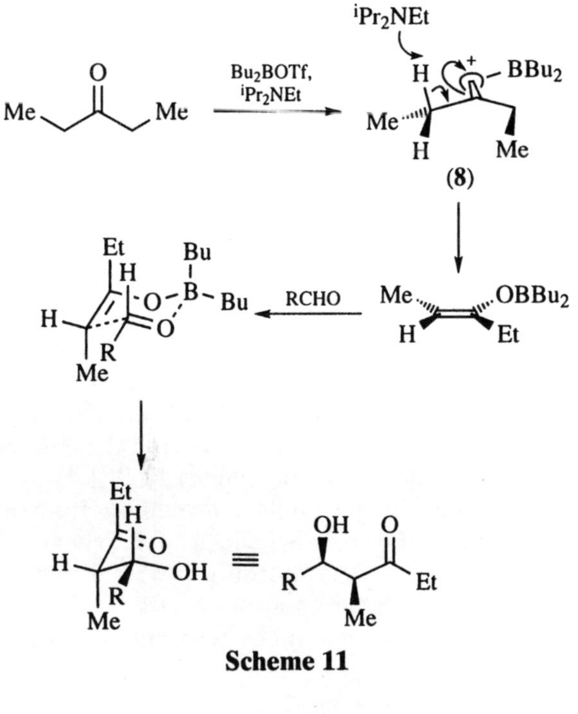

(8)

Scheme 11

The 'pinwheel'[†] conformation (9) (Scheme 12) from which proton removal takes place also accounts for (Z)-boron enolate formation from the boron-complexed thioester (10). Highly diastereoselective formation of *trans*-aldols takes place on reaction of the resulting enolate (11) with aldehydes (Scheme 12).[12b]

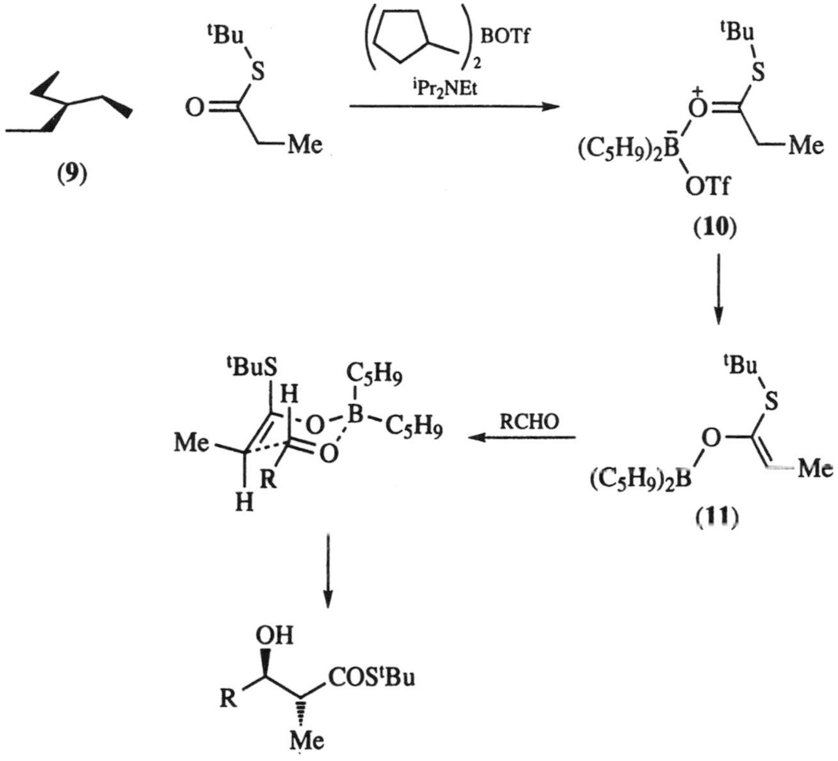

Scheme 12

However, as Corey and Kim[13] have shown, with thioester substituents other than *tert*-butyl, deprotonation of the boron-complexed thioester (12) using a hindered base (Scheme 13) may be slower than ionisation of bromide and conversion to (13); loss of a proton preferentially from (13) leads to the (E)-enolate and, consequently, the derived alcohol product is *cis*. The product in this reaction was in fact produced highly enantioselectively (97% e.e.) as well as highly diastereoselectively since the boron reagent R₂BBr used was the enantiopure *B*-bromodiazaborolane shown, i.e. the reaction is a Type II/III (see Chapter 13).

[†] A more correct description of this conformation is a triskelion (viz. the three-legged motif on the coat of arms of the Isle of Man); see Noe and Raban.[12a]

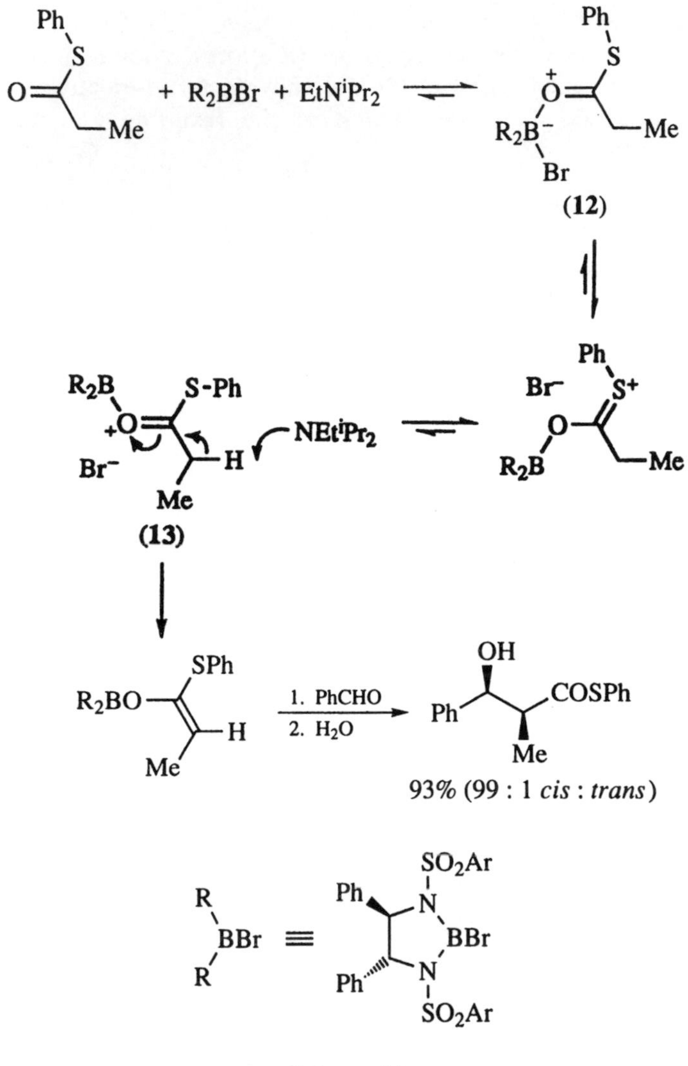

Scheme 13

Proton removal from the α-position of a carbonyl-containing substrate is not the only means by which enolates can be formed. Conjugate addition to α,β-unsaturated ketones, aldehydes, esters or amides provides a regiospecific route to enolates.[14] In Scheme 14, addition of a Grignard reagent to the α,β-unsaturated thioamide delivers the (Z)-enolate (**14**), probably as a result of coordination of the Grignard reagent to the sulphur and intramolecular delivery of the group R. High yields of one diastereoisomer (**15**) of the quaternary centre-containing product are obtained using alkyl Grignard reagents and alkyl aldehydes.

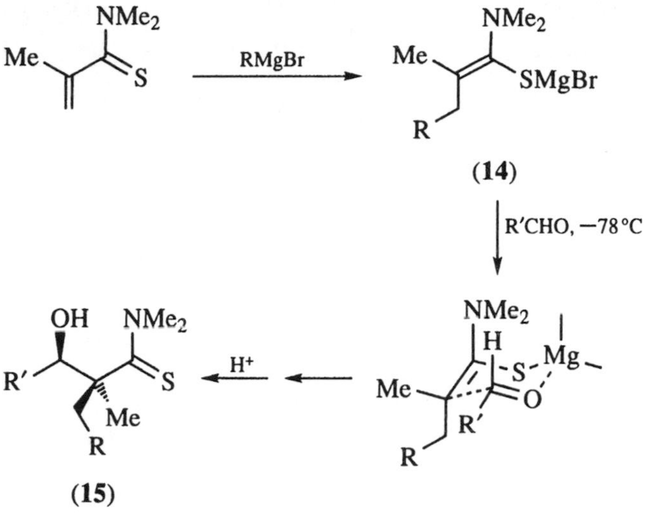

(14)

R′CHO, −78 °C

(15)

Scheme 14

The synthesis of configurationally defined enol silyl ethers (**16**) and silyl ketene acetals (**17**) is important because, unlike enolates, they are sufficiently stable to be isolated and purified. Also, they not only react with aldehydes in aldol-type reactions (see below) but also they are converted into lithium enolates by alkyllithiums with retention of the double bond configuration (Scheme 15). Enol silyl ethers of defined configuration are also important in Claisen rearrangements.

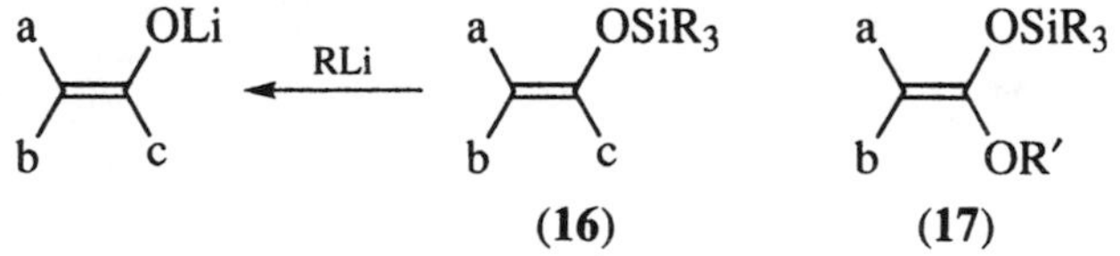

(16)   **(17)**

Scheme 15

Kuwajima and co-workers[15] have established conditions for highly diastereoselective formation of enol silyl ethers in a number of cases, e.g. Scheme 16. A combination of HMPA and phenyldimethylsilyl chloride activates α,β-unsaturated aldehydes and ketones towards attack by cuprates (RMgX + Cu⁺) and brings about trapping of the kinetically formed enolate [Scheme 17 (a)].[16]

The stereochemistry of these reactions is consistent with addition to α,β-unsaturated ketones in their *s-cis* conformations and to α,β-unsaturated aldehydes in their *s-trans* forms, as shown in Scheme 17 (a). By starting with

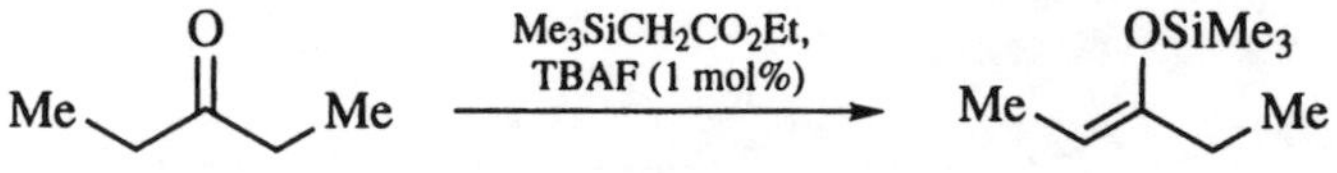

Scheme 16

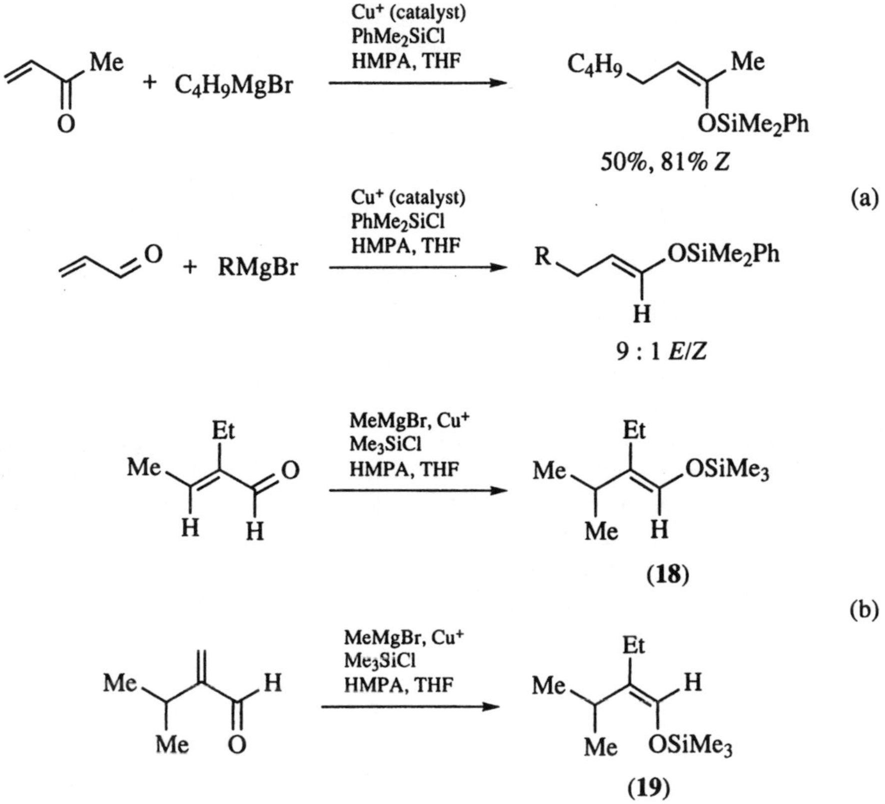

(a)

(b)

Scheme 17

the appropriate double bond isomer, access to (E) or (Z)-aldehyde enol silyl ethers **(18)** and **(19)** is available [Scheme 17 (b)]. In these reactions the silyl chloride is not simply a trap for the enolate but assists in the enolisation by coordinating with the aldehyde oxygen.

VIA OPEN (EXTENDED) TRANSITION STATES

Silyl enol ethers are unreactive towards aldehydes at ordinary temperatures but will react when the aldehyde is coordinated with Lewis acid such as BF_3, $SnCl_4$ or $TiCl_4$ (Scheme 18) to form the aldol product after work-up (the Mukaiyama reaction).[17]

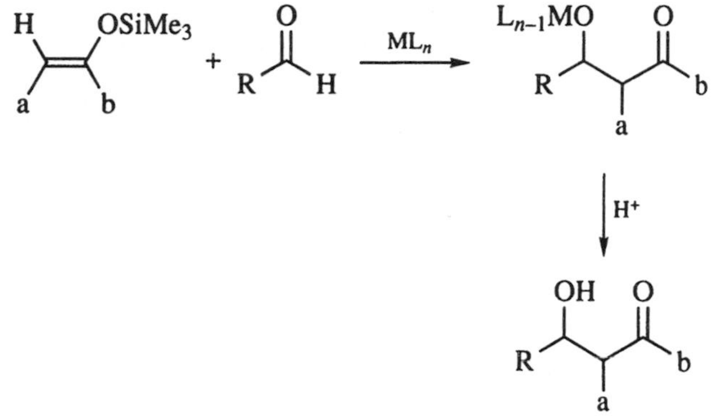

$ML_n = BF_3, SnCl_4, TiCl_4$

Scheme 18

Using acyclic substrates, the Mukaiyama reaction does not normally show high diastereoselectivity but, exceptionally, the version in Scheme 19 (a) gives a better than 95:5 ratio of *trans:cis* diastereoisomers.[18]

In this reaction, extended or open transition states are believed to be involved [Scheme 19 (b)] with (i) a staggered arrangement of bonds around the interacting sp^2–sp^2 centres, (ii) an *anti* relationship of the enol and carbonyl double bonds to maximise the distance between oxygen atoms which, in the transition state, will both have partial positive charge, (iii) coordination of the Lewis acid *syn* to the aldehyde proton and (iv) arrangement of the other substituents on the enol and the aldehyde so as to minimise steric interactions overall. In Scheme 19 (b) the preferred open transition state is (**20**); attack on the opposite face of the benzaldehyde as in (**21**) would involve Ph–^{t}Bu and Me–ML_{n-1} interactions of which the former is more severe.

(Z) and (E)-thioester-derived silyl ketene acetals (**22**) and (**23**) *both* react with benzaldehyde in the presence of boron trifluoride to give the *trans*-thioester with high diastereoselectivity, i.e. the reaction is *stereoconvergent* (Scheme 20).[19]

This stereoconvergence, or lack of dependence of the stereostructure of the aldol product on the configuration of the enolate double bond, is a frequent though not necessary accompaniment of extended transition states. It contrasts with the effect of a change in enolate geometry on the *cis:trans* aldol ratio when chair transition states are involved. The two transition states (**24**) and (**25**) (Scheme 20) have 'pinwheel' conformations for the ketene monothioacetals (**22**) and (**23**). Although in these transition states there is some steric interaction between the Ph and Me groups, this is likely to be less than the sum of the interactions that would arise were the opposite face of the BF_3-coordinated benzaldehyde to be attacked, e.g. as in (**26**) for the reaction of (**23**).[20]

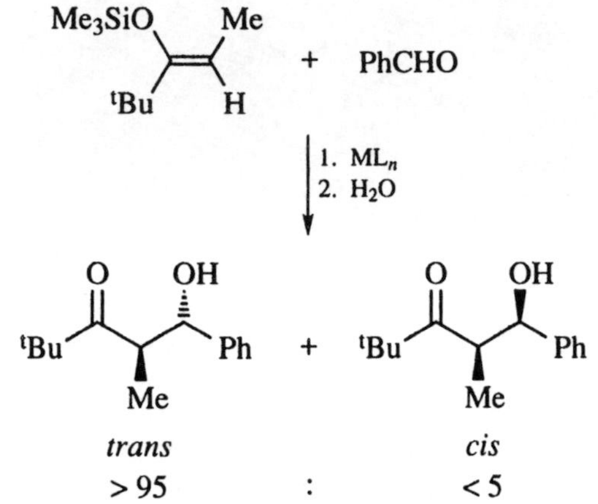

Scheme 19

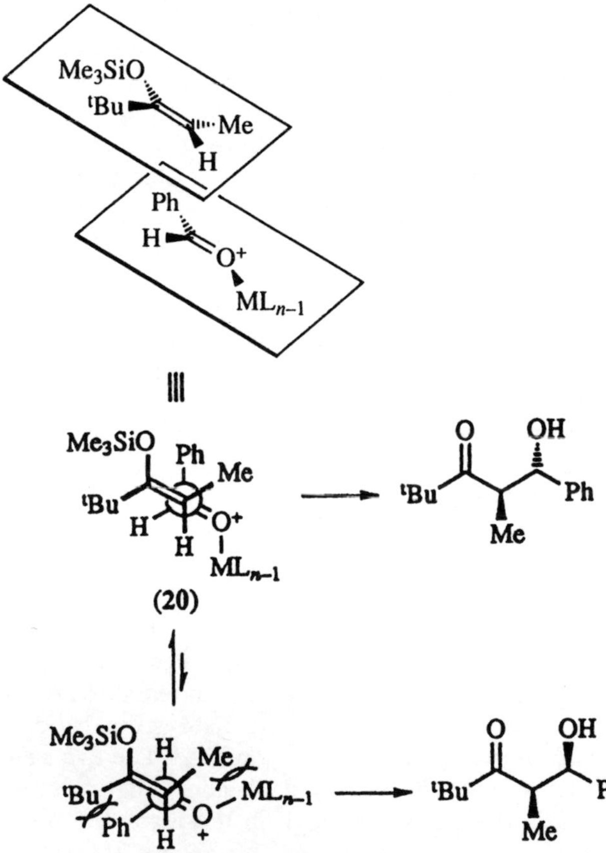

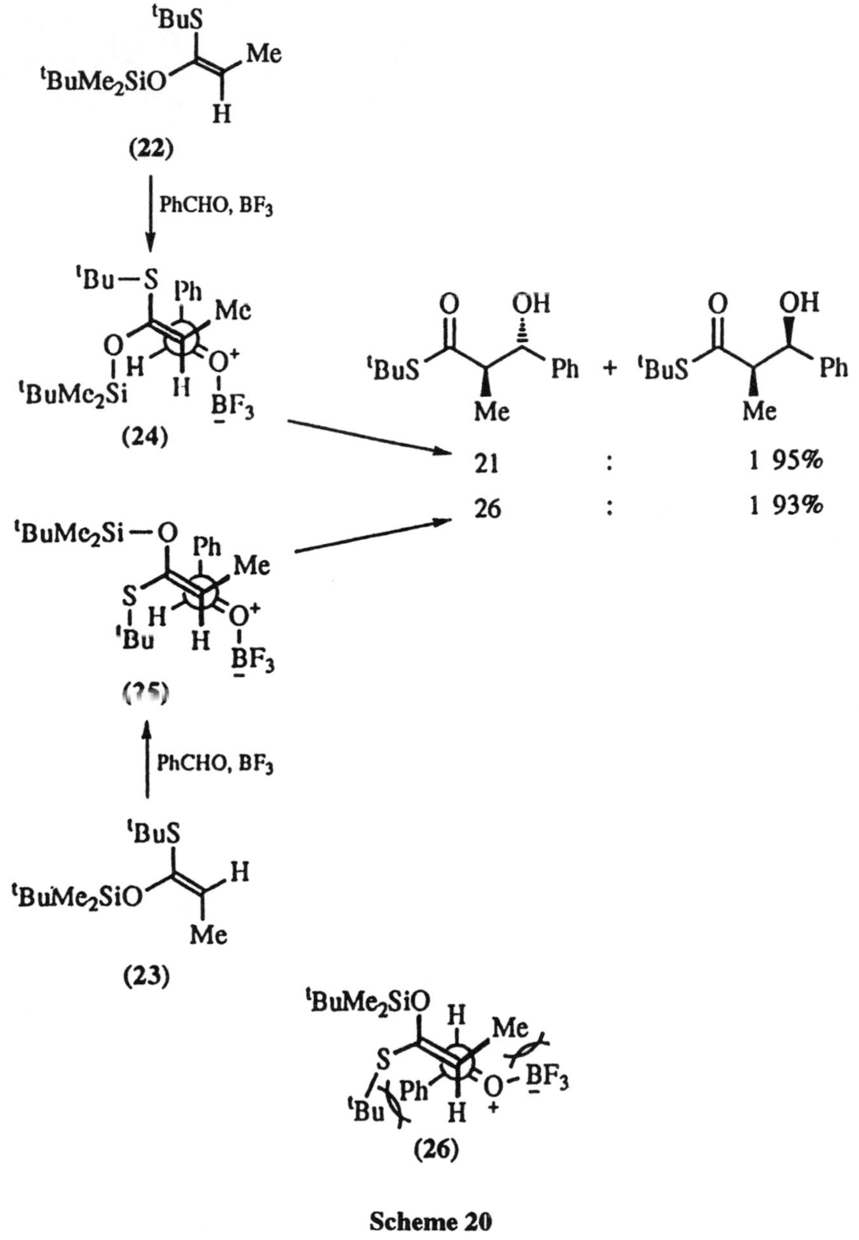

Scheme 20

A variation on the Mukaiyama reaction is the use of acetals in place of aldehydes. An efficient catalyst for this form of reaction is trimethylsilyl triflate (TMSOTf), which is able to generate a reactive carbocation (O-alkylated carbonyl) intermediate (27) (Scheme 21).[20]

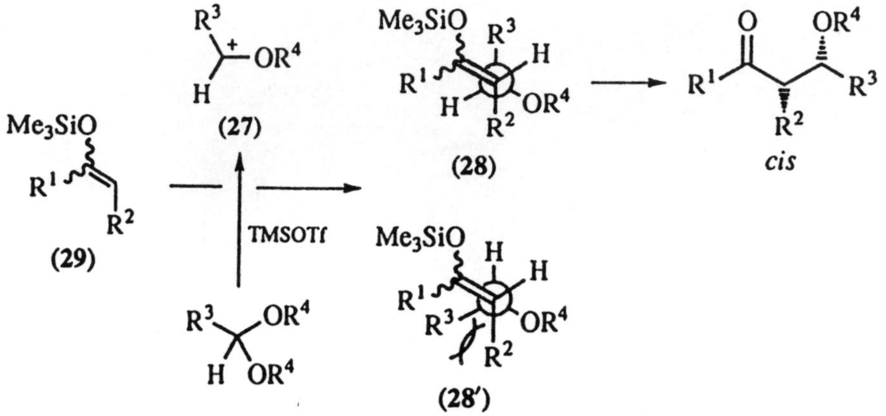

Scheme 21

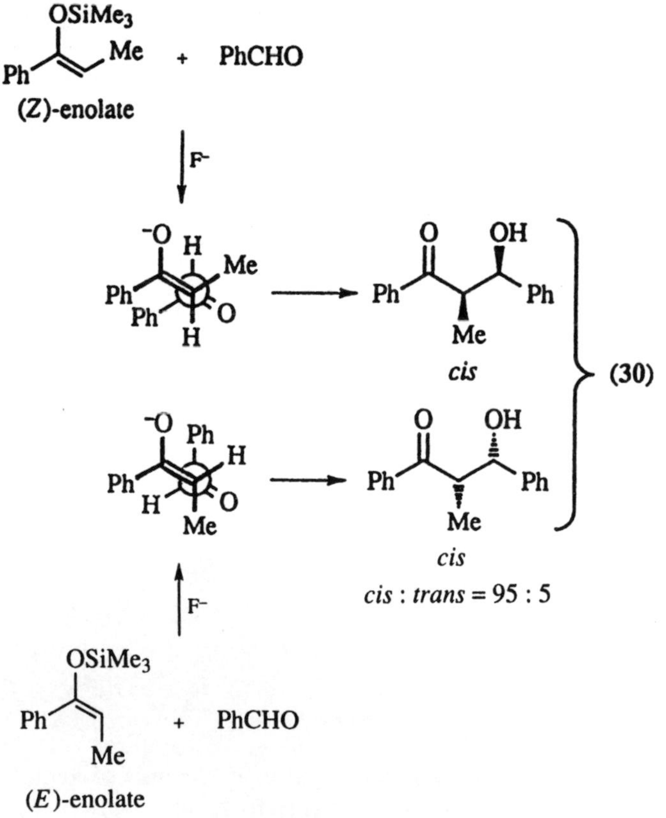

Scheme 22

Here *cis:trans* ratios of the products obtained are also interpreted in terms of extended transition states as in Scheme 21; formation of *cis* products via **(28)** is favoured because of the *gauche* interaction between R^2 and R^3 in the alternative **(28')**, irrespective of the configuration of the silyl enol ether.

Open transition states are also thought to be involved in the reaction of silyl enol ethers with aldehydes where the enolate is generated by fluoride ion removal of the silyl group (Scheme 22). Here also, the stereostructure of the product **(30)** (95:5 *cis:trans*) is independent of the configuration of the silyl enol ether used[21] [avoidance of a *gauche* interaction between Ph and Me is more important than avoidance of any Ph–Ph interaction in the case of the (Z)-enolate].

Type II reactions of alkenylmetals with aldehydes

A preference for a chair transition state can also account for the high diastereoselectivity which is obtained in the reaction of some alkenylmetals and alkenylmetalloids with aldehydes.[22a] Thus the (E)- and (Z)-crotylboronates in Scheme 23 each react diastereoselectively with aldehydes to give stereoisomeric homoallylic alcohols.[22b]

Crotyllithium and crotylmagnesium, however, do not show useful levels of diastereoselectivity in these reactions probably because the E- and Z-isomers interconvert at rates comparable to the rates of their reactions with aldehydes.[23]

Crotylchromium compounds give excellent diastereoselectivity but the product is the *trans*-alcohol (Scheme 24), irrespective of which crotyl bromide double bond isomer is used initially. Interconversion of the Z- and E-isomers in the alkenylchromium and reaction of only the E-isomer via a chair transition state accounts for this stereochemistry.

Other alkenylmetals, including those of Ti, Zr and Sn, undergo some reactions with aldehydes that are stereochemically consistent with chair transition states but it seems likely that as the metal–carbon bond increases in length, so the difference in energy between the chair and boat transition states will diminish, as it does with aldolisations (see above).

Alkenylstannanes react readily only with activated aldehydes, e.g. chloral, but they do so with high diastereoselectivity (Scheme 25), evidently via chair transition states.[24] Reaction with normal aldehydes is sluggish but can be accelerated by addition of Lewis acids; these also bring about a change in the transition state geometry and, consequently, in the stereochemistry.[25] Thus the reactions of both (Z)- and (E)-crotylstannanes with benzaldehyde catalysed by boron trifluoride give the *same cis*-homoallylic alcohol **(31)** (Scheme 26).[22a]

To account for the stereoconvergence, an extended transition state has been proposed similar to that for the Mukaiyama reaction (Scheme 19), with

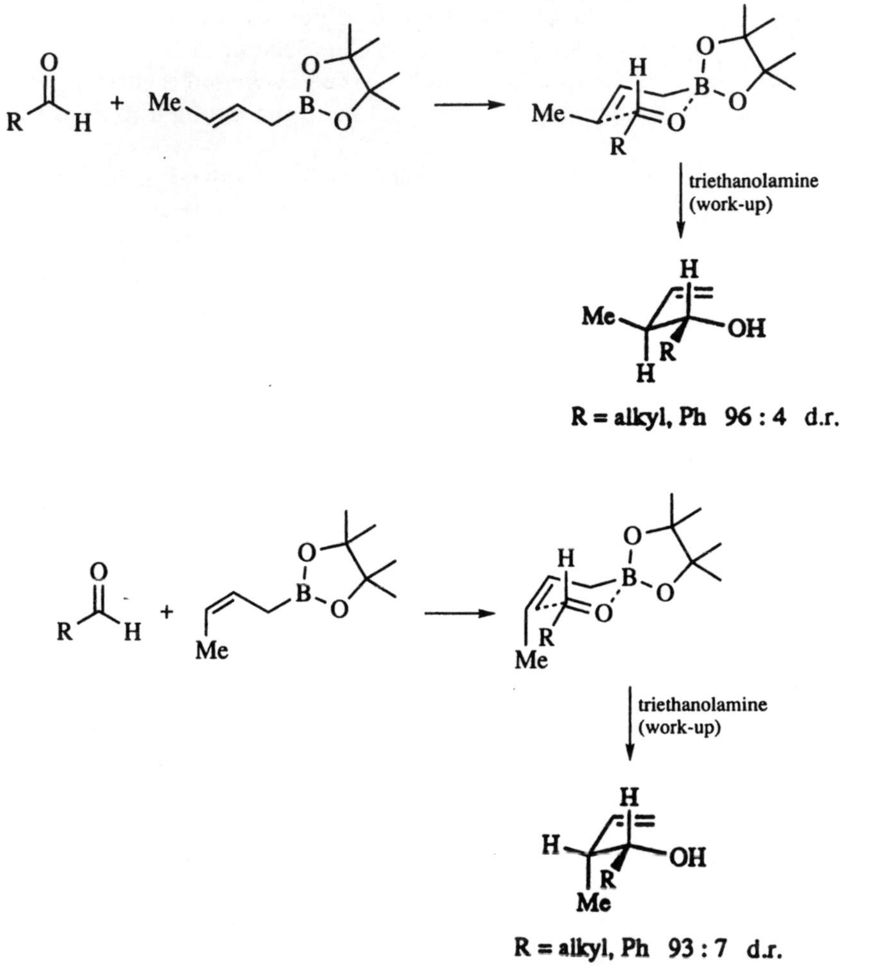

R = alkyl, Ph   96 : 4   d.r.

R = alkyl, Ph   93 : 7   d.r.

Scheme 23

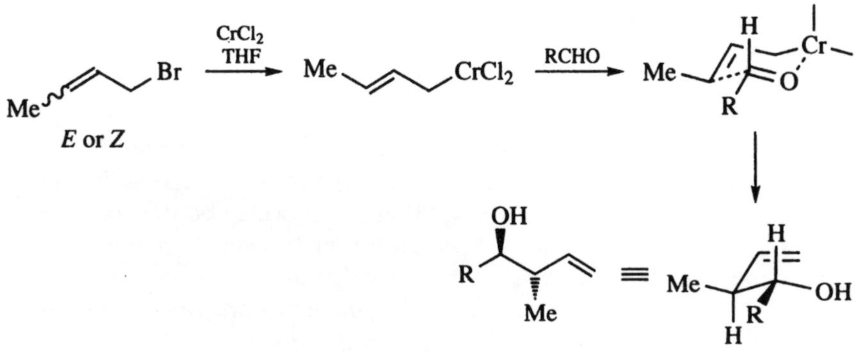

R = Ph, 96%, > 100 : 1 *trans* : *cis*

Scheme 24

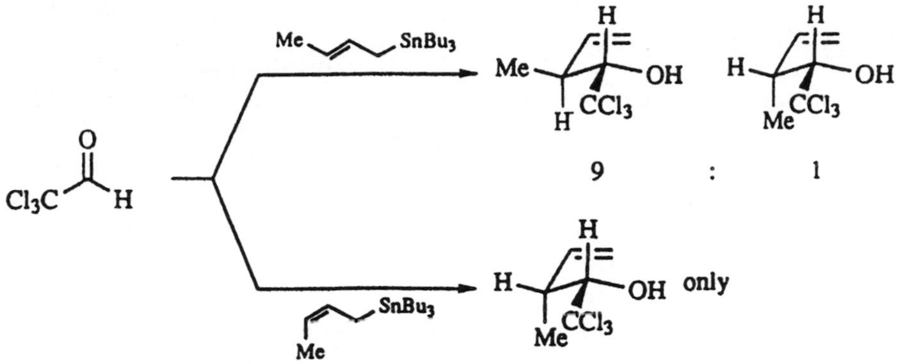

Scheme 25

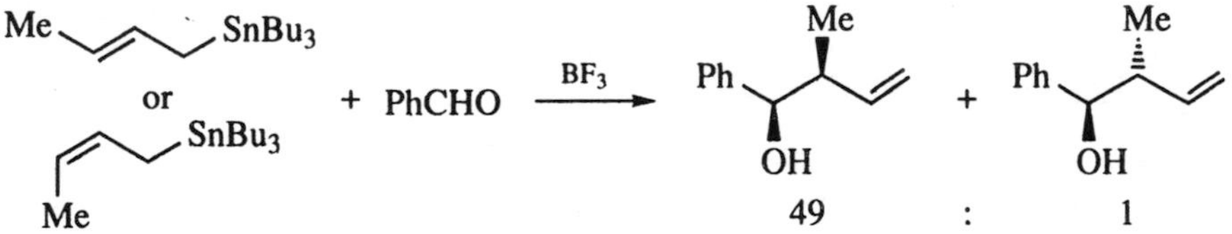

(31)

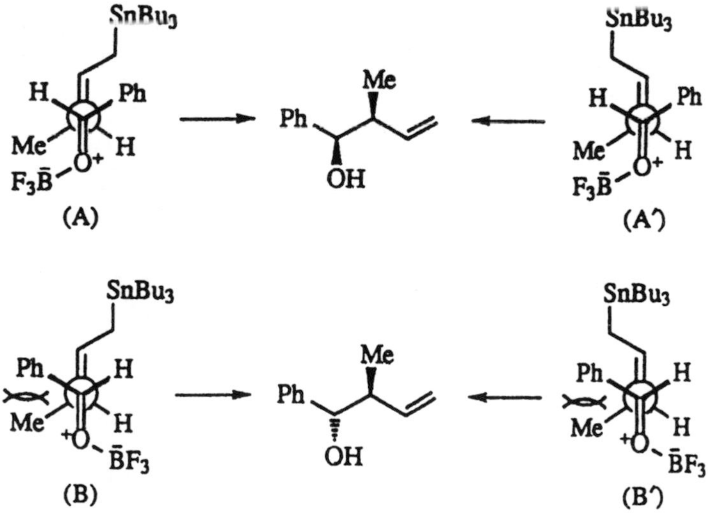

Scheme 26

a preferred *anti* relationship between the alkene double bond and the complexed aldehyde. The structure of the benzaldehyde–boron trifluoride complex has been shown to have the *trans* configuration, as in Scheme 26, both in the crystalline state and in solution.[26] Transition states A and A′ in

Scheme 26 are preferred because of the destabilising Ph, Me *gauche* interactions in B and B'.

This stereoconvergence in the Lewis acid-catalysed reactions of alkenyltins is of obvious preparative value because separation of *Z*- and *E*-double bond isomers of the alkenyltin is not now required. On the other hand, only the *cis* diastereoisomer is available by this route. Fortunately, as indicated above, preparation of the *trans* diastereoisomer can be accomplished by other means, e.g. by using alkenylchromium or alkenylboronate reagents.

Although open transition states having an extended *anti* relationship of enolate and carbonyl double bonds seem likely to be responsible for the stereochemistry of the reactions outlined above, it is also probable that the corresponding transition states in which the two interacting double bonds have a *syn* relationship are little different in energy and may be preferred in some cases (see below). This will be reflected in the stereochemistry of the reactions.

Type II Michael additions[27]

Michael addition of enolates, enamines and other nucleophiles to the β-positions of α,β-unsaturated carbonyl, nitro or sulphonyl systems creates two new chiral centres when the substitution is appropriate. High diastereoselectivity results from the kinetically controlled Michael addition of certain ketone-derived enamines to nitroalkenes. The relative configuration of the two chiral centres in the product is consistent with a *gauche* disposition of enamine and nitroalkene double bonds in the transition state (Scheme 27).[28]

Seebach and Golinski[28] suggested that this *gauche* relationship of donor

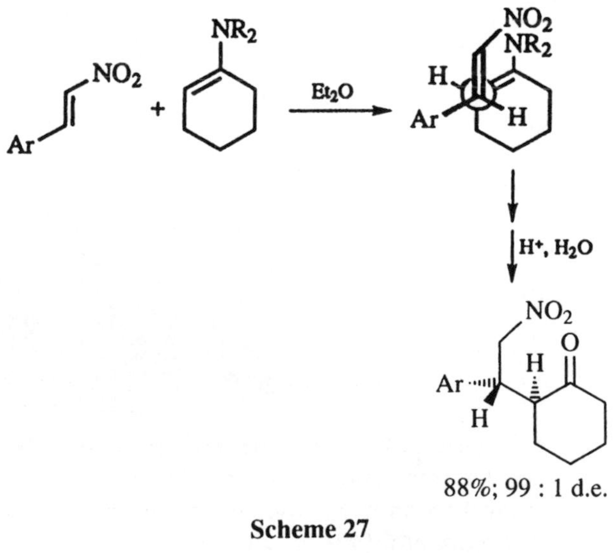

88%; 99 : 1 d.e.

Scheme 27

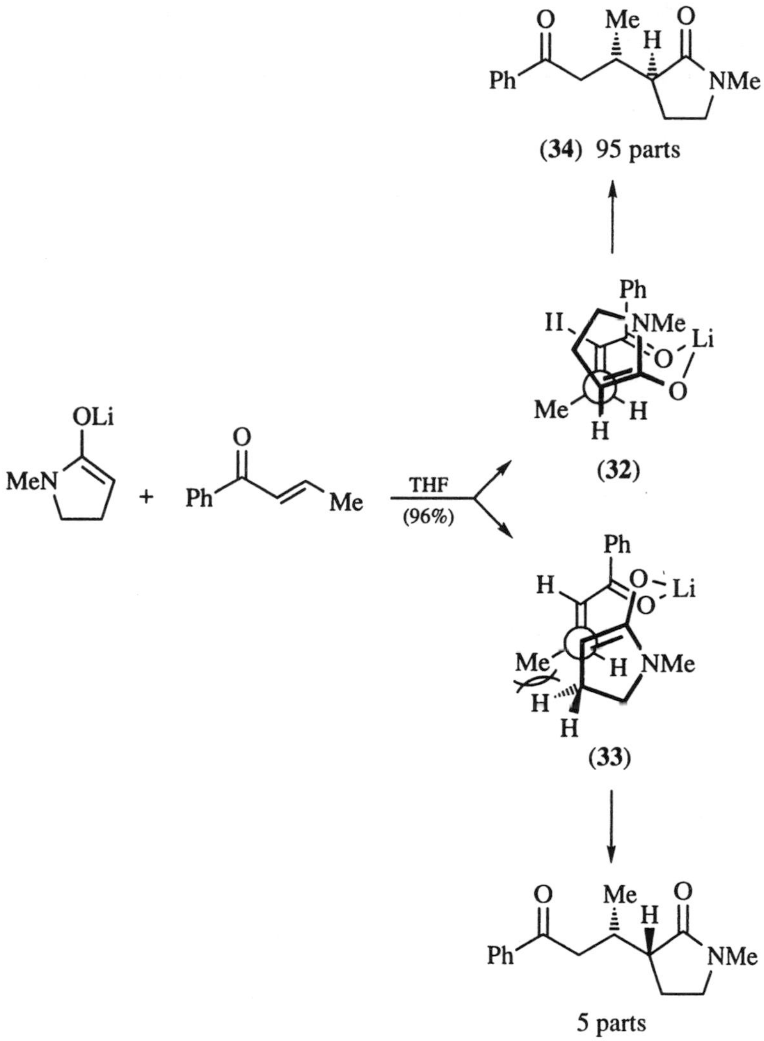

Scheme 28

and acceptor double bonds is an important stereochemical feature of a variety of reactions of differing mechanistic type (including Michael addition) in aprotic media under kinetic control.

Heathcock and co-workers[29] have studied the stereochemistry of a large number of Michael additions including the reaction of N,N-disubstituted amide and thioamide enolates with α,β-unsaturated ketones. Chelation of the oxygens of enolate and ketone by lithium favours a *gauche* disposition of double bonds; using the N-methylpyrrolidone-derived enolate (Scheme 28), transition state (32) leading to (34) is thought to be favoured over (33)

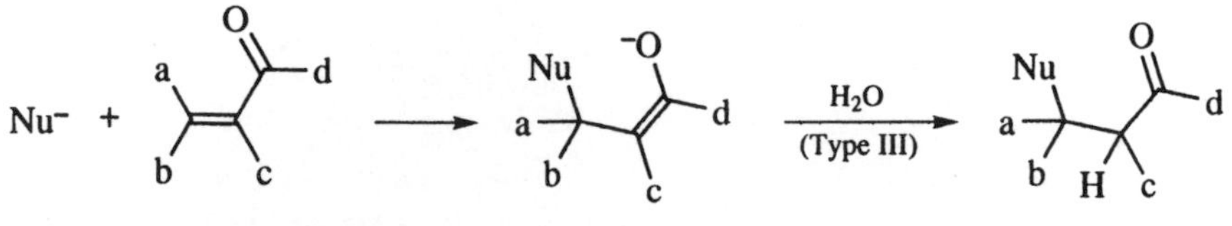

Scheme 29

possibly because of the greater interaction of the pyrroline ring methylene group with the methyl group in the latter.

Michael additions in which a second chiral centre results from protonation at the carbon of an enolate (Scheme 29) under the influence of a first-formed chiral centre are Type III reactions.

Type II carbocation-mediated diastereoselectivity

The biogenesis of steroids and other polyisoprenoids is believed to occur by enzyme-catalysed cyclisation of appropriately substituted polyenes.[30]

Work done by Johnson and co-workers over a number of years has shown that highly stereoselective (non-enzymic) cationic cyclisation of non-conjugated polyenes can be accomplished under carefully controlled conditions. The cation which initiates cyclisation is often a more stable one (allylic), the solvent of low nucleophilicity and the double bonds involved not substituted with electron-withdrawing groups.[31]

A 1,5-diene contained in an acyclic substrate can cyclise via chair or boat transition states (35) or (36) with antiperiplanar addition taking place at both double bonds (Scheme 30). The presence of an additional substituent in the

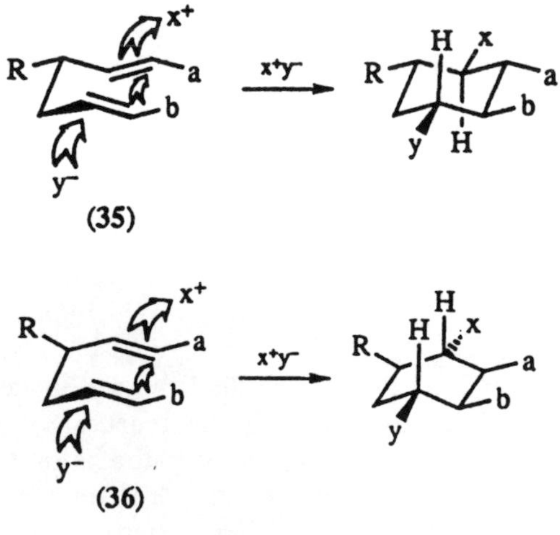

(35)

(36)

Scheme 30

1,5-diene [(35), (36); R≠H], with its predilection for an equatorial location, will result in asymmetric induction in the formation of up to four additional chiral centres (Type III).

As would be predicted, the chair transition state in Scheme 30 is favoured over the boat form and leads to complementary diastereoselectivity when addition takes place across *cis*- and *trans*-configured double bonds, respectively (Scheme 31).[32]

In these cyclisations of 1,5-dienes, a six-membered ring is formed in preference to a five-membered ring except when the diene is electronically biased towards the formation of the smaller ring.

Although the yields of bicyclised products in Scheme 31 are low (and epimers are produced at the hydroxyl-bearing carbon), the reaction is stereospecific in the sense that the *trans* isomer (37) gives only (38) and the *cis* isomer (39) gives only (40), with no apparent crossover.

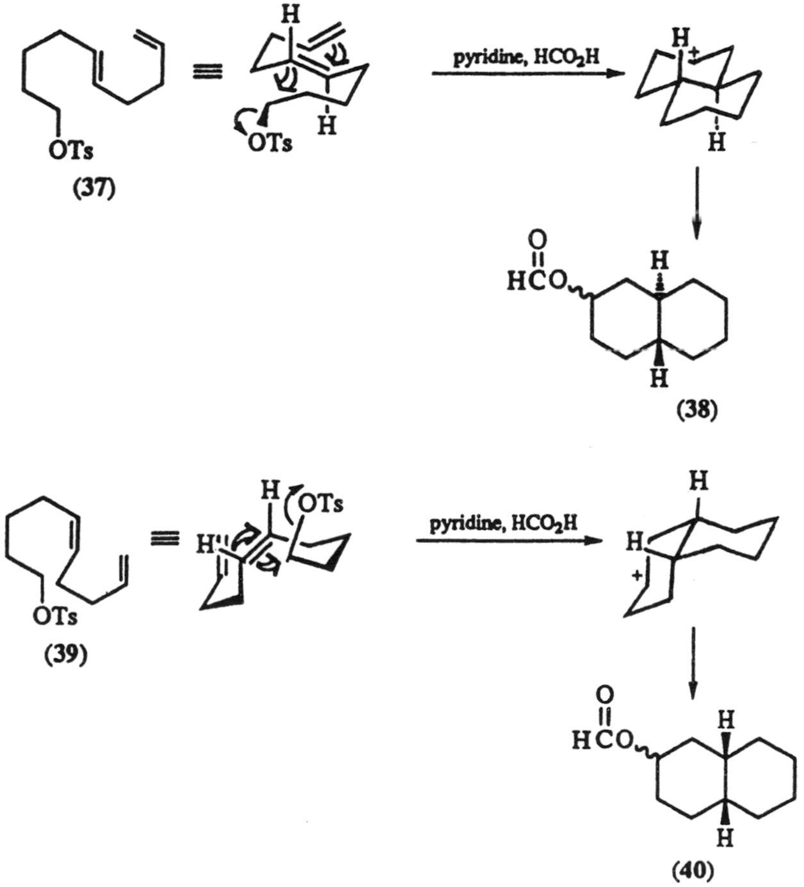

Scheme 31

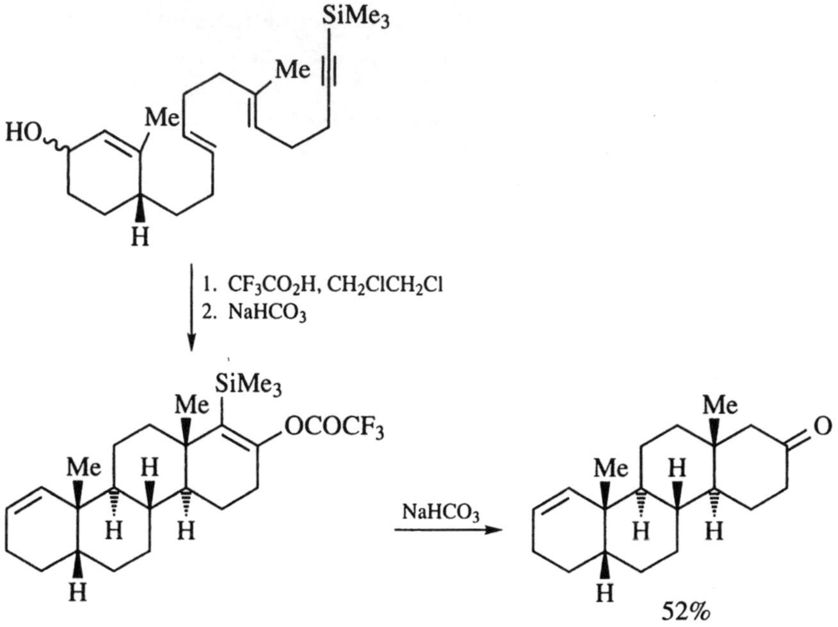

Scheme 32

Some elegant syntheses of steroid-related products have been devised based on this cyclisation, an example being that in Scheme 32.[33]

Summary

Many Type II aldol reactions of aldehydes RCHO and metal enolates proceed via closed transition states in which the chair shape is preferred. In this chair transition state the aldehyde group R is assumed to occupy the equatorial position and the sense of the diastereoselectivity is dependent on the configuration of the enolate double bond. Boat transition states may be preferred for aldol reactions of enolates having longer metal–oxygen bonds and/or particular substitution patterns.

Enolates having a particular configuration are usually prepared and used *in situ* but silyl enol ethers and silyl ketene acetals are isolable and can be purified. Both of the latter react with aldehydes in the presence of Lewis acids via open (extended) transition states. Some highly diastereoselective reactions of this type are known in which the sense of diastereoselectivity is independent of the double bond configuration of the enol ether, i.e. the reactions are stereoconvergent.

The reactions of alkenylmetals with aldehydes are similar in many ways to the aldol reaction. Chair transition states are preferred in many cases but open transition states are also believed to be involved in reactions showing stereoconvergence (see above).

Diastereoselectivity in some Type II Michael additions of enolates to double bonds has been rationalised using a *gauche* approach of the two carbon–carbon double bonds.

Type II cyclisations of polyenes via carbocation intermediates can be used to construct polycycles, e.g. steroid derivatives, diastereoselectively in reactions which resemble those by which these products are formed in nature.

References

1. H.E. Zimmerman and M.D. Traxler, *J. Am. Chem. Soc.*, 1957, **79**, 1920; W.A. Kleschick, C.T. Buse, and C.H. Heathcock, *J. Am. Chem. Soc.*, 1977, **99**, 247; for a summary of hypotheses regarding transition state geometries in the aldol, see J.D. Morrison (Ed.), *Asymmetric Synthesis*, Academic Press, New York, 1983, Vol. 3, Chapt. 2, pp. 154–161; D.A. Evans, *Top. Stereochem.*, 1982, **13**, 1.
2. D.A. Evans, E. Vogel and J.V. Nelson, *J. Am. Chem. Soc.*, 1979, **101**, 6120; D.A. Evans, J.V. Nelson, E. Vogel and T.R. Taber, *J. Am. Chem. Soc.*, 1981, **103**, 3099.
3. Y. Wei and R. Bakthavatchalam, *Tetrahedron Lett.*, 1991, **32**, 1535; see also K.A. Swiss, W-B. Choi, D.C. Liotta, A.F. Abdel-Magid and C.A. Maryanoff, *J. Org. Chem.*, 1991, **56**, 5978.
4. J.E. Dubois and P. Fellmann, *Tetrahedron Lett.*, 1975, 1225.
5. H.C. Brown, R.K. Dhar, R.K. Bakshi, P.K. Pandiarajan and B. Singaram, *J. Am. Chem. Soc.*, 1989, **111**, 3441; S. Masamune, S. Mori, D. Van Horn and D.W. Brooks, *Tetrahedron Lett.*, 1979, 1665.
6. D. Seebach, *Angew. Chem., Int. Ed. Engl.*, 1988, **27**, 1624.
7. E. Nakamura, and I. Kuwajima, *Tetrahedron Lett.*, 1983, **24**, 3343; I. Kuwajima and E. Nakamura, *Acc. Chem. Res.*, 1985, **18**, 181. D.A. Evans, *Top. Stereochem.*, 1982, **13**, 1.
8. S. Yamago, D. Machii and E. Nakamura, *J. Org. Chem.*, 1991, **56**, 2098.
9. R.E. Ireland, R.H. Mueller, and A.K. Willard. *J. Am. Chem. Soc.*, 1976, **98**, 2868; Z.A. Fataftah, I.E. Kopka and M.W. Rathke, *J. Am. Chem. Soc.*, 1980, **102**, 3959.
10. C.H. Heathcock, C.T. Buse, W.A. Kleschick, M.C. Pirrung, J.E. Sohn and J. Lampe, *J. Org. Chem.*, 1980, **45**, 1066.
11. C.T. Buse and C.H. Heathcock, *J. Am. Chem. Soc.*, 1977, **99**, 8109.
12. (a) E.A. Noe and M. Raban, *J. Am. Chem. Soc.*, 1975, **99**, 5811; (b) M. Hirama and S. Masamune, *Tetrahedron Lett.*, 1979, 2225.
13. E.J. Corey and S.S. Kim, *J. Am. Chem. Soc.*, 1990, **112**, 4976.
14. Y. Tamaru, T. Hioki, S. Kawamura, H. Satomi and Z. Yoshida, *J. Am. Chem. Soc.*, 1984, **106**, 3876.
15. E. Nakamura, K. Hashimoto and I. Kuwajima, *Tetrahedron Lett.*, 1978, 2079; *Org. Synth.*, 1983, **61**, 122.
16. S. Matsuzawa, Y. Horiguchi, E. Nakamura and I. Kuwajima, *Tetrahedron*, 1989, **45**, 349.
17. T. Mukaiyama, K. Banno and K. Narasaka, *J. Am. Chem. Soc.*, 1974, **96**, 7503; T. Makaiyama, *Org. React.*, 1982, **28**, 203.
18. C.H. Heathcock, S.K. Davidsen, K.T. Hug and L.A. Flippin, *J. Org. Chem.*, 1986, **51**, 3027.

19. C. Gennari, M.G. Beretta, A. Bernardi, G. Moro, C. Scolastico and R. Todeschini, *Tetrahedron*, 1986, **42**, 893.
20. S. Murata, M. Suzuki and R. Noyori, *Tetrahedron*, 1988, **44**, 4259.
21. R. Noyori, I. Nishida, and J. Sakata, *J. Am. Chem. Soc.*, 1981, **103**, 2106.
22. (a) Y. Yamamokto, *Acc. Chem. Res.*, 1987, **20**, 243; (b) R.W. Hoffman, *Angew. Chem., Int. Ed. Engl.*, 1982, **21**, 555.
23. T. Hiyama, K. Kimura and H. Nozaki, *Tetrahedron Lett.*, 1981, **22**, 1037.
24. C. Servens and M. Pereyre, *J. Organomet. Chem.*, 1972, **35**, C20; H. Yatagai, Y. Yamamoto and K. Maruyama, *J. Am. Chem. Soc.*, 1980, **102**, 4548.
25. S.E. Denmark, E.J. Weber, T.M. Wilson and T.M. Willson, *Tetrahedron*, 1989, **45**, 1053.
26. M.T. Reetz, M. Hüllmann, W. Massa, S. Berger, P. Rademacher and P. Heymanns, *J. Am. Chem. Soc.*, 1986, **108**, 2405.
27. Review of the stereochemistry of the Michael addition; D.A. Oare and C.H. Heathcock, *Top. Stereochem.*, 1989, **19**, 227.
28. D. Seebach and J. Golinski, *Helv. Chim. Acta*, 1981, **64**, 1413; S. Fabrissin and S. Fatutta and A. Risallti, *J. Chem. Soc., Perkin Trans. 1*, 1981, 109, and earlier papers.
29. D.A. Oare, M.A. Henderson, M.A. Sanner and C.H. Heathcock, *J. Org. Chem.*, 1990, **55**, 132.
30. A. Eschenmoser, L. Ruzicka, O. Jeger and D. Arigoni, *Helv. Chim. Acta*, 1955, **38**, 1890; G. Stork and A.W. Burgstahler, *J. Am. Chem. Soc.*, 1955, **77**, 5068.
31. J.K. Sutherland, *Chem. Soc. Rev.*, 1980, **9**, 265.
32. W.S. Johnson and J.K. Crandall, *J. Org. Chem.*, 1965, **30**, 1785.
33. W.S. Johnson, T.M. Yarnell, R.F. Meyers and D.R. Morton, *Tetrahedron Lett.*, 1978, 2549.

9 TYPE III REACTIONS: THOSE INVOLVING ASYMMETRIC INDUCTION

We have seen (Chapter 2) that Type III reactions involve the creation of one or more new chiral centres under the influence of an existing or parent chiral centre. Thus in Scheme 1 (a), the formation of an excess of one epoxide over the other is a Type III reactions with 1,2-asymmetric induction.

The parent chiral centre responsible for the induction does not have to be located in the same molecule as the prochiral element which gives rise to the new chiral centre: it may be present in any part of the reaction ensemble. Thus in Scheme 1 (b), asymmetric induction occurs in the addition of a chiral hydroborating agent to a prochiral double bond if one of the two diastereoisomers is produced in excess (cf. Chapter 1, Scheme 24). Alternatively, the chiral centre (chiral element) may be present in a catalyst or even in the solvent.

In Scheme 1 we have illustrated these Type III reactions using the parent chiral centres in enantiopure form, and the diastereoisomeric products will therefore also be enantiopure. However, it should be clear that the starting material may be enantioimpure or even racemic; the resulting diastereoisomeric products in that case will be enantioimpure to the same degree.

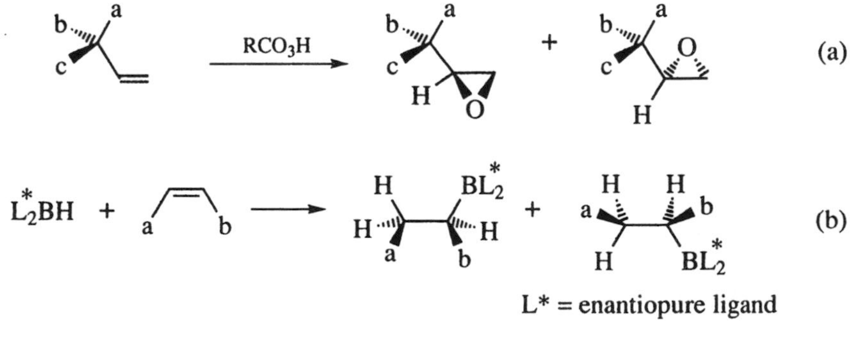

$L^* = $ enantiopure ligand

Scheme 1

In Scheme 1 also, the starting material must contain at least one chiral centre: in practice it may contain any number of chiral centres but will normally be a single diastereoisomer, although this may be a single enantiomer or enantioimpure as above.

Type III classification

Type III reactions involving additions to double bonds can, for our purposes, be sub-divided into three types:

(a) those, like the example in Scheme 1 (a) in which the parent chiral centre(s) and the prochiral double bond undergoing addition are contained within the same molecule; in this case the resulting diastereoselectivity is often described as *substrate-controlled* and we shall refer to these as Type $III_{s.c.}$;

(b) those in which the existing chiral centre(s) in the reagent and the prochiral double bond undergoing addition are in different molecules; the resulting diastereoselectivity is described as *reagent-controlled* and will be referred to as Type $III_{r.c.}$;

(c) those in which the substrate contains one or more chiral centres and a prochiral double bond undergoing addition and the reagent also contains one or more chiral centres; these involve double asymmetric induction and will be referred to as Type $III_{s.c./r.c.}$.

Most Type III reactions fall into one of these categories although a chiral centre can also be created by reaction of one of two identical but enantiotopic or diastereotopic atoms or groups on a prochiral sp^3-centre (Chapter 1). The above classification of Type III reactions should therefore be extended to include those in which 'prochiral centre undergoing reaction of one of its enantiotopic or diastereotopic atoms or groups' replaces 'prochiral double bond undergoing addition.' In practice, the majority of synthetically useful reactions involving Type III reactions other than addition to prochiral double bonds are enzyme catalysed (see Chapter 16).

TYPE III'$_{s.c.}$ REACTIONS

With the exception of simple chirality transfer reactions (see below), Type $III_{s.c.}$ reactions use starting materials in which at least one chiral centre is present initially and result in products in which at least one additional chiral centre is created. It is sometimes the case that because of symmetry, no chiral centres are present and yet the reaction can give rise to diastereoisomers. For example, reduction of the carbonyl group of 4-*tert*-butylcyclohexanone [Scheme 2 (a)] can give rise to *cis*- or *trans*-4-*tert*-butylcyclohexanol which are related as diastereoisomers (the carbonyl group has diastereofaces); in this case neither the starting material nor the product contain chiral centres.

The factors which bring about diastereoselectivity in this reaction are much

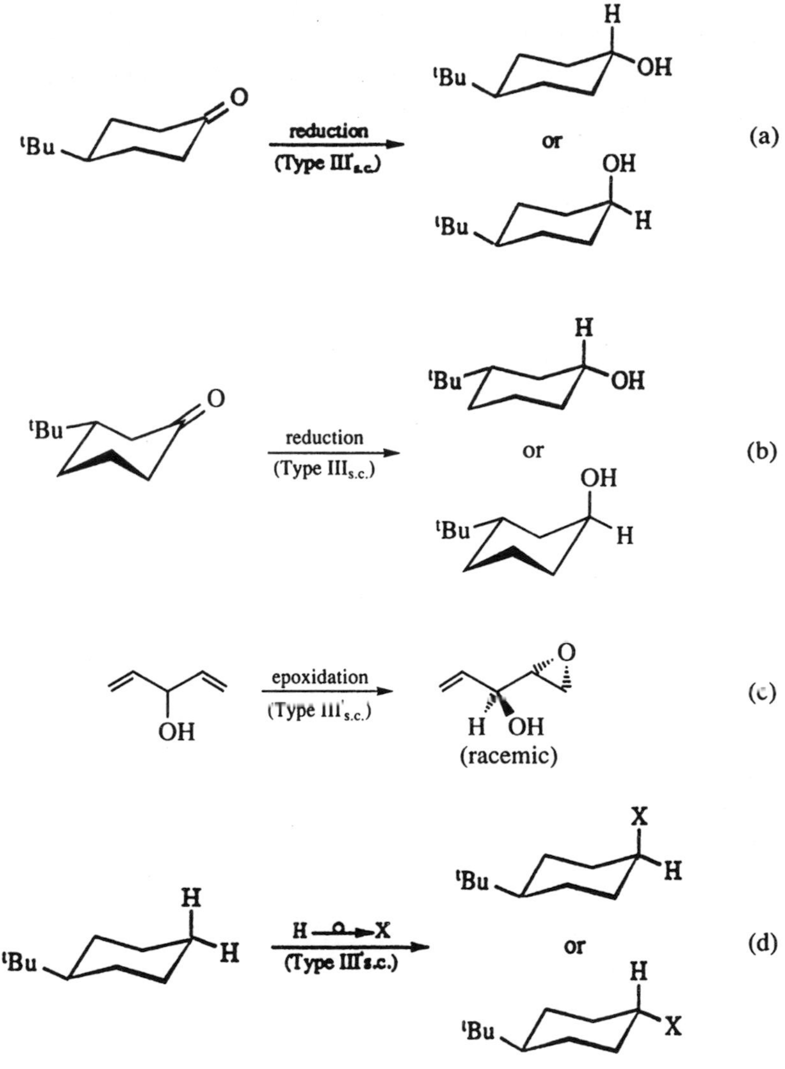

Scheme 2

the same as those which bring about diastereoselectivity in closely related Type $III_{s.c.}$ reactions, e.g. the reduction of 3-*tert*-butylcyclohexanone [Scheme 2 (b)] and so we shall refer to them as Type $III'_{s.c.}$. The products from a Type $III'_{s.c.}$ reaction will not always lack chiral centres; in Scheme 2 (c), the epoxide product contains two.

Likewise, one can conceive of a Type $III'_{s.c.}$ reaction in which one of a pair of diastereotopic atoms or groups is selectively reacted in a molecule containing no chiral centres [Scheme 2 (d)].

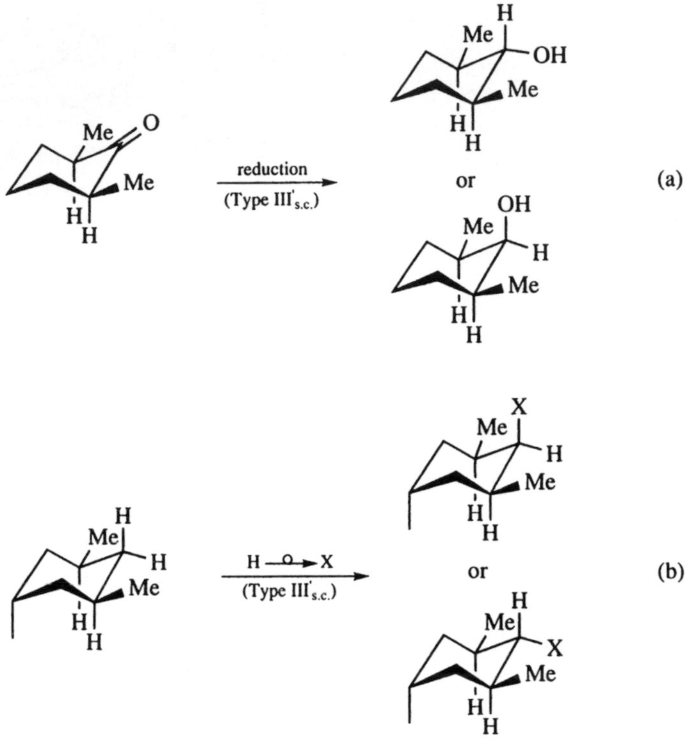

Scheme 3

Since *meso* compounds always contain pairs of chiral centres (chiral elements) which are of opposite configuration but otherwise identical, we may regard them as having no net chiral centres and therefore similar to substrates such as 4-*tert*-butylcyclohexanone in Schemes 2 (a), (c) and (d). Neither the diastereoselective reduction in Scheme 3 (a) nor the diastereoselective substitution in Scheme 3 (b) produce an additional chiral centre.[†]

Type III$'_{s.c.}$ reactions, therefore, are those in which addition to one diastereoface of a double bond occurs or reaction of one diastereotopic atom or group occurs in a substrate containing no net chiral centres; the product may or may not contain net chiral centres.

SIMPLE CHIRALITY TRANSFER REACTIONS

Simple chirality transfer reactions have already been considered in Chapter 5 and are a sub-class of Type III$_{s.c.}$ reactions. In these simple

[†]A chiral centre is commonly assumed to be an atom bearing four different tetrahedrally disposed substituents, but the IUPAC definition is 'an atom holding a set of ligands in a spatial arrangement not superimposable on its mirror image'. According to this definition, therefore, $(R)(S)$Cab centres such as those in the products in Scheme 3 (a) or (b) are not chiral.

chirality transfer reactions, a single new chiral centre is created under the influence of an existing chiral centre in the molecule but the parent centre is lost in the product. Exceptionally, therefore, in this sub-class of Type III reaction there is no net increase in the number of chiral centres.

Other (non-simple) chirality transfer reactions, in which more than one configured double bond is involved, may in fact give rise to an increase in the number of chiral centres. These will be considered in Chapter 13.

The sub-classification of Type III reactions used in this and succeeding Chapters is summarised in Figure 1.

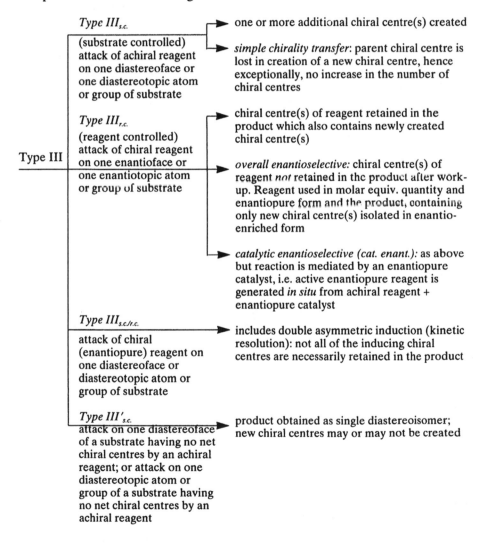

Figure 1. Sub-classification of Type III reactions

When two or more chiral centres are created (which excludes simple chirality transfer reactions), Type II reactions may be involved, i.e. the reactions become Type II/III$_{s.c.}$ or Type II/III$_{r.c.}$ (see Chapter 13).

Diastereoselective substrate-controlled reactions (Type III$_{s.c.}$)

How can an existing chiral centre in a molecule influence or control the configuration of a chiral centre which is being formed by addition to a double bond within the same molecule? In other words, what are the factors which bring about asymmetric induction in this reaction? This is the question which this and the following two chapters will seek to answer.

Asymmetric induction in all kinetically controlled Type III reactions is brought about by minimisation of adverse steric, torsional, angle and other strain factors and by optimisation of stereoelectronic effects, leading to a lowering of the energy of one diastereoisomeric transition state by comparison with another. There are, however, a multitude of ways in which these strain factors and electronic effects combine to mediate Type III reactions; the importance of conformational control and involvement of rings in this mediation will become apparent.

It is helpful to consider steric and electronic effects separately although in most cases they operate interdependently.

Asymmetric induction mediated by steric effects in Type III$_{s.c.}$: the importance of conformational control

CYCLIC COMPOUNDS

The fact that two atoms or groups cannot occupy the same space at the same time is an important and familiar concept and an ubiquitous means by which diastereoface discrimination can occur. In Scheme 4 are a number of monocyclic substrates which react by Type III$_{s.c.}$ reactions with high or complete diastereoselectivity.[1-5]

In all these examples, the bulky group (alkyl or trimethylsilyl) resides next to the double bond and, it is assumed, hinders approach of the attacking reagent to one face of the double bond. In Scheme 4 a variety of reaction types are represented, including those proceeding via carbanion, carbocation, radical and cycloaddition mechanisms.

In Scheme 5, 'serial' substitution adjacent to the carbonyl group gives the disubstituted products (1) and (2) in a ratio of 8:1.[6] When the order of addition of the alkyl groups is reversed, the major product obtained is (2). This tactic for reversal of the sense of diastereoselectivity is commonly used since it is always the second alkyl group which is directed *anti* by the bulky group present in the starting material.

Attack on dienes can also be directed to one diastereoface by steric effects.

The starting complex (3) in Scheme 6 is prepared from arabinose in enantiopure form and the complexed diene (4) is derived by the action of tetrafluoroboric acid. Attack of the nucleophile is directed to the less hindered face of (4).[7] The substrate (4) differs from those in Schemes 4 and 5 in that it does not contain a chiral centre; rather, the molecule (complex) as a whole is chiral.

Provided that a reaction is irreversible, the products will be formed under kinetic control. This may be *steric approach control*, if the transition state is early, or *product-development control*, if the transition state is late on the

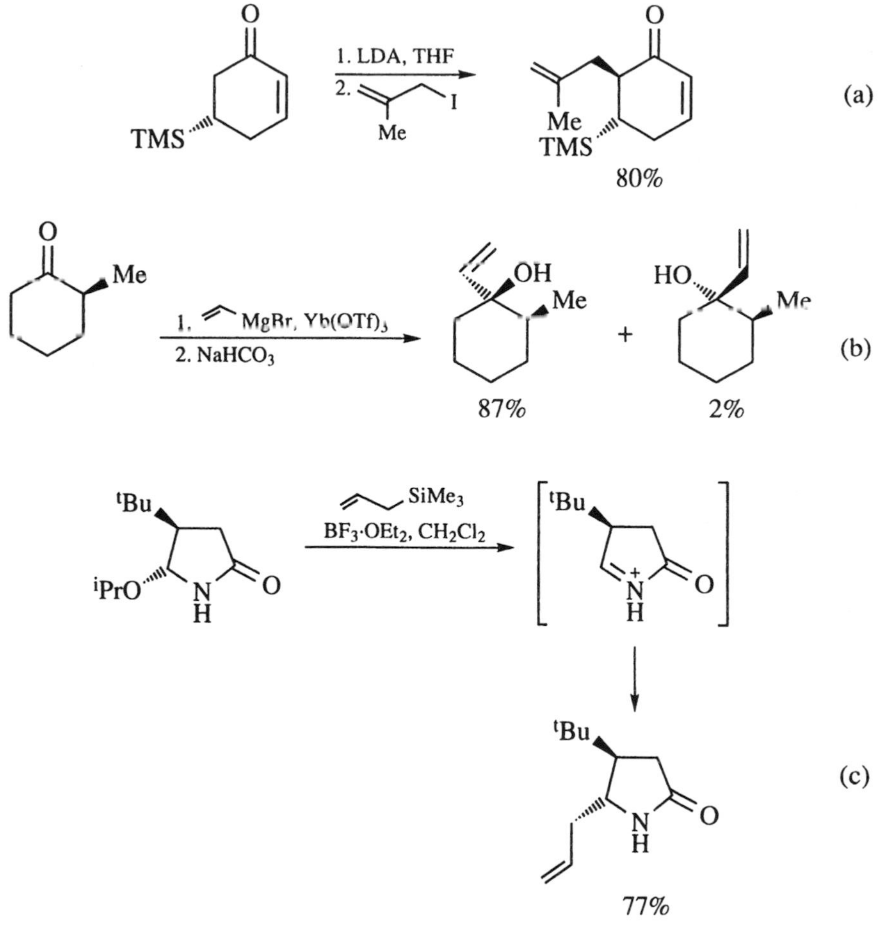

Scheme 4 *(continued)*

Scheme 4 *(continued)*

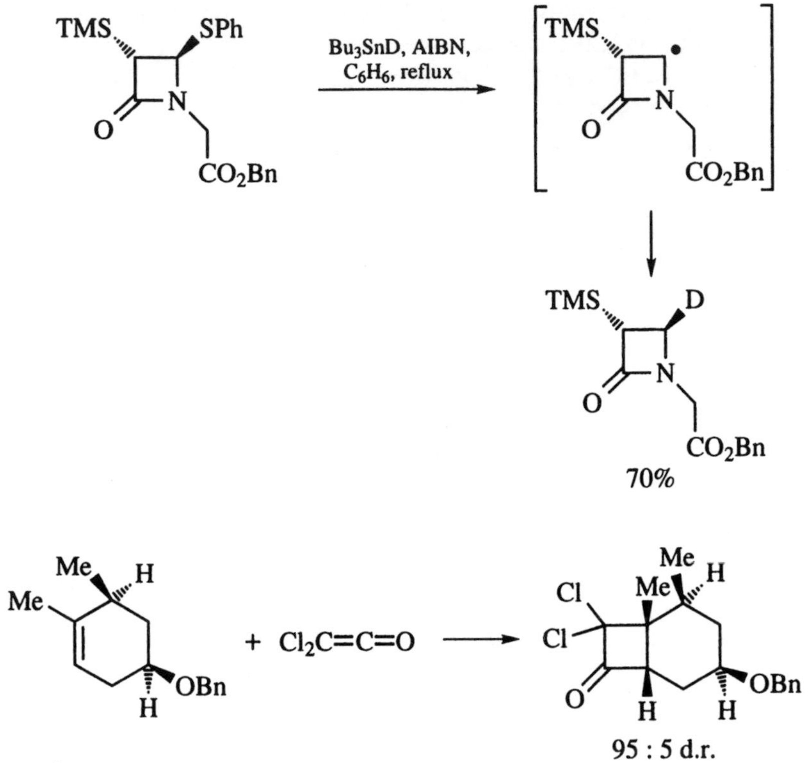
TMS
SPh
O
N
CO2Bn
Bu3SnD, AIBN,
C6H6, reflux
TMS
D
70%
Me
H
OBn
+ Cl2C=C=O
Cl
95 : 5 d.r.

Scheme 4

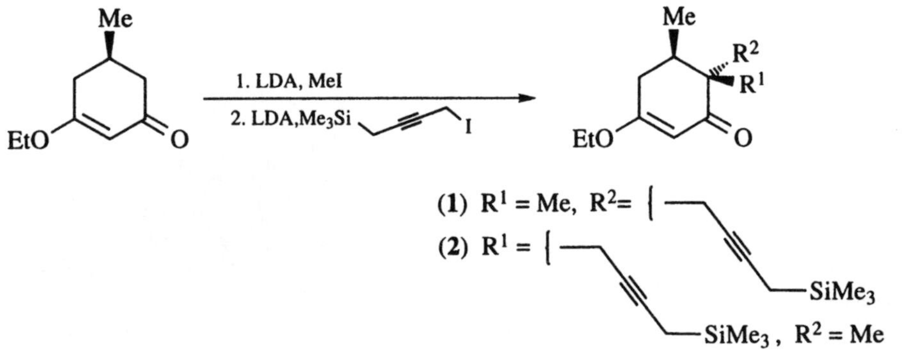
Me
EtO
O
1. LDA, MeI
2. LDA,Me3Si
I
R2
R1
(1) R1 = Me, R2=
(2) R1 =
SiMe3
SiMe3 , R2 = Me

Scheme 5

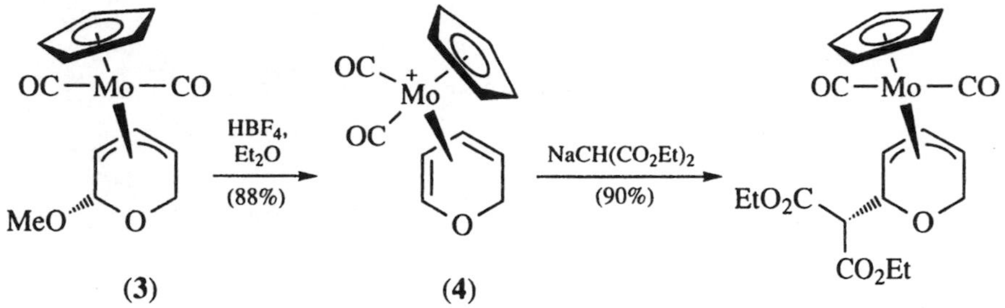

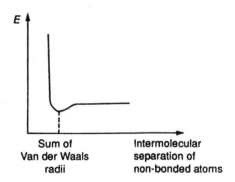

Scheme 6

reaction coordinate. The examples in Schemes 4–6 are those in which steric approach control is operative.

As two non-bonded atoms in a molecule approach one another there is a weak attractive interaction which is transformed into a strongly repulsive one when the internuclear distance closes to less than the sum of the van der Waals radii (Figure 2). In the examples in Schemes 4–6 we can infer that in each case the transition state for attack on the more hindered face requires non bonding atoms in the reagent and the allylic substituent to be closer than the sum of their van der Waals radii.

Figure 2

In spite of the many examples of diastereoselectivity apparently brought about by steric effects, as in Schemes 4–6, one cannot assume that addition will always be *anti* with respect to an existing (bulky) substituent adjacent the reaction centre. Thus reduction of the substituted cyclopentanone (5) with sodium tetrahydroborate [Scheme 7 (a)] takes place preferentially with addition of hydride to the *same* face as the adjacent *p*-methoxyphenyl ring,[8] albeit *anti* to the more distant phenyl group.

Similarly, reaction of the γ-silyloxy-substituted cycloalkenone (6) with the allylsilane (7) gives the product corresponding to attack *syn* to the γ-silyloxy

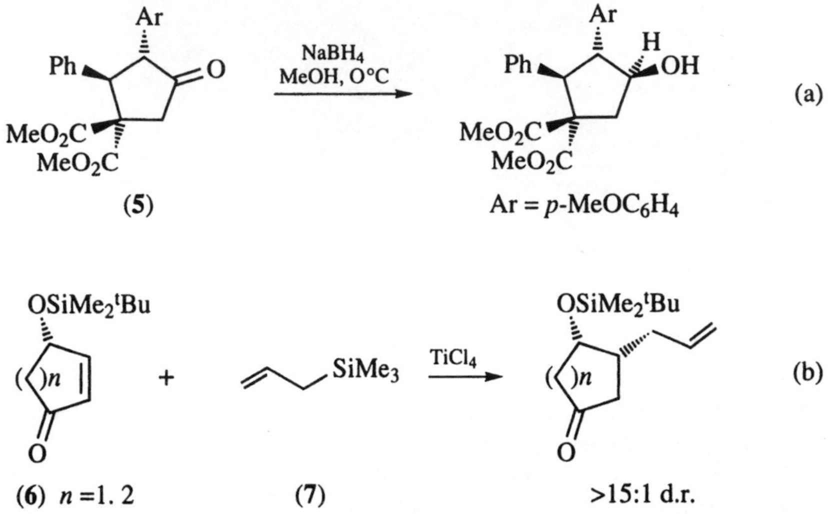

Ar
Ph   O   NaBH₄   Ph   H
                                 MeOH, 0°C   OH   (a)
MeO₂C   MeO₂C
MeO₂C   MeO₂C
        (5)   Ar = p-MeOC₆H₄

OSiMe₂ᵗBu   OSiMe₂ᵗBu

( )n   +   SiMe₃   TiCl₄   ( )n   (b)

O   O
(6) n =1. 2   (7)   >15:1 d.r.

Scheme 7

group [Scheme 7 (b)];[9] here, however, there is no more distant substituent to influence the reagent approach, making the result more surprising.

Clearly, in these cases factors other than the obvious steric repulsion control the diastereoselectivity.

Carbene additions to alkenes are also reactions which show 'contrasteric' diastereoselectivity: the major diastereoisomer formed in the (Type II) cyclopropanation in Scheme 8 is the more sterically hindered one.[10]

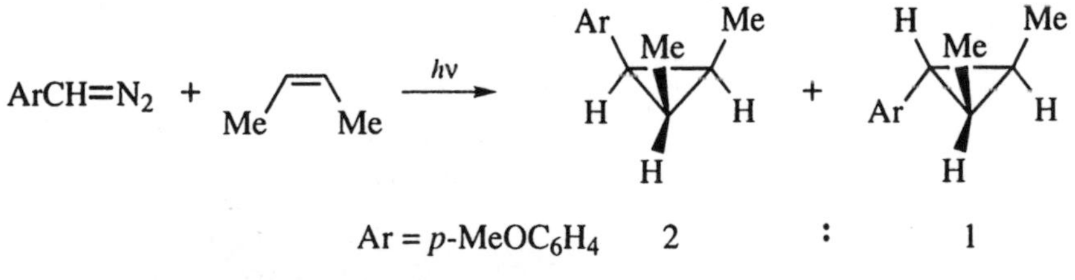

ArCH=N₂ +   Me   Me   hv   Ar   Me   Me   H   Me   Me
                                       H   H   +   Ar   H
                                           H   H

Ar = p-MeOC₆H₄   2   :   1

Scheme 8

Hammond[11] postulated that highly exothermic reactions, like the carbene addition in Scheme 8, will have reactant-like or 'early' transition states. In these transition states, the larger than average internuclear distance between interacting centres will result in a weakening of steric repulsion which may be present between substituents in the product; this repulsion will only develop fully at a later point on the reaction coordinate beyond the transition state, i.e. too late to influence the ratio of diastereoisomers formed. With this attenuation of steric repulsion in the transition state, other factors can manifest themselves; for example, two non-bonded atoms may find themselves just outside their combined van der Waals radii (Figure 2) and experience a mild attractive interaction, the result of which may be 'contrasteric' in its effect on the diastereoselectivity.

In product-development control, the diastereoisomer ratio is more likely to reflect the thermodynamic stability of the products. This is because the transition states, being further along the reaction coordinate, will more likely resemble the respective product diastereoisomers and hence reflect their relative energies.

Product-development control is relatively uncommon in kinetically controlled Type III$_{s.c.}$ reactions.[12] However, epoxidation of the dimethylcyclohexene (8) [Scheme 9 (a)] gives an excess of the cis-epoxide and so an explanation based on steric repulsion used in Schemes 4–6 is inadequate here.[13]

To rationalise this syn-epoxidation we can identify a further way in which steric effects can favour the formation of one diastereoisomer over another, other than by direct steric shielding of one diastereoface from attack by the

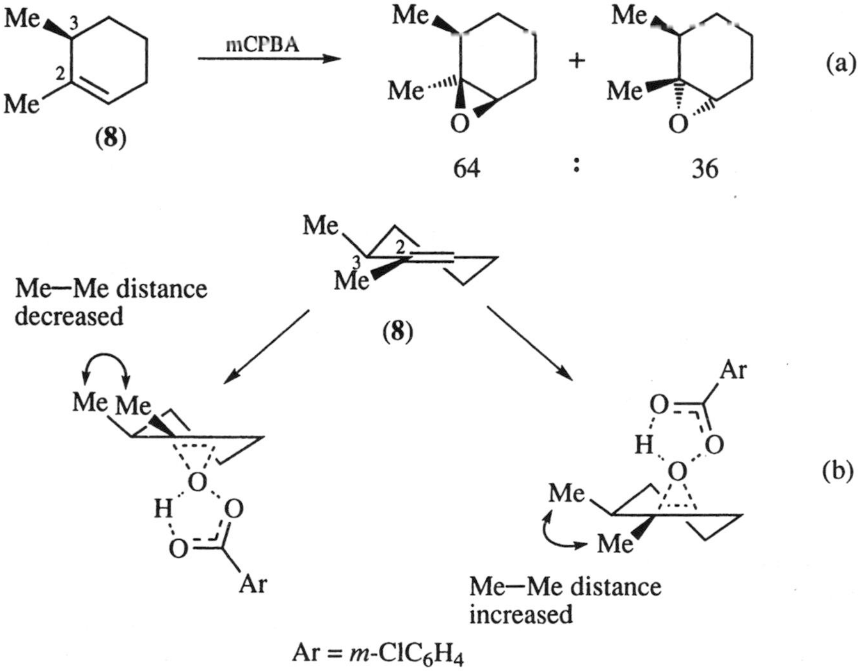

Scheme 9

reagent. This is the development of greater steric effects by interaction between *existing* groups in the molecule as a result of attack on one diastereoface rather than the other. Thus, as attack takes place from the top face of the double bond of (**8**) [Scheme 9 (b)] *syn* to the C-3 methyl group, the distance between the two methyl groups *increases* as the hybridisation at C-2 changes from sp^2 to (formally) sp^3. Conversely, attack from the bottom face *anti* to the methyl group *decreases* the distance between these two methyl groups.

It is assumed here that it is the half-chair conformation with the allylic methyl group pseudo-equatorial that undergoes reaction and that the allylic methyl group therefore offers little steric discouragement to attack from the top face by what is a reagent of small bulk.

Recognition of steric effects of this type will clearly require conformational analysis of the substrate and product—in this case the cyclohexene and its derived epoxide—together with an understanding of the reaction mechanism. The analysis in Scheme 9 (b) assumes that, in the transition state, significant sp$^2 \rightarrow$ sp^3 re-hybridisation has taken place and bonds to the epoxide oxygen are partially made. In other words it is assumed that this reaction does not have a particularly early transition state, i.e. it proceeds with product-development control.

The third, less direct, way in which steric effects can bring about diastereoselectivity is by suppressing the conformational freedom of a molecule. Probably one of the most familiar examples of this is the conformational anchoring that the *tert*-butyl group imposes on the cyclohexane ring (Scheme 10).

In a 4-substituted cyclohexanone there will be an equilibrium between the two chair conformations (**9**) and (**10**) (Scheme 10), and this must be taken into account when attack on the two faces of the carbonyl group is considered. A cyclohexanone normally has a preference for axial attack by non-bulky nucleophiles (see Chapter 10). If both chair forms are present, axial attack on each of them will lead to diastereoisomers (Scheme 10). With R = ^{t}Bu, however, the equilibrium is completely on the side of the chair having the ^{t}Bu group equatorial (**9**; R = ^{t}Bu); steric interaction between the ^{t}Bu group and the *cis*-1,3-axial hydrogens in (**10**; R = ^{t}Bu) is responsible for this. Consequently, the highest diastereoselectivity in axial nucleophilic attack on 4-substituted cyclohexanones is found with 4-*tert*-butylcyclohexanone e.g. using LiAlH$_4$ (Scheme 10).

Generally, the greater the number of conformations significantly populated by a molecule the more likely is it that reactions will be taking place from more than one conformation.

The Curtin–Hammett principle reminds us that the rate of reaction of a molecule is a function not only of the concentration of any reacting conformation but also of its transition state energy in the reaction.[14] Thus a reaction *may* take place through a little-populated conformation if the

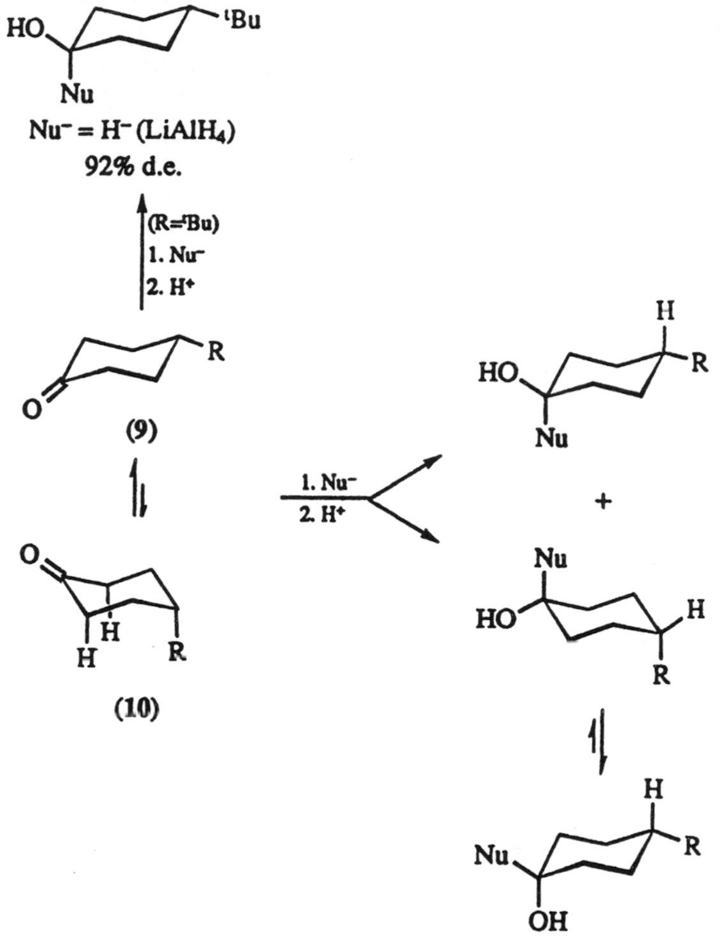

Scheme 10

transition state for reaction of that conformation happens to be significantly lower in energy than that for reaction of a major conformation. Nevertheless, it remains generally true that the best way to encourage reaction from a single conformation is to suppress the concentrations of all other conformations of the molecule if necessary by the introduction of nicely placed substituents. Conversely, in Type III$_{s.c.}$ reactions, reaction from more than one conformation is likely to result in a loss of diastereoselectivity, as would be the case in Scheme 10 (R ≠ ^{t}Bu) if both chairs react by axial attack.

Clearly, the nature of the cyclic compound undergoing reaction will be important here: cyclohexane has a well defined conformational minimum— the chair—and its reactions via other conformations can usually be discounted. Diastereoselectivity in Type $III_{s.c.}$ reactions will therefore be more likely with cyclohexane derivatives than, for example, analogous medium-ring compounds, for which a greater number of reactive conformations of similar energy is probable. The greatest degree of conformational freedom will be present in acyclic compounds and it is in the latter that stereocontrol is most difficult.

Our consideration of steric effects so far has focused on the effect of substituents within the substrate but an increase in size of the reagent may also lead to greater diastereoselectivity as in the reduction of (11) in Scheme 11. The effect of the two (equatorial) *cis*-methyl groups in the 3- and 5-positions in (11) is much the same as a *tert*-butyl group, i.e. they anchor the cyclohexane ring in one particular chair form. The axial methyl group at position 3 opposes the normally preferred axial attack by $LiAlH_4$ (see above) and increasing the bulk of the hydride-supplying reagent greatly increases the preference for equatorial attack.[15]

(11)				
	$LiAlH_4$	42	:	58
	$LiAl(O^tBu)_3H$	4	:	96

Scheme 11

ACYCLIC COMPOUNDS

In order to achieve diastereoselectivity in addition to a double bond in an acyclic compound, ways must be found of limiting the conformational freedom of the molecule.

$A_{1,3}$-strain[16]

$A_{1,3}$-strain is the steric strain associated with some substituted allyl systems. Minimisation of $A_{1,3}$-strain can be an important factor in limiting the conformational freedom of an allyl system, thus increasing the likelihood of diastereoselectivity in attack on the double bond.

In a Z-configured allyl system such as that in Scheme 12, conformation (A) is much preferred over (B) or (C), largely as a result of the 1,3-methyl groups being closer than their van der Waals radii in (B) and (C). Conformation (C) has the two methyl groups eclipsed thus raising its energy above that of (B); $A_{1,3}$-strain is minimised in (A).

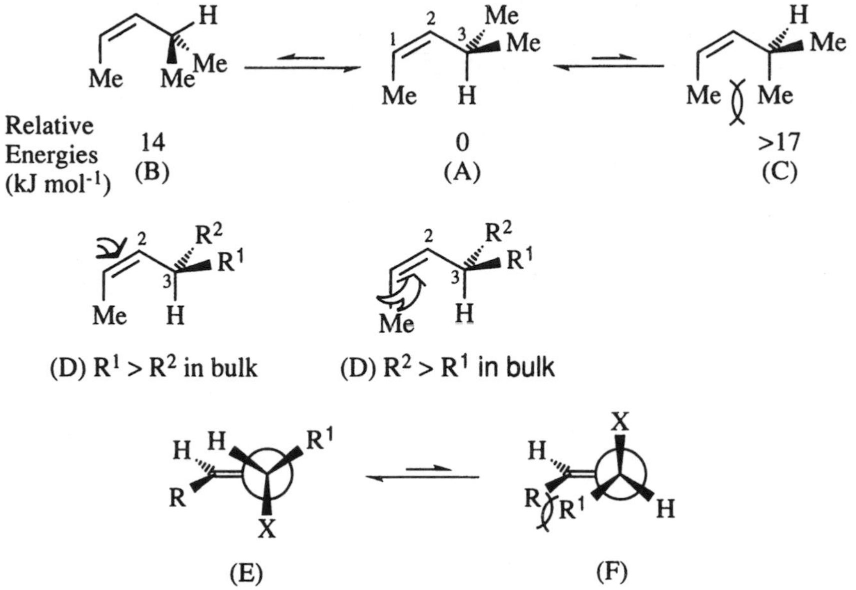

Scheme 12

When the geminal methyl groups in (A) are replaced by groups R^1 and R^2 of differing bulk, as in (D), the two diastereofaces of the double bond will be differentially shielded, and preferential attack on the less hindered face can be anticipated, as shown.

It is relatively inexpensive, in terms of energy, to rotate around the 2,3-bond in (D) by up to ~30°. This is important when the σ-bond to the allylic substituent R^1 or R^2 interacts with the π-bond (a σ–π interaction; see Chapter 10), as such an interaction will usually be maximised when the C—R^1 or C—R^2 σ-bond is parallel to the π-orbital of the double bond.

Assuming that the σ–π interaction with an allylic σ-bond C—X ($R^2 = X$) activates the double bonds towards attack, the two conformations from which reaction can occur are shown as Newman projections in (E) and (F). Because the probable steric interaction in (F) is absent from (E), the latter will be present in higher concentration; attack on E will be easier from the top face which is not sterically shielded by X, i.e. the reaction will be diastereoselective.

$A_{1,3}$-strain was recognised as a general property of appropriately substituted allyl systems by Johnson[17] and applied to explain the stereochemistry of, e.g., cyclohexylidene derivatives (Scheme 13). In (12) the Z-configured exocyclic double bond has R' *cis* to the substituted α-carbon and the R,R' steric interaction will, if sufficient, favour the alternative chair conformation (12') in which R assumes the axial position. The axial location

(12)   **(12′)**

Scheme 13

of R will give rise to considerably greater steric shielding in (12′) of the bottom face of the double bond and this can be expected to direct facial attack on the double bond accordingly.

Avoidance of $A_{1,3}$-strain can be important in bringing about conformational control in the vicinity of a reacting double bond in *any* molecule, but it is of particular importance for acyclic compounds. Cyclic compounds have relatively fewer degrees of freedom and conformational control can be mediated by the steric and/or electronic effects of substituents remote from the double bond [e.g. the *tert*-butyl group of 4-*tert*-butylcyclohexanone (Scheme 10)]. For acyclic compounds, however, any conformational control must usually be exercised by substituents adjacent to the reacting double bond and $A_{1,3}$-strain is one of the few ways in which this can be accomplished. $A_{1,3}$-strain may also arise in allyl systems containing heteroatoms in the double bond and/or at the allyl position. In Scheme 14 the arrows indicate the more accessible face of the preferred conformer resulting from $A_{1,3}$-strain, in such heteroatom-substituted allyl systems.

We have seen already that $A_{1,3}$-strain is an important factor in the diastereoselectivity of some 2,3-sigmatropic rearrangements and S_N2' reactions (Chapter 5). Further examples in which alleviation of this strain is the major cause of diastereoselectivity will be given later. There will also be

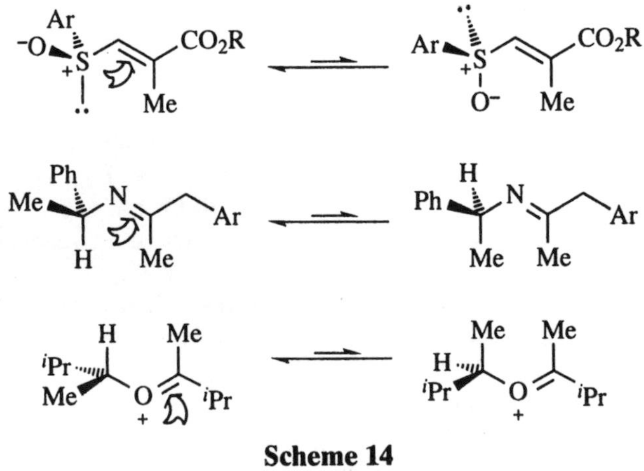

Scheme 14

examples of the interplay between steric and electronic effects of the $\sigma-\pi$ type referred to above, such that the stereoelectronic effect of an allylic substituent *and* the minimisation of $A_{1,3}$-strain both tend to direct attack to the same diastereoface of the double bond.

$A_{1,2}$-strain

The existence of $A_{1,2}$-strain as a general property of appropriately substituted allyl systems was also explicitly recognised by Johnson and Malhotra and applied by them particularly to α-substituted cyclohexenyl derivatives (Scheme 15).[17]

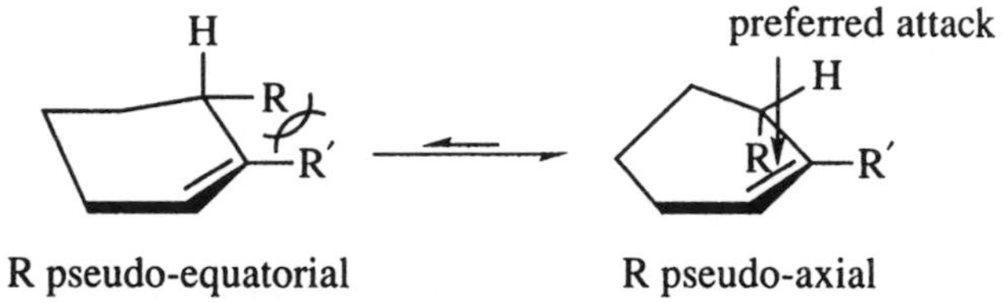

R pseudo-equatorial   R pseudo-axial

Scheme 15

When R and R' are of sufficient bulk, the favoured half-chair may be that in which the R group occupies the pseudo-axial rather than the pseudo-equatorial position; by this means, the steric interaction between R and R' is minimised. The consequence of this pseudo-axial location for R is that attack on the opposite face of the cyclohexenyl double bond will be less hindered, as shown in Scheme 15.

$A^{1,2}$-strain is particularly prevalent in the stereochemistry of α-substituted enamines and enolates of cyclohexanone (Scheme 16).

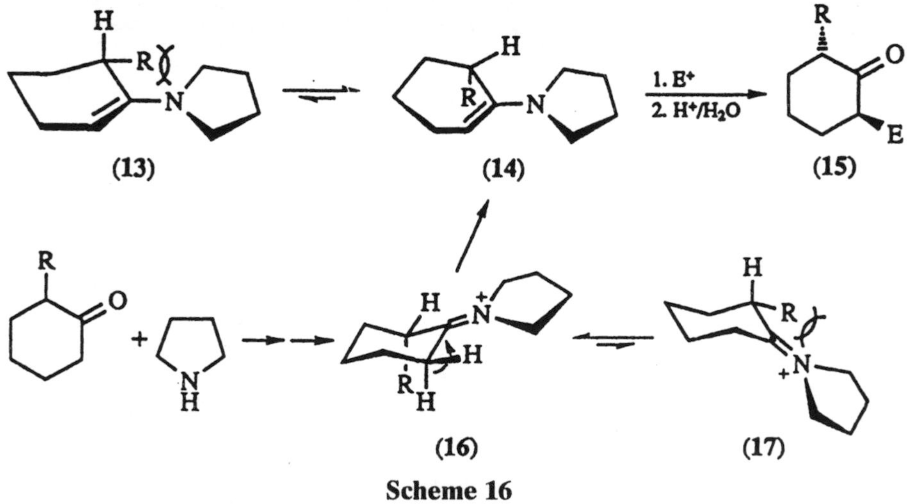

(13)   (14)   (15)

(16)   (17)

Scheme 16

For maximum activation of the enamine double bond towards attack by electrophiles, overlap of its π-orbital with the nitrogen lone pair is required. This is the conformation around the C_{sp^2}—N bond which maximises the $A_{1,2}$-strain in conformer (13). Attack on the more abundant conformer (14), on its less hindered (top) face, is to be expected leading to the 1,3-*trans*-substituted cyclohexanone (15) after acid hydrolysis.

Formation of the enamine itself occurs by loss of an axial proton (stereoelectronic control) from the first-formed iminium salt. Preferential formation of the less-substituted enamine double bond regioisomer (14) is a consequence of $A_{1,3}$-strain in (17), which means that loss of an axial proton is more likely from (16). The lesser strain in the less substituted enamine double bond means that this will be the preferred isomer even if enamine formation is under thermodynamic control.

Dummy substituents

Sometimes it may be expedient to introduce steric strain deliberately so as to favour one of two diastereoisomeric transition states for a reaction. This can be accomplished by the selection and location of a bulky substituent which, having served its purpose, can easily be removed after the diastereoselective reaction.

The role of bulky substituent is often assumed by the trimethylsilyl (TMS) group. Thus the enantiopure deuteriated β-lactam (19) (Scheme 17) is obtained by desilylation of the product (18) from Scheme 4 (d). The corresponding deuterium-labelled β-lactam was not available by direct reduction of (20).[4]

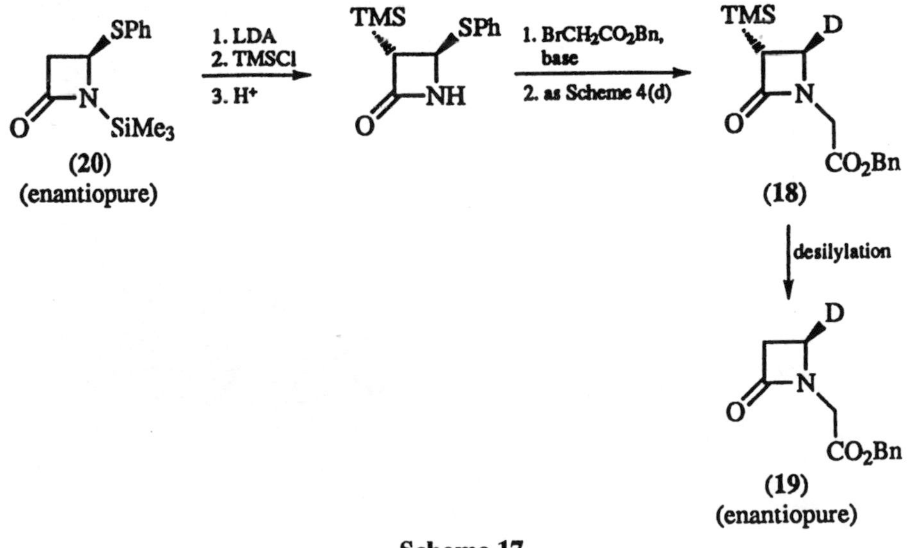

(20)
(enantiopure)

1. LDA
2. TMSCl
3. H+

1. BrCH₂CO₂Bn, base
2. as Scheme 4(d)

(18)

desilylation

(19)
(enantiopure)

Scheme 17

BICYCLIC COMPOUNDS

Just as a cyclic molecule has less conformational freedom than its acyclic analogue, so can the conformational freedom of a ring be further reduced by its incorporation into a bicyclic skeleton. In some bicyclic compounds, the shape or molecular architecture of the skeleton may of itself give rise to shielding of one diastereoface of, e.g., a double bond in one of the rings. Thus a number of bicyclo[$m.n.0$] systems, where $m + n$ is usually less than 7, have wedge-shaped contours; this means that the *exo* and *endo* faces are clearly differentiated, with the *exo* face more exposed to attack. In the examples in Scheme 18, attack is in each case completely diastereoselectively *exo* as indicated.[18-20]

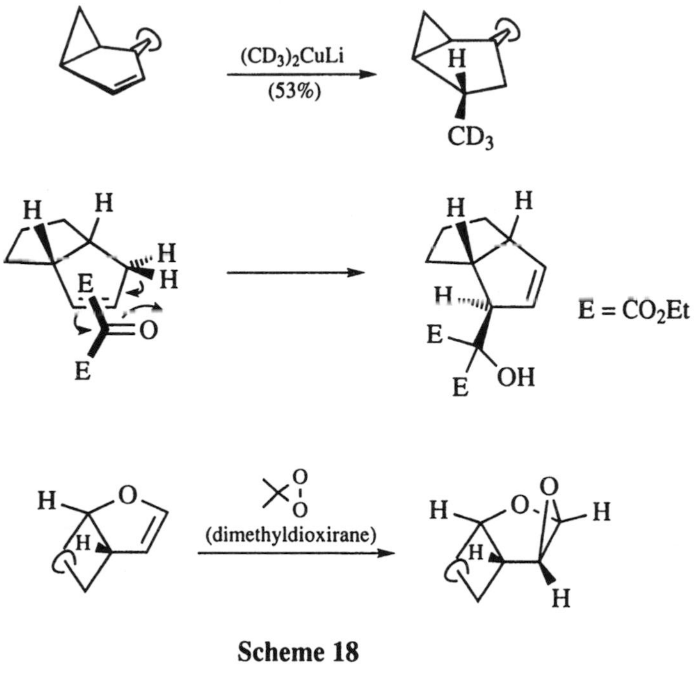

Scheme 18

Of course, the extent of this predilection for *exo* attack will depend on the type and location of substituents in or on the bicyclic skeleton, on the stereoelectronic effects associated with attack on the functional group, and on the bulk of the attacking reagent.

In the examples in Scheme 18, one ring can be regarded as a pair of adjacent *cis*-substituents on the other ring, which is undergoing addition; attack is directed to the opposite (*exo*) face of the second ring, in the same way as the allylic substituents direct attack in Scheme 4.

The decomposition of esters of *N*-hydroxypyridine-2-thione, e.g. **(21)**

(Scheme 19), is now a standard route to carbon-centered radicals by homolysis of the N—O bond followed by decarboxylation. In the present case, reaction of the radical (22) takes place on the *exo* face. The derived α-sulphonyl radical (23) then attacks the thione sulphur of the starting material and homolysis of the N—O bond sets up a chain reaction.[21]

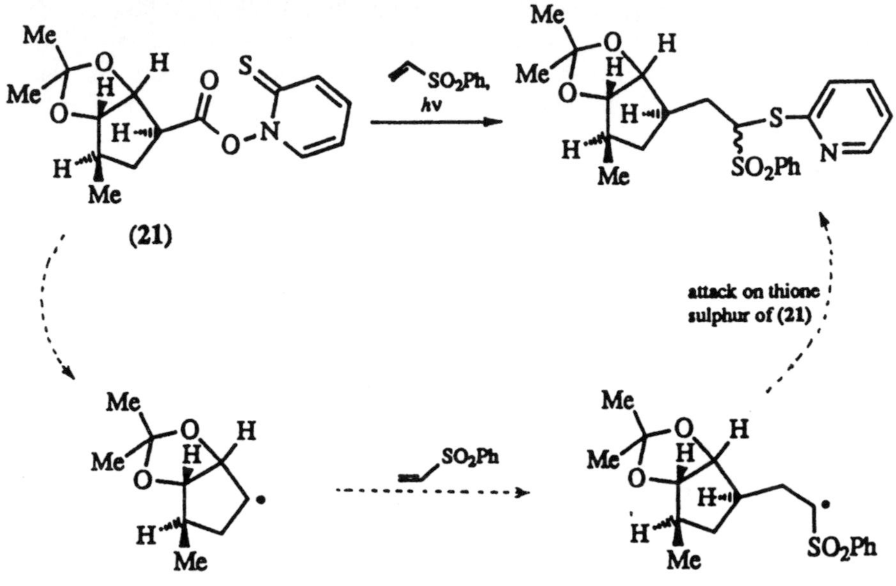

Scheme 19

Bicyclo[2.2.1] and bicyclo[3.2.1] systems also have shapes which predispose them to attack on the *exo* face (Scheme 20). The pair of *endo*-hydrogens on the two-carbon bridge in these skeletons (see (a)) assist in this *exo*-directing steric effect.[22-25]

In example (b) in Scheme 20, the *exo* or *endo* orientation of the groups introduced can be reversed by changing the order in which the alkylations are performed (cf. Scheme 5), thus adding to the usefulness of the *exo*-directing effect.

In example (d) in Scheme 20, the regioselectivity of the selenenium ion ring opening is controlled by the electronegative CN and OAc substituents; these discourage a build-up of positive charge at C-6, and so direct the attack of chloride to C-5.

Bicycle[2.2.1] substrates are particularly important in synthesis because of their accessibility by Diels–Alder reactions and because of the high *exo* diastereoselectivity of addition to their double bonds. Type 0 ring opening of

[2.2.1] substrates, diastereoselectively modified in this way, can then lead to single diastereoisomers of the corresponding cyclopentane or cyclohexane derivatives (Chapter 3).

The predilection for *exo* attack [which can be overidden by the steric effects of substituents at the 7-position (see below)] can only in part be attributed to the *endo*-hydrogens on the two-carbon bridge, because even

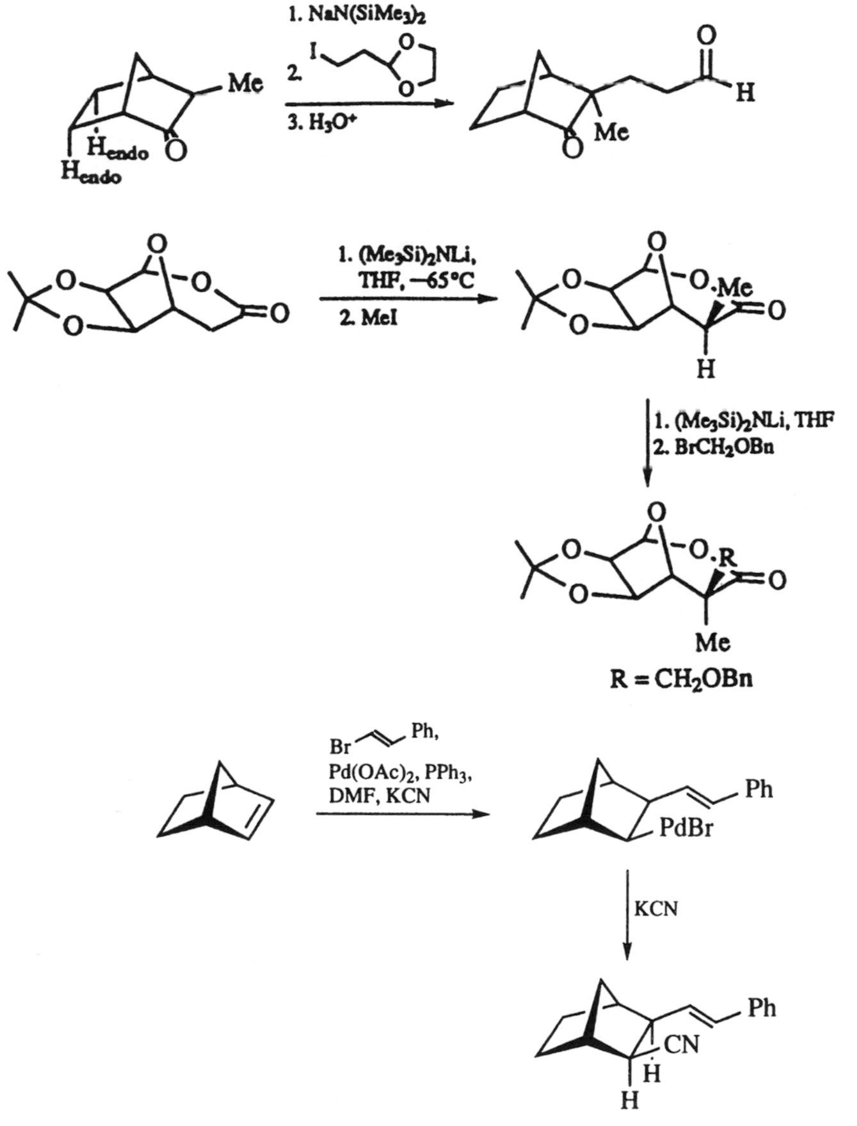

(a)

(b)

R = CH₂OBn

(c)

Scheme 20   *(continued)*

Scheme 20 *(continued)*

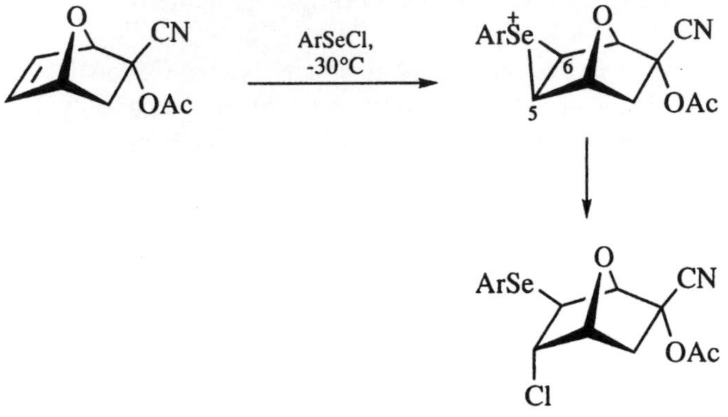

Scheme 20

bicyclo[2.2.1] substrates lacking these hydrogens also undergo preferential *exo* attack. Considerable effort has been made to understand the origin of this preference for *exo* selectivity, which does not have its origin in steric effects (see Chapter 10).

It is predictable that the inherent tendency for *exo* attack in most bicyclo[2.2.1] derivatives will be eroded when the bridging carbon atom (C-7) is substituted, as is seen with camphor where *endo* attack can intrude and may even become exclusive (Scheme 21).[26]

Scheme 21 illustrates an advantage of using compounds that have little conformational freedom: the effect of introducing substituents on the diastereoselectivity of their reactions is more reliably predictable.

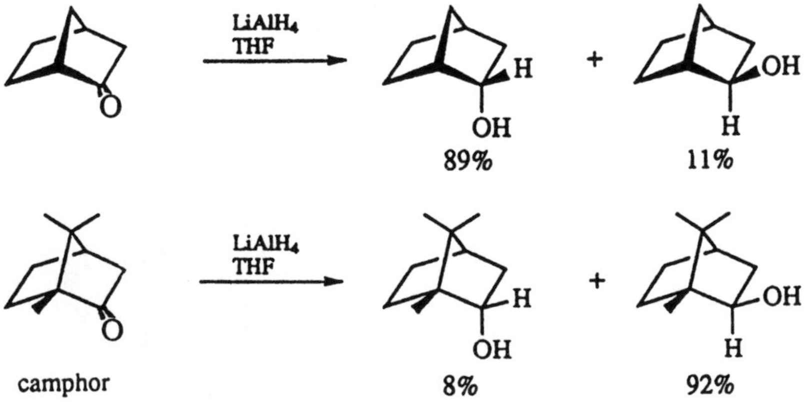

Scheme 21

In the bicyclic systems considered so far there has either been no conformational freedom or so little that the inherent diastereoface preference is not affected. Many other bicyclic systems, although they may exist largely as single conformers, show little inherent difference in reactivity between the two diastereofaces of their double bonds. Here again the introduction of substituents can be used to manipulate the diastereoselectivity in a predictable way in a Type III$_{s.c.}$ reaction, always assuming that the introduction of this substituent does not alter the preferred conformation of the molecule.

For example, *trans*-decalin is a relatively flat molecule, and its derived alkenes undergo addition with almost equal facility at both faces of the double bond, e.g. the epoxidation of (24) (Scheme 22).[27]

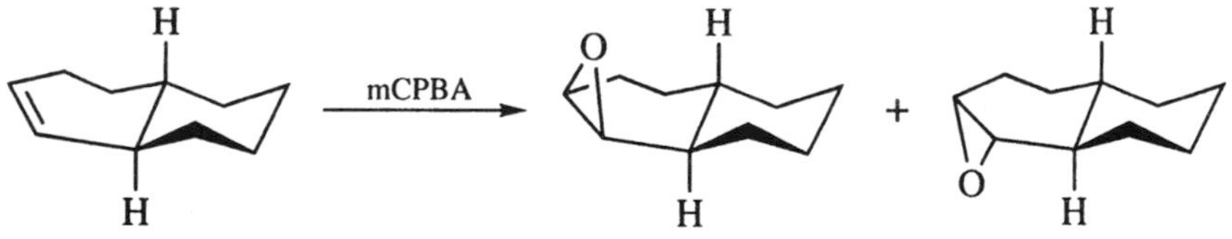

(24)   55   :   45

Scheme 22

Introduction of a methyl group at the ring junction has a pronounced effect on the relative accessibility of the diastereofaces of double bonds contained in the *trans*-decalin. This has been investigated in detail in steroids, where rings A and B constitute such a system substituted by a methyl group (C_{19}) at the ring junction. The stereochemistry of steroid reactions is generally dominated by the presence of the 18- and 19-methyl groups which stand guard over the top (β) face of the molecule [see (25)] and direct attack onto the lower (α) face. Thus in the epoxidation of cholest-2-ene (25) (Scheme 23), the α-epoxide (26) is produced with high diastereoselectivity.[28]

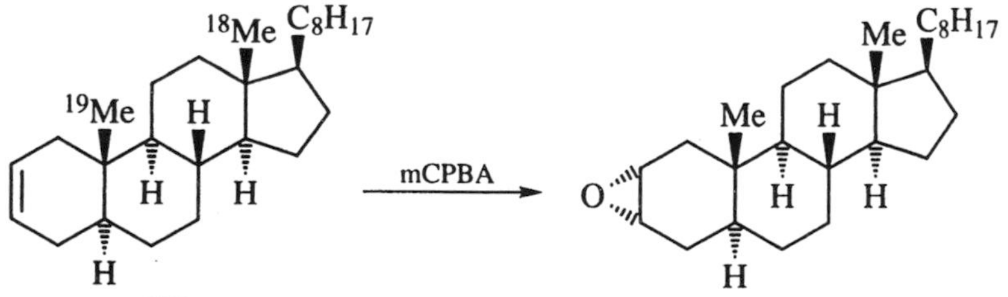

(25)   **(26)**

Scheme 23

However, the degree to which the 19-methyl group hinders attack on the β-face is less in cholest-4-ene (**27**) and cholest-5-ene (**28**) (Scheme 24) than in cholest-2-ene (**25**), in spite of the fact that their double bonds are closer to the 19-methyl group.[29]

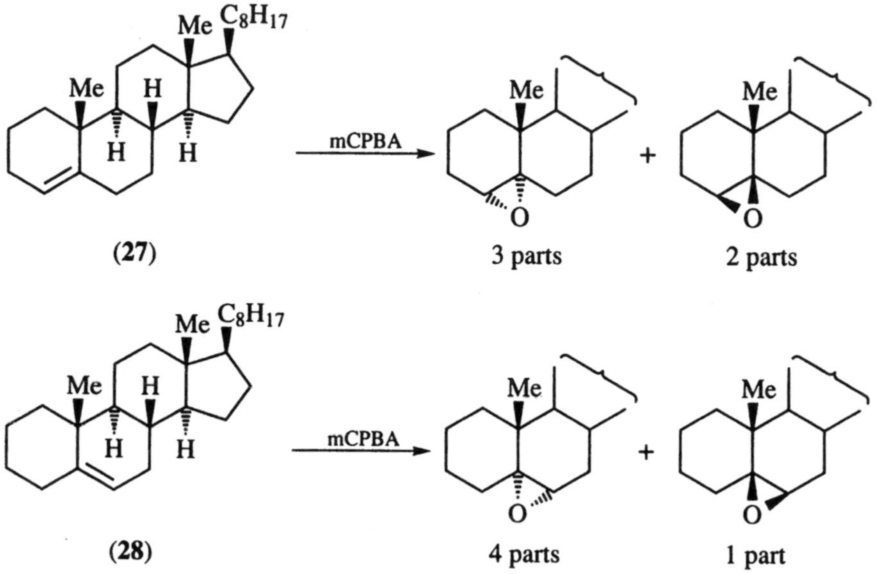

Scheme 24

Rationalisation of these disparities in α-:β-epoxide ratios requires an examination of the conformations of rings A and B in the three isomeric alkenes (Figure 3).

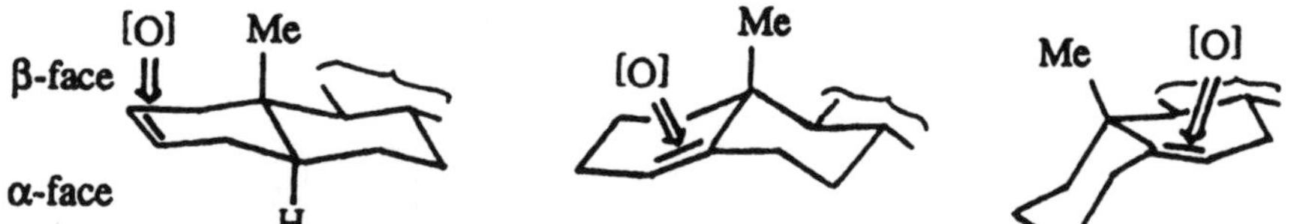

Figure 3

It can be seen that the origin of the apparently greater steric directing effect which operates in (**25**) is the result, at least in part, of the proximity of the 19-methyl group to the trajectory which must be followed by the *m*-chloroperoxybenzoic acid when it engages on the β-face with the π-orbital of the 2,3-double bond. For both (27) and (28), the 19-methyl group will be less effective at hindering this attack on the β-face because it is tilted away from the double bond.

As it stands, this is a superficial analysis of the problem of why, and by how much, the α:β ratios differ in epoxidation of the double bonds. It ignores, for example, the relative ease of attack on the α-faces in (25), (27) and (28).

Transition-state energies for reactions at these α- and β-faces are only accessible by calculation and while considerable progress continues to be made, we are still not routinely able to predict quantitatively small differences in transition state energy for reactions such as the epoxidations in Scheme 24.

STERIC EFFECTS IN REACTIONS OF POLYCYCLIC COMPOUNDS

Diastereoface differentiation of a double bond can be accomplished by its incorporation into a polycyclic structure. Thus the double bond in (29) (Scheme 25) is attacked by deuterodiborane from the more accessible top face to give (30) after the normal oxidative work-up followed by acetylation.[30]

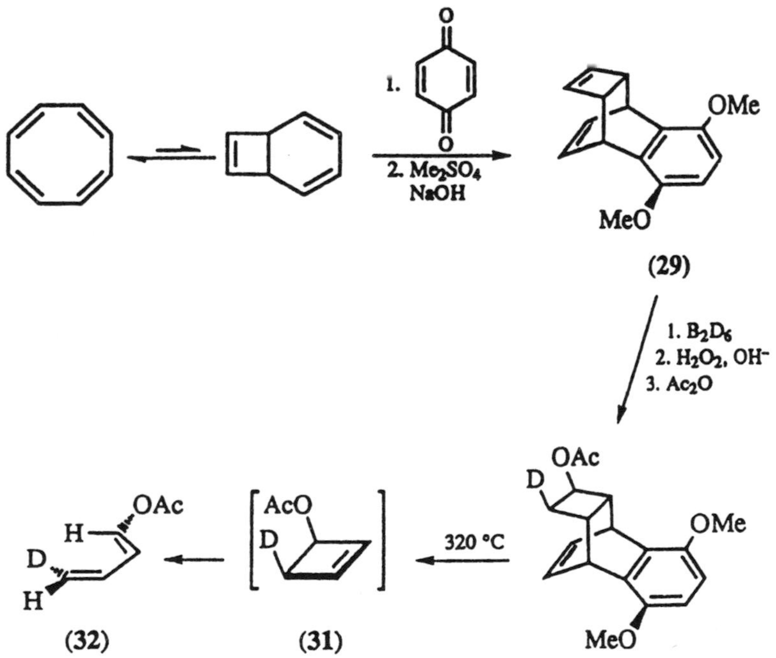

Scheme 25

In this reaction we have Type II diastereoselectivity associated with the *syn* addition of deuterodiborane and Type III (Type III')diastereoselectivity in selective attack on the top face, i.e. the reaction is a Type II/III hybrid (Chapter 13). Compound (**30**) was used to prepare the specifically labelled diene (**32**) via retro-Diels–Alder reaction followed by electrocyclic ring opening of (**31**) and so the Type III diastereoselectivity is of no consequence in this case.

Thermodynamic equilibration

In our discussion of asymmetric induction in Type III$_{s.c.}$ reactions so far, formation of the products has been assumed to be under kinetic control so that diastereoselectivity arises from the lower energy of one of the diastereoisomeric transition states.

In general, diastereoisomeric products will also have different thermodynamic stabilities. If an equilibrium between two diastereoisomers can be established, then if their thermodynamic stabilities are sufficiently disparate there can be complete conversion of the less stable into the more stable via a Type III$_{s.c.}$ reaction on an intermediate which usually contains a prochiral double bond (often an enolate).

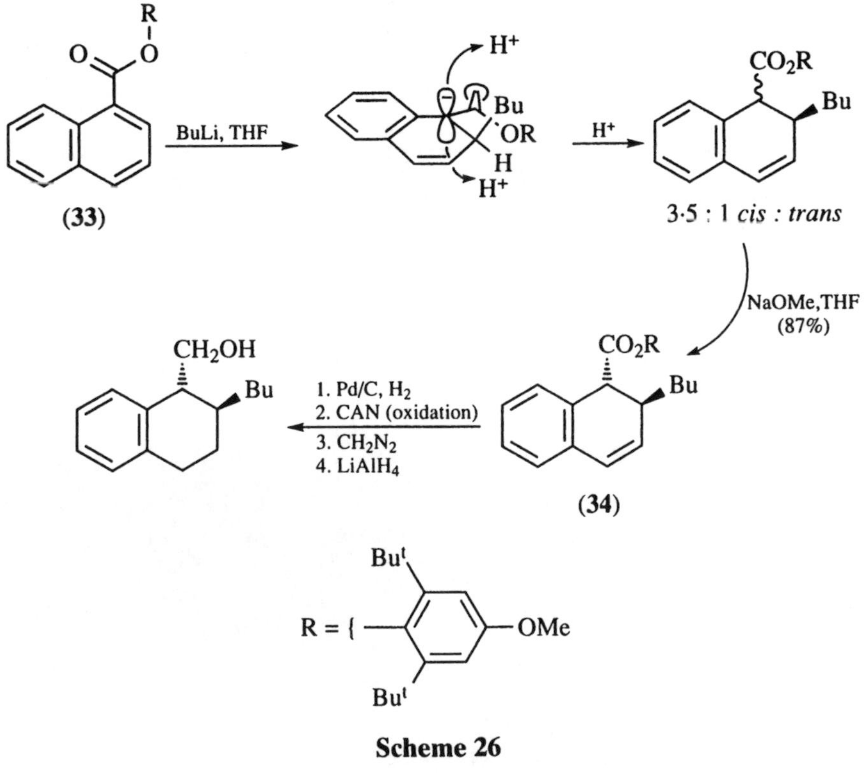

Scheme 26

By and large, the thermodynamically preferred diastereoisomer can usually be predicted from the influence of local steric effects near the chiral centre in question. Thus in Scheme 26, the mixture of *cis*- and *trans*-dihydronaphthalenes arising from organolithium Michael addition to the bulky ester (33) is converted into the *trans* isomer (34) by equilibration with base. The bulky ester is converted into the acid using cerium(IV) ammonium nitrate (CAN).[31]

Likewise, the all-*cis*-substituted cyclopropane (35) in Scheme 27 is converted into the acid (36) by thermodynamic equilibration via the enolate followed by oxidation.[32]

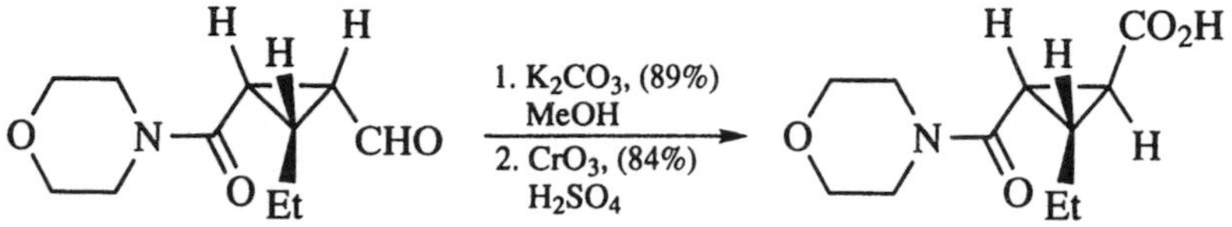

1. K_2CO_3, (89%) MeOH

2. CrO_3, (84%) H_2SO_4

(35)   (36)

Scheme 27

A common setting for thermodynamic equilibration is in or on six-membered rings where the greater stability of an equatorially located substituent determines the equilibrium position. Thus in Scheme 28, the axial formyl group in (37) is relocated equatorially on treatment with mild base via the enolate.[33] Note that the thermodynamic driving force for the overall inversion of configuration at C-2 in conversion of (37) into (38) is unlikely to be present in the acyclic diol (39) from which the cyclic acetal is formally derived.

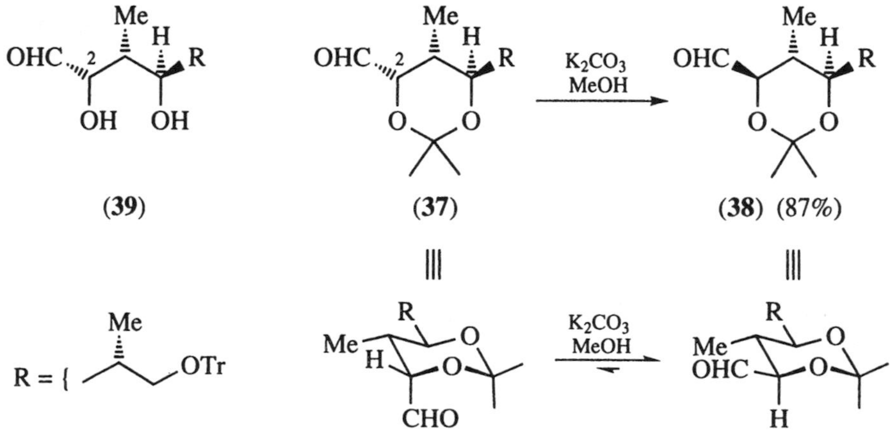

(39)   (37)   (38) (87%)

K_2CO_3 MeOH

$R = \{$

K_2CO_3 MeOH

Scheme 28

Conversion of the *trans*-substituted tetrahydropyran in Scheme 29 into the *cis*-isomer on standing with base is an example of thermodynamic equilibration via (reversible) Michael addition. Underpinning this isomerisation is, as before, the thermodynamic preference for an equatorially disposed substituent on a chair-shaped six-membered ring.[34]

The product of any reaction like that in Scheme 29, in which the step involving creation of the new chiral centre is reversible, may result from thermodynamic control.

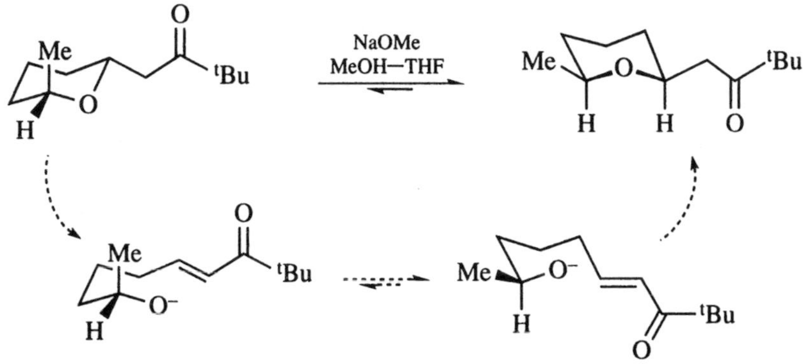

Scheme 29

Summary

A knowledge of the preferred conformation and the conformational freedom available to a molecule is indispensable for an understanding of steric effects in Type III$_{s.c.}$ reactions. There are four common ways in which steric effects can help to bring about diastereoselectivity in Type III$_{s.c.}$ reactions on a double bond: (a) shielding one face by substituent(s) such that non-bonded atoms in the incoming reagent and the substituent(s) would have to approach within the sum of their van der Waals radii; (b) directing attack to the diastereoface that minimises the close approach of existing substituents in the substrate during the ensuing bond angle and/or bond length changes; or (c) restricting the conformational freedom available to the substrate, ideally to the point where reaction at the prochiral double bond takes place from a single conformation. In a number of double bond-containing bicyclic compounds a single conformation is present, and this has contours that preferentially expose one diastereoface to attack. In other double bond-containing mono- or bicyclic compounds which exist in single conformations, the diastereoface of the double bond may not be so differentiated; however, the effect of substituents on the diastereoselectivity of addition to the double bond is more readily predicted. Further, (d) all other things being equal, an increase in the size of the attacking reagent will lead to an increase in diastereoselectivity.

Minimisation of local steric strain is often the driving force for thermodynamic equilibration, e.g. the greater stability of *trans*-1,2- versus *cis*-1,2-disubstitution in cyclic systems. Conversion of axially into equatorially located substituents in cyclohexane derivatives also constitutes such a driving force.

Finally, it is because steric effects are reasonably well understood that they are routinely used in the design of chiral auxiliaries, almost all of which function, at least in part, from the screening of one diastereoface by steric effects resulting from an appropriately located substituent or substituents in the auxiliary.

References

1. M. Asaoka, T. Aida, S. Sonoda and H. Takei, *Tetrahedron Lett.*, 1989, **30**, 7075.
2. G.A. Molander, E.R. Burkhardt and P. Weinig, *J. Org. Chem.*, 1990, **55**, 4990.
3. W.J. Koot, H. Hiemstra and W.N. Speckamp, *Tetrahedron Lett.*, 1992, **33**, 7969.
4. A. Basak, S.P. Salowe and C.A. Townsend, *J. Am. Chem. Soc.*, 1990, **112**, 1654.
5. B. Hartmann, A.M. Kanazawa, J.-P. Deprés and A.E. Greene, *Tetrahedron Lett.*, 1991, **32**, 767.
6. G. Majetich, D. Lowery and V. Khetani, *Tetrahedron Lett.*, 1990, **31**, 51.
7. S. Hansson, J.F. Miller and L.S. Liebeskind, *J. Am. Chem. Soc.*, 1990, **112**, 9660.
8. B.M. Trost, B. Yang and M.L. Miller, *J. Am. Chem. Soc.*, 1989, **111**, 6482.
9. L.O. Jeroncic, M.-P. Cabal, S.J. Danishefsky and G.M. Shulte, *J. Org. Chem.*, 1991, **56**, 387.
10. G.L. Closs and R.A. Moss, *J. Am. Chem. Soc.*, 1964, **86**, 4042; R.A. Moss, in *Selective Organic Transformations*, ed. B.S. Thyagarajan, Wiley–Interscience, New York, 1970; for similar *syn* selectivity in reaction of carbenoids, see R.A. Moss, *J. Org. Chem.*, 1965, **30**, 3261.
11. J. March, *Advanced Organic Chemistry*, McGraw-Hill, Kogakusha, Tokyo, 2nd edn, 1977, p. 307; G.S. Hammond, *J. Am. Chem. Soc.*, 1955, **77**, 334.
12. S. Chandrasekhar, *Chem. Soc. Rev.*, 1987, **16**, 313.
13. P.M. McCurry, *Tetrahedron Lett.*, 1971, 1841; see also E.W. Garbisch *et al.*, *J. Am. Chem. Soc.*, 1965, **87**, 2932.
14. E.L. Eliel, *Stereochemistry of Carbon Compounds*, McGraw-Hill, New York, 1962.
15. E.L. Eliel and Y. Senda, *Tetrahedron*, 1970, **26**, 2411.
16. R.W. Hoffmann, *Chem. Rev.*, 1989, **89**, 1841.
17. F. Johnson, *Chem. Rev.*, 1968, **68**, 375.
18. P.A. Parziale and J. Berson, *J. Am. Chem. Soc.*, 1990, **112**, 1650.
19. M.F. Salomon, S.N. Pardo and R.G. Salomon, *J. Org. Chem.*, 1984, **49**, 2446.
20. R. Hambalek and G. Just, *Tetrahedron Lett.*, 1990, **31**, 4693.
21. D.H.R. Barton, S.D. Géro, B. Q-Sire and M. Samadi, *Chem. Commun.*, 1988, 1372.
22. A. Krotz and G. Helmchen, *Tetrahedron: Asymmetry*, 1990, **1**, 537.
23. J. Wagner and P. Vogel, *Chem. Commun.*, 1989, 1634.
24. S. Torii, M. Okumoto, H. Ozaki, S. Nakayasu and T. Kotani, *Tetrahedron Lett.*, 1990, **31**, 5319.
25. A. Warm and P. Vogel, *Tetrahedron Lett.*, 1985, **26**, 5127.
26. H.C. Brown and H.R. Deck, *J. Am. Chem. Soc.*, 1965, **87**, 5620; see also H.C. Brown, J.H. Kawakami and K. Liu, *J. Am. Chem. Soc.*, 1973, **95**, 2209.

27. A. Casedevall, E. Casedevall and M. Mion, *Bull. Soc. Chim. Fr.*, 1968, 4498.
28. A. Fürst and P.A. Plattner, *Helv. Chim. Acta*, 1949, **32**, 275; P. Kocovsky, University of Leicester, personal communication.
29. Y. Houminer, *J. Chem. Soc., Perkin Trans. 1*, 1975, 1663.
30. R.K. Hill and M.G. Bock, *J. Am. Chem. Soc.*, 1978, **100**, 637.
31. K. Tomioka, M. Shindo and K. Koga, *Tetrahedron Lett.*, 1990, **31**, 1739.
32. S.F. Martin, R.E. Austin and C.J. Oalmann, *Tetrahedron Lett.*, 1990, **31**, 4731.
33. S. Hanessian, N.G. Cooke, B. DeHoff and Y. Sakito, *J. Am. Chem. Soc.*, 1990, **112**, 5276.
34. D. Seebach and M. Pohmakotr, *Helv. Chim. Acta*, 1979, **62**, 843.

10 SUBSTRATE-CONTROLLED DIASTEREOSELECTIVE REACTIONS (TYPE III$_{s.c.}$) MEDIATED BY STEREOELECTRONIC EFFECTS: TORSIONAL STRAIN AND σ–π INTERACTIONS

For many reactions there exists a preferred spatial relationship between the bonds made and broken. Such reactions are often described as proceeding under stereoelectronic control.[1] Preferred spatial relationships may also exist between bonds made or broken in the reaction and bonds or orbitals in the starting material that are retained in the product. This includes orbitals containing a pair of electrons or a single electron.

Even in a molecule not undergoing reaction there exist a preferred spatial relationships between bonds. Torsional strain arises from the repulsion between electrons which are contained in σ-bonds that are coplanar (Figure 1). Thus as the dihedral angle φ between two σ-bonds decreases, the torsional strain increases and is at a maximum when the bonds are eclipsed as in (1). It is this torsional strain which is responsible for the rotational barrier in ethane, when three pairs of σ C—H bonds are coplanar in the higher energy eclipsed conformation. The same strain is present for non-bonded pairs of electrons in orbitals which are eclipsed but its magnitude is greater.

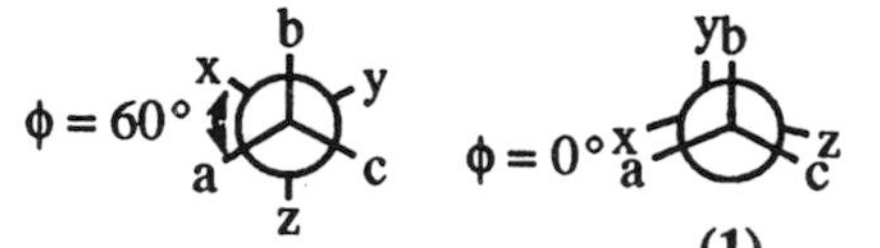

$$\phi = 60° \qquad \phi = 0°$$

(1)

Figure 1

Thus the barrier to rotation around the N—N bond between the two sp^2-hybridised nitrogens in (2) is $\sim 98\,kJ\,mol^{-1}$ and the eclipsed nitrogen lone pairs in (2') are responsible for the major part of this barrier[2]; by comparison, the rotational barrier in ethane is $12.5\,kJ\,mol^{-1}$.

A minimum of torsional strain is one of the major factors which stabilises the chair form of cyclohexane relative to the boat (and other even less stable conformations) (Figure 2). The chair is not completely free from torsional strain since the three axial hydrogens shown on each side of the chair are eclipsed.

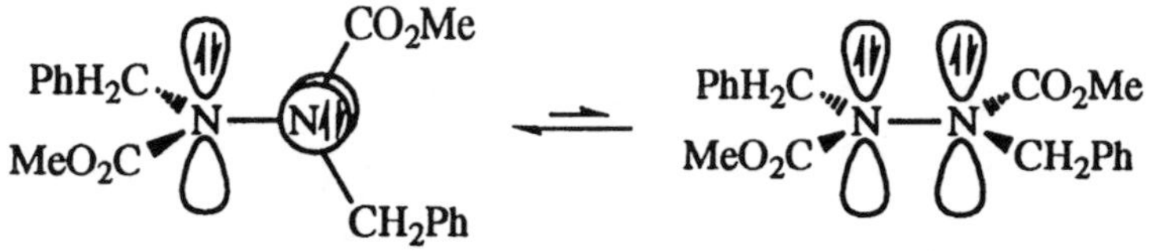

(2)   (2')

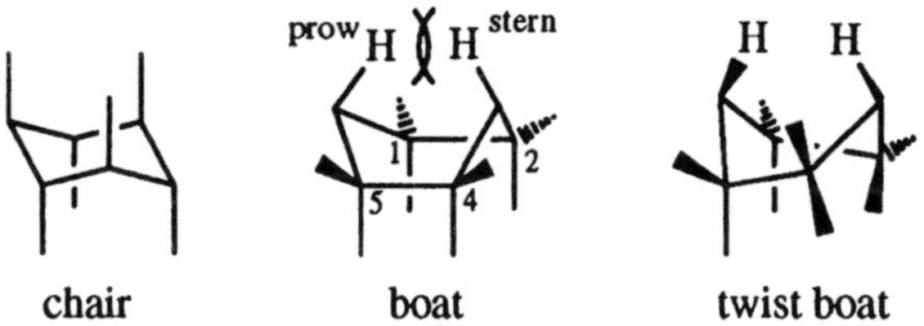

chair   boat   twist boat

Figure 2

In the boat, the torsional strain between bonds on C-1 and C-2 and between those on C-4 and C-5 is a maximum; some relief of this torsional strain is afforded by deformation to the *twist boat*, and this also relieves the H(stern)–H(prow) steric interaction.

Similarly, the half-chair conformation of cyclohexene is more stable than the half-boat (Figure 3), eclipsing of bonds to C-4 and C-5 being minimised in the half-chair and maximised in the half-boat.

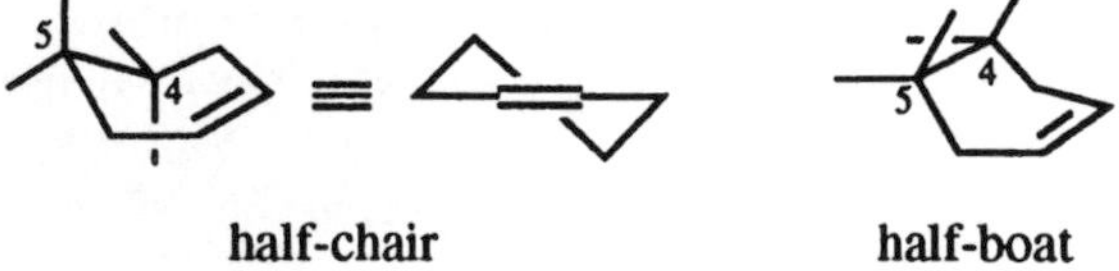

half-chair   half-boat

Figure 3

Minimisation of all the strain elements, and in particular torsional strain, angle strain and steric strain also determines the preferred conformations of other rings. These three important strain elements also play a major role in the preferred conformations of saturated carbon chains. This is important when the chain bears functional groups on its terminal carbon atoms that may undergo intramolecular reaction. The reaction of two functional groups X and Y separated by four methylene groups (Scheme 1) in a molecule is a common occurrence. Just as substituents on cyclohexane prefer equatorial positions, so a substituent R in the four-carbon tether will prefer an equatorial site on the chair motif.

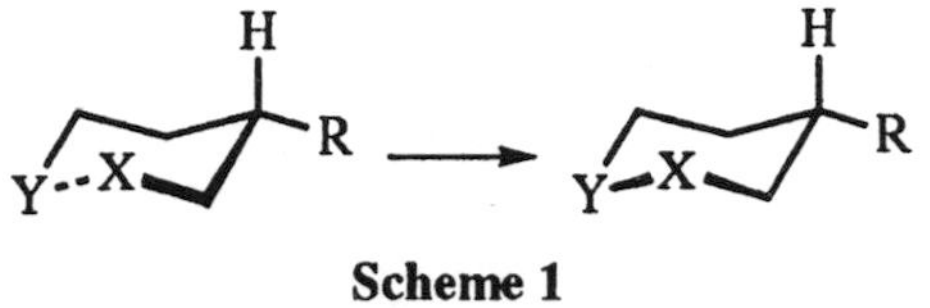

Scheme 1

As will be seen later, this constitutes an important means for an existing chiral centre to influence the configuration of a nascent chiral centre resulting from the X–Y union in a Type III$_{s.c.}$ reaction, i.e. for bringing about asymmetric induction.

Likewise, in an analogous reaction between functional groups separated by three methylenes, the five-membered cyclic transition state will take up conformations resembling (3), (4) or (5) to minimise the torsional strain. In these possible transition-state conformations (which resemble closely some of the stable conformations of cyclopentane), (4) may appear to have less torsional strain than (3) or (5) but the nature of the tether (e.g. sp² substitution within it) may favour reaction via one of the latter. In any event,

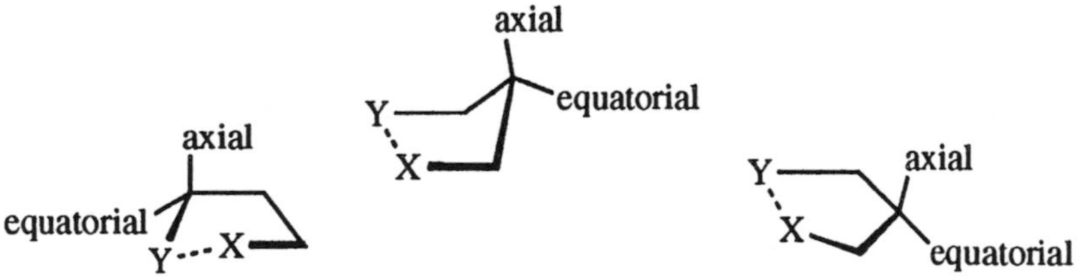

(3)   (4)   (5)

there are equatorial-like sites which will be preferred by substituents in the tether, and this may also give rise to asymmetric induction in a ring-closure reaction (see Chapter 11).

Stereoelectronic effects in Type III$_{s.c.}$ reactions

ADDITION TO UNSATURATED SIX-MEMBERED RINGS

Reactions of the α,β-unsaturated-δ-lactone (6) [Scheme 2(a)] with divinyl cuprate gives the *trans*-substituted diastereoisomer (7) with high selectivity.[3]

This is unlikely to be the result of a shielding effect on the top face of the double bond by the isopropenyl group in **6** since it is equatorial. Instead, the diastereoselectivity in this conjugate addition is believed to be the result of the stability of the developing half-chair (with its lesser torsional strain) resulting from axial attack on the bottom face of the double bond, as shown in Scheme 2 (a), compared with the half-boat which results from axial attack on the top face as in Scheme 2 (b). Axial attack, i.e. orthogonal to the plane of the ring, is necessary in either Scheme 2 (a) or (b) since attack takes place via the π^*-orbital of the double bond.

This Type III$_{s.c.}$ reaction is predictable in additions to other substituted cyclohexenones as in the Sakurai reaction in Scheme 2 (c)[4] provided that bulky

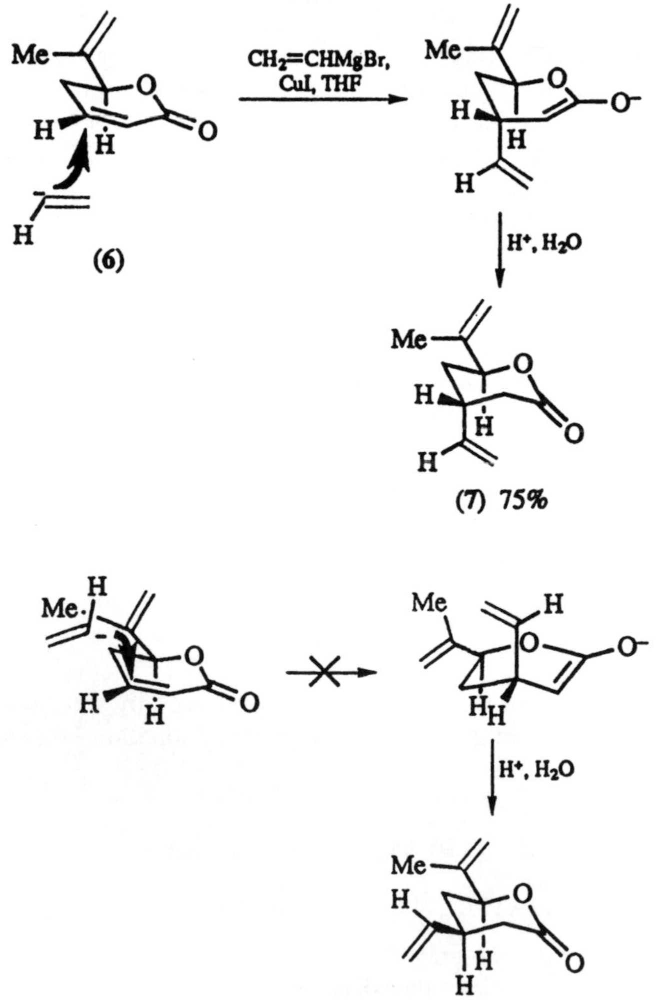

(a)

(7) 75%

(b)

t present to discourage attack in the sense shown. The
arising from addition to double bonds in tetrahydropyran
interpreted similarly (see Schemes 37 and 38).

r hand, *intramolecular* attack on the double bond by a
ay be constrained to proceed via a boat-shaped transition state
amination in Scheme 2 (d).[5]

NS TO BICYCLO[2.2.1] DERIVATIVES

rable effort has been made to understand the preference for *exo*
in reactions of bicyclo[2.2.1]heptene derivatives, as referred to in
er 9. This preference does not have its origin solely in steric effects.[6]
s *endo* attack proceeds on the 2,3-double bond in Scheme 3, torsional
in increases between the exocyclic σ-bonds on C-1 and C-2 as a
nsequence of the developing sp³ character of C-2. Moreover, there is
dditional torsional strain between the developing nucleophile—C-2 bond
and the C-1—C-6 bond in the ring.[7]

In *exo* attack, by contrast, these torsional strains are both reduced since the
exocyclic σ-bond at C-2 moves away from the bridgehead σ-bond, and the
developing bond to C-2 is not eclipsed by a bond in the ring.[7]

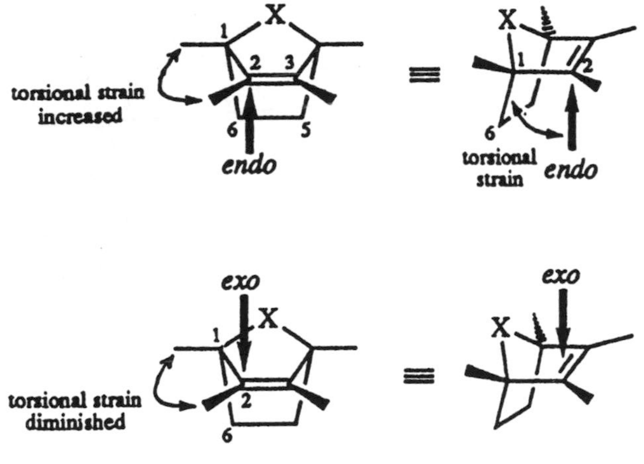

Scheme 3

There appear to be additional factors involved, however, arising from σ–π
interactions between the π-bond and the σ-bonds of the skeleton (see
below). In some of these strained bicyclo[2.2.1]heptene derivatives, a small
pyramidalisation of the formally sp²-hybridised carbon atoms of the double
bond has been found in the solid state by X-ray crystallography. In (**8**), for
example, the 14,13 and 5,6 vinylic σ-bonds are bent out of the 1,14,5-plane,

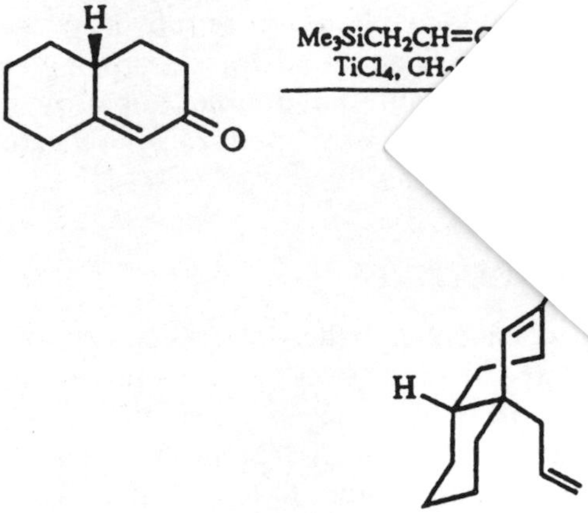

Me₃SiCH₂CH=C
TiCl₄, CH-

substituents are n
diastereoselectivit
derivatives can b
On the oth
nucleophile m
as in the iod

ADDITI

Conside
attack
Chap
A
str
c

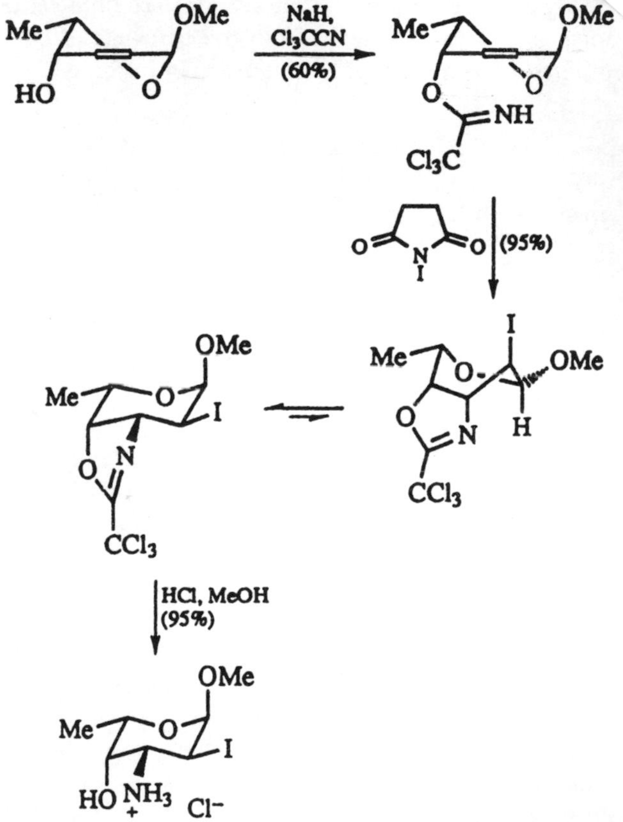

Me—⟨...⟩—OMe
HO   O
$\xrightarrow[\text{(60\%)}]{\substack{\text{NaH,} \\ \text{Cl}_3\text{CCN}}}$
Me—⟨...⟩—OMe
O   O
=NH
Cl₃C

(95%)

Me—⟨...⟩—OMe
       I
O   N
   CCl₃
$\rightleftharpoons$
Me—⟨...⟩
   I
   ⁙OMe
O   N   H
   CCl₃

HCl, MeOH
(95%)

Me—⟨...⟩—OMe
       I
HO NH₃ Cl⁻
   +

(d)

Scheme 2

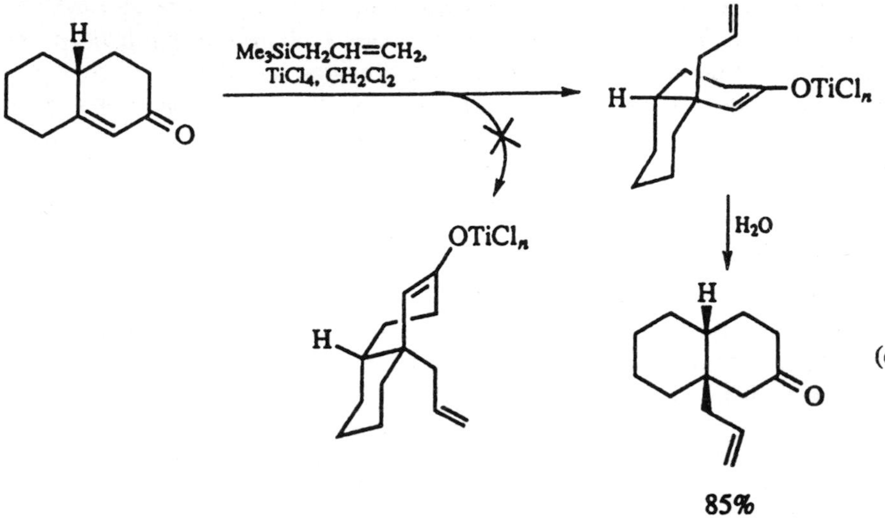

(c)

85%

(d)

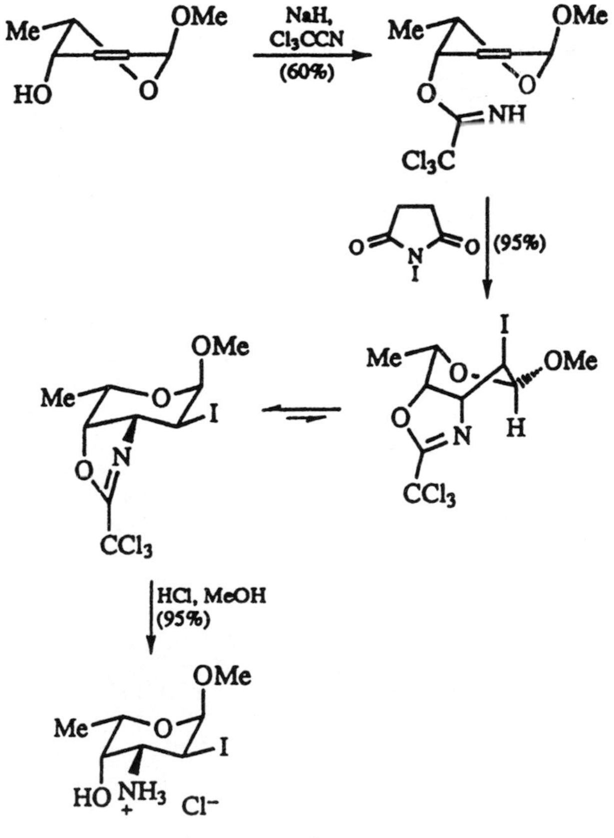

Scheme 2

substituents are not present to discourage attack in the sense shown. The diastereoselectivity arising from addition to double bonds in tetrahydropyran derivatives can be interpreted similarly (see Schemes 37 and 38).

On the other hand, *intramolecular* attack on the double bond by a nucleophile may be constrained to proceed via a boat-shaped transition state as in the iodoamination in Scheme 2 (d).[5]

ADDITIONS TO BICYCLO[2.2.1] DERIVATIVES

Considerable effort has been made to understand the preference for *exo* attack in reactions of bicyclo[2.2.1]heptene derivatives, as referred to in Chapter 9. This preference does not have its origin solely in steric effects.[6]

As *endo* attack proceeds on the 2,3-double bond in Scheme 3, torsional strain increases between the exocyclic σ-bonds on C-1 and C-2 as a consequence of the developing sp³ character of C-2. Moreover, there is additional torsional strain between the developing nucleophile—C-2 bond and the C-1—C-6 bond in the ring.[7]

In *exo* attack, by contrast, these torsional strains are both reduced since the exocyclic σ-bond at C-2 moves away from the bridgehead σ-bond, and the developing bond to C-2 is not eclipsed by a bond in the ring.[7]

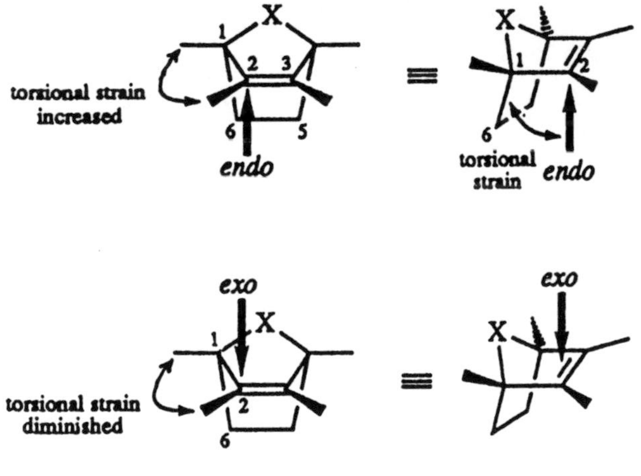

Scheme 3

There appear to be additional factors involved, however, arising from σ–π interactions between the π-bond and the σ-bonds of the skeleton (see below). In some of these strained bicyclo[2.2.1]heptene derivatives, a small pyramidalisation of the formally sp²-hybridised carbon atoms of the double bond has been found in the solid state by X-ray crystallography. In (**8**), for example, the 14,13 and 5,6 vinylic σ-bonds are bent out of the 1,14,5-plane, away from the oxygen bridge, by ~10° [see (8′)].[8]

Although such pyramidalisation[9] may be small, it makes attack more likely from the *exo* direction: small distortions in the ground state can lead to significant energy differences between competing transition states.

A small number of bicyclo[2.2.1] derivatives are known in which, exceptionally, attack takes place preferentially on the *endo* face. These include 2,3-dimethylene-substituted derivatives such as (9) and (10).[10] Thus (10) adds dienophiles predominantly on the *endo* face. It is suggested that in these cases, σ–π interactions (see below) are dominant and responsible for the *endo* face reactivity.

The high *endo* face preference in the (double) Diels–Alder reaction of the bis-diene (9), coupled with the high *exo* preference of (11) (Scheme 4) acting as a bis-dienophile, has been used by Stoddart *et al.*[6] to prepare the adduct (12) and thence the macrocycle (13).

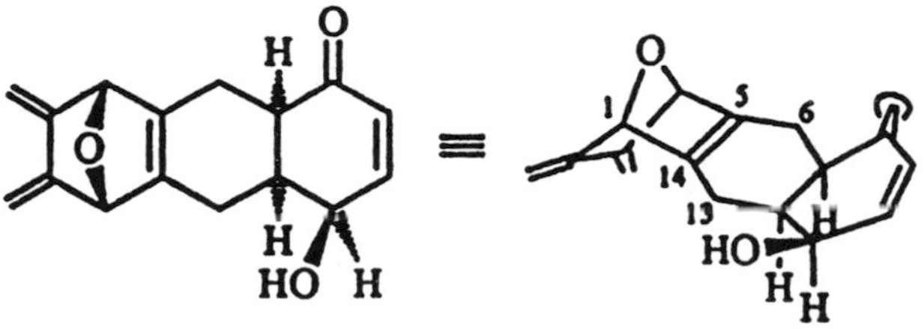

(8) 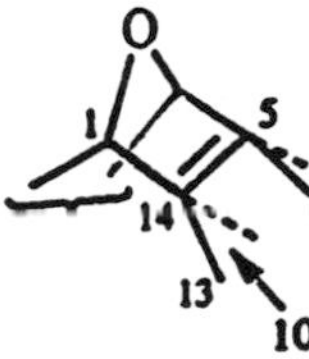　　　(8′)

(9) 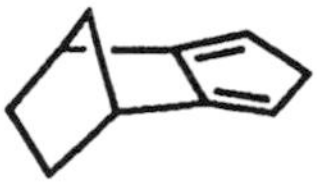　　　(10)

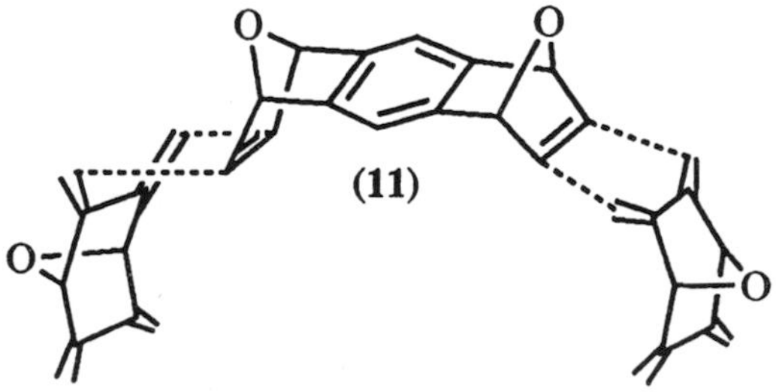

(11)

Scheme 4　　　*(continued)*

Scheme 4 *(continued)*

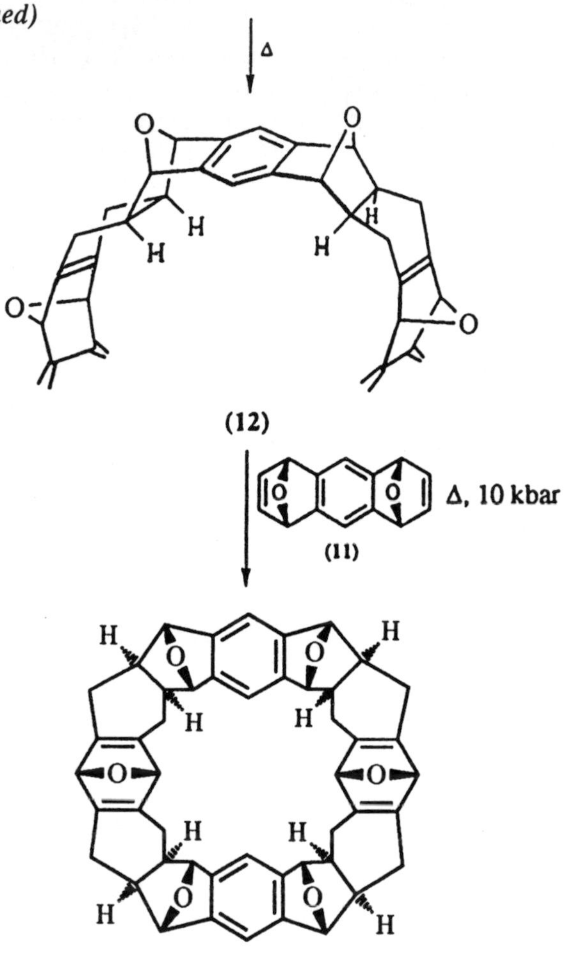

(12)

(11)

Δ, 10 kbar

(13)

Scheme 4

EFFECT OF σ-BONDED SUBSTITUENTS ADJACENT TO DOUBLE BONDS

The reaction of double bonds with electrophiles is greatly facilitated by the presence of a heteroatom X (e.g. OR, NR$_2$) as a substituent on the double bond. This is because the developing positive charge in the transition state can be stabilised by a non-bonding electron pair on the heteroatom (n–π interaction) (Figure 4a).

Likewise, the π-electrons in an adjacent C=C bond can stabilise the developing positive charge by delocalisation (π–π interaction) (Figure 4b). It is not surprising, therefore, that an adjacent σ-bond can similarly stabilise the

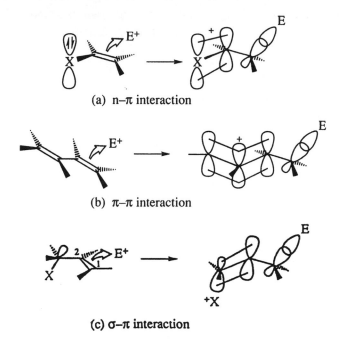

(a) n–π interaction

(b) π–π interaction

(c) σ–π interaction

Figure 4

development of positive charge leading in resonance terms to Figure 4c (σ–π interaction). In frontier molecular orbital terms we would describe this as a σ–π interaction which raises the energy of the HOMO of the alkene and results in a smaller HOMO (alkene)–LUMO (electrophile) energy gap and hence a faster reaction (Figure 5).[11]

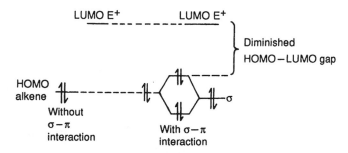

Figure 5

However, just as an adjacent C=O bond is unable to stabilise positive charge as a C=C does in Figure 4b, so some C—X bonds are not good σ-donors. C—H and C—C bonds are σ-donors and the concept of hyperconjugation, developed to account for the greater reactivity of alkenes substituted with a greater number of alkyl groups, uses the σ-donor properties of allylic C—H and C—C bonds. C—Si bonds are excellent σ-donors (see below), but when X is an electronegative atom, the C—X bond becomes a σ-acceptor rather than a donor.

Stabilisation of the positive charge as in Figure 4a or b requires that the overlapping orbitals be aligned as illustrated. Similarly, maximum σ–π interaction arises only when the C—X bond is parallel to the developing p-orbital on C-2 as in Figure 4c. Conversely, this interaction will be minimal when the C—X bond is orthogonal to the developing p-orbital as in Figure 6a or b.

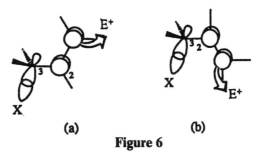

(a)   (b)

Figure 6

Provided that rotation around the C_2—C_3 bond is possible, a σ-*acceptor* C—X can take up an orientation as in Figure 6a or b, leaving a σ-donor to interact more favourably with the π-bond undergoing electrophilic attack.

The interaction of a σ-bond with an adjacent π-bond as in Figure 4c can be expected to increase the electron availability at each of the faces of the alkene to different extents, and it differs in this respect from the n–π and π–π interactions in Figure 4a or b. In practice, it may be difficult to separate directionality resulting from σ–π interaction of a donor C—X bond and the steric effect associated with X since both direct attack of the electrophile on the double bond in the same *anti* (to X) sense.

Silicon is more electropositive than carbon and it is this, together with its high polarisability, which makes the Si—C bond such a good σ-donor. The directionality associated with the σ–π interaction of the Si—C bond and the steric effect of the R_3Si group both conspire to direct electrophilic attack on allylsilanes *anti* (to Si) and in the S_E2' reaction (Scheme 5) the silicon is lost.

As Scheme 5 shows, however, in electrophilic attack on acyclic allylsilanes, the outcome of the reaction is not wholly predictable there is an ambiguity even though attack by the electrophile is invariably *anti* to the C—Si bond. This is because there are two conformations of the substrate, (**14**) and (**15**), in which the σ–π interactions can operate. In fact, allylsilanes with R^1 = Me react in both conformations. Thus hydroboration of the (*E*)-alkene (**16**) (Scheme 6) with diborane gives a 50:50 mixture of diastereoisomeric alcohols (**17**) and (**18**) after oxidative work-up, presumably as a result of reaction via both allylsilane conformers, as shown.[12]

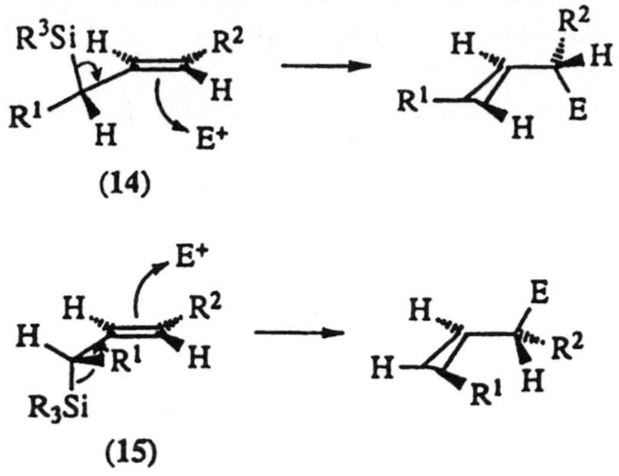

(14)

(15)

Scheme 5

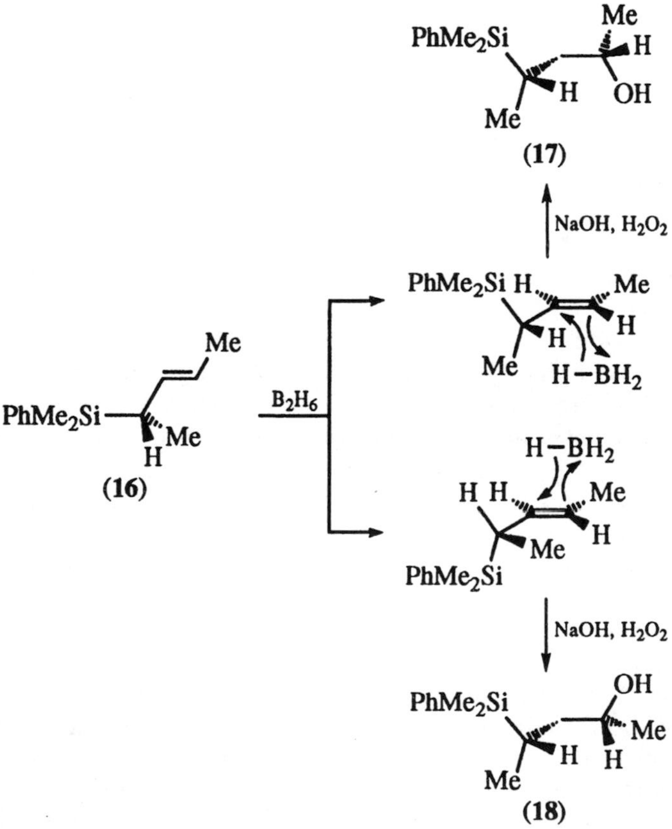

(16)

(17)

(18)

(17) + (18) 80%
(17) : (18) 50 : 50

Scheme 6

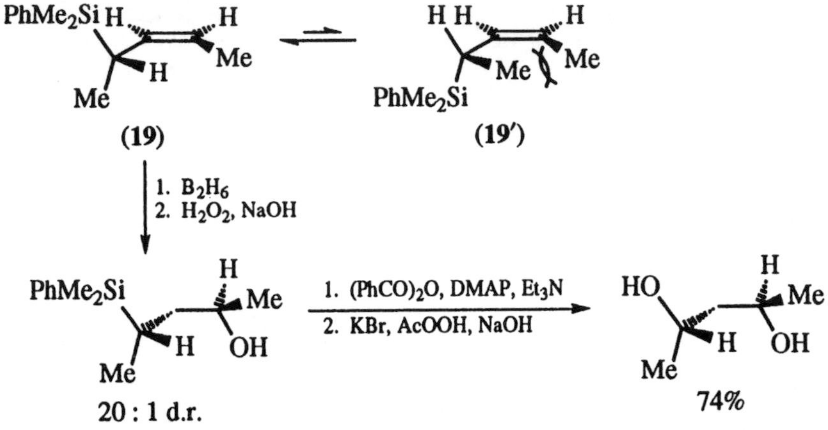

Scheme 7

By contrast, hydroboration of (**19**), the Z-double bond isomer of (**16**) (Scheme 7), is highly diastereoselective. The effect of the methyl group when *cis* is to favour reaction from (**19**) because of the $A_{1,3}$-strain present in conformation (**19'**).

Attack of electrophiles via conformation (**15**) in Scheme 5 can also be discouraged by increasing the size of R^1. Thus hydroxylation (via osmylation), epoxidation and methylenation all respond to an increase in size of R^1 from Me to Ph by giving higher diastereoselectivity in the sense expressed by (**14**).[13]

An attribute of the C—Si bond is that, like the C—B bond in the reactions in Schemes 6 and 7, the heteroatom can be replaced by OH with complete retention of configuration at carbon (see Scheme 7 and Chapter 4).

The *anti* face-directing effect of an C—Si bond can similarly be harnessed in the Diels–Alder reaction. In a normal Diels–Alder reaction, the diene may be considered nucleophilic, and to be undergoing attack by an electrophilic dienophile; in frontier orbital terms the dominant interaction is HOMO(diene)–LUMO(dienophile).[14] The effect of a σ–π interaction is, as seen previously, to raise the HOMO(diene) energy level, thus diminishing the HOMO–LUMO gap and facilitating the reaction (Figure 5).

When C-1 of the diene bears a chiral centre having a C—SiR_3 bond, attack of the dienophile can take place via two different transition states in both of which the C—Si bond is *anti* to the developing C—C bond to the (electrophilic) dienophile (Scheme 8). Diastereoselectivity in this Diels–Alder reaction, therefore, hinges on the selectivity of the groups a and b for sites 'inside' and 'outside' relative to the diene residue.

The major diastereoisomer (**20**) which results from reaction of *N*-phenyl-maleimide with such a substituted diene (Scheme 9) is that which

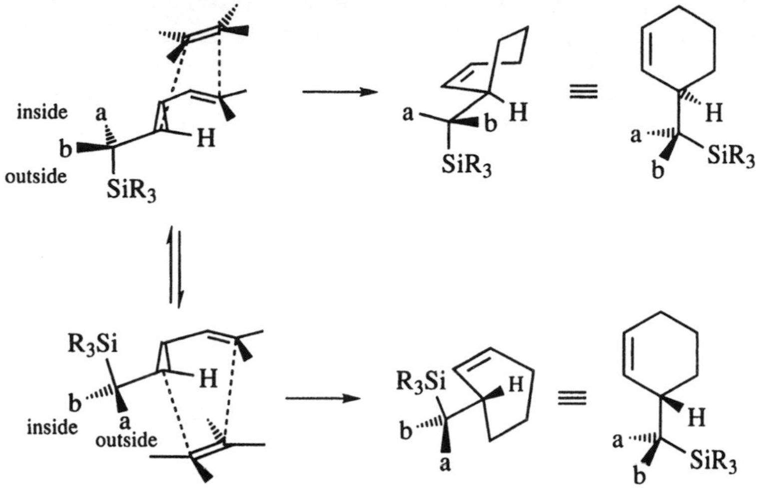

Scheme 8

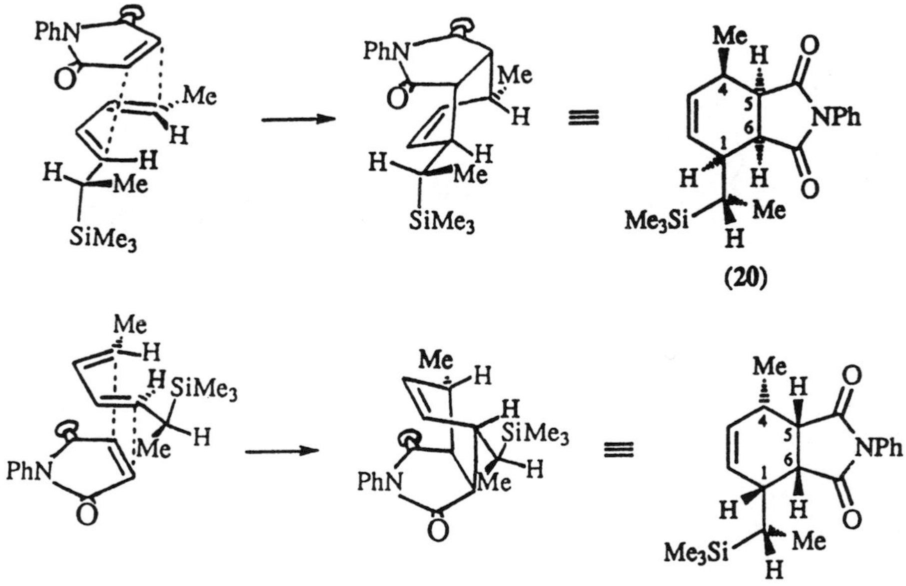

(20)

(21)

(20) + (21) 96%
(20) : (21) 82 : 18

Scheme 9

corresponds to a methyl group occupying the outside position and H the inside position, although the diastereoselectivity is modest.[15] Note that (20) and (21) are clearly diastereoisomeric since they have the same configuration at the silicon-bearing chiral centre but opposite configurations at all other chiral centres.

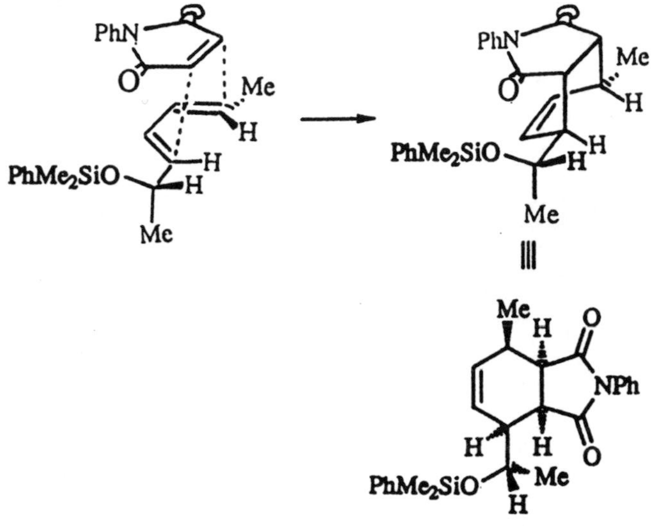

(22)

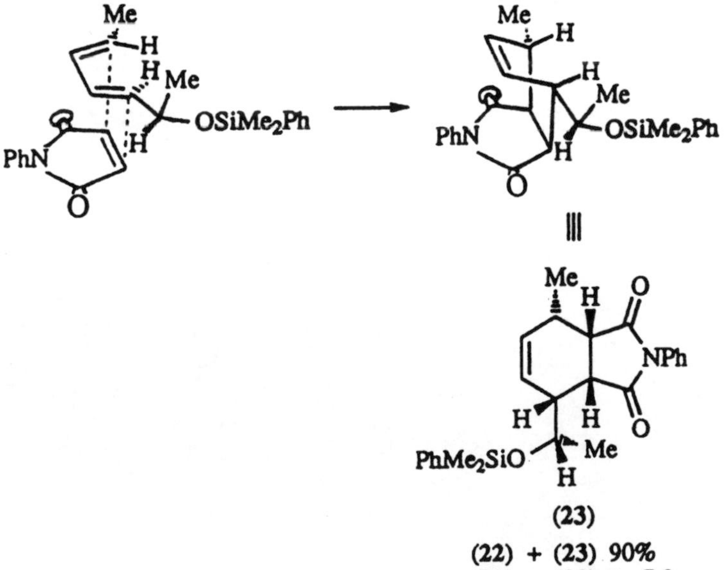

(23)

(22) + (23) 90%
(22) : (23) 1 : 7·3

Scheme 10

Without the chiral centre in the substituent at C-1 this would be a Type II Diels–Alder reaction with *endo* overlap leading necessarily to a racemic mixture of enantiomers. As a result of the chiral centre, the diene has diastereofaces and one of these is attacked in preference to the other. This reaction is therefore a Type II/III reactions, further examples of which will be considered in Chapter 13. For the present, note that the *relative* configuration at the chiral centres created at C-1, C-4, C-5 and C-6 is as would be expected for the Type II component of the reaction were the chiral substituent absent. Although Scheme 9 depicts the reaction of a single enantiomer of the diene, the same diastereoselectivity would obtain if the latter were used in racemic form (as in fact was the case); the diastereoisomeric products would then both have been racemic.

In contrast to the C—Si bond, a C—OR bond is a σ-acceptor. A double bond bearing an OR bond in the allylic position and undergoing electrophilic attack will mitigate the σ–π electron-withdrawing effects of this C—OR bond by undergoing reaction from conformations in which OR is 'inside' or 'outside' but not *anti*. A comparison of the Diels–Alder reaction in Scheme 10 with that in Scheme 9 where an OSiMe$_2$Ph group has replaced the SiMe$_3$ group is instructive.[15]

The *major* diastereoisomers produced in Schemes 9 and 10 [(**20**) and (**23**)] have the same configuration for the chiral substituent at C-1 of the diene (with OSiMe$_2$Ph replacing SiMe$_3$) but opposite configurations at all other centres. In other words, the substitution of SiMe$_3$ by OSiMe$_2$Ph reverses the diastereoface preference of attack on the diene.

To account for the preferred sense of asymmetric induction in Schemes 9 and 10, it is assumed that the 'inside' position is more congested than the 'outside' one and so in both cases this inside position will be occupied by H rather than Me (Scheme 9) or OSiMe$_2$Ph (Scheme 10).[16]

The transition-state models in Schemes 9 and 10 account satisfactorily for the stereostructures of the products, but they are not the only models that can do this. For example, attack on the diene of Scheme 10 with the substituents on the chiral centre oriented as shown in Scheme 11 would also account for the formation of (**23**) as the major diastereoisomer. This is the model used by Kahn and Hehre[17] in which, it is assumed, attack of the dienophile is directed largely by electrostatic effects, i.e. that the positive portion of the electrophilic

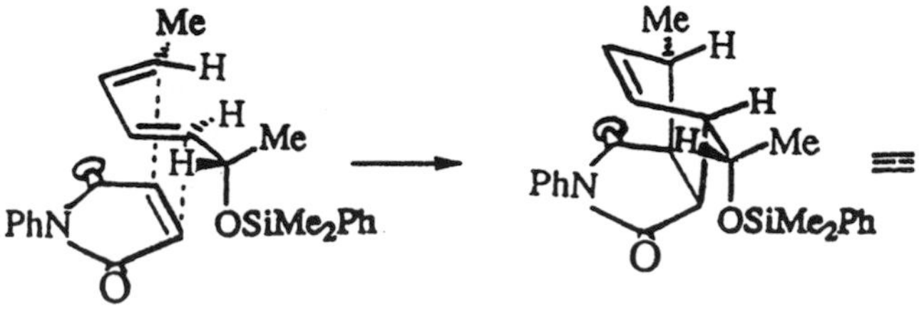

Scheme 11

dienophile adds *cis* to the allylic oxygen which, by virtue of its two lone pairs and electron-withdrawing inductive effect, has negative charge.

INTRAMOLECULAR NUCLEOPHILIC ATTACK ON ALLYLICALLY SUBSTITUTED DOUBLE BONDS MEDIATED BY 'ONIUM IONS

The inherent lack of reactivity of alkyl-substituted alkenes towards nucleophiles can be overcome by the formation of cyclic 'onium ions. Thus cyclisation of 3-substituted pent-4-enols (24) (Scheme 12) is mediated by a number of electrophiles capable of forming such 'onium ions.

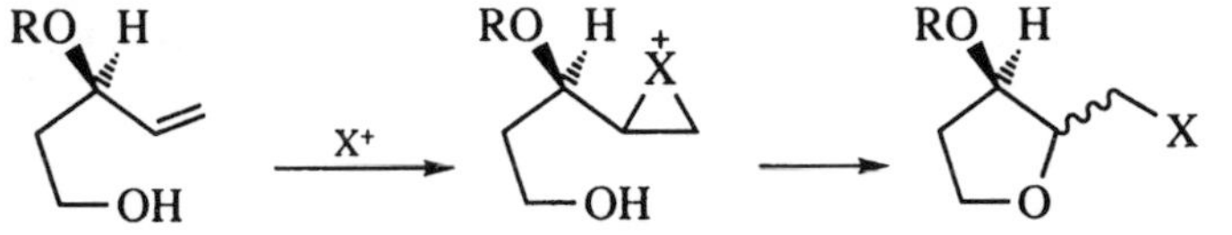

X = I, Br, ArSe, HgY

Scheme 12

In these cyclisations, regioselectivity favours five-membered ring formation with attack on the more substituted carbon of the 'onium ion. Diastereoselection favours a 2,3-*cis*-relationship of substituents on the tetrahydrofuran ring (Scheme 13).[18]

Note that this cyclisation can also be carried out [Scheme 13 (b)] using a benzyl ether analogue (25) with the benzyl group being easily lost from an intermediate tetrahydrofuranonium species (26).

The asymmetric induction in these cyclisations is not the result of steric effects since analogues bearing allylic methyl groups are *less cis*-selective or even weakly *trans*-selective, depending on the electrophile and substrate (see below).

The same stereochemical pattern emerges in the iodolactonisation,[19] iodolactamisation[20] and iodoetherification[21] reactions in Scheme 14. Here also the (non-reacting) allylic OH group brings about high 2,3-*cis*-diastereoselectivity which is eroded when it is replaced by an allylic methyl group and the same effect is observed even when the double bond is substituted, as in (a) and (c) in Scheme 14.

Since in (c) in Scheme 14 the preferences for formation of the *cis*-2,3-diastereoisomers are of similar magnitude for X = OH, F and OMe, and since hydrogen bonding would be expected to be much stronger for fluorine than oxygen, an explanation solely in terms of hydrogen bonding is unsatisfactory.

What is the origin of this useful *cis*-2,3-diastereoselectivity? In Schemes 13 and 14, the fact that *cis*-diastereoselectivity is present when the substituent in the allylic position is oxygen or fluorine but absent when it is methyl suggests

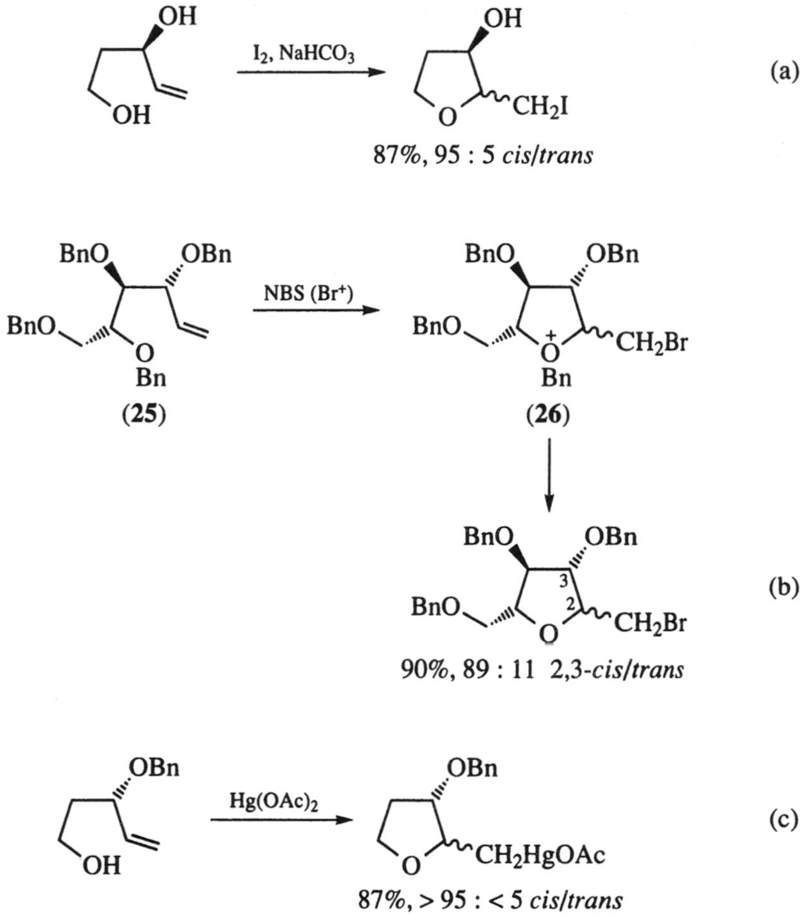

87%, 95 : 5 *cis/trans* (a)

(25)

(26)

90%, 89 : 11 2,3-*cis/trans* (b)

87%, > 95 : < 5 *cis/trans* (c)

Scheme 13

that σ–π interactions are implicated, with the σ-bond acting as acceptor.

In these cyclisations, formation of the 'onium ion is thought to be reversible. High diastereoselectivity is likely to be *dependent* on this reversibility since (non-reversible) epoxidation of, e.g., (27) (Scheme 15) gives a mixture of diastereoisomeric epoxides (28) which spontaneously cyclises to a 45:55 mixture of the corresponding tetrahydrofurans (29),[18] i.e. the stereostructure of the product is determined by the stereochemistry of formation of the intermediate epoxide.

Diastereoselectivity in the cyclisations in Schemes 13 and 14 will then be determined by the relative transition state energies for attack by the internal nucleophile on the two diastereoisomeric 'onium ions.

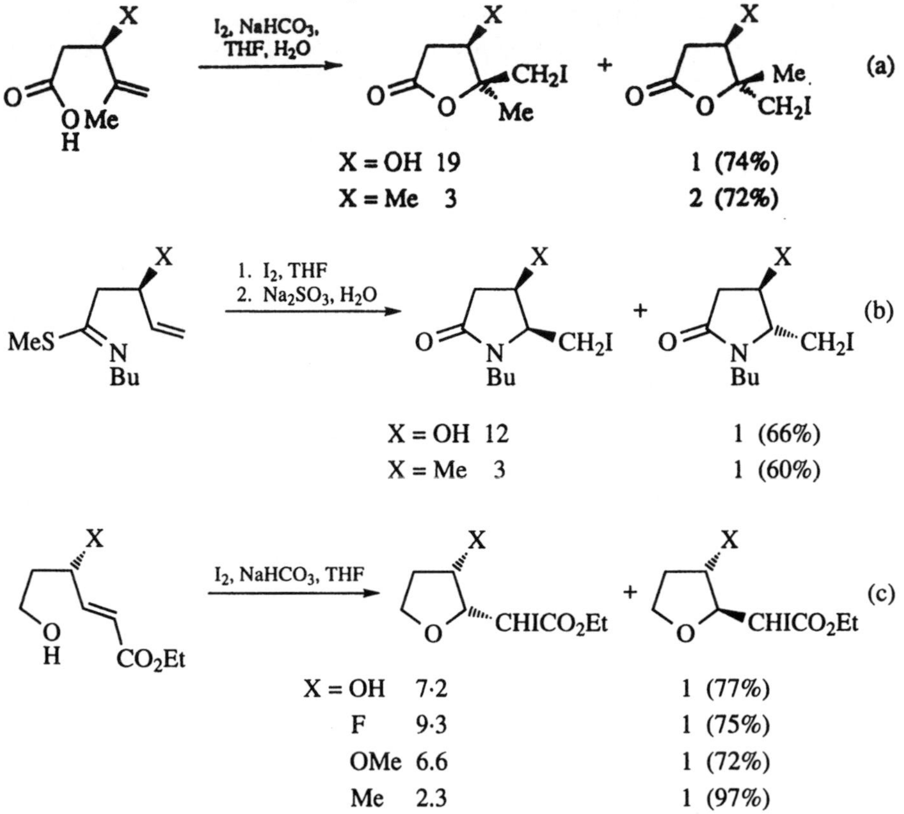

I_2, $NaHCO_3$, THF, H_2O
(a)
X = OH 19 1 (74%)
X = Me 3 2 (72%)
1. I_2, THF
2. Na_2SO_3, H_2O
(b)
X = OH 12 1 (66%)
X = Me 3 1 (60%)
I_2, $NaHCO_3$, THF
(c)
X = OH 7·2 1 (77%)
F 9·3 1 (75%)
OMe 6.6 1 (72%)
Me 2.3 1 (97%)

Scheme 14

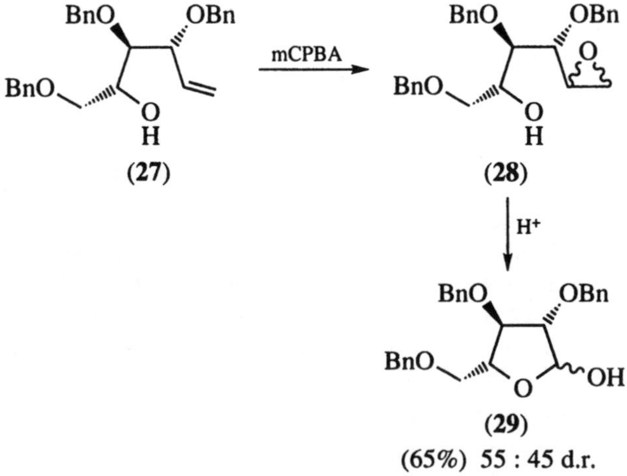

mCPBA
(27)
(28)
H+
(29)
(65%) 55 : 45 d.r.

Scheme 15

A transition state model which has been proposed to account for the diastereoselectivity in (c) in Scheme 14 but which can be extended to all examples in Schemes 13 and 14 is shown in Scheme 16.

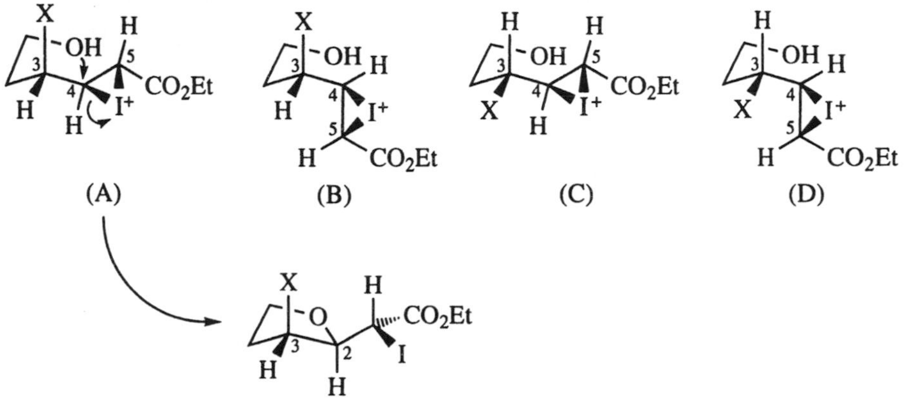

Scheme 16

Cyclisation is possible via four envelope chair-like transition states [(A)–(D)] in all of which torsional strain in the C_1–C_4 segment is minimised and the stereoelectronic requirement for the S_N2 is satisfied. However, (B) and (D) are less stable than (A) and (C) because C_5 occupies an axial position in (B) and (D) but an equatorial position in (A) and (C).[21] Attack on the iodonium ion by the OH in Scheme 16 is facilitated by the fractional positive charge at C-4. The allylic C—X bond with its σ-acceptor properties (X = OH, OMe, F) is best located in the plane of the original double bond allowing stabilization by the allylic hydrogen of development of this positive charge at C-4.

Thus cyclisation via (A) with the C—X bond axial, leading to cis-2,3-diastereoselectivity is preferred. Note that in Scheme 16, (C) and (D) have been used in which the configuration at C-3 is opposite to that in (A) and (B) to show more clearly the relationships between these four transition states although these experiments were in fact carried out on enantiopure materials having absolute configurations as in (A) and (B).

In contrast to the examples in Schemes 13 and 14, cyclisation of Z-configured 3-substituted hex-4-enols (30) (Scheme 17) results in the formation of 2,3-trans-substituted tetrahydrofurans.[18a] Here the sense of asymmetric induction derives from the steric interaction between the allylic substituent X and the methyl group ($A_{1,3}$-strain) when X is axial [conformation (31)] which means that cyclisation takes place exclusively via conformation (32). This is the case even when X = OBn, i.e. any σ–π electronic effects are overridden with this substitution pattern for the double bond.

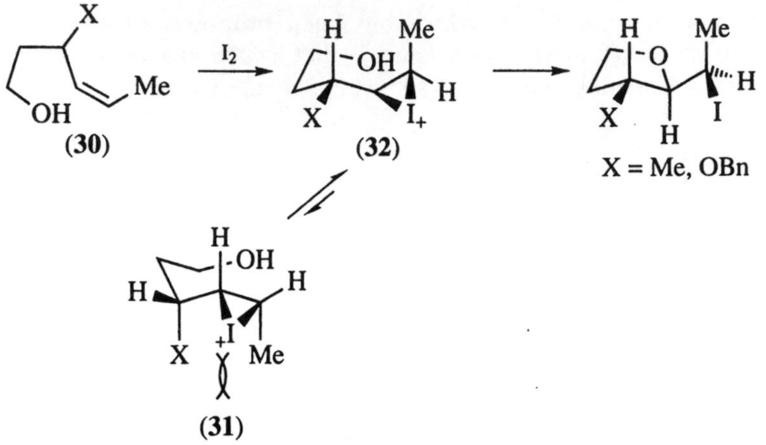

Scheme 17

Diastereoselectivity in attack on carbonyl groups

AXIAL VERSUS EQUATORIAL ATTACK IN CYCLOHEXANONES

Attack on cyclohexanones by bulky nucleophiles takes place preferentially to give axial alcohols [Scheme 18 (a)], and the presence of substituents at 3β or 5β of the cyclohexanone ring increases the diastereoselectivity (Chapter 9, Scheme 11).

However, using small nucleophiles and unhindered cyclohexanones, axial attack is favoured [Scheme 18 (b)].[22] Clearly, the *tert*-butyl group in the equatorial position, which is *cis* to the developing axial bond, offers no appreciable *direct* shielding towards attack, but it has an indirect effect in anchoring the chair as discussed in Chapter 9.

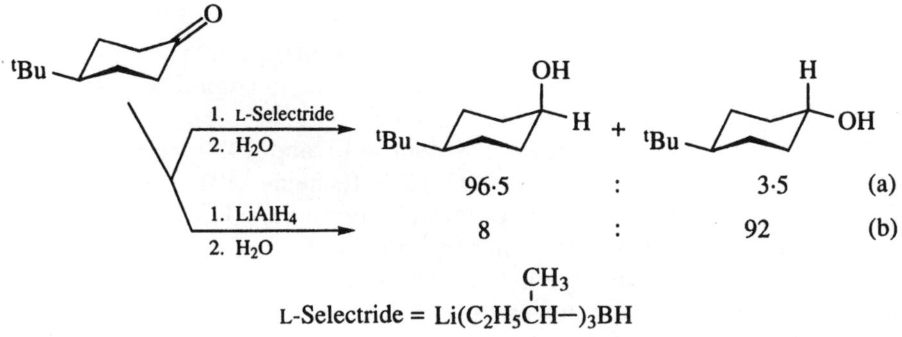

Scheme 18

What is the origin of the preferential axial attack in Scheme 18 (b)? In cyclohexanones some torsional strain arises from the near-eclipsing of the carbonyl group and the neighbouring equatorial C—H bonds (dihedral angle ~ 4°). At least part of the reason why axial attack by (small) nucleophiles is favoured may be the changes in this torsional strain as the hybridisation of the carbonyl group changes from sp^2 to sp^3. This is illustrated for 4-*tert*-butylcyclohexanone in Scheme 19.

In this Type $III'_{s.c.}$ reaction (see Chapter 9), the torsional strain between the equatorial $H_{2\beta}$ and the carbonyl is initially *increased* as equatorial attack (33) proceeds, since the C—O bond must pass through the plane containing this C—H bond. There will, of course, be an identical torsional strain between the carbonyl and the equatorial 6-H($H_{6\beta}$) (not shown).[23]

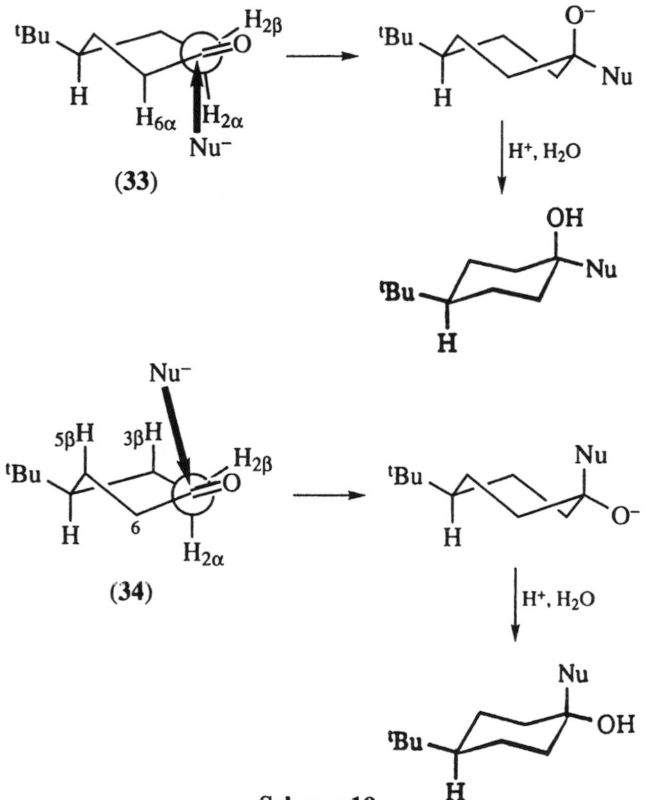

Scheme 19

By contrast, axial attack (34) will alleviate the torsional strain between these bonds since the C—O bond moves further away from C—$H_{2\beta}$ and C—$H_{6\beta}$ bonds as the reaction proceeds.

A further important factor to be considered is the trajectory of attack of the nucleophile. Bürgi and Dunitz[24] examined the crystal structures of a number of compounds each containing an amine and a carbonyl group in proximity in the molecule. They assumed that the relative position of the amine and carbonyl groups in space could be correlated with the trajectory taken in nucleophilic attack by an amine on a carbonyl group. The preferred direction of attack was concluded to be that having an N—C—O angle of 105°, i.e. close to tetrahedral. Equatorial attack by a nucleophile on 4-*tert*-butylcyclohexanone along this Bürgi–Dunitz trajectory [see (33)] will result in greater torsional strain between the axial bonds C—H$_{2\alpha}$ and C—H$_{6\alpha}$ and the developing C—Nu bond than between axial bonds C—H$_{3\beta}$ and C—H$_{5\beta}$ and the developing C—Nu bond in axial attack [as in (34)].

On the other hand, axial attack will be very sensitive to substitution of the axial hydrogens H$_{3\beta}$ and H$_{5\beta}$ by larger atoms or groups since the trajectory of the nucleophile takes it close to these hydrogens. For this reason also, axial attack will be more affected than equatorial attack by increased bulk of the nucleophile.

Diastereoselectivity resulting from unhindered nucleophilic attack on the carbonyl group of cyclohexenones by small nucleophiles is even higher

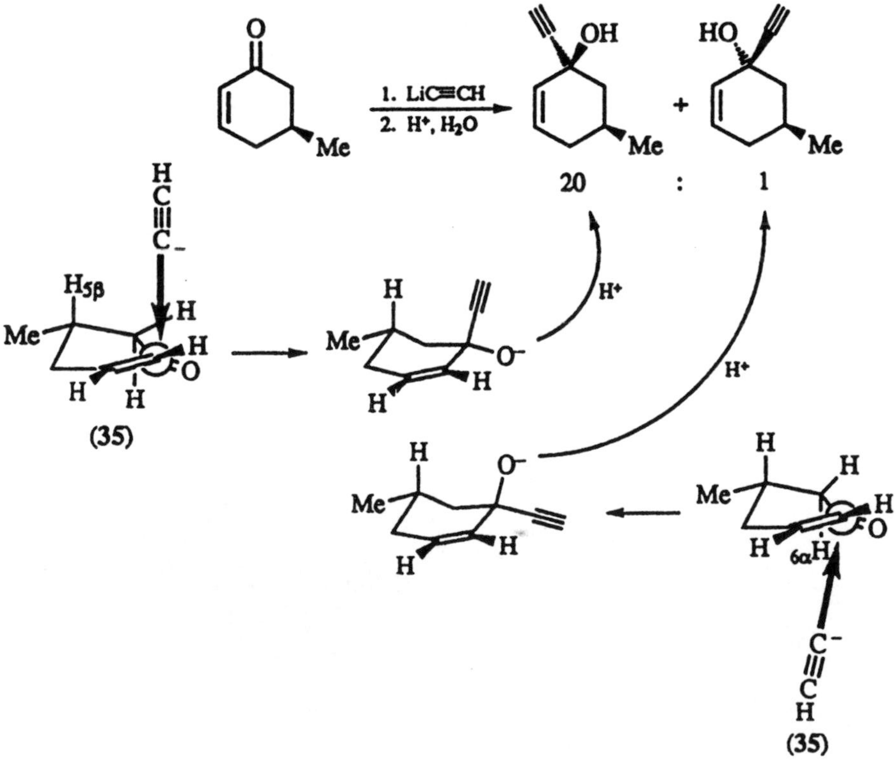

Scheme 20

in some cases than for attack on cyclohexanones (Scheme 20).[25] Here again the developing torsional strain is minimised by pseudoaxial attack; in particular, torsional strain between the developing bond and $H_{6\alpha}$ for equatorial attack (**35**) is greater than between the developing bond and the $H_{5\beta}$ for axial attack (**36**). Here the role of the methyl group is simply to occupy an equatorial position and hence dictate that the substrate reacts in the half-chair conformation shown.

CIEPLAK'S MODEL FOR DIASTEREOFACE SELECTION

A different model to account for the preference in addition to diastereofaces has been advanced by Cieplak.[26] It has been applied particularly to the cases of nucleophilic attack on cyclohexanones and electrophilic attack on methylenecyclohexanes.

Cieplak's model focuses on the relative stabilities of the two diastereoisomeric transition states. It is based on the premise that the principal factor differentiating the reactivity of the two diastereofaces is the nature of the σ-bond(s) interacting with the σ*-orbital of the developing σ-bond.

For the case of attack by a nucleophile Nu⁻ on cyclohexanone, it is postulated that axial attack is facilitated by interaction of the σ*-orbital of

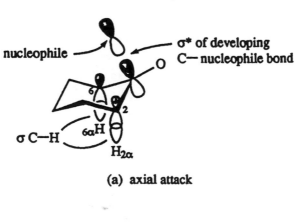

(a) axial attack

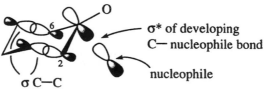

(b) equatorial attack

Figure 7

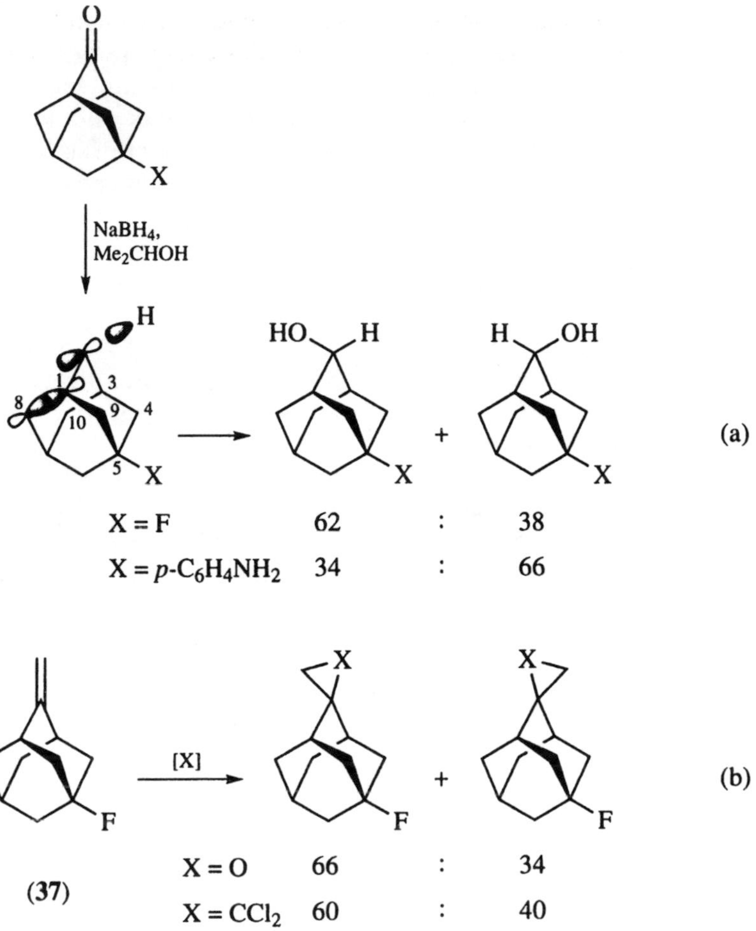

Scheme 21

the developing C—Nu bond by 'hyperconjugative' (σ–σ^*) interaction with the C—$H_{2\alpha}$ and C—$H_{6\alpha}$ bonds (Figure 7a). Likewise, equatorial attack is assisted by a corresponding hyperconjugative interaction with the ring C—C bonds shown (Figure 7b).

The preference for axial attack on cyclohexanones, in the absence of overriding steric effects, is ascribed to the greater hyperconjugative interaction of a C—H bond in comparison with a C—C bond.[†] This model, therefore, provides an alternative rationalisation of the preference for axial attack to that given above which was based primarily on torsional strain

---

†Cieplak's ordering of σ-donating substituents is C—Si > C—S > C—H > C—C > C-Hal; the relative ordering of C—H and C—C is still the subject of some debate.

considerations. It also predicts that *electrophilic* attack on methylenecyclohexanes should also be preferentially from the axial direction (see below).

Although this model has aroused considerable controversy, it does account for the disastereoselectivity of addition reactions to 5-substituted adamantan-2-ones and 5-substituted 2-methyleneadamantanes studies by Le Noble and co-workers.[27]

In Scheme 21 (a) it is proposed that attack by the nucleophile H⁻ takes place *anti* to the better C—C σ-donor, i.e. the more nucleophilic C—C σ-bond. The effect of fluorine substitution (X = F) will be to make the C_1—C_8 and C_3—C_{10} bonds better electron donors than the C_1—C_9 and C_3—C_4 bonds so attack takes place preferentially *syn* to the fluorine. Conversely, when X = $C_6H_4NH_2$-*p*, attack *anti* to X is preferred.

Similarly, electrophilic attack on the double bond in (37) [Scheme 21 (b)] by electrophiles is also *syn* selective.[28]

Even cycloadditions respond to these hyperconjugative effects as in the *syn*-selective Diels–Alder addition (Scheme 22).[29]

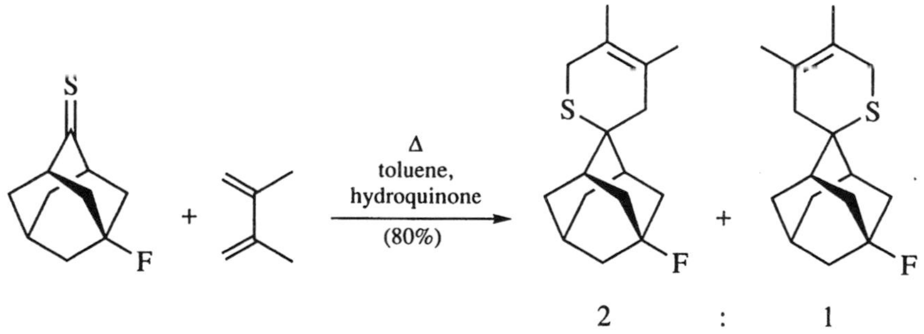

Scheme 22

CRAM AND FELKIN–ANH DIASTEREOSELECTIVITY IN NUCLEOPHILIC ADDITION TO THE CARBONYL GROUP

As mentioned previously, stereocontrol in acyclic molecules is complicated by the large number of conformations from which reaction can occur. In Type $III_{s.c.}$ reactions, competitive reaction from different conformations will invariably result in erosion of diastereoselectivity since the same diastereoface is unlikely to be attacked in each of the conformations.

Almost all highly diastereoselective Type $III_{s.c.}$ reactions on acyclic compounds are successful because they involve reducing to one the number of conformations through which reaction takes place. For an aldehyde or ketone, one way this can be achieved, using a combination of steric and electronic effects, is by having the chiral centre and the prochiral carbonyl

double bond adjacent to one another and by having at the chiral centre the
three substituents L, M and S of disparate size (L = large, M = medium, S =
small) (Scheme 23).

The Felkin–Anh interpretation[30] of the diastereoselectivity resulting from
nucleophilic addition to the carbonyl group uses the model (38) in Scheme 23
and makes the following assumptions: (a) the reactive conformation has the
bonds to L, M and S staggered relative to the carbonyl group as in (38) (and
(39)), with L located in the least sterically hindered site; (b) attack by the
nucleophile takes place along the Bürgi–Dunitz trajectory taking it close to S
in (38) but M in (39); formation of (40) via (38) is thus favoured.

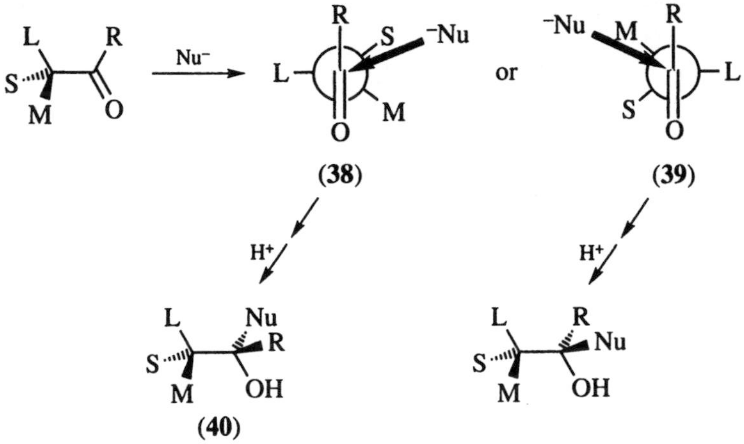

Scheme 23

Remember that we are dealing here with a model for the transition state,
with a view to predicting correctly the configuration of the alcohol resulting
from nucleophilic attack. There are other models which can do the same,
including the Cram model. The latter has been superseded by the
Felkin–Anh model, although the predicted sense of diastereoselectivity using
this model is still referred to as Cram selectivity (*anti*-Cram selectivity is
where the sense is the opposite to that predicted).

Examples of Cram selectivity which can be rationalised by the Felkin–Anh
model (38) are given in Scheme 24.[31,32]

Because of both its bulk and its easy subsequent elimination, the role of L
in Scheme 23 is frequently allotted to a trimethylsilyl group. For example,
addition of lithium alkyls to α-trimethylsilyl ketones is a highly
diastereoselective route to alcohols which are precursors to either (Z)- or
(E)-alkenes (Scheme 25).[33]

The basic Felkin–Anh (and Cram) model, which considers only the steric

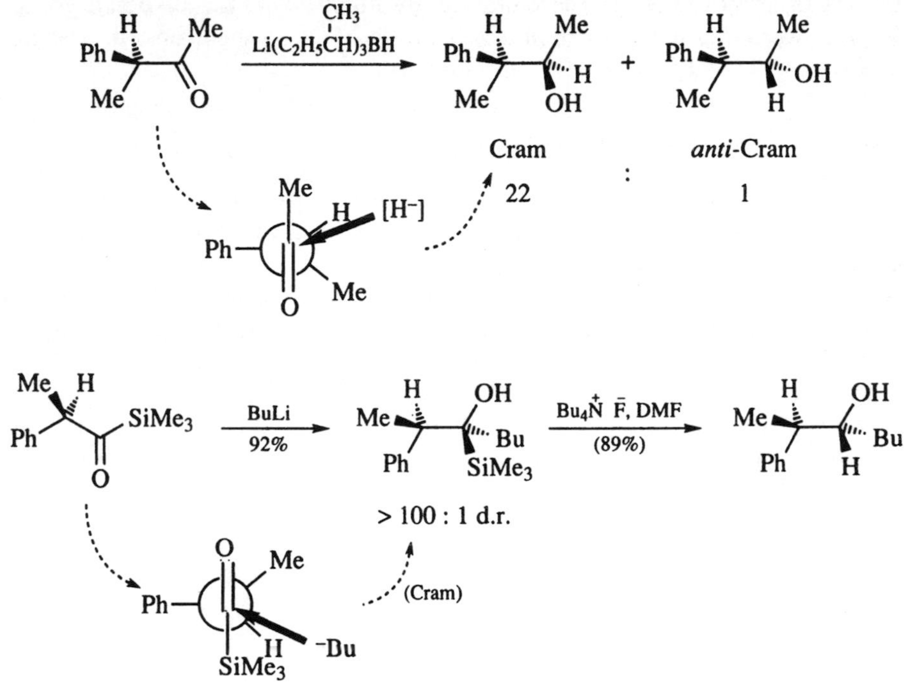

Scheme 24

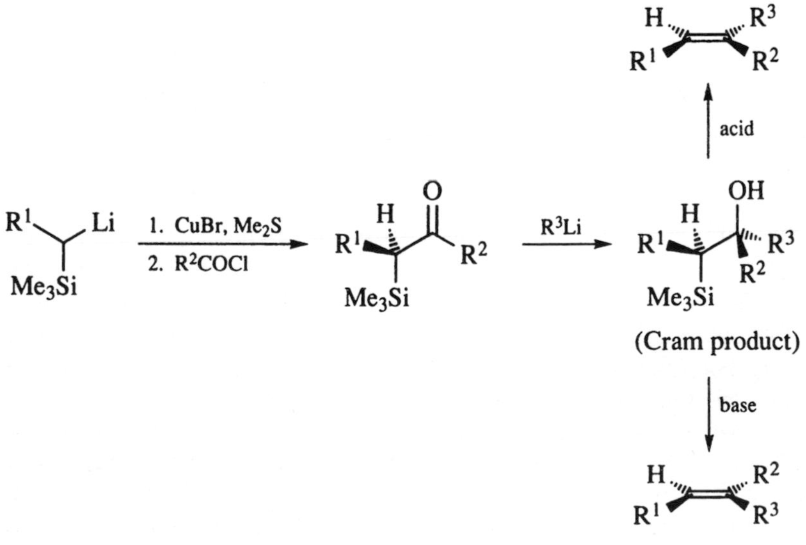

Scheme 25

effects of substituents on the chiral centre adjacent to the carbonyl group, breaks down when the medium-sized group M is an electron-withdrawing substituent (usually NR^1R^2, OR or halogen).

The σ^*–π^*-Dominated Felkin–Anh Model

The effect of substituents in the allylic position of a double bond on its reactivity towards *electrophiles* has been considered previously. What is the effect of substituents in the α-position on the reactivity of a carbonyl group undergoing nucleophilic addition? How does facial attack on the carbonyl double bond respond to the presence of electron-withdrawing or electron-donating σ-bonds at an adjacent chiral centre?

In frontier orbital terms, the LUMO of the carbonyl group (the π^*-orbital) is lowered by interaction with the σ^*-orbital of the α-substituent's σ-bond (Figure 8). This decreases the energy gap between it and the HOMO of the nucleophile and hence facilitates reaction.

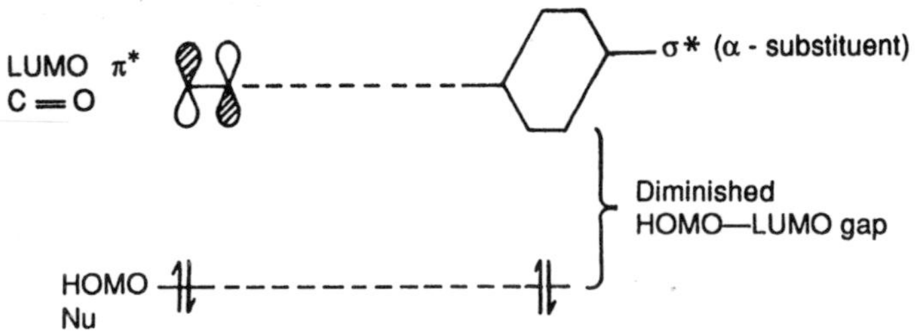

Figure 8

Which bonds will have σ^*-orbitals which will interact best with the π^*-orbital in Figure 8? They will be the best σ-acceptor bonds since their σ^*-orbitals will be lowest in energy and hence closest to π^*. Consequently, for facilitating nucleophilic attack, the best σ-acceptor bond at the position α to the carbonyl should be aligned *anti* to the face undergoing attack by the nucleophile. However, this position *anti* to the incoming nucleophile is that occupied by the largest group L in the 'steric' Felkin–Anh model in Scheme 23.

It was recognised early on by Cornforth et al.[34] that the presence of chlorine as the medium-sized substituent could change the sense of diastereoselectivity in nucleophilic additions to carbonyl groups from Cram to *anti*-Cram. The σ^*–π^*-dominated Felkin–Anh model in Scheme 26 is preferred over that in Scheme 23 when the medium-sized substituents on the chiral centre is electron withdrawing (M=EWG). In Scheme 26, (**41**) is preferred over (**42**) because the nucleophile passes close to the smaller group S rather than the larger group L.

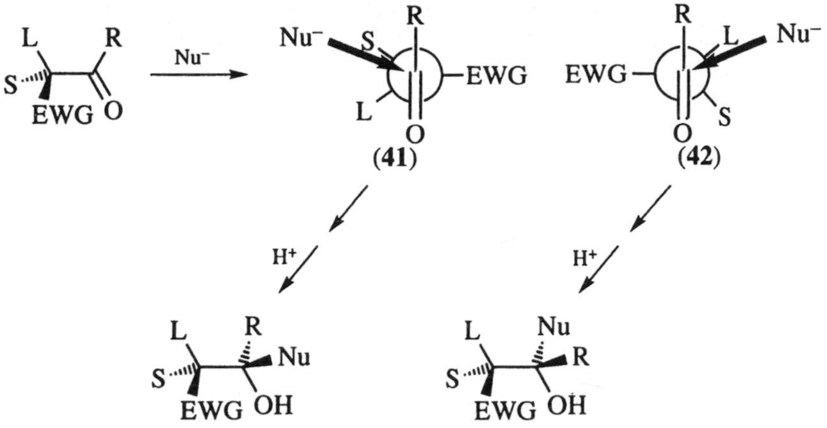

(41)   (42)

EWG = inductively electron-withdrawing group
(NR^{1}R^2, OR^1, halogen)

Scheme 26

In the examples in Scheme 27, the electron-withdrawing group in each case is oriented *anti* to the incoming nucleophile and attack on the carbonyl group is in accord with **(41)** (Scheme 26).[35,36]

The work of Dondoni *et al.*[36] in Scheme 27 (b) has shown that 2-trimethylsilylthiazole (a formyl anion synthetic equivalent) adds to glyceraldehyde acetonide **(43)** in a $\sigma^*-\pi^*$-dominated Felkin–Anh sense. The opposite configuration at this new chiral centre can be obtained by sodium tetrahydroborate reduction of the derived ketone **(44)** in the same Felkin–Anh sense.

The highest levels of diastereoselectivity are likely to arise when the large substituent L is also the σ-acceptor. This is often the case when L is a substituted nitrogen atom as in Scheme 28. Reetz *et al.*[37] have used the *N,N*-dibenzylamino group in this way [Scheme 28 (a)] in substrates, e.g. **(45)**, derived from readily available enantiopure amino acids. The benzyl groups are eventually removed by hydrogenolysis. Diastereoselectivity in the complementary sense can again be achieved by sodium tetrahydroborate reduction of the appropriate ketones [Scheme 28 (b)].

Conversely, the lowest levels of diastereoselectivity are likely to arise when the chiral centre bears a σ-acceptor as one substituent and a bulky group as another. The loss of diastereoselectivity in these cases may well be the result of reaction via Felkin–Anh steric and $\sigma^*-\pi^*$-dominated transition states **(38)** and **(41)** in competition.[30]

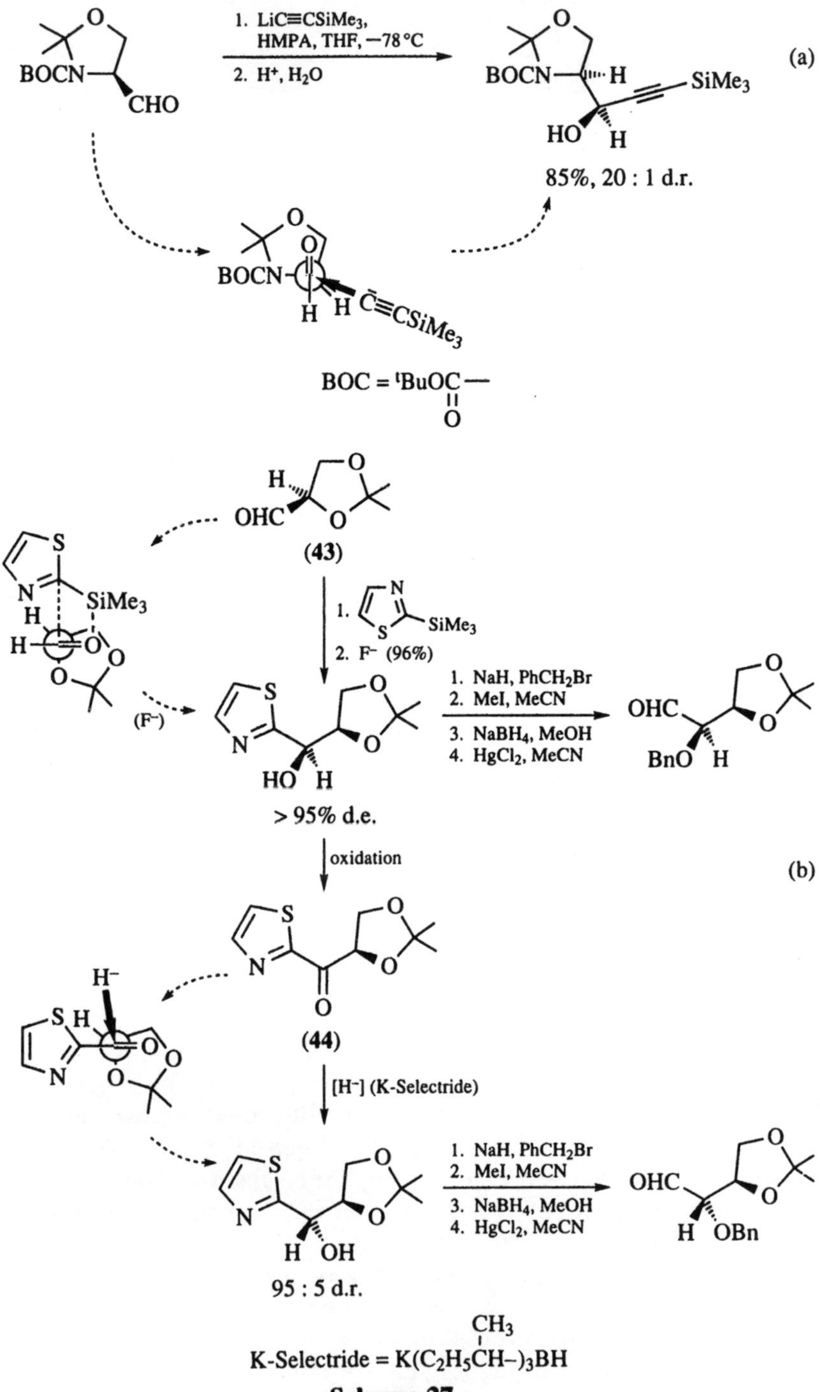

BOC = ^{t}BuOC—

(a)

85%, 20 : 1 d.r.

(43)

> 95% d.e.

(44)

95 : 5 d.r.

(b)

K-Selectride = K(C$_2$H$_5$CH–)$_3$BH

Scheme 27

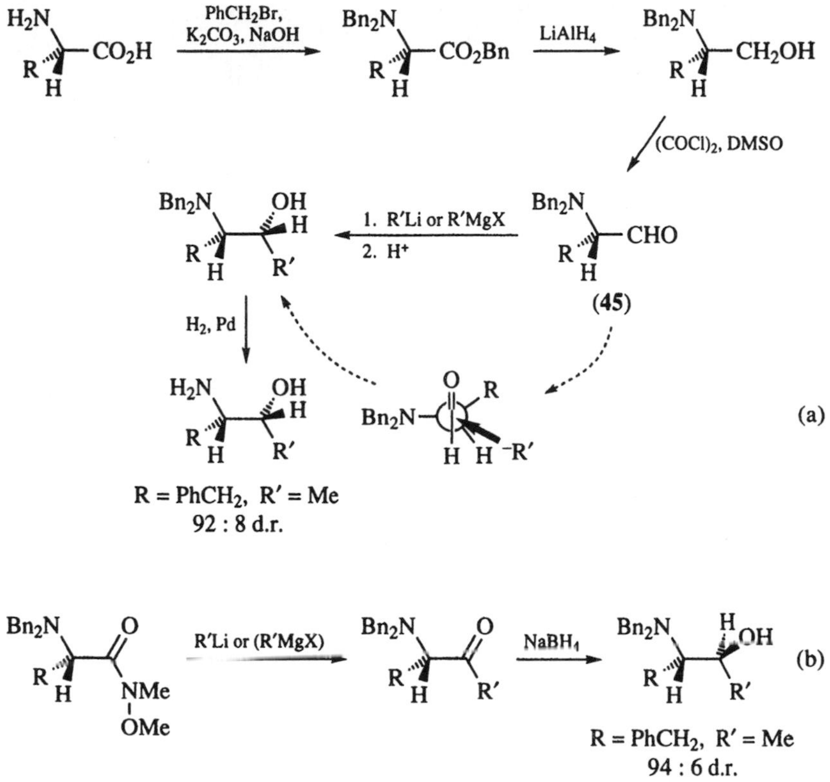

Scheme 28

The importance of ring formation in bringing about diastereoselectivity

DIASTEREOSELECTIVITY IN ADDITION TO ACYCLIC CARBONYL-CONTAINING COMPOUNDS

A complementary means to those above for defining the conformer that undergoes reaction is the linking one of the substituents L, M or S to the carbonyl oxygen via an ion X, usually a metal. As a consequence of this *chelation*, the carbonyl group is fixed in a more or less planar five-membered ring; attack on one diastereoface rather than the other is then controlled by the steric effects between the non-chelating substituents and the incoming nucleophile, as indicated in Scheme 29.

In this chelate model, first introduced by Cram and Kopecky,[38] the substituent at the chiral centre is usually chelated via oxygen or nitrogen, and the reactivity of the carbonyl group towards nucleophilic attack is enhanced by this chelation.[39]

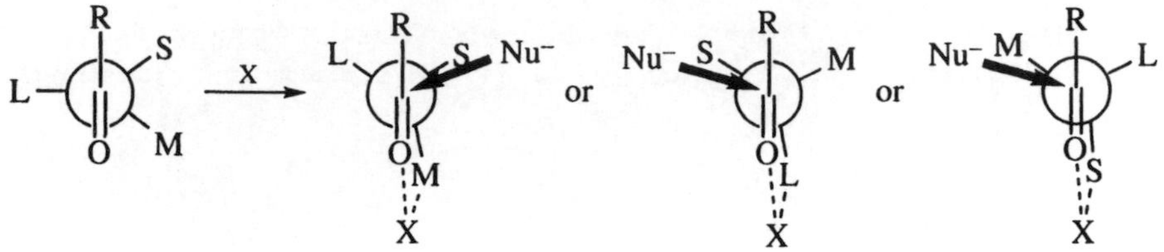

Scheme 29

When the chelated atom is the oxygen of a trialkylsilyloxy group, the chelation is disfavoured by an increase in size of the alkyl groups on silicon. Eliel and co-workers[40] have shown that as the silyloxy group in (46) increases in size, the diastereoselectivity of the reaction in Scheme 30 decreases.

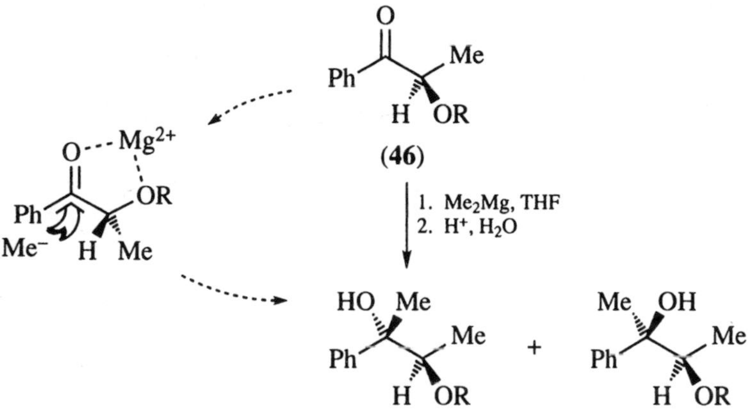

				relative 2nd-order rate constants
R = Me	99	:	1	~1000
R = Me$_3$Si	99	:	1	100
R = ^{t}BuMe$_2$Si	88	:	12	2.5
R = ^{i}Pr$_3$Si	42	:	58	0.45

Scheme 30

It seems likely that steric effects are responsible for this trend by reducing or preventing the chelation but the change in substitution on silicon may also affect the bascity of the coordinating oxygen lone pair.

A necessary accompaniment of chelation control of these nucleophilic addition reactions is that the rate of reaction of the chelated substrate should be greater than that of the non-chelated substrate. The relative second-order rate constants in the additions in Scheme 30 reveal that this is the case, i.e. that the more diastereoselective chelation-controlled reaction is also the faster reaction. Some examples of chelation-controlled addition to aldehydes are shown in Scheme 31.[41]

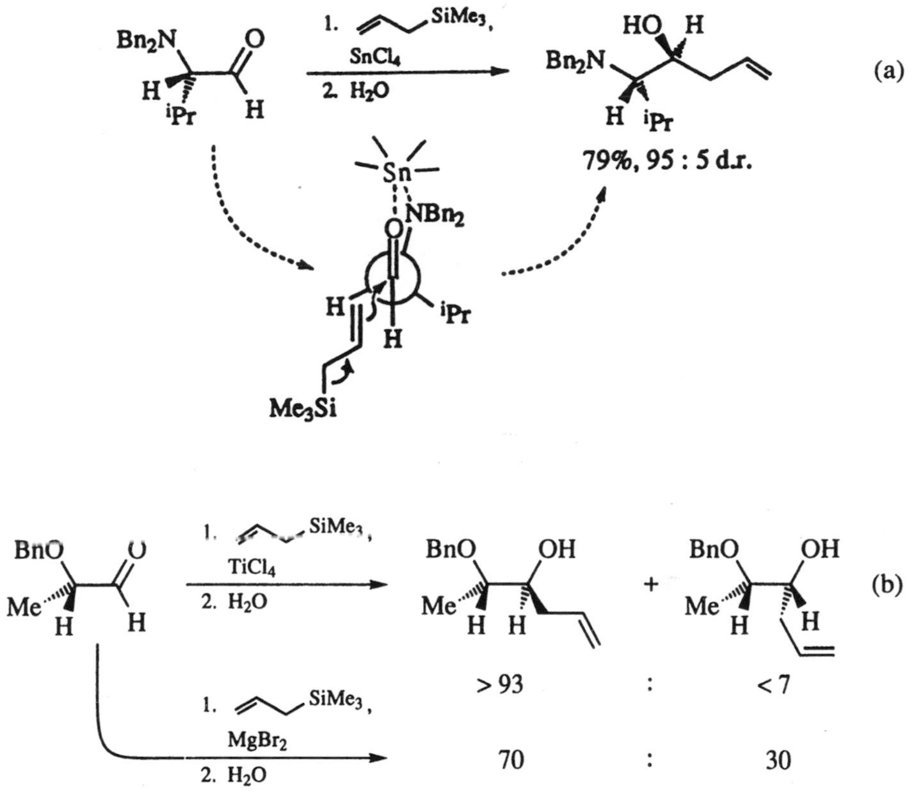

Scheme 31

The work of Reetz[39] in particular has shown that $TiCl_4$ or $SnCl_4$ may be used to form a chelate from an α-alkoxycarbonyl compound, and that this can then be attacked by carbon nucleophiles which do not disrupt the chelate. Note that in Scheme 31 (a) the sense of diastereoselectivity is opposite to that obtaining in Scheme 28 (a) where the reaction is of the σ^*–π^*-dominated Felkin–Anh type.

In chelation-controlled reactions, the choice of chelating metal is often critical if high levels of diastereoselectivity are to be achieved: in Scheme 31 (b) the titanium-catalysed allylsilane addition is highly diastereoselective but the corresponding magnesium-catalysed reaction is not.

Diastereoselective nucleophilic attack on a carbonyl group in a six-membered chelate is also possible.[42] The ring in this case is almost certainly not planar but the sense of asymmetric induction corresponds to attack *anti* to the bulkier unchelated substituent α to the carbonyl group (Scheme 32) i.e. it is still in the same sense as attack on a 5-membered chelate. Here also, a high level of chelation-controlled diastereoselectivity is metal dependent; replacement of the lithium dimethylcuprate by methylmagnesium bromide results in a loss of diastereoselectivity.[39]

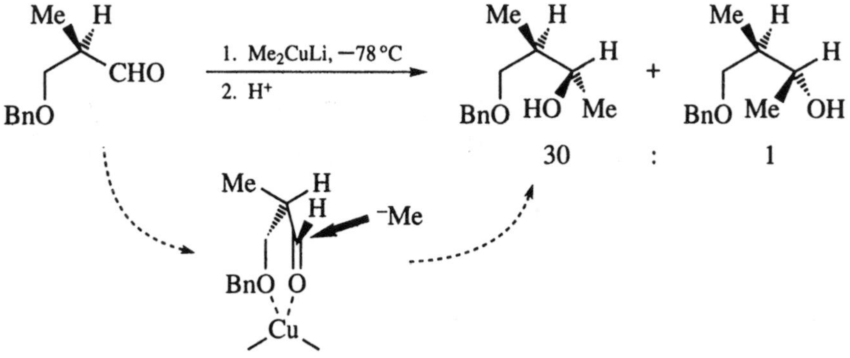

Scheme 32

1,3-Asymmetric induction in chelation-controlled addition to carbonyl groups

In the foregoing nucleophilic additions, the newly created chiral centre is adjacent to the parent chiral centre and the result is 1,2-asymmetric induction.

Reetz and Jung[43] have also shown that the aldehyde (**47**), in which the chiral centre is one carbon remove (β) from the carbonyl group, can also react highly diastereoselectively with allylsilanes in the presence of titanium tetrachloride (Scheme 33).

If the presumed chelated intermediate (**48**) takes up a half-chair conformation in which the methyl group occupies an equatorial position, attack of the allylsilane on the top face, as indicated, will deliver the chair conformation of the titanium-complexed product with the observed sense of asymmetric induction. However, one should be cautious in applying this analysis to chelated six-membered ring systems such as (**48**); the greater Ti—O bond length in comparison with C—C, and uncertainty as to the coordination state of the metal may mean that conformations other than the half-chair are those by which the reaction proceeds.

Chelation control can be used to advantage in other reactions of α-heteroatom-substituted carbonyl compounds. Thus Diels–Alder reaction of α-benzyloxyaldehyde (**49**) with the diene (**50**) in the presence of magnesium bromide proceeds as indicated in Scheme 34, with attack of the diene taking place on the less hindered face of the chelated aldehyde.[44]

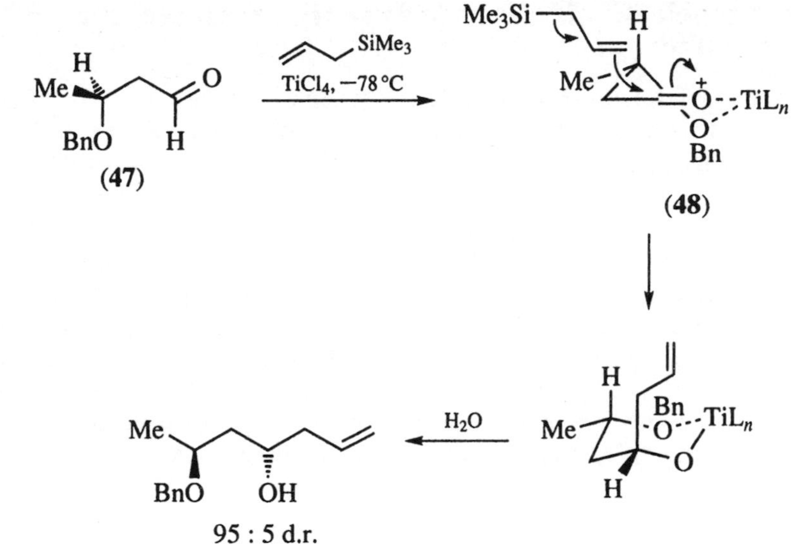

Scheme 33

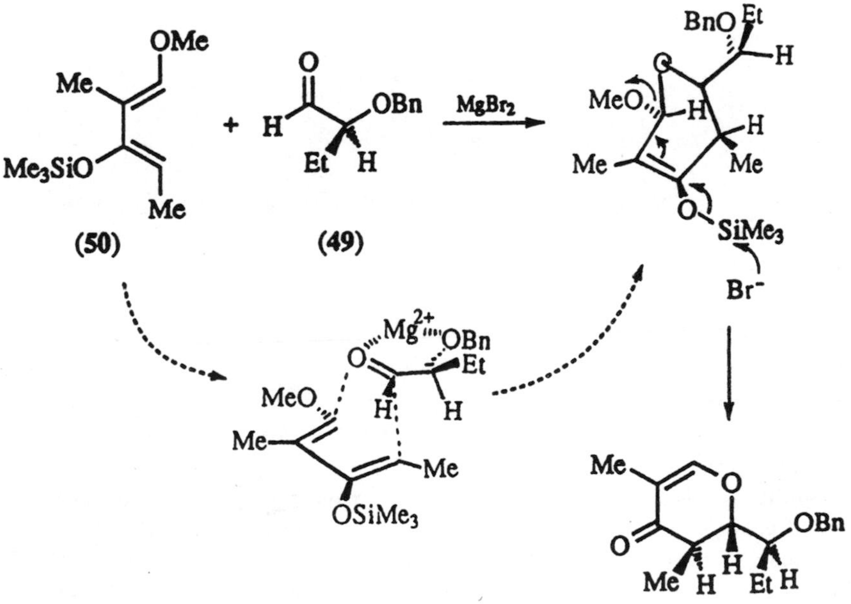

Scheme 34

Other stereoelectronic effects in Type III$_{s.c.}$ reactions: the anomeric effect

Minimisation of torsional strain and the presence of σ–π interactions are important in bringing about diastereoselectivity in many additions to double bonds, as discussed in this chapter. However, there are other stereoelectronic effects which may become dominant in particular classes of compounds. One of the best known of these is the anomeric effect.[1,45]

In 2-alkoxytetrahydropyrans (Scheme 35), for example, the alkoxy group has a preference for the axial over the equatorial position, the opposite of what would have been expected by analogy with the corresponding cyclohexane. This is the result of the anomeric effect, i.e. the stabilising interaction of the σ*-orbital of the exocyclic C$_2$—OR bond and the antiperiplanar non-bonding electron pair on the ring oxygen in the chair conformer having the OR group axial (51) (n–σ* interaction).

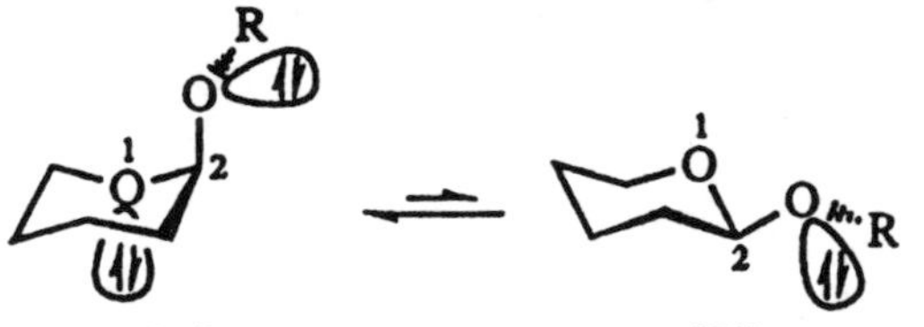

(51)   (52)

Scheme 35

A similar stabilising n–σ* interaction (the *exo*-anomeric effect) also exists between the (antiperiplanar) lone pair on the exocyclic oxygen and the O—C$_2$ σ-bond, but this is present in both conformations (51) *and* (52), and so does not influence their relative stability (Scheme 35).

These stabilising interactions lead to smaller bond lengths than expected for *both* O—C—O bonds in compounds bearing an axial OR group, but only for the exocyclic C$_2$—OR bond (the glycosidic bond) in compounds bearing an equatorial OR group. This is revealed in the X-ray crystal structures of a number of carbohydrate derivatives.[46]

Anomeric effects are important in stabilising particular conformations not only of acetals (Scheme 35) but also of related functional groups which have electronegative atoms (e.g. Cl, F) and heteroatoms bearing lone pair(s) (e.g. S, N, O) linked to the same atom. For example, α-fluoropyranose derivatives are obtained by treating protected pyranose derivatives with pyridinium polyhydrogen fluoride (Scheme 36) in a reaction which is under thermodynamic control.[47]

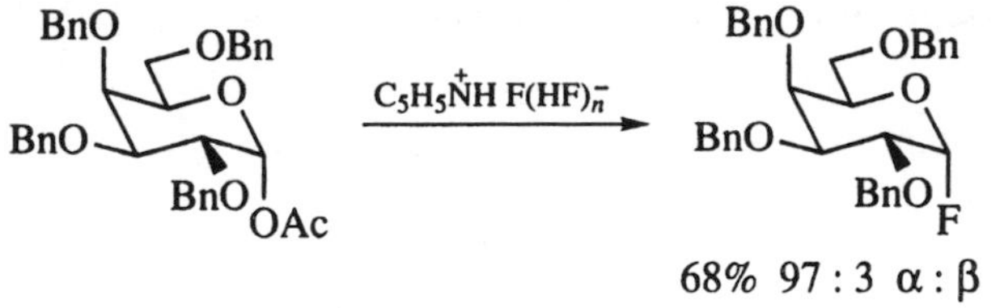

$$68\% \quad 97:3 \quad \alpha:\beta$$

Scheme 36

KINETIC ANOMERIC EFFECTS

Some stereochemistry of acetals, orthoesters and related compounds suggests that the n–σ* interactions, which give rise to ground-state anomeric effects, may also operate in the transition states for their reactions.[1,45]

Diastereoselectivity in reactions at the anomeric centre of pyranose derivatives often gives the axial anomer as the kinetically favoured product. Some of these reactions [Scheme 37 (a) and (c)] may lead to stereoconvergence. Suitable choice of the resident group in the anomeric position of the substrate and the attacking nucleophile can lead to one anomer or the other being the dominant product [Scheme 37 (b) and (c)].[48,49] As example (d) in Scheme 37 indicates, the formation of the axial product is also favoured in reactions involving a radical at the anomeric position.[50] However, it is not clear in the examples in Scheme 37 that a kinetic anomeric effect is necessarily responsible for the diastereoselectivity. The ground-state anomeric effects associated with O—C—C and O—C—H systems, which are

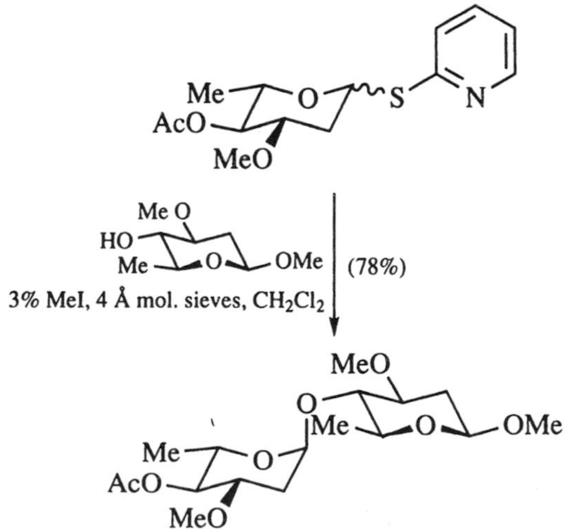

(a)

Scheme 37 *(continued)*

Scheme 37 *(continued)*

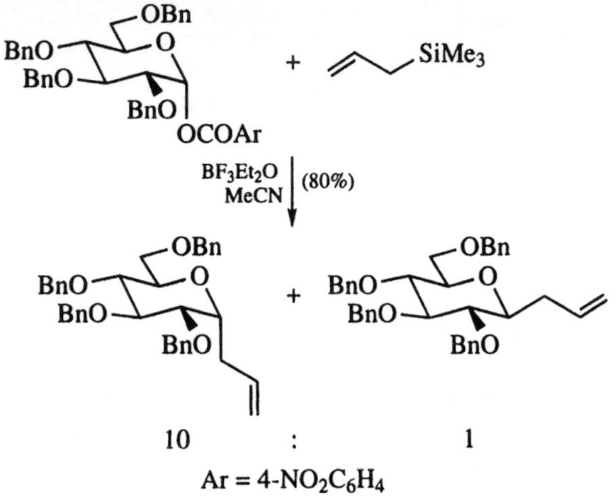

(b)

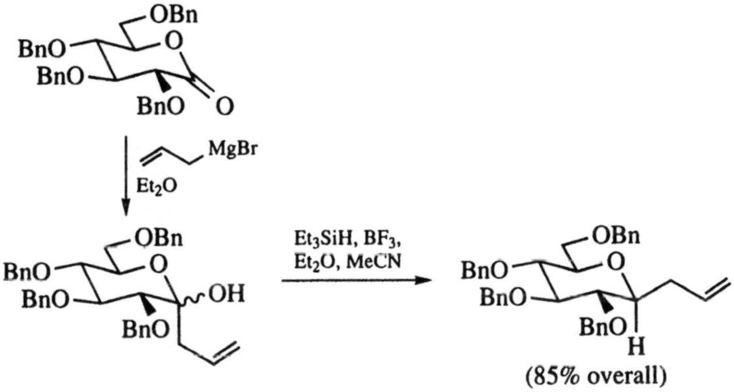

10 : 1

Ar = 4-NO$_2$C$_6$H$_4$

(c)

(85% overall)

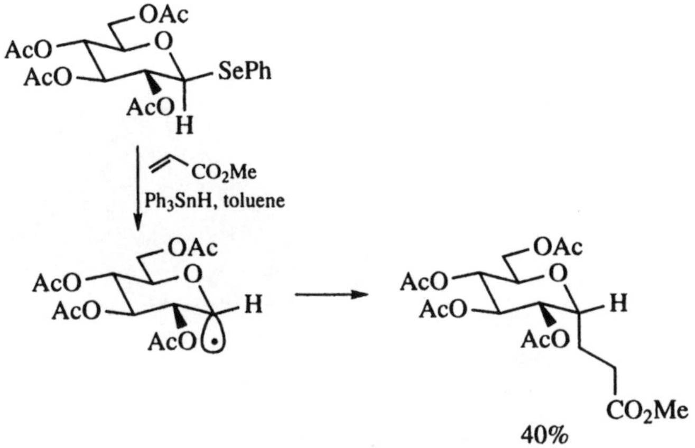

(d)

40%

Scheme 37

those being formed in (b)–(d), are small or non-existent. Attack on the intermediate carbocation or radical to give the equatorial anomer would require the product to be formed initially in a boat conformation, with its associated torsional strain, whereas the axial anomer can be formed directly in the chair conformation (Scheme 38; cf. Scheme 2).

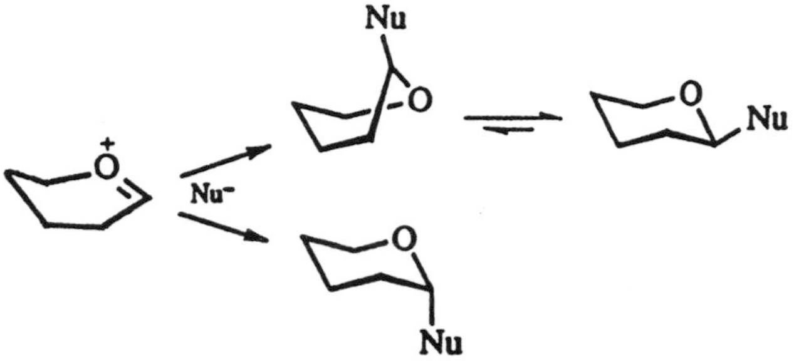

Scheme 38

Summary

Minimisation of torsional strain is one of the major reasons why cyclohexane assumes a chair conformation and why cyclopentane prefers an envelope conformation. Intramolecular reactions which occur between functional groups linked by three or four (carbon) atoms will also prefer to have the tether in a chair-like or envelope-like motif, respectively, in their transition states.

The greater stability of the developing the half-chair over the half-boat form accounts for the diastereoselectivity in many additions to six-membered α,β-unsaturated carbonyl-containing rings.

Substituents on the allylic position of a double bond may be either σ-donors, e.g. SiR$_3$, H, alkyl, or σ-acceptors, e.g. OR, F. For maximum activation of the double bond towards electrophiles, the σ-donor bond is aligned parallel to the plane of the π-electrons and electrophiles are directed *anti* to this σ-bond. On the other hand, σ-acceptor bonds align themselves perpendicular to the plane of the π-electrons, thus allowing activation by σ-donor bonds (if present). Similarly, activation of double bonds towards nucleophilic attack, by the formation and reaction of 'onium ions, requires electron-withdrawing allylic substituents (F, OH) to be located orthogonal to the plane of the π-electrons of the original double bond.

Nucleophilic attack on unhindered cyclohexanones by non-bulky nucleophiles takes place axially to give the equatorial alcohol diastereoselectively. In one explanation, the selectivity is attributed to the lesser torsional strain which results from axial attack. Another explanation

(Cieplak) focuses on the σ^*-orbital of the developing σ-bond to the nucleophile; this is assumed to be better stabilised by hyperconjugative interaction with the axial C—H bonds on the 2- and 6-positions.

The 'steric' Felkin–Anh model (38) (Scheme 23) predicts the sense of asymmetric induction in nucleophilic addition to a carbonyl group having chirality at the α-carbon, based on the size of the substituents at the chiral centre. This model has superseded the earlier Cram model, although the predicted sense of asymmetric induction is still referred to as Cram selectivity. When the medium-sized substituent M is electron withdrawing (Cl, O, N), M is assumed to take the place of L in model (38) [cf. (41), M=EWG, Scheme 26] in order to maximise its σ^*–π^* interaction with the carbonyl group; the sense of asymmetric induction is then the opposite of that predicted from model (38).

Chelation-controlled nucleophilic additions to carbonyl groups can result in high diastereoselectivity. In this case, one of the substituents on the α-carbon (L, M or S as above) becomes chelated to the carbonyl group via a metal ion. The chelate exists in a single conformation and contains an activated carbonyl group; it has diastereofaces which are clearly differentiated by the two flanking unchelated substituents (Scheme 29).

Similar chelation control can sometimes be effected in substrates with the chiral centre β to the carbonyl group.

The anomeric effect arises from n–σ^* interaction in molecules containing X—Y—Z where Z is electron withdrawing (OR, halogen) and X (usually OR) has a lone pair of electrons which interact with the Y—Z σ^*-orbital. This effect is manifest in pyranose derivatives by a greater stability of the conformer with an axial OR substituent on the anomeric position than would otherwise be expected.

Preferential formation of axial bonds to carbon or hydrogen by a carbocation or radical at the anomeric centre of pyranose derivatives is more likely to be the result of direct formation of the chair conformation of the product (cf. addition to six-membered α,β-unsaturated carbonyl-containing rings above), rather than a kinetic anomeric effect.

References

1. P. Deslongchamps, *Stereoelectronic Effects in Organic Chemistry*, Pergamon Press, Oxford, 1983.
2. G.J. Bishop, B.J. Price and I.O. Sutherland, *Chem. Commun.*, 1967, 672; Y. Shvo, in *The Chemistry of the Hydrazo, Azo and Azoxy Groups Part II*, ed. S. Patai, Wiley–Interscience, New York, 1974.
3. J.A. Marshall, M.J. Coghlan and M. Watanabe, *J. Org. Chem.*, 1984, **49**, 747.
4. A. Hosomi and H. Sakurai, *J. Am. Chem. Soc.*, 1977, **99**, 1673.
5. G. Cardillo, M. Orena, S. Sandri and C. Tomasini, *J. Org. Chem.*, 1984, **49**, 3951.
6. J.F. Stoddard, *et al. J. Am. Chem. Soc.*, 1992, **114**, 6330, and references cited therein.

7. P.R. Schleyer, *J. Am. Chem. Soc.*, 1967, **89**, 699.
8. A.A. Pinkerton, D. Schwartzenbach, J.H.A. Stibbard, P.A. Carrupt and P. Vogel, *J. Am. Chem. Soc.*, 1981, **103**, 2095.
9. For further references to pyramidalisation and reactivity of trigonal carbons, see D. Seebach, T. Maetzke, W. Petter, B. Klötzer and D.A. Plattner, *J. Am. Chem. Soc.*, 1991, **113**, 1781; W. Luef and R. Keese, *Top. Stereochem.*, 1990, **20**, 231.
10. R. Gleiter and L.A. Paquette, *Acc. Chem. Res.*, 1983, **16**, 328.
11. K. Burgess, J. Cassidy and M.J. Ohlmeyer, *J. Org. Chem.*, 1991, **56**, 1020.
12. I. Fleming and N.J. Lawrence, *Tetrahedron Lett.*, 1988, **29**, 2077.
13. I. Fleming, A.K. Sarkar and A.P. Thomas, *Chem. Commun.*, 1987, 157.
14. I. Fleming, *Frontier Orbitals and Organic Chemical Reactions*, Wiley, New York, 1976.
15. I. Fleming, A.K. Sarkar, M.J. Doyle and P.R. Raithby, *J. Chem. Soc., Perkin Trans. 1*, 1989, 2023.
16. The nature of the dienophile can affect this face selectivity; see R. Tripathy, R.W. Frank and K.D. Onan, *J. Am. Chem. Soc.*, 1988, **110**, 3257.
17. S.D. Kahn and W.J. Hehre, *J. Am. Chem. Soc.*, 1987, **109**, 663.
18. (a) A.B. Reitz, S.O. Nortey, B.E. Maryanoff, D. Liotta and R. Monahan, *J. Org. Chem.*, 1987, **52**, 4191; (b) Y. Tamaru, S. Kawamura and Z. Yoshida, *Tetrahedron Lett.*, 1985, **26**, 2885.
19. A.R. Chamberlin, M. Dezube, P. Dussault and M.C. McMills, *J. Am. Chem. Soc.*, 1983, **105**, 5819.
20. H. Takahata, T. Takamatsu and T. Yamazaki, *J. Org. Chem.*, 1989, **54**, 4812.
21. M. Labelle and Y. Guindon, *J. Am. Chem. Soc.*, 1989, **111**, 2204.
22. M. Cherest, H. Felkin and N. Prudent, *Tetrahedron Lett.*, 1968, 2199; Y.-D. Wu and K.N. Houk, *J. Am. Chem. Soc.*, 1987, **109**, 908.
23. F.A. Carey and R.J. Sundberg, *Advanced Organic Chemistry, Part A*, Plenum Press, New York, 1977.
24. H.B. Bürgi and J.D. Dunitz, *Acc. Chem. Res.*, 1983, **16**, 153, and references cited therein.
25. Y.-D. Wu, K.N. Houk and B.M. Trost, *J. Am. Chem. Soc.*, 1987, **109**, 5560.
26. A.S. Cieplak, *J. Am. Chem. Soc.*, 1981, **103**, 4540.
27. C.K. Cheung, L.T. Tseng, M.-H. Lin, S. Srivastava and W.J. le Noble, *J. Am. Chem. Soc.*, 1976, **108**, 1598.
28. S. Srivastava and W.J. le Noble, *J. Am. Chem. Soc.*, 1987, **109**, 5874.
29. W.-S. Chung, N.J. Turro, S. Srivastava, H. Li and W.J. le Noble, *J. Am. Chem. Soc.*, 1988, **110**, 7882; H. Li, J.E. Silver, W.H. Watson, R.P. Kashyap and W.J. Le Noble, *J. Org. Chem.*, 1991, **56**, 5932.
30. E.P. Lodge and C.H. Heathcock, *J. Am. Chem. Soc.*, 1987, **109**, 3353 (for an account of earlier work in this area).
31. M.M. Midland and Y.C. Kwon, *J. Am. Chem. Soc.*, 1983, **105**, 3725; *anti-* Cram products are obtained preferentially using dicyclohexylborane.
32. M. Nakada, Y. Urano, S. Kobayashi and M. Ohno, *J. Am. Chem. Soc.*, 1988, **110**, 4826.
33. A.G.M. Barrett and J. Flygare, *J. Org. Chem.*, 1991, **56**, 638.
34. J.W. Cornforth, R.H. Cornforth and K.K. Mathew, *J. Chem. Soc.*, 1959, 112.
35. P. Herold, *Helv. Chim. Acta*, 1988, **71**, 354.
36. A. Dondoni, G. Fantin, M. Fogagnolo, A. Medici and P. Pedrini, *J. Org. Chem.*, 1989, **54**, 693, 702.
37. M.T. Reetz, M.W. Drewes, K. Lennick, A. Schmitz and X. Holdgrün, *Tetrahedron: Asymmetry*, 1990, **1**, 375; M.T. Reetz, *Angew. Chem., Int. Ed. Engl.*, 1991, **30**, 1531.

38. D.J. Cram and K.R. Kopecky, *J. Am. Chem. Soc.*, 1959, **81**, 2748.
39. Review: M.J. Reetz, *Angew. Chem., Int. Ed. Engl.*, 1984, **23**, 556.
40. X. Chen, E.R. Hortelano, E.L. Eliel and S.V. Frye, *J. Am. Chem. Soc.*, 1990, **112**, 6130.
41. M.T. Reetz, M.W. Drewes and A. Schmitz, *Angew. Chem., Int. Ed. Engl.*, 1987, **26**, 1141.
42. W. Clark Still and J.A. Schneider, *Tetrahedron Lett.*, 1980, **21**, 1035.
43. M.T. Reetz and A. Jung, *J. Am. Chem. Soc.*, 1983, **105**, 4833.
44. S.J. Danishefsky, W.H. Pearson and D.F. Harvey, *J. Am. Chem. Soc.*, 1984, **106**, 2546.
45. A.J. Kirby, *The Anomeric Effect and Related Stereoelectronic Effects at Oxygen*, Springer, Berlin, 1983; P.R. Graczyk and M. Mikalajczyk, *Top. Stereochem.*, 1900, **21**, 351.
46. G.A. Jeffrey, J.A. Pople, J.S. Binkley and S. Vishveshwara, *J. Am. Chem. Soc.*, 1978, **100**, 373.
47. M. Hayashi, S. Hashimoto and R. Noyori, *Chem. Lett.*, 1984, 1747.
48. G.V. Reddy, V.R. Kulkarni and H.B. Mereyala, *Tetrahedron Lett.*, 1989, **30**, 4283, 4287.
49. M.D. Lewis, J.K. Cha and Y. Kishi, *J. Am. Chem. Soc.*, 1982, **104**, 4976.
50. R.M. Adlington, J.E. Baldwin, A. Basak and R.P. Kozyrod, *Chem. Commun.*, 1983, 944.

11 SUBSTRATE-CONTROLLED DIASTEREOSELECTIVE REACTIONS (TYPE III$_{s.c.}$) MEDIATED BY RING FORMATION

1,n-Asymmetric induction in acyclic substrates

We have seen that steric and electronic effects from an adjacent chiral centre can lead to high levels of asymmetric induction in addition to C=O and C=C double bonds in (1) (Chapters 9 and 10). However, these effects decline rapidly in importance as the number of bonds separating the chiral centre from the prochiral double bond increases in (2).

$$X = O, C\diagdown$$

(1)

(2)

In (2; p > 0) there will be a larger number of accessible conformations and some uncertainty as to which of these will be preferred. In addition, there will be the ever-present possibility that the reacting conformation(s) may not correspond to the more stable one(s) (cf. Curtin–Hammett principle). The likelihood of reaction taking place from more than one conformation increases as the number of atoms p comprising the tether in (2) increases and this, as always, is a recipe for reduced diastereoselectivity. Ways need to be found of eliminating reactions from, ideally, all but one conformation.

Limitation of the conformational freedom of a chain can most easily be achieved by incorporating it, or part of it, into a ring. We have seen that this tactic can even be used for (2; p = 0) (Chapter 10) by using a metal atom to chelate the carbonyl group and one of the substituents a, b or c. The two diastereofaces in the resulting five-membered cyclic chelate are clearly differentiated because one of them must be *cis* to the more bulky of the two unchelated substituents on the chiral centre.

Diastereoselectivity originating in the cyclisation reaction itself

An alternative way in which ring formation can mediate diastereoselective addition to the double bond in (2) is by intramolecular attack with, e.g., c (Figure 1), making use of the conformational preferences of the connecting tether and the substituents on it.

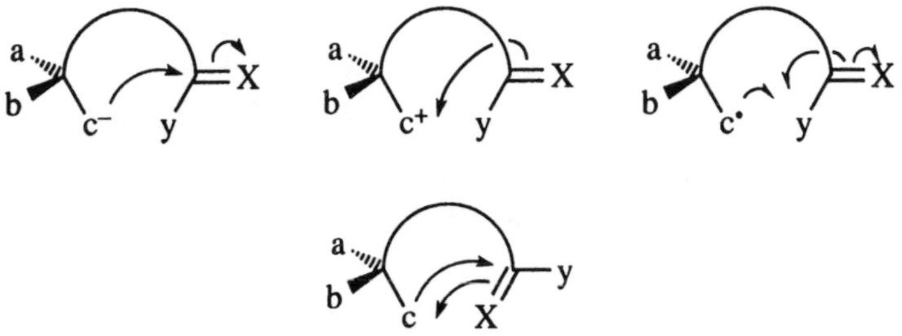

Figure 1

In this case, therefore, ring formation constitutes the diastereoselective reaction; it contrasts with chelation-controlled reactions referred to above where chelate ring formation is followed by the reaction which determines the diastereoselectivity.

As indicated in Figure 1, the intramolecular reaction may be nucleophile-, electrophile- or radical-mediated, or it may involve an intramolecular cycloaddition. The cyclic transition state sets up a new set of steric or electronic interactions and these can give rise to high diastereoselectivity.

Thus in Scheme 1, cyclisation takes place via nucleophilic attack on the iodonium ion (3) by the carbonate anion to give the 4,6-*cis*-disubstituted

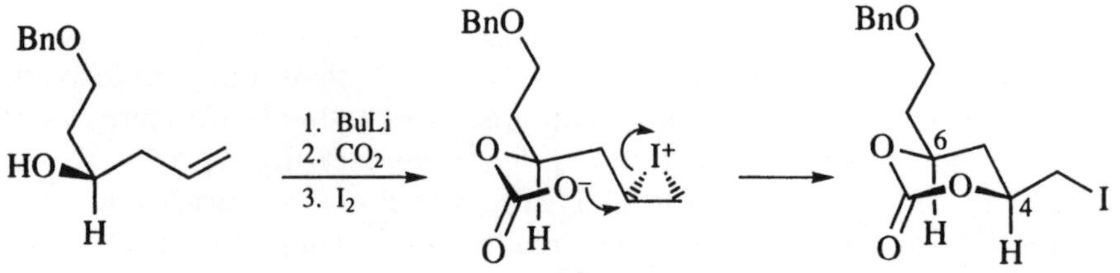

(3)

(4) 69%

9 : 1 d.r.

Scheme 1

product (4) diastereoselectively.[1] It is helpful to consider the stereochemistry of this cyclisation in some detail because it is typical of many other reactions of this type. Asymmetric induction is the result of a number of factors including (a) cyclisation via a chair conformation, (b) preference of the side-chain $(CH_2)_2OBn$ for an equatorial rather than an axial position on the chair, (c) preference of the CH_2 of the iodonium ion for an equatorial rather than an axial position and (d) reversibility of iodonium ion formation. The formation of the 4,6-*trans*-substituted diastereoisomer as a minor product can be ascribed to steric interactions between the pairs of protons indicated in the two transition states (5) and (6) (Scheme 2) which would lead to its formation.

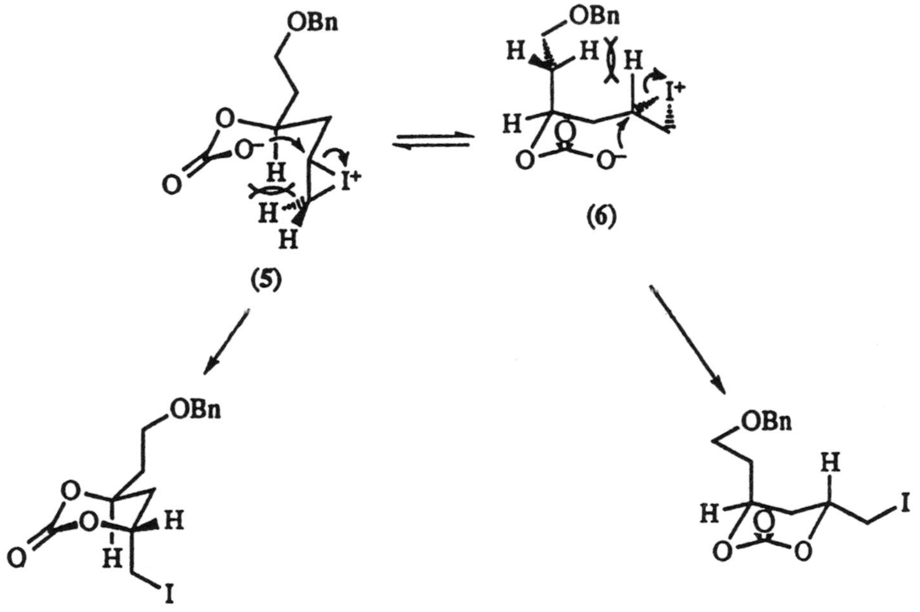

Scheme 2

Since intramolecular attack of the carbonate oxygen on the iodonium ion can only take place with inversion, the iodonium ion must have the configuration shown in (3) (Scheme 1). However, there is no compelling reason why only one diastereoface of the alkene should be attacked in the formation of the iodonium ion, and consequently attack on the C=C must be reversible if high yields of the final product are to be obtained.

It should be noted that the 4,6-*cis*-isomer, the product in Scheme 1, could, in principle, also result from a boat transition state (7) as shown in Scheme 3 (a). It should not automatically be assumed that the chair will always be favoured over the boat. When we draw the boat we are in fact referring to the twist boat. The all-carbon twist boat lies only ~21 kJ above the chair in energy and

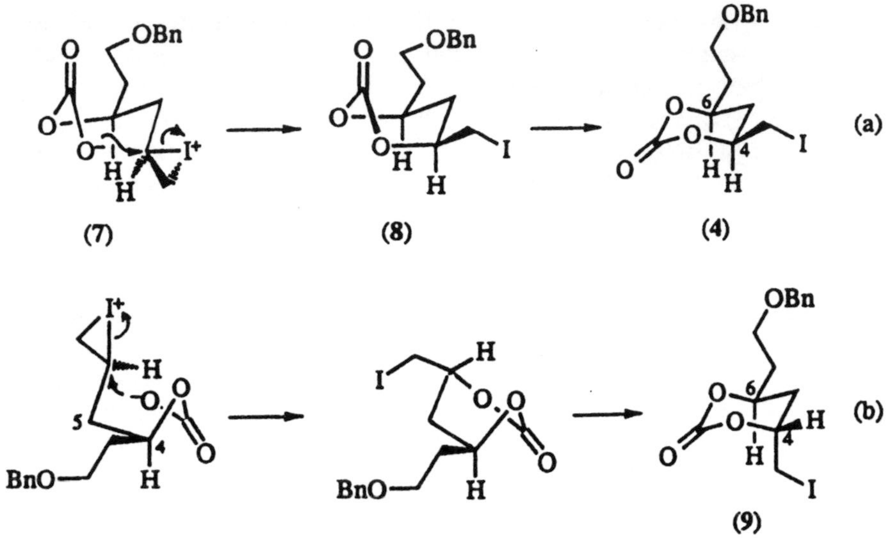

(7)   (8)   (4)

(a)

(b)

(9)

Scheme 3

this difference will be reduced in the cyclic carbonate in Scheme 3 (a): the stern–prow interaction and eclipsed-butane interactions which raise the energy of the all-carbon boat relative to the chair are both reduced in (8) by incorporation of the carbonate function into the ring. This reduction in strain would be expected to benefit the transition state (7) also. In the present case this is of no consequence because the boat transition state (7) and the chair transition state (3) (Scheme 1) both lead to the observed diastereoisomer (4).

When considering the possibility of boat-like transition states it is important to appreciate that whereas there are only two different chair conformations for a monosubstituted cyclohexane, there are four possible

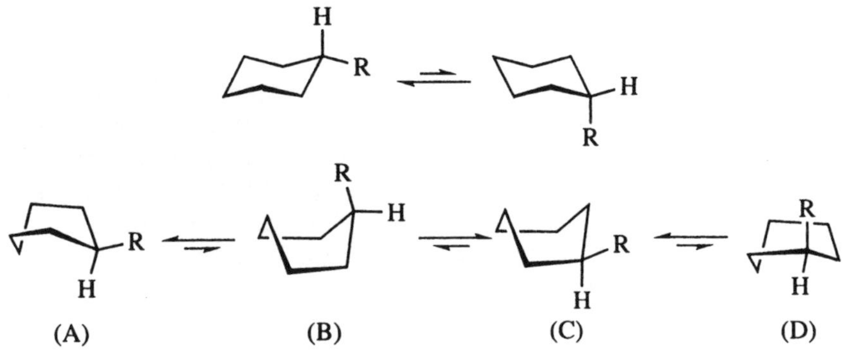

(A)   (B)   (C)   (D)

Scheme 4

boat conformations (Scheme 4). This is the case even though strictly the conformations should be shown as twist boats.

Normally one can discount involvement of the conformations (B) and (D) because of the stern (prow) location of R in the former and the axial position of R in the latter. This still leaves the possibility of two boats (A) and (C).

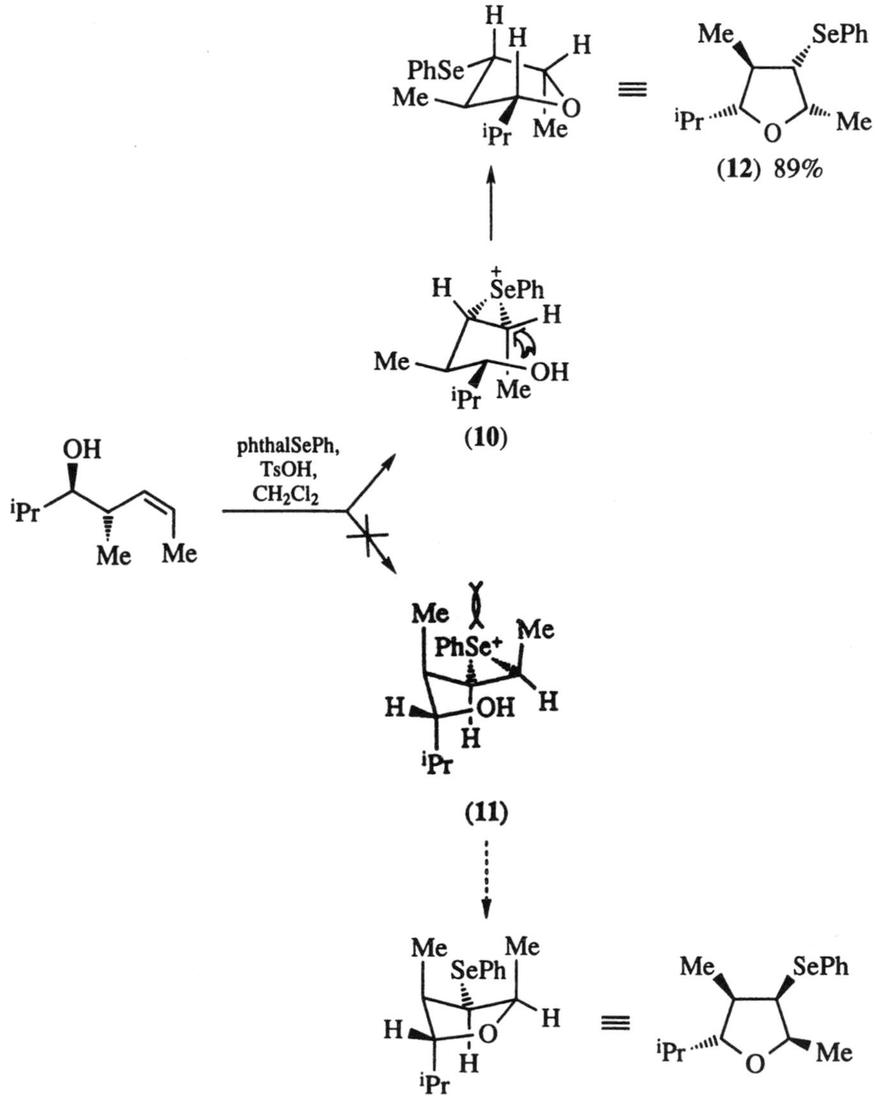

Scheme 5

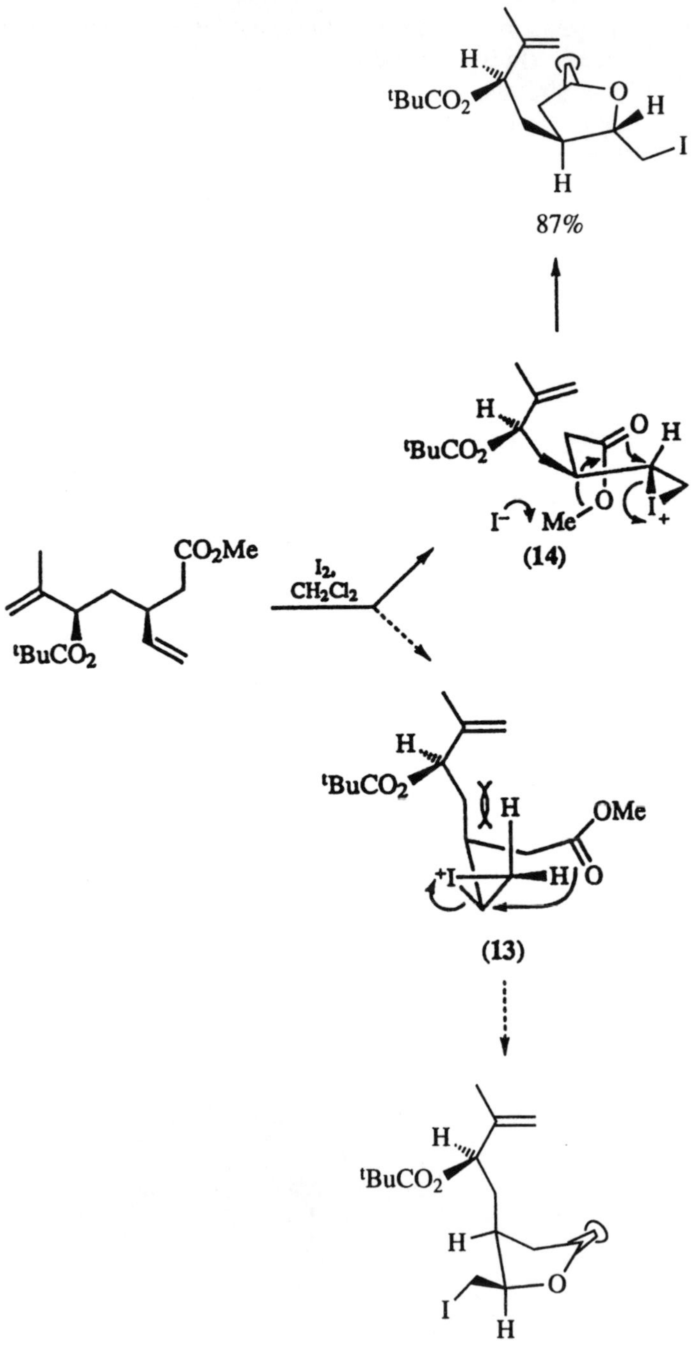

H
$^tBuCO_2$
O
H
I
H
87%
H
$^tBuCO_2$
O
H
O
I⁻
Me
I
+
(14)
CO_2Me
$^tBuCO_2$
I_2,
CH_2Cl_2
H
$^tBuCO_2$
H
OMe
+I
H
O
(13)
H
$^tBuCO_2$
H
O
I
H

Scheme 6

Likewise in cyclisation reactions, when the tether and interacting functional groups can assume a chair or boat motif, there will always be at least two boat conformations that should be examined.

In fact, this alternative boat in Scheme 3 (b) would not be expected to compete with that in (a) because one of the eclipsed-butane interactions is still present, between C-4 and C-5. Consideration of this boat, of course, is not just academic because it would lead to a 4,6-*trans*-disubstituted product (9), diastereoisomeric with the one obtained.

When five-membered rings are being formed, preference for an envelope-shape transition state is a recurring theme. Diastereoselectivity can again result when substitution on the tether is appropriate. In the selenolactonisation in Scheme 5, for example, the part of the envelope motif which the chain adopts gives rise to $A_{1,3}$-strain in (11) and destabilises it relative to (10) from which the product (12) is formed.[2] As in the previous example, selenenium ion formation must be reversible since high yields of products are obtained even though facial selectivity in its initial formation is unlikely to be complete in the required sense.

Even iodolactonisation of a terminally unsubstituted double bond with a single substituent on the tether may take place diastereoselectively (Scheme 6).[3] Here, the attacking nucleophile is the carbonyl group of an ester rather than a carboxylate anion. To the extent that the transition state resembles a chair, the 'axial' substituent in (13) will retard its reaction relative to that of (14), in which the substituent is 'equatorial.'

A useful property of both γ-lactone and dioxanone (cyclic carbonate) ring systems in synthesis is the ease with which they undergo ring opening under mildly basic conditions. The oxyanions generated in this way from the products in Scheme 1 and 6 bring about intramolecular displacement of iodide and formation of epoxides (Scheme 7). Overall, therefore, the ring-closure, ring-opening, ring-closure reactions in Schemes 1, 6 and 7 result in formation of epoxides (15) and (16) from the corresponding acyclic alkenes with high 1,3- and 1,2-asymmetric induction, respectively.

Examples of diastereoselective formation of five-membered ring derivatives via chair-like transition states in which stereoelectronic effects are in part responsible for the sense of induction are shown in Chapter 10, Schemes 14 and 16.

For radical cyclisations of hex-5-enyl and related radicals, as in Scheme 8, the model suggested by Beckwith *et al.*[4] again makes use of a chair-like transition state, with cyclisation actually taking place to give a five-membered ring. Substitution on the ring will favour formation of a *cis* or *trans* relative configuration in the product depending on the location of the substituent. Heating the bromide (17) with tributyltin hydride (Scheme 9) gives a modest preference for the 1,3-*cis* substituted cyclic product (18).

Likewise, cyclisation of substituted hex-5-enyl anions (Scheme 10) follows a similar stereochemistry: cyclisation via conformation (19) is favoured because the alternative (20) suffers from $A_{1,3}$-strain.[5]

The diastereoselectivity of intramolecular cycloadditions is often controlled by the location of a single substituent in an equatorial position on a connecting tether with a chair-like transition state (Scheme 11; see also Chapter 13).[6]

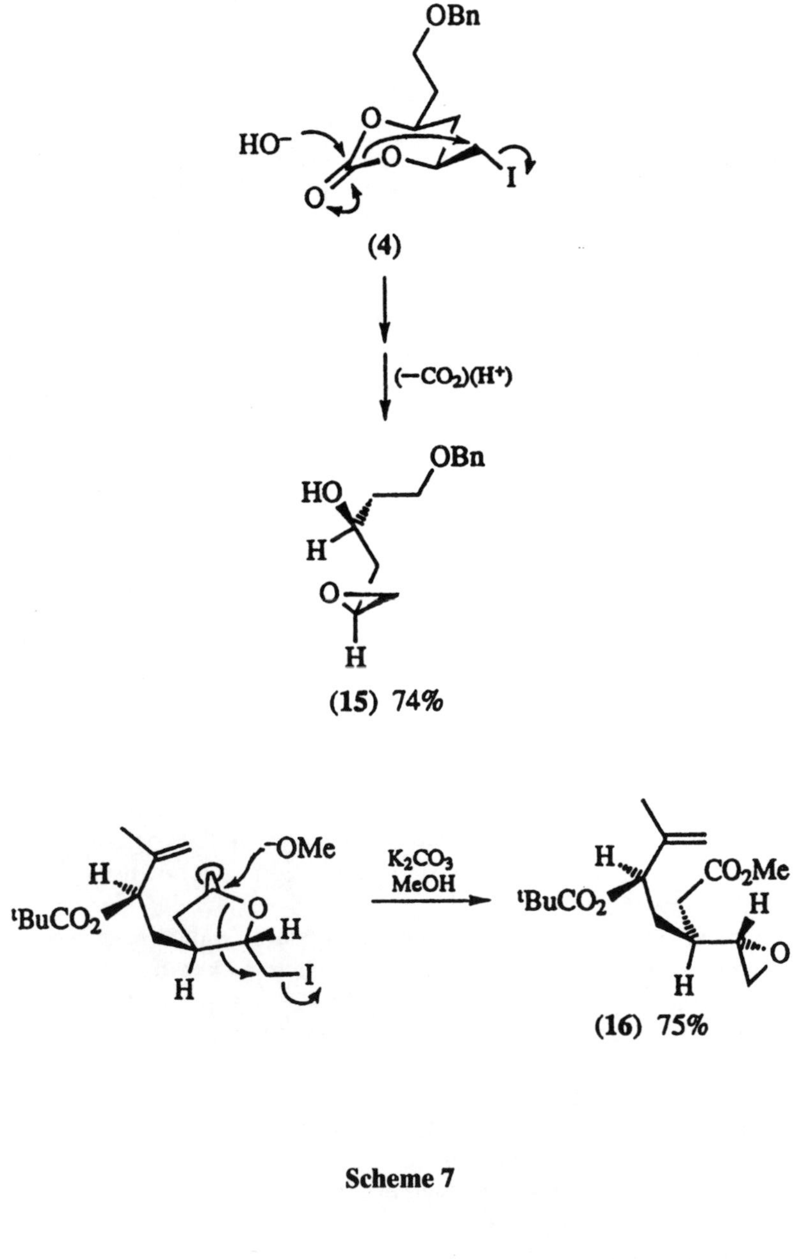

(4)

$(-CO_2)(H^+)$

(15) 74%

(16) 75%

Scheme 7

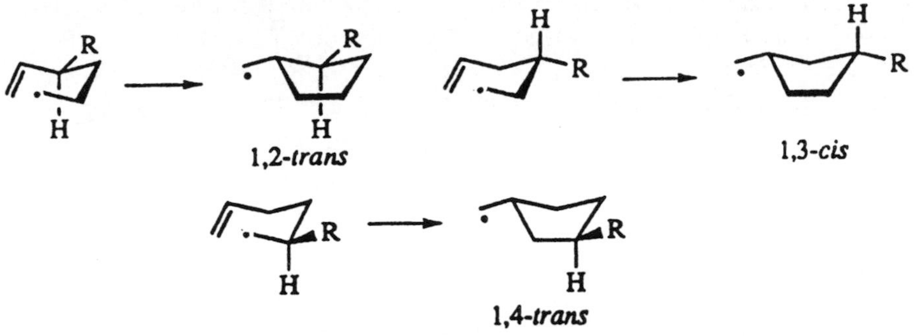

1,2-*trans*

1,3-*cis*

1,4-*trans*

Scheme 8

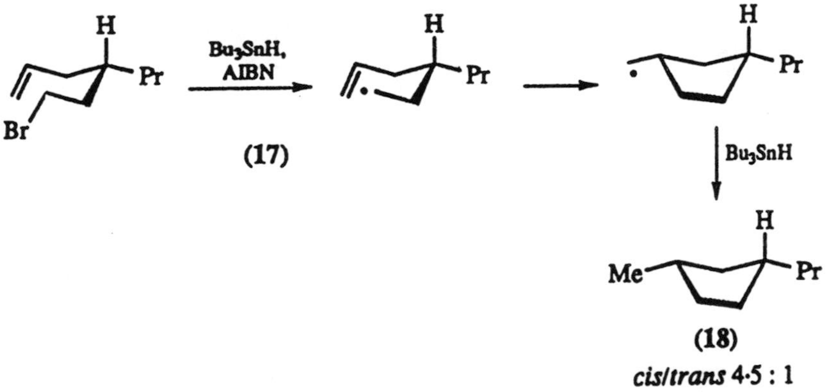

(17)

(18)

cis/trans 4·5 : 1

Scheme 9

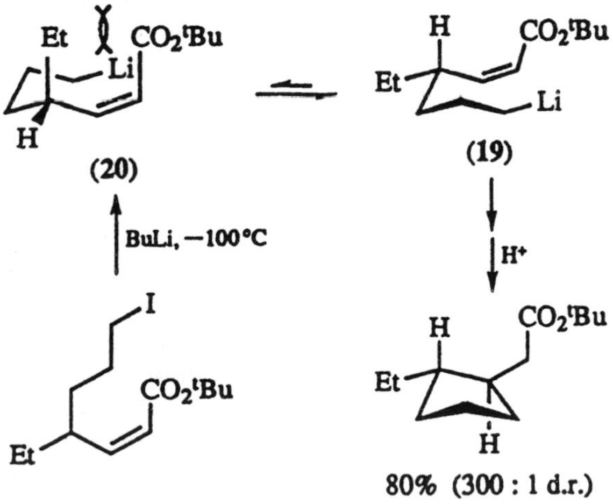

(20)

(19)

BuLi, −100 °C

H⁺

80% (300 : 1 d.r.)

Scheme 10

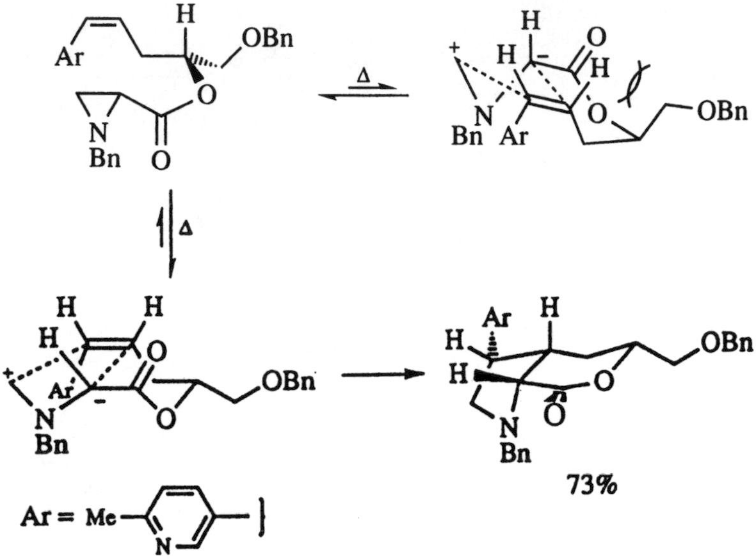

Scheme 11

Asymmetric induction dictated by bicyclic ring formation: geometrical imperatives

In an acyclic substrate containing a chiral centre and a prochiral double bond, high levels of asymmetric induction in addition to the double bond require limitation of the conformational freedom of the molecule. One way of imposing some conformational restraint is to carry out the reaction intramolecularly, as we have seen above, using one of the substituents on the chiral centre as the attacking species.

This tactic can be applied to the diastereoselective addition to a double bond contained in or on a ring, the chiral centre being one of the ring atoms. The nature of the linkage between the attacking functional group and the chiral centre may mean that in this case it is impossible to form anything other than a single diastereoisomer of the product. This is a common occurrence when the products are bridged bicyclics. Thus in a compound of the general type shown in Scheme 12, with a limitation on the total number of bridging atoms (say p + q + r ≤ 9),[7] only one of the two diastereofaces in (**21**) and (**22**) can conceivably be attacked (whatever the mechanism) because of the grossly different energies of the products (and the transition states which lead to them).

This *geometrical imperative* is present in, for example, the

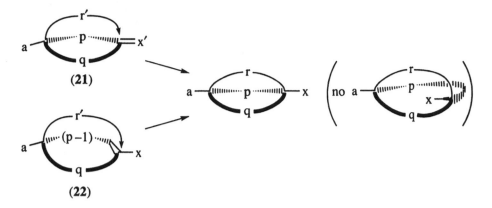

Scheme 12

selenolactonisation in Scheme 13.[8] As we have seen previously in cyclisations mediated by 'onium ions (Chapter 10), it is necessary that formation of the intermediate selenenium ion be reversible if good yields of the product are to be obtained since attack on both faces of the alkene (**23**) by phenylselenenyl chloride will undoubtedly occur.

It is noteworthy that in Scheme 13, there is little if any competitive formation of the isomer (**24**); this is an example of the faster formation of a five-membered relative to a six-membered ring when other factors are more or less equal.

Removal of the selenium in Scheme 13 and hydrolysis of the lactone results in a monocyclic acid (**25**) in which a new chiral centre has been formed with *complete* 1,3-asymmetric induction (and double bond shift) from the original monocyclic acid (**23**).

The geometrical imperative in reactions of this type is particularly valuable in synthesis because diastereoselectivity is necessarily complete. This derives, as mentioned above, from the impossibility of forming smaller bicyclic systems having anything other than a single defined relationship between the configurations at the bridgehead positions (Scheme 12). The diastereoselectivity in these reactions, therefore, is distinguishable from that in the Type III$_{s.c.}$ reactions previously considered in this chapter, in which the formation of both diastereoisomers was at least conceivable. This complete diastereoselectivity coupled with Type 0 ring opening of the bridged bicyclic product leads to a single diastereoisomer of a monocyclic product [e.g. (**25**); Scheme 13] for which, unlike its bicyclic precursor, two diastereoisomers *are* capable of existence.

Intramolecular cycloadditions to double bonds contained in rings will lead to tricyclic compounds. The existing chiral centre and the geometrical imperative of the reaction can again often lead to complete

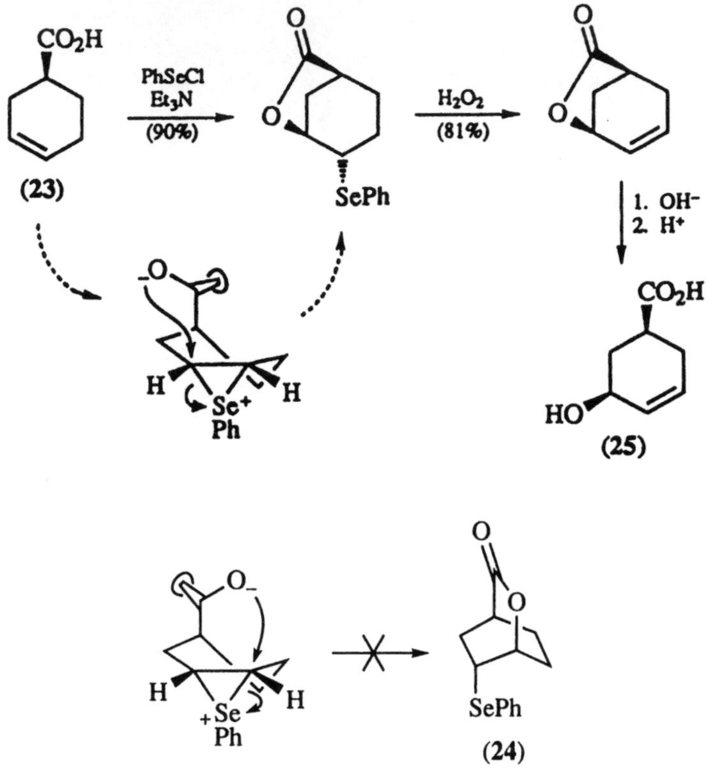

Scheme 13

diastereoselectivity. Thus Scheme 14 is part of a synthesis[9] of vernolepin, a natural product with anti-tumour activity. The intramolecular carbenoid addition to the cyclohexadiene double bond will result in the relative configuration at the four adjacent chiral centres shown in (26). Note that because of the symmetry in (27), either double bond can be attacked but the same relative configuration results in either case. Tricyclic to bicyclic conversion in Scheme 14 is accomplished by opening of the cyclopropane ring (with concomitant decarbomethoxylation). This acetic acid–sulphuric acid-mediated reaction proceeds with inversion of configuration at the cyclopropane ring carbon as shown.

A further example of intramolecular cycloaddition to a double bond contained within a ring, in which geometrical imperatives dictate the relative configuration of the existing and created chiral centres, is that in Scheme 15.[10] Dehydration of the primary alkylnitro group in (28) using the Mukaiyama method generates a nitrile oxide which adds as a 1,3-dipole to the cyclohexene double bond exclusively from one face. It should be clear, even

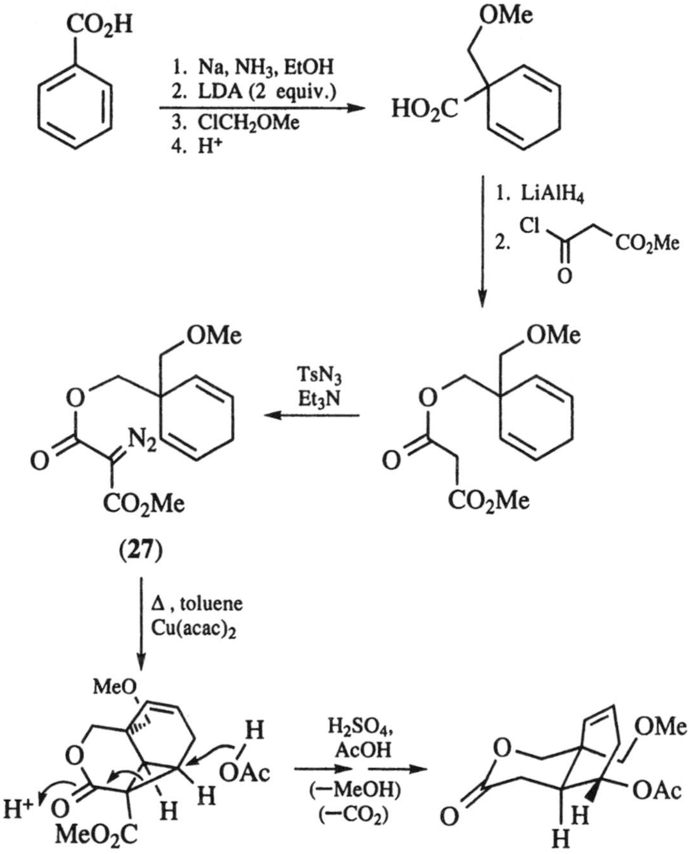

Scheme 14

without recourse to molecular models, that overlap of the orbitals required in this cycloaddition would be grossly inferior were cycloaddition to be attempted from the other face (as well as leading to a highly strained product). Tricyclic to bicyclic conversion is easily accomplished by reduction of the weak N—O bond and, overall, Scheme 15 involves annelation of a cyclopentanone completely diastereoselectively on to the original cyclohexenol double bond. In Scheme 15 the Claisen–Ireland rearrangement is also compelled to proceed on the bottom face of (**29**) as drawn.

When intramolecular reaction of a double bond in a cyclic precursor results in the formation of a bicyclo[*n*.1.0] compound, the geometrical

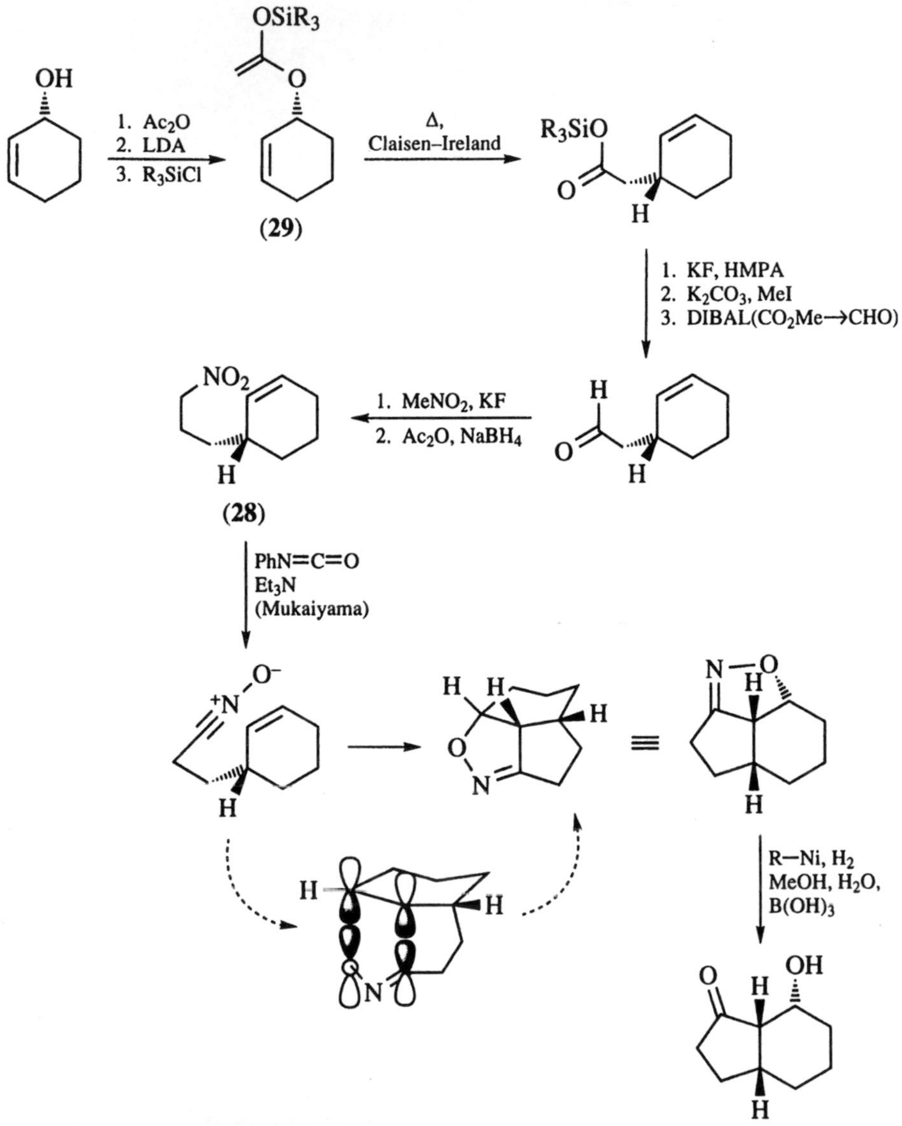

Scheme 15

imperative is for formation of a *cis* ring junction if $n \leq 6$. Thus Scheme 16 depicts a synthesis[11] of enantiopure chrysanthemic acid (**30**), a natural insecticide. Intramolecular S_N2 displacement of the mesylate by the ketone enolate (the double bond) can only result in the *cis*-fused bicyclic product (**31**) whose hydrolysis gives the required 1,2-*cis* configuration in (**30**). Although examples of the *trans*-bicyclo[4.1.0] skeleton are known, they are

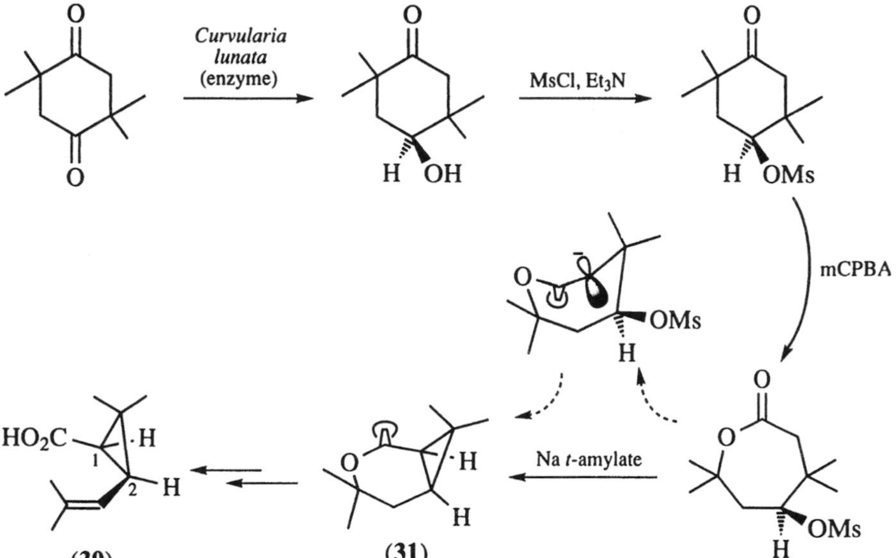

Scheme 16

highly strained and never formed in competition with their *cis* analogues. Note that in Scheme 16 the existing chiral centre is more actively involved in the asymmetric induction than heretofore since it is itself undergoing inversion of configuration (Type I/III).

Scheme 17 shows part of a synthesis[12] of (±)-juvabione (**32**), an insect moulting hormone. Formation of the bicyclo[3.3.1]nonane skeleton in (**33**) is crucial in establishing the required relationship between the two adjacent chiral centres in the product since the double bond in (**33**) is preferentially reduced from the less hindered *exo* face when incorporated into this [3.3.1] skeleton.

In Scheme 17, the relative configuration of the two chiral centres in (**33**) is ordained by their incorporation as bridgehead positions in the bicyclo[3.3.1] ring, i.e. there is a geometrical imperative in cyclisation of (**34**). This newly created chiral centre in (**33**) is, in fact, subsequently destroyed in the formation of (±)-juvabione, but not before it has served to allow differentiation between the two faces of the double bond in (**33**), as described above.

Not all cyclisations to form bicyclic compounds are necessarily completely diastereoselective. Where fused bicyclo[*m.n.*0] systems are formed, a geometrical imperative may be absent but nevertheless optimised orbital overlap may give rise to diastereoselectivity. Cyclisation of (**35**) gives only

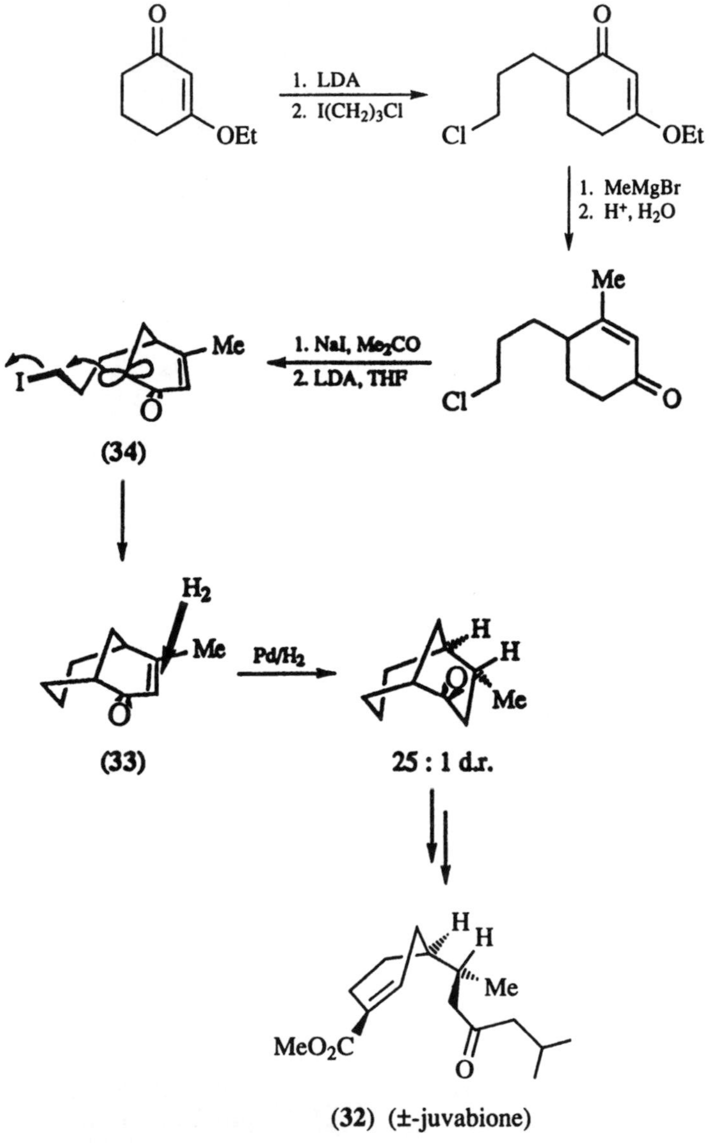

(34)

(33)   **25 : 1 d.r.**

(32) (±-juvabione)

Scheme 17

the *cis*-fused product (**36**)[13] (Scheme 18), although the *trans*-fused compound is certainly capable of existence. The introduction of the double bond into the chain which cyclises, stiffens it sufficiently for cyclisation via (**37**) to become much more favourable than cyclisation via (**38**) or (**39**). Cyclisation

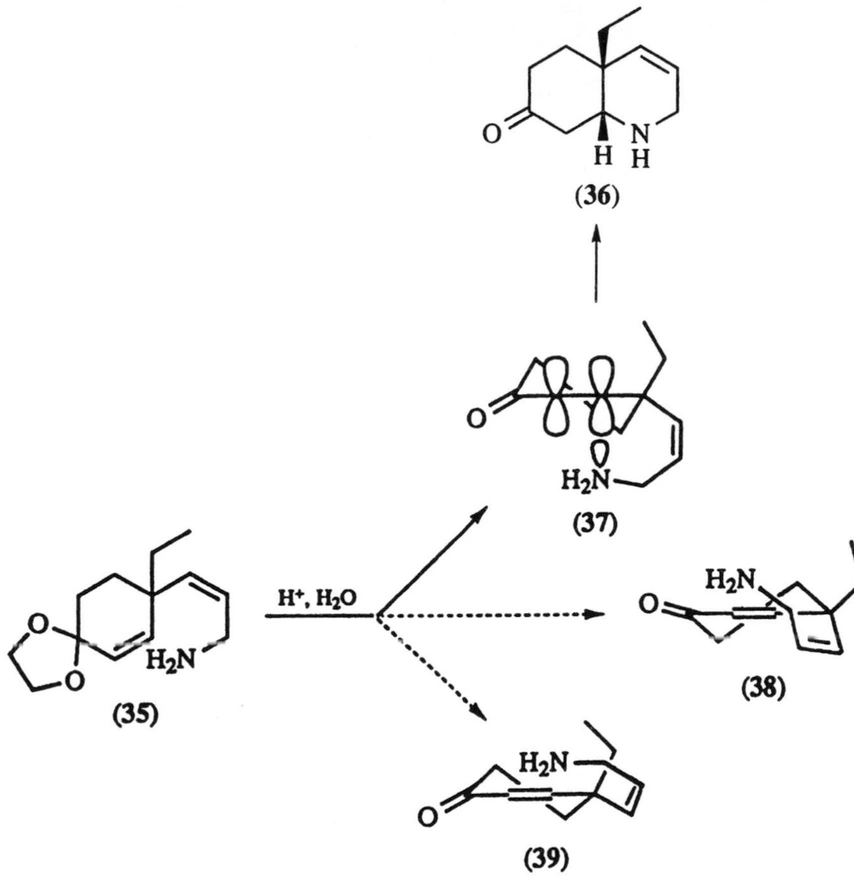

Scheme 18

via (38) does not allow orbital overlap [see (37)]; cyclisation via (39) maximises steric interaction with the adjacent ethyl group and delivers the enol in its half-boat conformation (see Chapter 10, Scheme 2).

Similarly, reducing the number of atoms in a chain involved in intramolecular Michael addition also encourages *cis* addition in bicyclo[4.3.0] formation [Scheme 19 (a)].[14] In this case, since the *cis*-fused product mesembrine (40), an alkaloid isolated from *Mesembryanthemum tortosum*, is likely to be thermodynamically more stable than the *trans* product, one cannot be certain that its formation is not the result of thermodynamic control. This is less likely to be the case in the related cyclisation in Scheme 19 (b)[15] and radical cyclisation in Scheme 19 (c).[16]

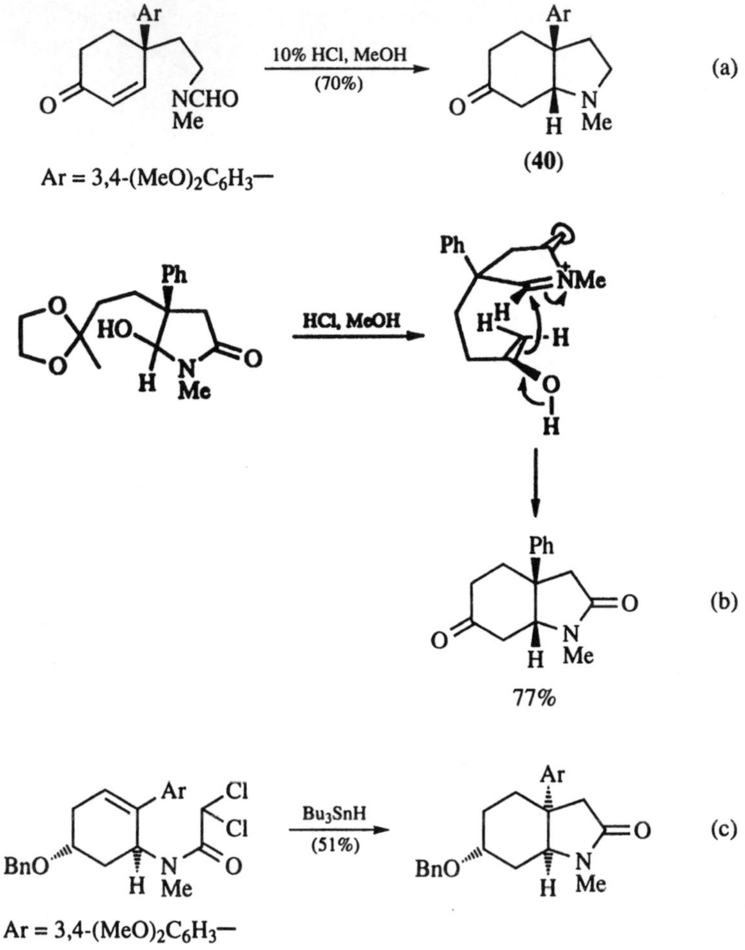

(a)

(40)

Ar = 3,4-(MeO)$_2$C$_6$H$_3$—

(b)

77%

(c)

Ar = 3,4-(MeO)$_2$C$_6$H$_3$—

Scheme 19

Intramolecular delivery from *in situ* appended reagents—diastereoface selection

In the previous section we saw how additions to double bonds contained in rings could be made completely diastereoselective by carrying out the reactions intramolecularly. Inevitably these reactions resulted in bicyclic or tricyclic ring formation although in many cases cleavage of one ring was practicable with conservation of the newly formed chiral centre(s).

A closely related class of reactions, also of great value in stereoselective synthesis, is that in which complexation or chelation of the reagent with some functional group in the substrate is followed by intramolecular delivery of atom(s) or group(s) to one double bond diastereoface (Scheme 20).

Decomplexation of the spent reagent from the product may take place spontaneously or in the reaction work-up.

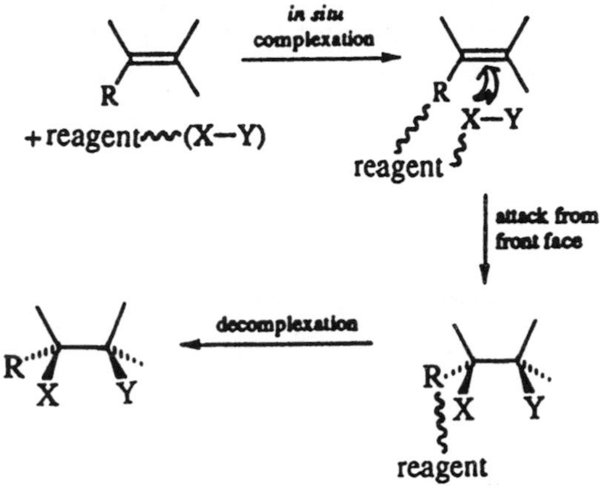

Scheme 20

INTRAMOLECULAR DELIVERY IN EPOXIDATION, AZIRIDINATION AND CYCLOPROPANATION

Oxidation of the steroid ring B allylic alcohol (41) with *tert*-butyl hydroperoxide in the presence of a vanadium(V) catalyst gives the *cis*-epoxide (42) completely diastereoselectively (Scheme 21).[17] [In the absence of the hydroxy group, epoxidation takes place preferentially from the

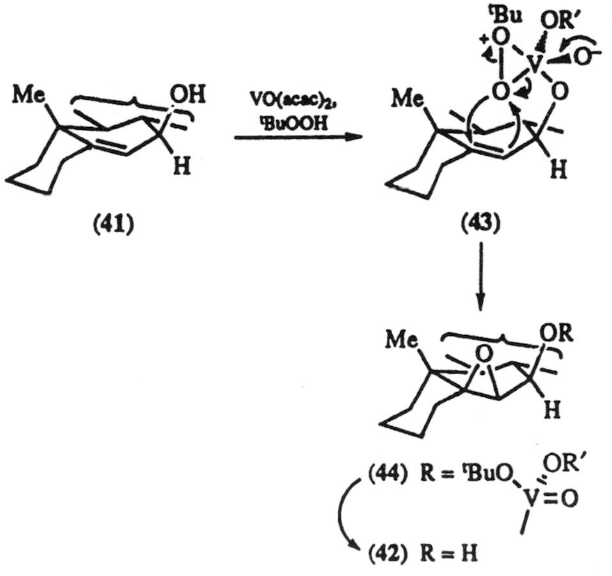

Scheme 21

opposite (α) face; see Chapter 9.] An intermediate complex (43) is assumed
to result from a combination of the allylic alcohol and *tert*-butyl
hydroperoxide with the vanadium by ligand exchange. In-line attack on the
O—O bond can be accomplished only from the top face of the double bond
in (43), giving the epoxide (44) and hence (42) as shown.[18]

These V^V-catalysed epoxidations can also be highly stereoselective with
acyclic alkenols in which the geometrical imperative in Scheme 21 is absent.

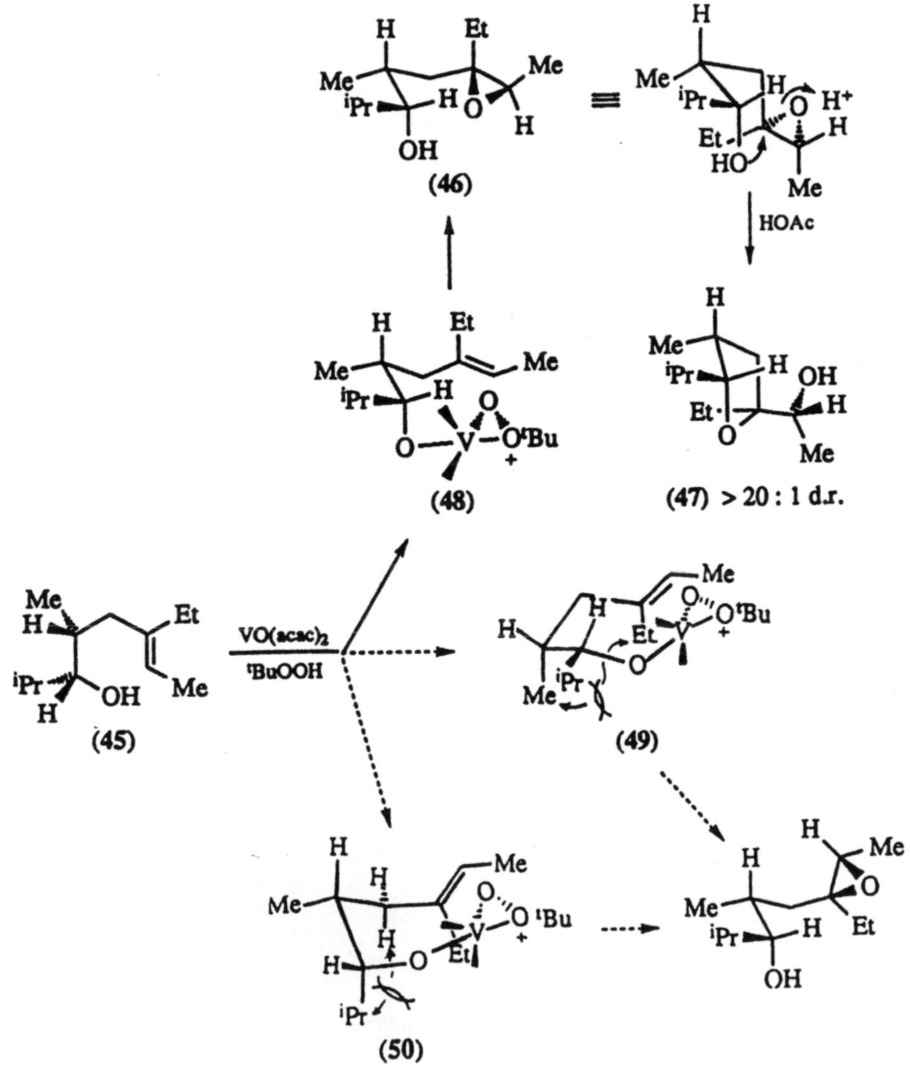

Scheme 22

Thus conversion of (45) into epoxide (46) and hence into tetrahydrofuran (47) (Scheme 22) is part of Kishi and co-workers' lasalocid synthesis.[19] The preferred diastereoisomer is that produced via a transition state resembling (48) rather than that which would result from (49) or (50) because of 1,3-'diaxial' strain in both of the latter.

Epoxidations directed by a hydroxyl group coordinated to other metals including molybdenum and, in particular, titanium (see Chapter 14) proceed by related mechanisms.

Since they are easily made and broken, hydrogen bonds are a useful way of temporarily linking a reagent to a substrate and thus directing attack on to one diastereoface. Epoxidation of 5- and 6-membered ring allylic alcohols with peroxy acids takes place largely *cis* to the hydroxyl group. This is believed to be the result of hydrogen bonding between one of the peroxy acid's oxygen atoms and the hydroxyl group proton as indicated in Scheme 23 (a).[20] There is evidence to suggest that this intramolecular delivery of the peroxy acid oxygen can be more comfortably accommodated when the hydroxyl group is pseudo-equatorial as shown in Scheme 23 (a).[21] A similar mechanism appears to obtain in aziridination of cyclohexenol [Scheme 23 (b)].[22]

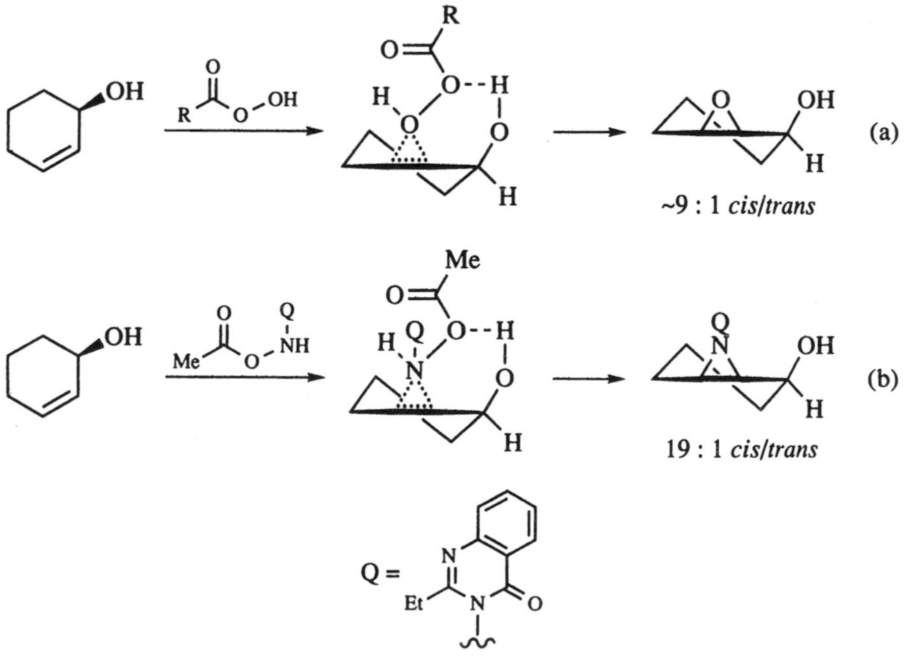

Scheme 23

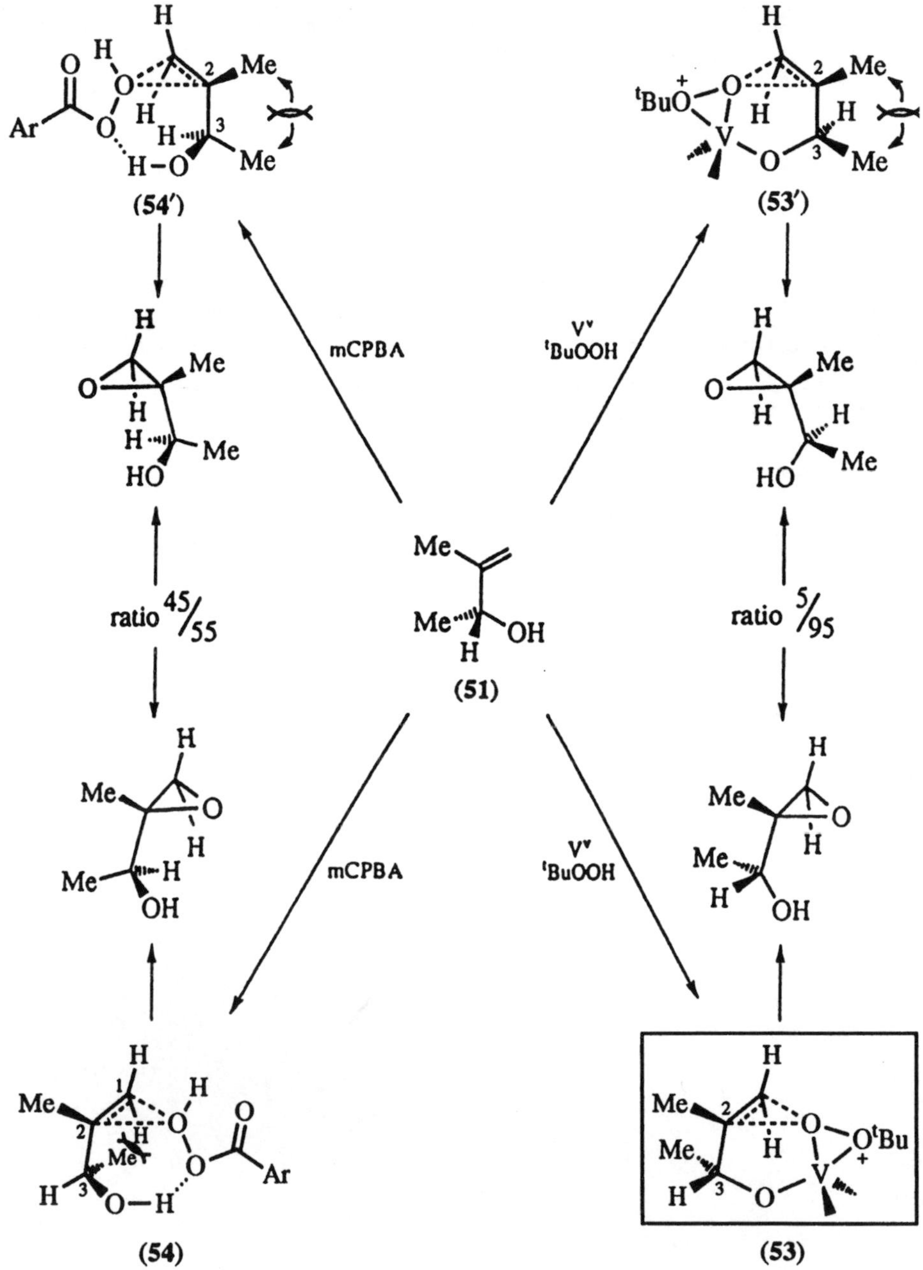

Scheme 24   *(continued)*

Scheme 24 *(continued)*

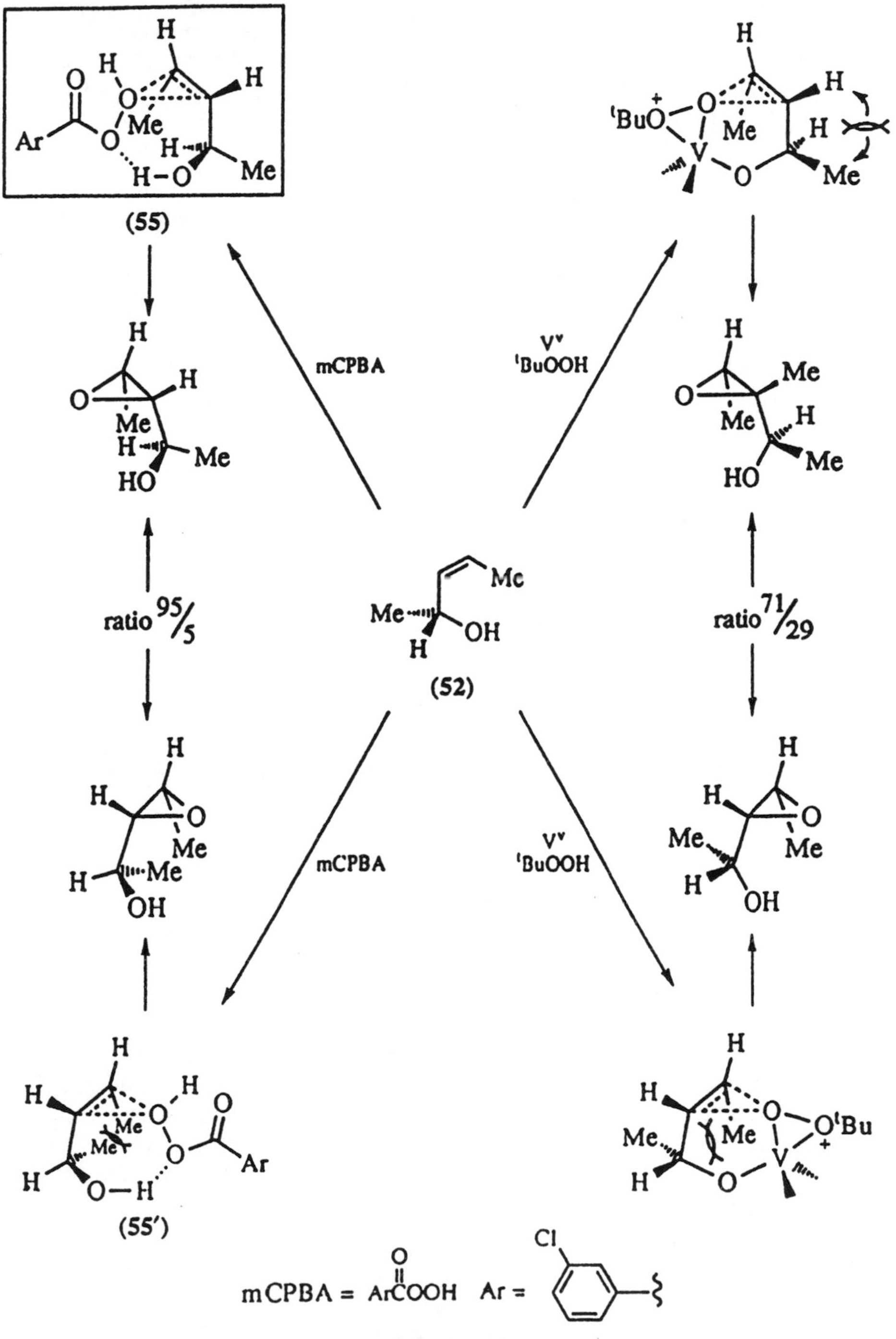

mCPBA = ArCOOH   Ar =

Scheme 24

As with V^V-catalysed epoxidations described above, high diastereo-selectivity can also accompany intramolecular delivery of the reagent using acyclic substrates, even though the geometrical imperative that may be present in cyclic substrates is not present in acyclic substrates.

A comparison of the diastereoselectivity of epoxidation using V^V–*tert*-butylhydroperoxide and *m*-chloroperoxybenzoic acid (mCPBA) for the two alkenes (51) and (52) is shown in Scheme 24. The differences can be accounted for by assuming that the two transition states giving rise to the major diastereoisomers in the diastereoselective epoxidations of (51) and (52) are represented by (53) and (55) respectively.[18]

In (53) and (53'), the O—O bond is required to be in-line with the π-orbital of the double bond. The relayed conformational consequence is that there is a clear preference for the methyl group at C-3 to take up residence as shown in (53) because of interaction between the two methyl groups in (53'). In (54), however, there is a difference in torsional angles around the 2,3-bond which means that a methyl group in the 3-position is closer to a hydrogen at C-1 and its preference for this site over that in (54') is accordingly less.

As a corollary, however, the diastereoselectivity in epoxidation of alkene (52) is higher with mCPBA since, in this case, Me–Me repulsion in (55') disfavours this transition state and leads to the formation of the alternative diastereoisomer via (55).

For the related Simmons–Smith reaction, *syn*-cyclopropanation of allylic alcohols, e.g. (56), by the reagent ICH$_2$ZnI occurs but not necessarily via hydrogen bonding or zincate salt formation. Thus the corresponding allyl ether (57) also shows a *syn*-directing effect and a similar reactivity (Scheme 25).[23] Chelation between the zinc atom and the oxygen lone pair is believed to be responsible, possibly involving a dimeric form of the reagent ICH$_2$ZnI.

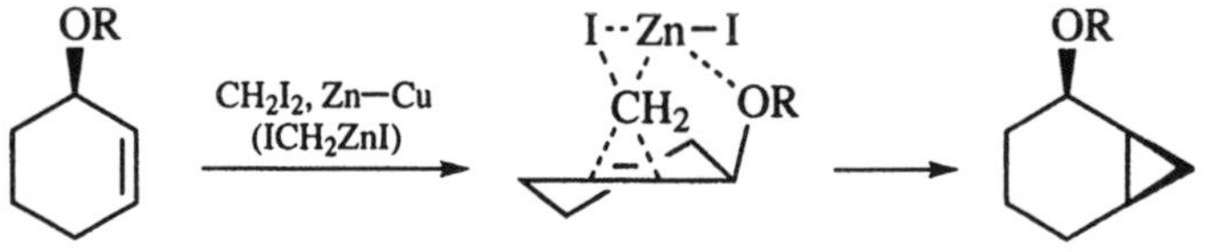

(56) R = H
(57) R = Me

Scheme 25

Acetal oxygens can also direct the Simmons–Smith reagent to the *syn* face of an allylic double bond (Scheme 26). Whereas the cyclopropanation of (58) is highly diastereoselective, that of (59) is not.[24] This difference in selectivity has been rationalised by assuming that intramolecular delivery takes place by coordination of the zinc with the lone pair of the oxygen indicated in (58), i.e. on that lone pair which is *syn* to the double bond and on the oxygen which is pseudoequatorial. Coordination to this same lone pair in (59) is less favourable because of the neighbouring *syn*-methyl and this allows competitive cyclopropanation from the top face.

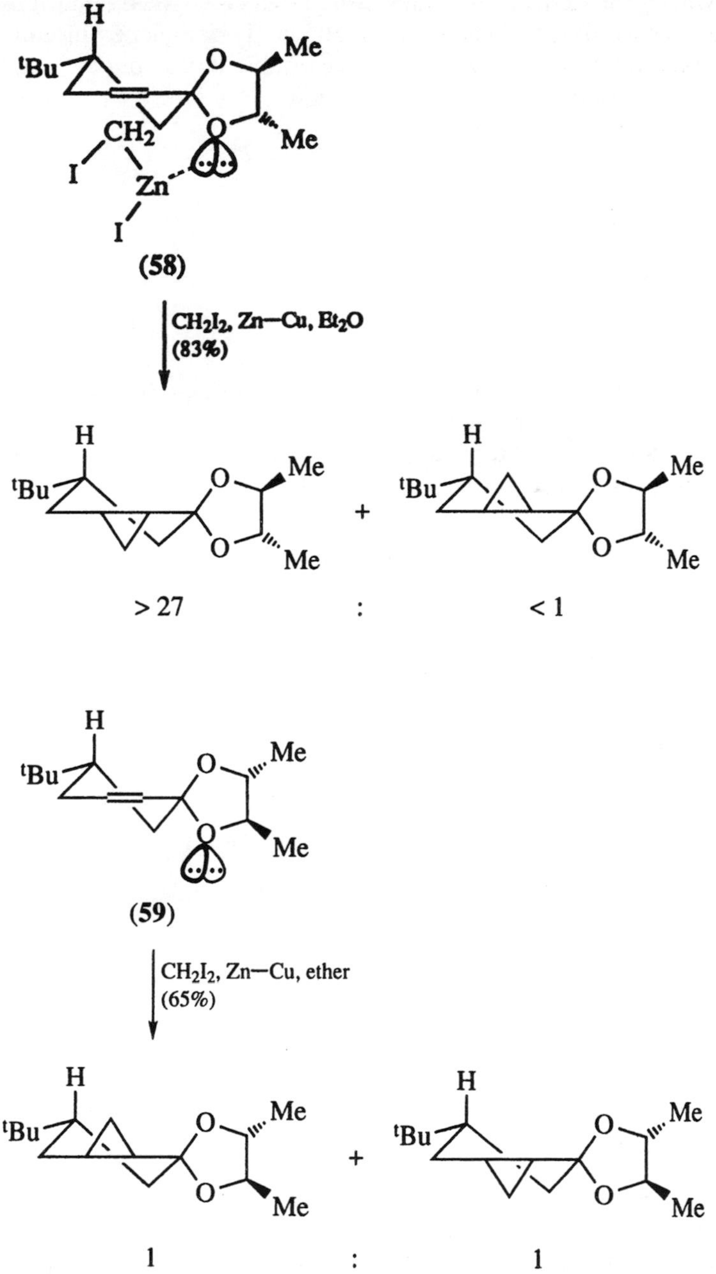
H
ᵗBu
O
Me
CH₂
O
Me
I
Zn
I
(58)
CH₂I₂, Zn—Cu, Et₂O
(83%)
H
ᵗBu
O
Me
O
Me
+
H
ᵗBu
O
Me
O
Me
> 27
:
< 1
H
ᵗBu
O
Me
O
Me
(59)
CH₂I₂, Zn—Cu, ether
(65%)
H
ᵗBu
O
Me
O
Me
+
H
ᵗBu
O
Me
O
Me
1
:
1

Scheme 26

Methylenation of double bonds using the analogous samarium reagent ICH$_2$SmI, derived from samarium metal and methylene diiodide, is also hydroxyl-directed (Scheme 27);[25] isolated double bonds and even homoallylic alcohols do not react.

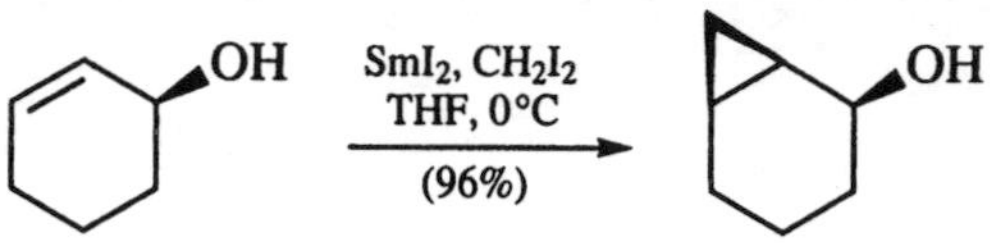

Scheme 27

Diastereoselectivity in intramolecular cyclopropanation of acyclic allylic alcohols can also be high (Scheme 28); the isopropyl group suffers less steric interaction in transition state (60) than it would do in (61), corresponding to attack on the other face of the double bond.

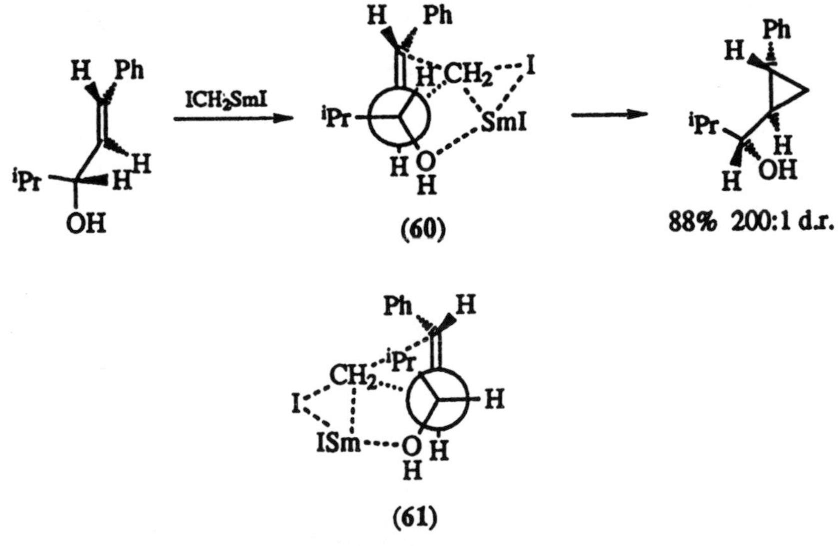

Scheme 28

In many cases, *syn*-epoxidation, *syn*-aziridination and *syn*-cyclopropanation reactions, analogous to those above, can be effected by a hydroxyl group that is not allylic but homoallylic or even more remote from the double bond undergoing reaction. Hydrogen bonding-directed epoxidation mediated by acidic groups other than OH, notably NH, is also possible.

INTRAMOLECULAR DELIVERY IN REDUCTION

Many natural products contain polyol segments (62) or (63) (Scheme 29) which, being polyketide derived, have the OH groups in a 1, 3, 5, etc., array; they are often 'stereoregular,' e.g. the adjacent OH groups are all *cis* or all *trans*.

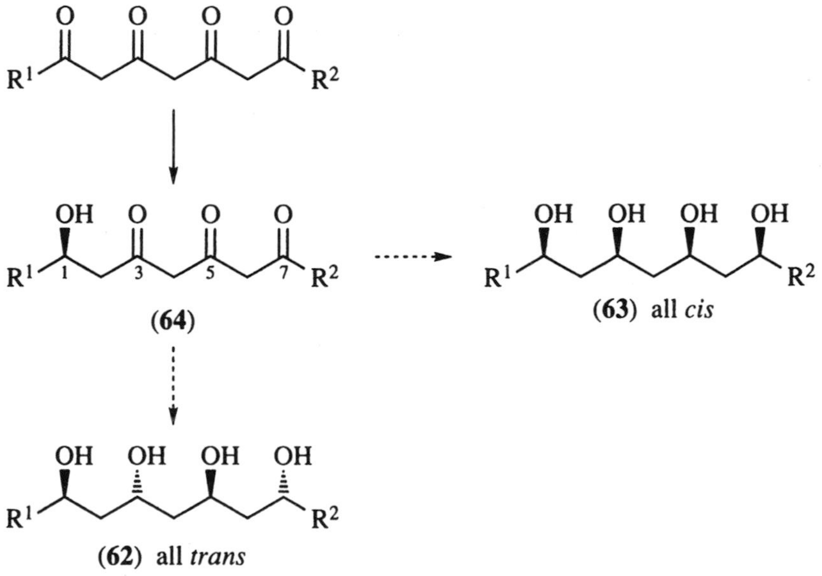

Scheme 29

One approach to the laboratory synthesis of such polyols is from a polyketo alcohol such as (64), by reduction at the C-3 carbonyl group using 1,3-asymmetric induction by the existing chiral alcohol (C-1) to control the relative or absolute configuration of the new alcohol. This process can be carried out iteratively, i.e. the newly created chiral alcohol centre (C-3) can be used to control the diastereoselectivity of the reduction at the C-5 carbonyl group, and so on.

A method of reduction of 3-keto alcohols that gives *trans*-1,3-diols with high diastereoselectivity makes use of tetramethylammonium triacetoxyhydroborate in the presence of acetic acid (Scheme 30).[26] Coordination of the reagent to the existing hydroxyl is followed by intramolecular hydride delivery. Diastereoselectivity arises from the more adverse R²–OAc interaction in the chair-like transition state (65) in comparison with the O(H)–OAc interaction in (66).

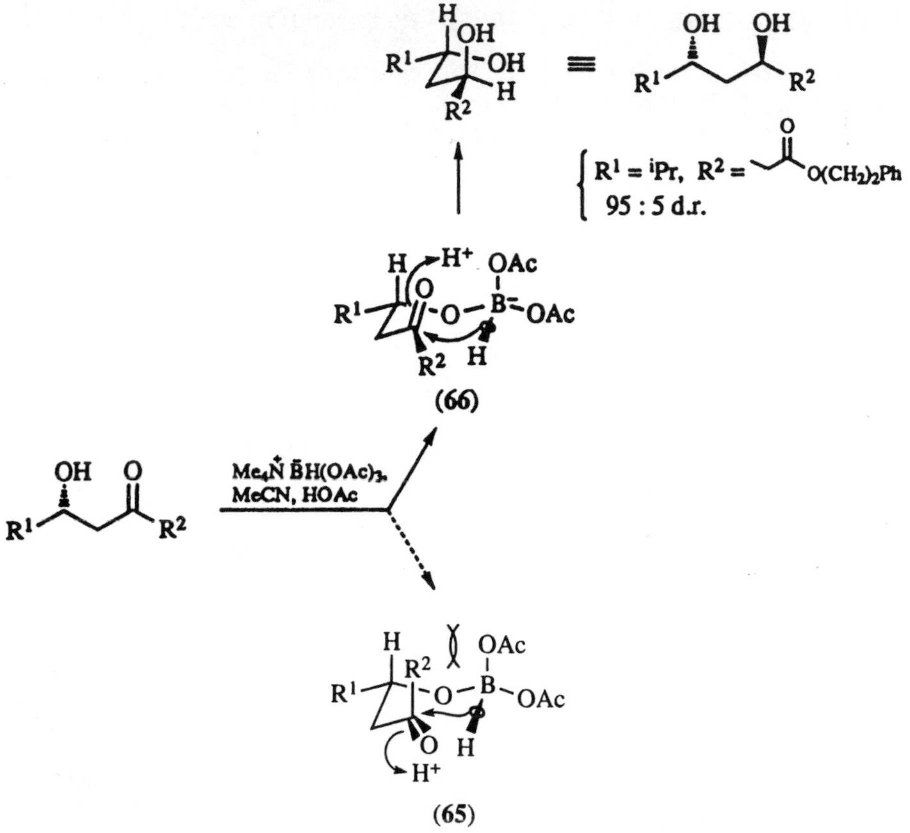

Scheme 30

It should be noted that this complexation–internal delivery (by a single reagent) is an intramolecular reaction and therefore differs from the previously encountered reactions in which chelate formation was followed by *intermolecular* diastereoselective attack on the chelate by another reagent (cf. Chapter 10, Scheme 29). The latter method, can, however, be applied to the reduction of β-hydroxy ketones to 1,3-*cis*-diols, i.e. reduction with the opposite sense of induction to that in Scheme 30, by treatment with a trialkylborane followed by sodium tetrahydroborate reduction (Scheme 31).[27] Note that axial attack on the half-chair (67) gives the chair conformation of the product directly (cf. Chapter 10, Scheme 2).

In reduction by zinc tetrahydroborate (Scheme 32)[28] the chelating agent and reducing agent are the same, but here too the diastereoselectivity is thought to be established in an intermolecular reaction analogous to that in Scheme 31.

The fact that the reactions in Schemes 30 and 31 give opposite senses of induction supports the view that the products in Scheme 30 are *not* formed by a chelation-intermolecular reduction mechanism as in Scheme 31.[26] This is further supported by the high stereoselectivity of reduction of the cyclic

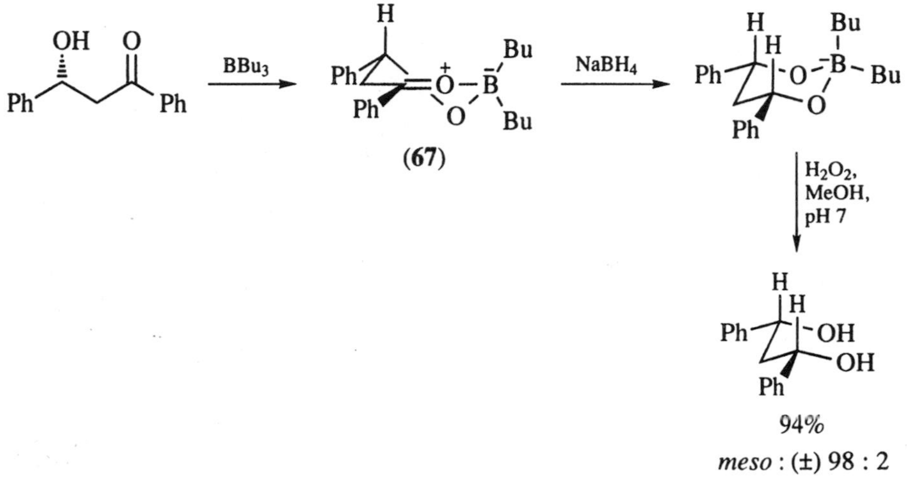

(67)

94%

meso : (±) 98 : 2

Scheme 31

(69%) 91 : 9 *cis* : *trans*

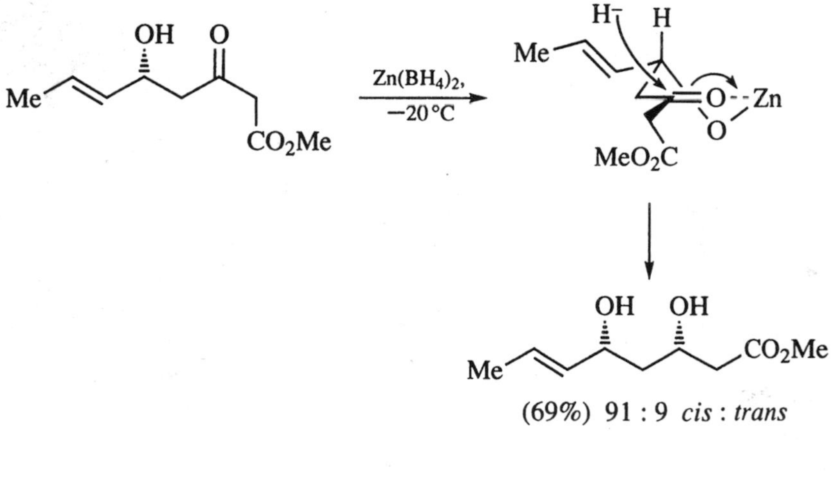

Scheme 32

hydroxy ketone (**68**) in Scheme 33 (a), where chelation between the hydroxyl and ketone groups is not possible. Unlike the reduction in Scheme 30, there is a geometrical imperative here that dictates that the sense of asymmetric induction must be as shown.

Interestingly, the corresponding equatorial alcohol (**69**) [Scheme 33 (b)] is also reduced with complete stereoselectivity, although at a slower rate

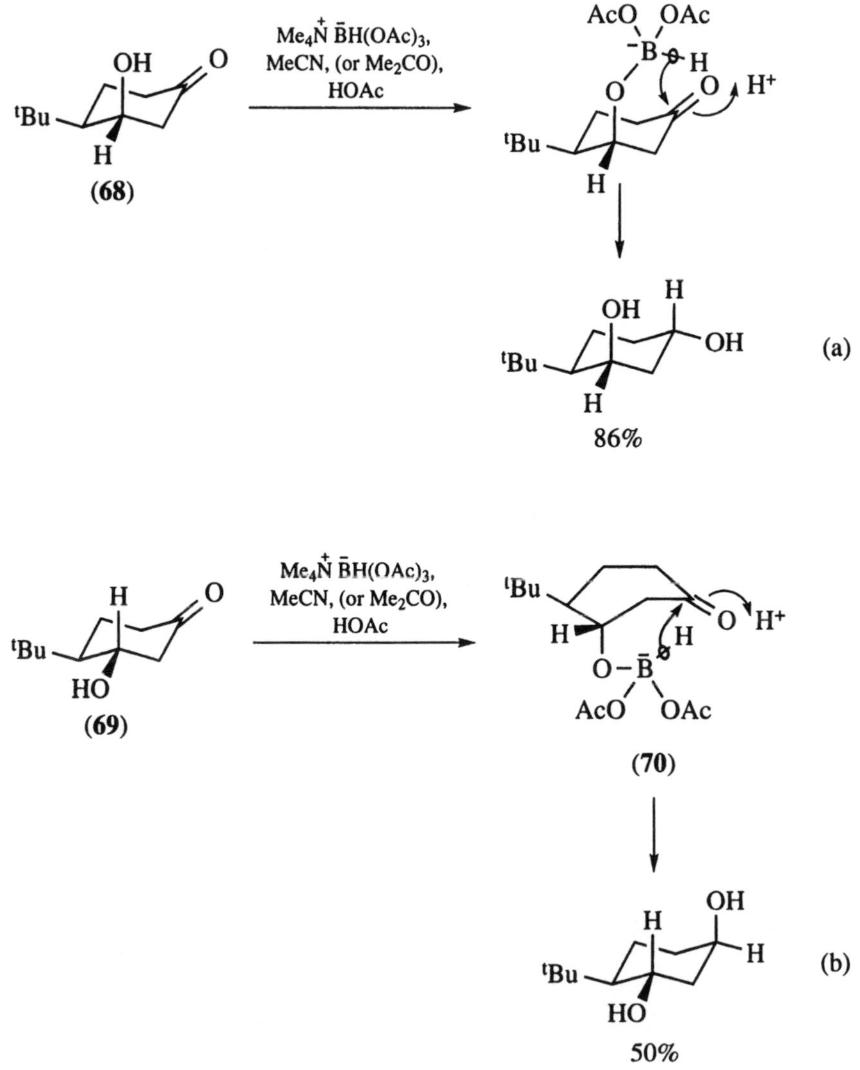

Scheme 33

than its axial counterpart (**68**). In this case the reaction is thought to proceed via the twist boat (**70**) in which again there is a geometrical imperative present.

Exchange of one acetoxy group of the reagent in the coordinated hydroborate intermediates, as indicated in Schemes 30 and 33, very likely makes the hydroborates more powerful reducing agents and, together with the intramolecularity of the reaction, accounts for the fact that acetone is not reduced and can be used as solvent.

INTRAMOLECULAR DELIVERY IN HYDROGENATION[29]

The diastereoselectivity of both heterogeneous and homogeneous hydrogenation of double bonds can be directed by coordination between the metal catalyst and a neighbouring group (OH, NHR, CO_2R), thus facilitating delivery of hydrogen to one diastereoface. With cyclic alkenes of seven members or less, *syn* delivery of hydrogen is the rule (Scheme 34) even when this gives the thermodynamically less stable product. In Scheme 34 (b) the ratio of *cis* to *trans* diastereoisomers is very dependent on the solvent polarity, consistent with there being competition between the solvent and the catalyst for coordination to the hydroxyl group. Thus there is much more intramolecular delivery in hexane than in DMF.[30]

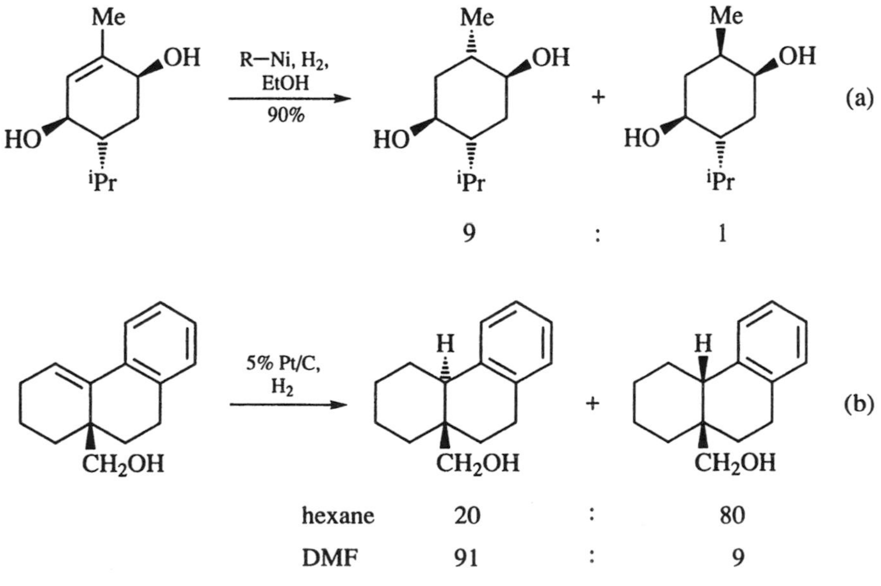

Scheme 34

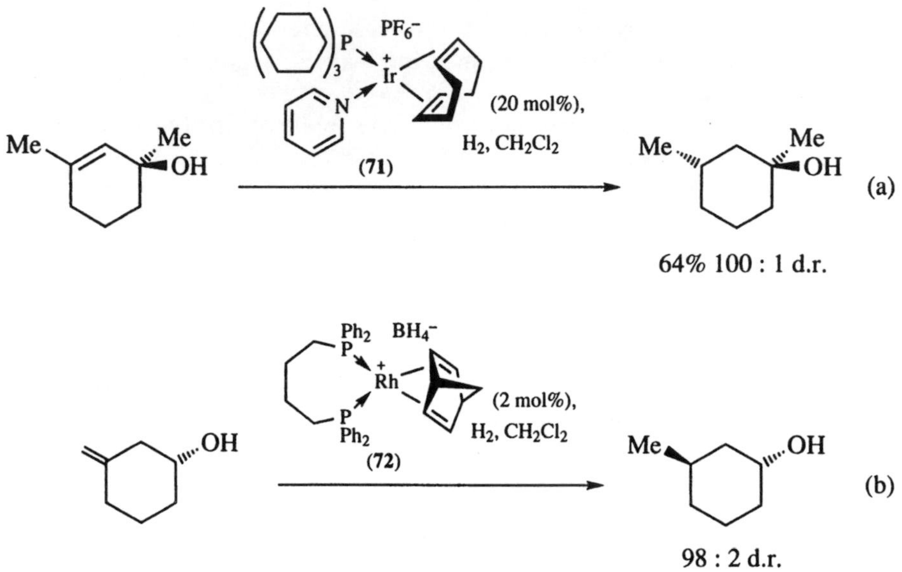

(a)

64% 100 : 1 d.r.

(b)

98 : 2 d.r.

Scheme 35

For homogeneous hydrogenations, the iridium catalyst (71) designed by Crabtree[31] or the rhodium catalyst (72)[32] is used (Scheme 35). For the hydrogenation in Scheme 35 (a), little diastereoselectivity is obtained using palladium on charcoal. An advantage of homogeneous catalysis is that reproducibility in catalyst preparation is not a problem, as it can be with heterogeneous catalysis.

In the cyclic substrates in Schemes 34 and 35, coordination between the metal and the neighbouring (hydroxyl) group, together with the limited length of the tether, are two of the factors which account for the *syn* delivery of hydrogen.

Homogeneous reduction of *acyclic* allylically substituted alkenes (usually allylic alcohols) is also often stereoselective, in spite of the greater conformational freedom which allows approach of the coordinated metal hydride to either face in a way not possible with a cyclic substrate. In Scheme 36,[33] it is assumed to be a steric interaction, between the methyl group and the *ortho*-hydrogen atom of the phenyl ring, which destabilises (73) relative to (74) and leads to the product (75).

Homogeneous reduction of acyclic homoallylic alcohols can also achieved stereoselectively (Scheme 37).[34] In (a) it is $A_{1,3}$-strain which destabilises (76) relative to (77), with a chair transition state and *syn*

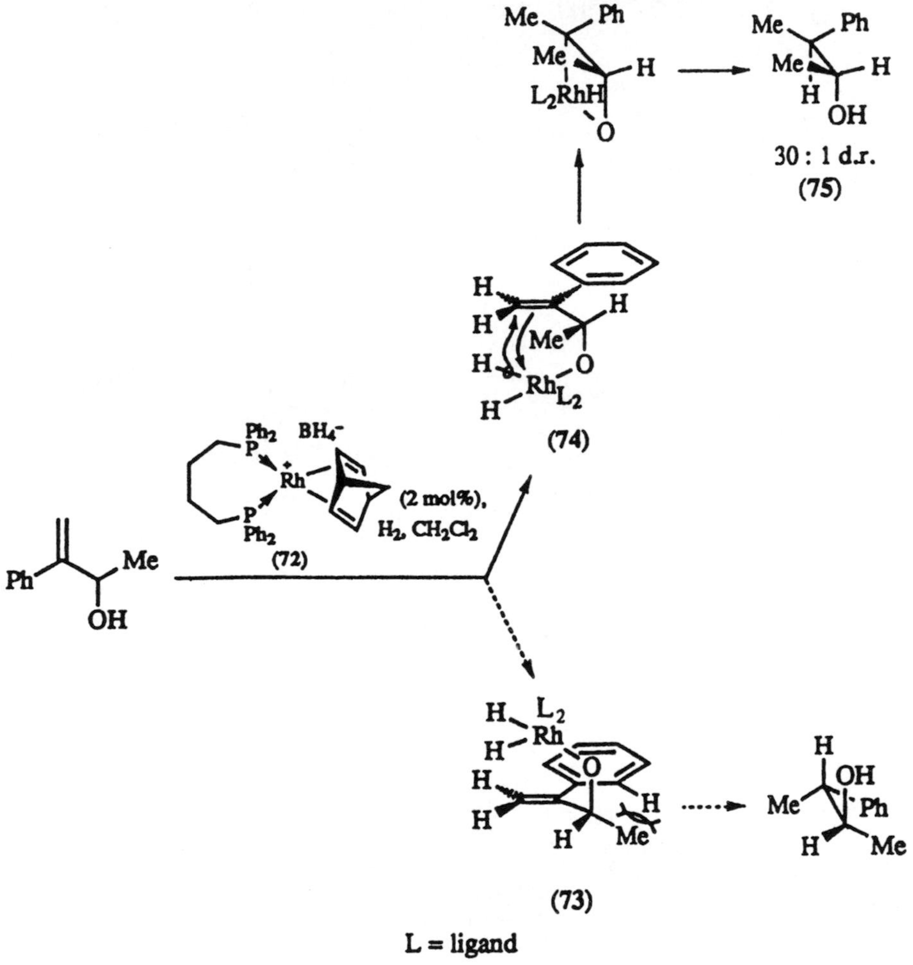

Scheme 36

addition of the Rh—H to the C=C bond. Note that in Schemes 36 and 37, substitution of the Rh—C bond by H—C goes with *retention* of configuration at carbon.

In Scheme 37 (b), the preferred conformation of the side-chain of the dihydronaphthalene (78) undergoing hydrogenation with Wilkinson's catalyst leads to (79).[35]

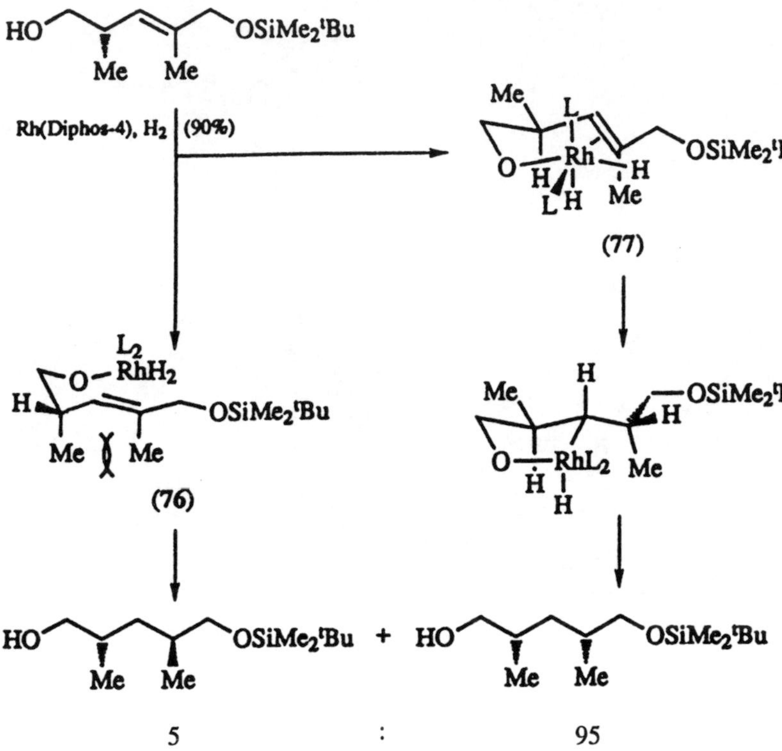

(a)

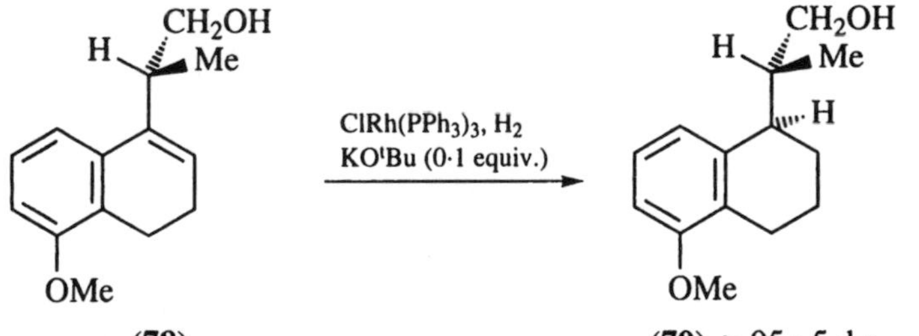

(78) (79) >95 : 5 d.r.

(b)

Scheme 37

INTRAMOLECULAR DELIVERY IN MICHAEL ADDITIONS

Addition of methyllithium to the Michael acceptor (**80**) in Scheme 38 gives (**81**) after desilylation.[36] Intramolecular chelation of the methyllithium by the acetal (MOM) group delivers the methyl anion from the top face of the double bond in (**82**); $A_{1,3}$-strain in the alternative (**83**) (Gt-SiMe$_3$) means that it hardly comes into contention.

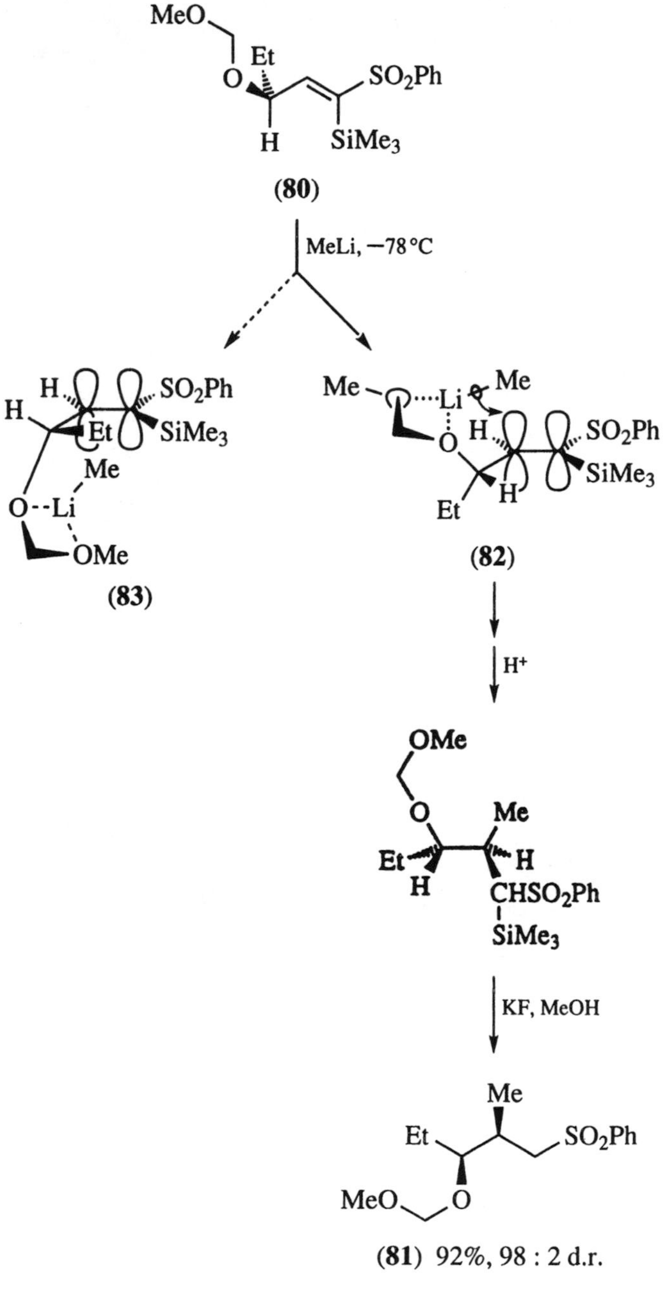

(80)

MeLi, −78 °C

(83)

(82)

H⁺

KF, MeOH

(81) 92%, 98 : 2 d.r.

Scheme 38

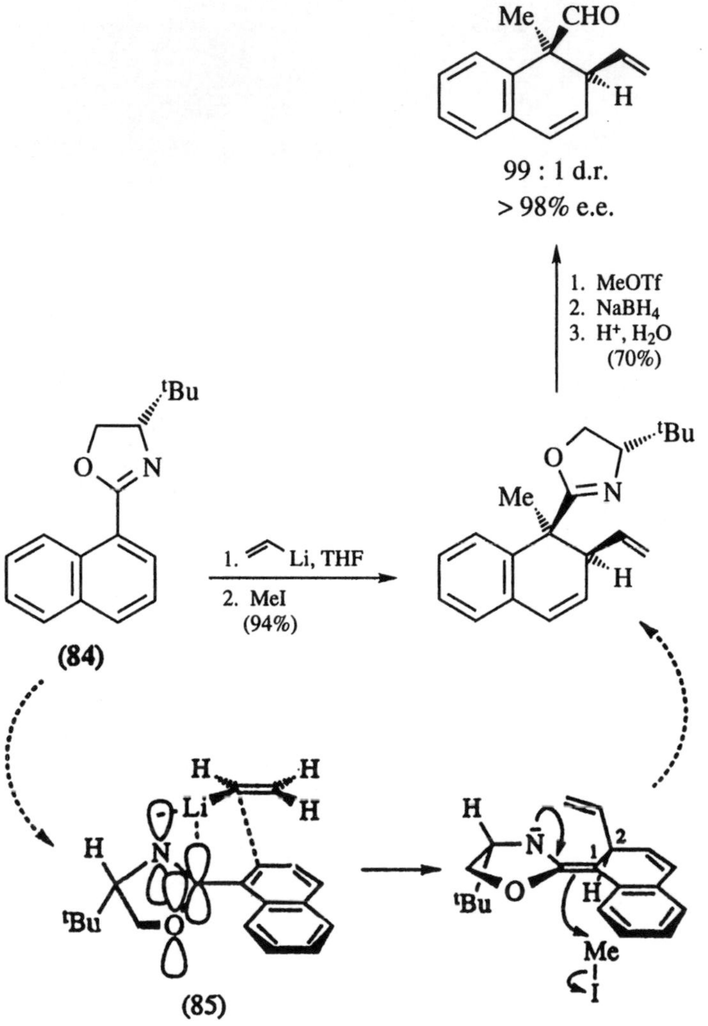

Scheme 39

Face-selective attack on the naphthalene 1,2-double bond in **(84)** (Scheme 39) has been suggested by Rawson and Meyers[37] to result from π-electrons in the oxazoline complexing with the vinyllithium as in **(85)**. The *tert*-butyl group discourages complexation from the bottom face and hence delivery of the vinyl group is on to the top face. Reaction of the resulting anion at C-1 is exclusively from the bottom face, directed thereto by the vinyl group.

Intramolecular delivery and diastereoface selection from covalently bound reagents

In the previous section we have shown how *in situ* linkage of a reagent to a neighbouring functional group can result in diastereoselectivity in the subsequent reaction. In a cyclic substrate, geometric imperatives will often bring about complete diastereoselectivity; in an acyclic substrate, the intramolecularity of the reaction will bring into play a new set of steric/electronic influences that will often increase the diastereoselectivity/regioselectivity.

There is, of course, no reason why the linkage need be only transient; it may be expedient to form a covalent bond between the substrate and the reagent and to isolate this product, then to allow the intramolecular reaction to proceed, and finally to disconnect the (residue of) the reagent. The advantage of *in situ* complexation of reagent and substrate is that the three steps take place in one stage without the need for isolation of intermediates.

Examples of the strategy of first covalently binding the reagent are often ingenious, and usually rely for their diastereoselectivity on geometrical imperatives operating in the reaction. Thus in Scheme 40, an α-oriented methyl group was required from reduction of the double bond in (86). Catalytic reduction gave exclusively the unwanted β-oriented methyl group [cf. Scheme 37 (b)]. Tethering of a silicon hydride to the primary hydroxyl

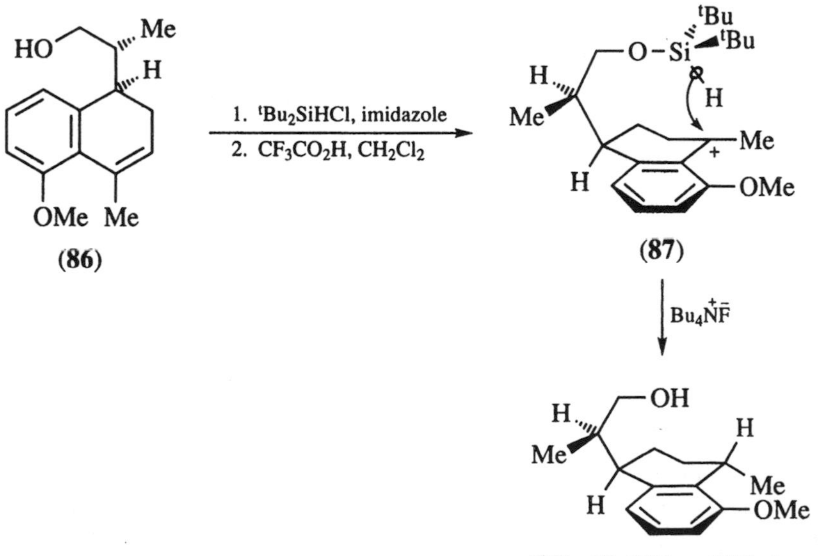

(88) 65–75%, > 95% d.e.

Scheme 40

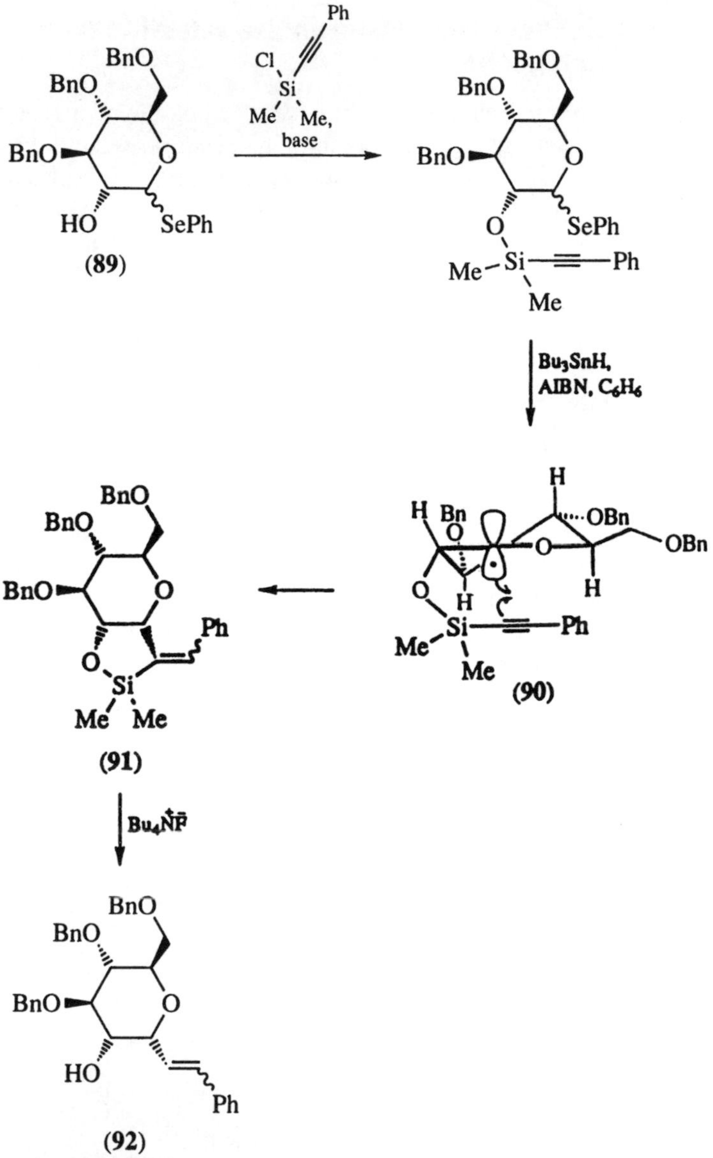

Scheme 41

and intramolecular delivery of the hydride to a benzylic carbocation gave exclusively the required diastereoisomer (88) via a transition state resembling (87).[35]

Stork *et al.*[38] used the easy formation and cleavage of the silicon–oxygen bond as a basis for the diastereoselective radical-mediated C—C bond formation shown in Scheme 41. The radical generated by deselenylation of

(89) is either planar or rapidly inverting on the time-scale of the reaction, and so both epimers of (89) can be used in the reaction. Attack of this radical on the triple bond occurs from one side, as in (90), leading to a *cis* ring fusion in (91). It is noteworthy that the free hydroxyl group in (92), liberated in the last step, is available for further manipulation and that the styrenyl group can serve as a formyl equivalent, e.g. by ozonolysis.

Diastereoselectivity in reactions of acyclic compounds mediated by medium and larger ring formation

Diastereoselective reactions on acyclic compounds pose particular problems because of the conformational freedom they enjoy. Although some of the methods outlined in this chapter can be applied to acyclic substrates, success depends on the presence of adventitious local substitution which favours one diastereoisomeric transition state over others.

Medium- and large-ring saturated hydrocarbons C_nH_{2n} have a larger number of conformations of similar energy available to them than is the case for cyclohexane.[39] However, just as a single *tert*-butyl group can anchor the cyclohexane ring in one conformation, so appropriate substitution on larger rings can limit the number of accessible conformers.

Still *et al.*[40] and Vedejs *et al.*[41] have exploited the limitation of conformational freedom resulting from cyclisation of a substituted chain to a medium-ring compound. Thus, in Scheme 42, cyclisation of hydroxy acid (93) to lactone (94) and diastereoselective Michael addition to the derived α,β-unsaturated lactone (95) constitutes, after ring opening, overall diastereoselective alkylation of an acyclic substrate.[42]

In this synthesis of (96), diastereoselectivity is believed to result from cuprate attack on the more exposed (upper) face of the double bond of the nine-membered cyclic α,β-unsaturated ester in the conformation shown (97). Needless to say, little diastereoselectivity could have been anticipated had this cuprate addition been carried out on the acyclic (protected) hydroxy α,β-unsaturated ester corresponding to (95).

For the highly diastereoselective reaction of, for example, a carbonyl group on a large ring it is not necessary that the whole molecule be constrained to react via a single conformation. It is only necessary for that part of the ring in the locality of the carbonyl group to exist in a single conformation in all those conformations of the entire molecule that undergo reaction at comparable rates.

For example, in Scheme 43, reduction of the substituted tetradecanolide (98) with a bulky reducing agent (L-Selectride) gives an 89:11 ratio of diastereoisomers in which the *trans* diastereoisomer (99) predominates.[43] In principle, therefore, ring closure of hydroxy acid (100) to tetradecanolide (98) and ring opening of the reduced lactone (99) would constitute a method for diastereoselective reduction of the 9-oxo group in (100).

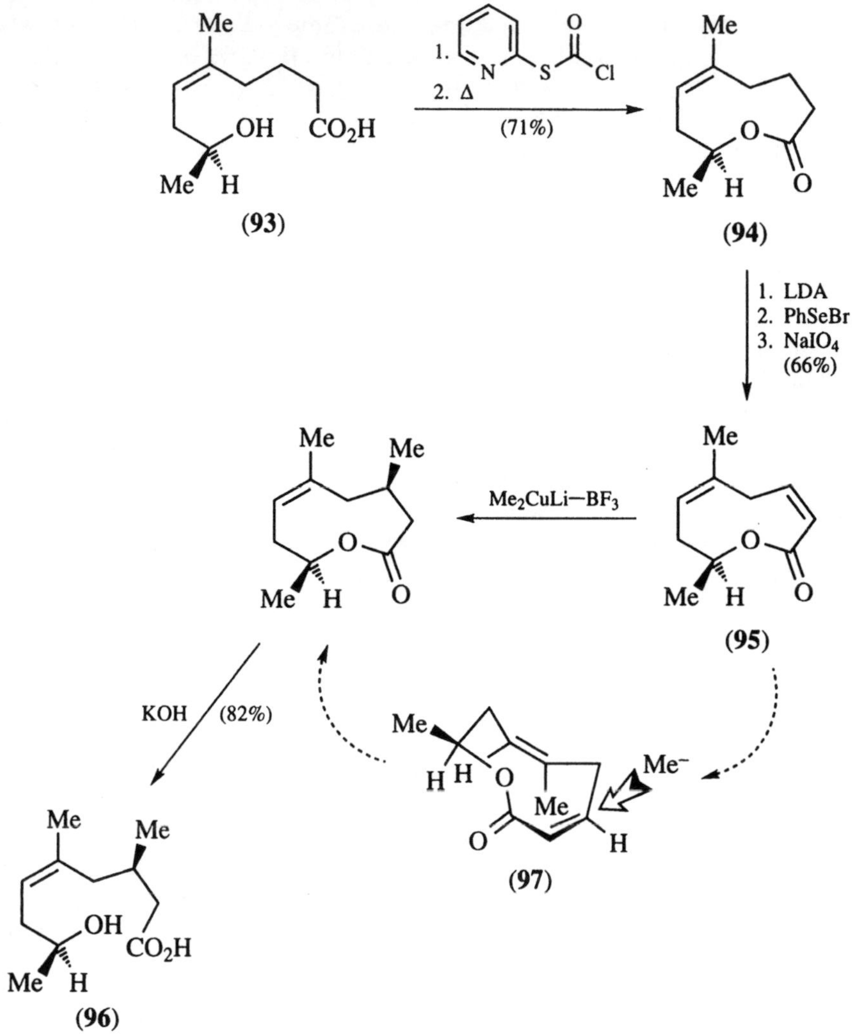

Scheme 42

The conformations accessible to compounds such as (98), together with their relative energies, can be computed using molecular mechanics computer programs such as MM2.[44] This program, which is particularly suitable for medium- or large-ring compounds such as (98), affords descriptions of the conformational energy minima as terms for steric, torsional and angle strain are varied iteratively. Conformations with energies >17 kJ mol⁻¹ (4 kcal mol⁻¹) greater than the minimum are usually neglected. The preferred reacting conformation (97) in Scheme 42 was derived in this way.

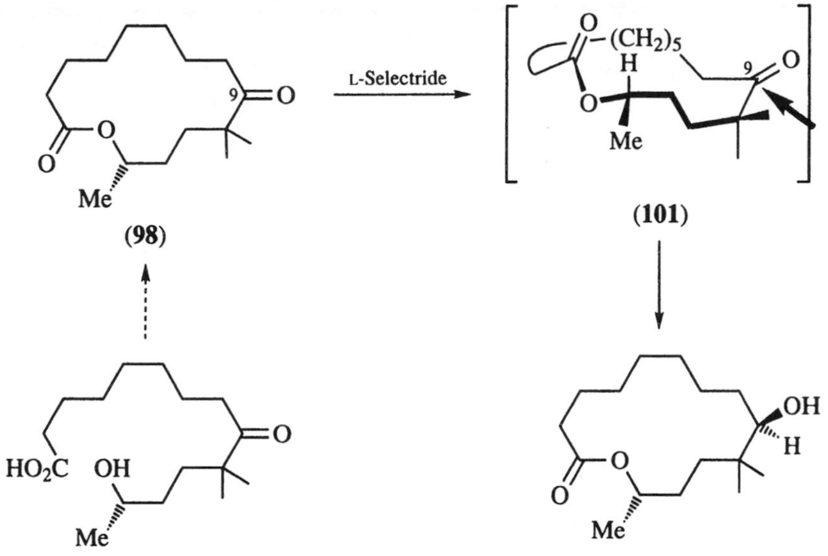

(98)

L-Selectride

(101)

(100)

(99) (89 : 11 d.r.)

$$\text{L-Selectride} \equiv \underset{\overset{\displaystyle CH_3}{\displaystyle |}}{Li(C_2H_5CH-)_3BH}$$

Scheme 43

Application of the MM2 program to **(98)** predicts that all conformations within $17\,kJ\,mol^{-1}$ of the minimum have the same local conformation, shown in **(101)** around the carbonyl group being reduced, irrespective of the conformation of the $(CH_2)_5$ unit. Attack on the 9-oxo group from outside the periphery will, as shown, deliver the diastereoisomer actually isolated. Attack from the interior of the ring is likely to be of higher energy, particularly with a bulky nucleophile and a tetrahedral angle of attack (the Burgi–Dunitz angle; see Chapter 10).

Summary

In situ ring formation is of great value for conformational control in acyclic compounds. The resulting ring is more likely to experience attack by an external reagent diastereoselectively than is the acyclic substrate (cf. the Cram–Kopecky chelation and reaction of α-substituted carbonyl compounds, Chapter 10).

Alternatively, diastereoselectivity may arise in the ring-forming reaction itself, e.g. in the intramolecular reaction between two functional groups at the termini of a chain. If one of these functional groups is a double bond having diastereofaces, the new set of steric and electronic interactions set up in the cyclisation may favour attack on one diastereoface rather than the other. Ring opening of the product so formed can result in a single diasteroisomer of an acyclic product in which overall 1,n (usually 1,2 or 1,3) asymmetric induction has been accomplished.

In some cases, notably when the product is a bridged bicyclic molecule, there is a *geometrical imperative*, which means that only one diastereoisomer of the product can be formed, i.e. diastereoselectivity is complete. Similarly, intramolecular cycloadditions result in tricyclic structures and geometrical imperatives can dictate the diastereoselectivity of the reaction.

Type 0 ring opening of these bicyclic or tricyclic single diastereoisomers regenerates mono- or bicyclic products as single diastereoisomers, even though the products may be capable of existing in diastereoisomeric forms.

A related class of reaction, that makes use of the increased diastereoselectivity afforded by intramolecular reaction, uses *in situ* appended reagents. Here, a reagent becomes attached to one functional group in the substrate and then attacks a second functional group intramolecularly. The spent reagent becomes detached from the product in the reaction or on work-up.

Finally, a further extension which takes advantage of geometrical imperatives is covalent binding of the reagent to the substrate in a preliminary step. Diastereoselective reaction is then followed by removal of the reagent residue, sometimes simply in the work-up of the reaction.

Many reactions with *in situ* appended reagents can be applied to acyclic substrates. Although in these cases the geometrical imperative is absent, the intramolecularity of the reaction again brings a set of steric/electronic influences into play that may favour attack on one diastereoface of the double bond rather than the other.

The conformational freedom of an acyclic compound is reduced when it is cyclised to a medium or large ring. The presence of a single conformation of the whole molecule can lead to diastereoselective addition; ring opening then leads to a single diastereoisomer of the acyclic product. Alternatively, the low-energy conformations of the molecule may all have the same local conformation around the functional group undergoing addition, and this can be sufficient to give diastereoselectivity. Molecular mechanics programs are useful for calculating the relative energies of the stable conformations of these compounds.

References

1. M. Majewski, D.L.J. Clive and P.C. Anderson, *Tetrahedron Lett.*, 1984, **25**, 2101.
2. E.D. Mehelich and G.A. Hite, *J. Am. Chem. Soc.*, 1992, **114**, 7318.
3. J.A. Marshall, M.J. Coghlan and M. Watanabe, *J. Org. Chem.*, 1984, **49**, 747.

4. A.L.J. Beckwith, T. Lawrence and A.K. Serelis, *Chem. Commun.*, 1980, 484.
5. M.P. Cooke, *J. Org. Chem.*, 1992, **57**, 1495, and references cited therein.
6. S. Takano, Y. Iwabuchi and K. Ogasawara, *J. Am. Chem. Soc.*, 1987, **109**, 5523.
7. 'Inside–outside' isomerism: see J.D. Winkler, B.C. Hong, J.P. Hey and P.G. Williard, *J. Am. Chem. Soc.*, 1991, **113**, 8839.
8. K.C. Nicolaou and Z. Lysenko, *J. Am. Chem. Soc.*, 1977, **99**, 3185; K.C. Nicolaou, S.P. Seitz, W.J. Sipio and J.F. Blount, *J. Am. Chem. Soc.*, 1979, **101**, 3884.
9. F. Zutterman, P. de Clercq and M. Vandewalle, *Tetrahedron Lett.*, 1977, 3191.
10. D.P. Curran and P.B. Jacobs, *Tetrahedron Lett.*, 1985, **26**, 2031.
11. J. d'Angelo and G. Revial, *Tetrahedron Lett.*, 1983, **24**, 2103; D. Buisson, R. Azerad, G. Revial and J. d'Angelo, *Tetrahedron Lett.*, 1984, **25**, 6005.
12. A.G. Shultz and J.P. Dittami, *J. Org. Chem.*, 1984, **49**, 2615.
13. S.S. Klioze and F.P. Darmory, *J. Org. Chem.*, 1975, **40**, 1588.
14. S. Yamada and G. Otani, *Tetrahedron Lett.*, 1971, 1133.
15. J.B.P.A. Wijnberg and W.N. Speckamp, *Tetrahedron Lett.*, 1975, 3963.
16. H. Ishibashi, T.S. So, K. Okochi, T. Sato, N. Nakamura, H. Nakatani and M. Ikeda, *J. Org. Chem.*, 1991, **56**, 95.
17. D. Baldwin and J.R. Hanson, *J. Chem. Soc., Perkin Trans. 1*, 1975, 1941.
18. K.B. Sharpless and T.R. Verhoeven, *Aldrichim. Acta*, 1979, **12**, 63.
19. T. Fukuyama, B. Vranesic, D.P. Negri and Y. Kishi, *Tetrahedron Lett.*, 1978, 2741.
20. G. Berti, *Top. Stereochem.*, 1973, **7**, 93.
21. P. Chamberlain, M.L. Roberts and G.H. Whitham, *J. Chem. Soc. B*, 1970, 1374.
22. R.S. Atkinson and B.J. Kelly, *J. Chem. Soc., Perkin Trans. 1*, 1989, 1515.
23. J.H.-H. Chan and B. Rickborn, *J. Am. Chem. Soc.*, 1968, **90**, 6406; F. Mohamadi and W.C. Still, *Tetrahedron Lett.*, 1986, **27**, 893, and references cited therein.
24. E.A. Mash, S.B. Hemperly, K.A. Nelson, P.C. Heidt and S. van Deusen, *J. Org. Chem.*, 1990, **55**, 2045; E.A. Mash and S.B. Hemperley, *J. Org. Chem.*, 1990, **55**, 2055.
25. G.A. Molander and L.S. Harring, *J. Org. Chem.*, 1989, **54**, 3525; M. Lautens and P.H.M. Delanghe, *J. Org. Chem.*, 1992, **57**, 798.
26. D.A. Evans, K.T. Chapman and E.M. Carreira, *J. Am. Chem. Soc.*, 1988, **110**, 3560.
27. K. Narasaka and F.-C. Pai, *Tetrahedron*, 1984, **40**, 2233.
28. F.G. Kathawala, B. Prager, K. Prasad, O. Repic, M.J. Shapiro, R.S. Stabler and L. Widler, *Helv. Chim. Acta*, 1986, **69**, 803.
29. J.M. Brown, *Angew Chem., Int. Ed. Engl.*, 1987, **26**, 190.
30. H.W. Thompson, E. McPherson and B.L. Lences, *J. Org. Chem.*, 1976, **41**, 2903.
31. G. Stork and D.E. Kahne, *J. Am. Chem. Soc.*, 1983, **105**, 1072.
32. J.M. Brown and S.A. Hall, *Tetrahedron Lett.*, 1984, **25**, 1393.
33. J.M. Brown and R.G. Naik, *Chem. Commun.*, 1982, 348.
34. D.A. Evans, M.M. Morrissey and R.L. Dow, *Tetrahedron Lett.*, 1985, **26**, 6005.
35. S.W. McCombie, B. Cox, S.-I. Lin, A.K. Ganguly and A.T. McPhail, *Tetrahedron Lett.*, 1991, **32**, 2083.
36. C. Alcaraz, J.C. Carretero and E. Dominguez, *Tetrahedron Lett.*, 1991, **32**, 1385; M. Isobe, Y. Ichikawa, Y. Funabashi, S. Mio and T. Goto, *Tetrahedron*, 1986, **42**, 2863.
37. D.J. Rawson and A.I. Meyers, *J. Org. Chem.*, 1991, **56**, 2292.
38. G. Stork, H.S. Suh and G. Kim, *J. Am. Chem. Soc.*, 1991, **113**, 7054.
39. J. Dale, *Top. Stereochem.*, 1976, **9**, 199.
40. W.C. Still, L.J. MacPherson, T. Harada and J.F. Callahan, *Tetrahedron*, 1984, **40**, 2275.

41. E. Vedejs, W.H. Dent, D.M. Gapinski and C.K. McClure, *J. Am. Chem. Soc.*, 1987, **109**, 5437, and references cited therein.
42. W.C. Still and I. Galynker, *J. Am. Chem. Soc.*, 1982, **104**, 1774.
43. T.H. Keller and L. Weiler, *J. Am. Chem. Soc.*, 1990, **112**, 450.
44. N.L. Allinger, *Adv. Phys. Org. Chem.*, 1976, **13**, 1; U. Burkert and N.L. Allinger, *Molecular Mechanics*, American Chemical Society, Washington, DC, 1982.

12 TYPE III REAGENT-CONTROLLED REACTIONS (TYPE III$_{r.c.}$)

In substrate-controlled diastereoselective reactions, the chiral centre and the double bond undergoing addition are contained in the *same* molecule (Chapters 9–11). In reagent-controlled diastereoselective reactions (Type III$_{r.c.}$), the chiral centre(s) and the double bond undergoing addition are contained in *different* molecules.

Suppose we have a carbonyl group which is undergoing addition by a reagent x–y [Scheme 1 (a)]. The product will be racemic because (1) and $_{enant.}$(1) will be produced in equal amounts by addition to the two *enantiofaces* of the carbonyl group. When addition of a chiral reagent x*–y to the same carbonyl group takes place [Scheme 1 (b)], the two products are diastereoisomers and in favourable cases one may be produced in large excess over the other. This is because in the transition states for addition of x*–y, the two enantiofaces of the carbonyl group become diastereofaces and one diastereoisomeric transition state can be favoured over another (see Chapter 1, Figure 2).

The differences between Type III$_{s.c.}$ and III$_{r.c.}$ reactions are not as great as might be supposed; in both cases, asymmetric induction takes place in the transition state of the reaction and in both similar steric and electronic effects bring about diastereoselectivity.

The difference between Type III$_{s.c.}$ and Type III$_{r.c.}$ is, for the majority of Type III$_{r.c.}$ reactions, even further reduced because, in this majority, some part of the reagent x*–y becomes attached (appended) to the substrate in a preliminary step e.g. to the carbonyl oxygen in Scheme 1(c) forming a complex (2). The two faces of the carbonyl group are now diastereofaces and the intramolecular attack to give say (3) is indistinguishable from a Type III$_{s.c.}$ reaction. Type III$_{r.c.}$ reactions of this kind, or rather the Type III$_{s.c.}$ component that they contain, have clearly much in common with Type III$_{s.c.}$ reactions proceeding via appended reagents, discussed in Chapter 11.

Type III$_{r.c.}$ reactions of C=C bonds can also frequently occur via an initially formed complex of the substrate with the reagent, but in this case the complexing must take place with a substituent on the double bond, e.g. d (Scheme 2; cf. Chapter 1, Figure 4).

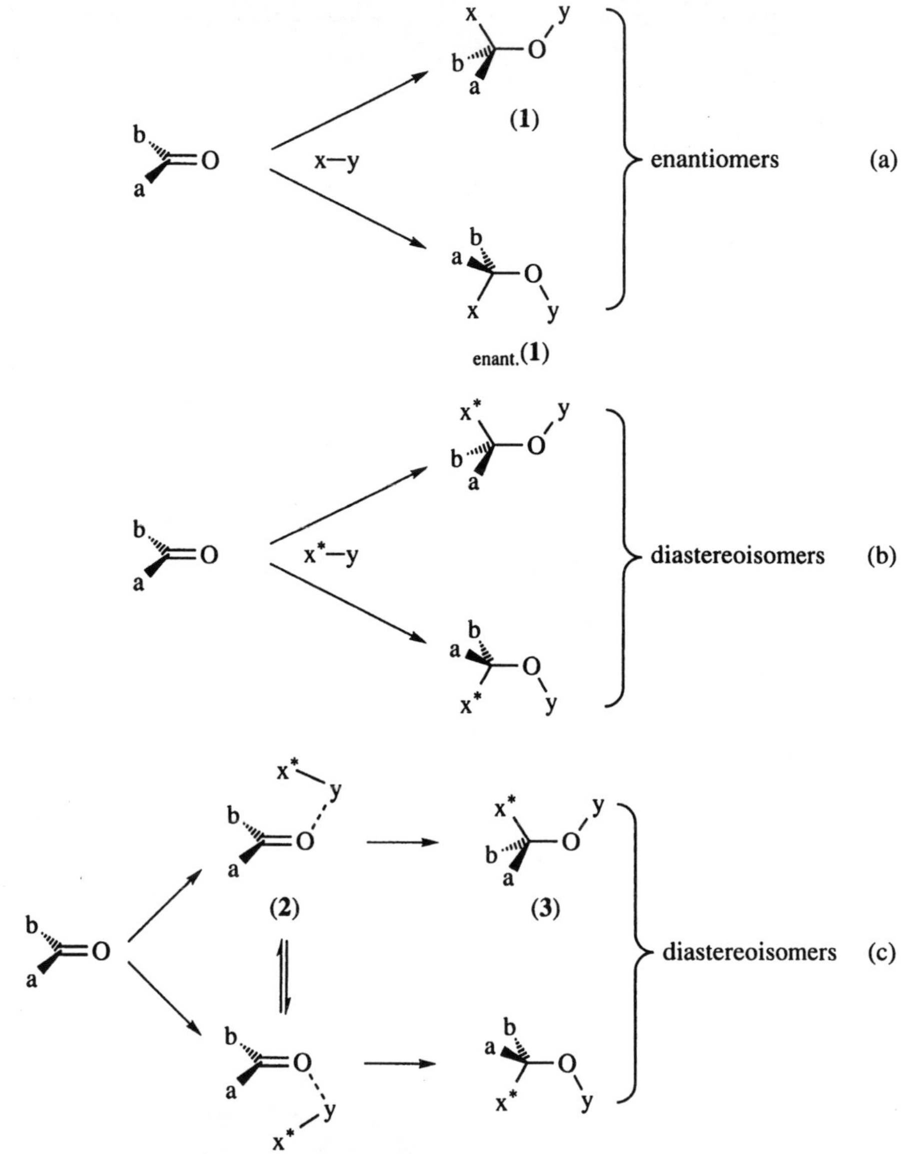

Scheme 1

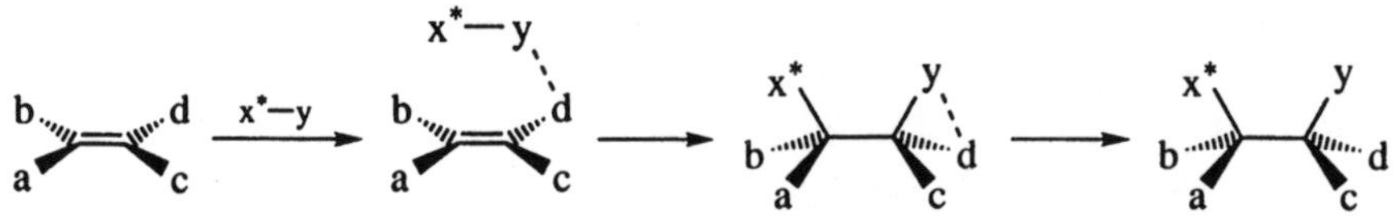

Scheme 2

The temporary link between d and y is broken either spontaneously or in the reaction work up, as it is in Type III$_{s.c.}$ reactions via appended reagents (Chapter 11).

Design of Type III$_{r.c.}$ reactions

There is a problem which arises in the design of *echt* Type III$_{r.c.}$ reactions, i.e. those which proceed via a single transition state as in Scheme 1 (b). In the simplest version of this Type III$_{r.c.}$ reaction (Scheme 3), a double bond undergoes attack by a reagent containing a single chiral centre. To maximise the opportunity for asymmetric induction, it should be advantageous to locate the anion, cation or radical that attacks the double bond *on* the chiral centre, so that the distance between the existing and developing chiral centres is minimised [Scheme 3 (a)].

Unfortunately, carbocations are almost invariably planar and radicals and carbanions are, with a few exceptions, either planar or not configurationally stable (rapidly inverting). Furthermore, for a common situation in Scheme 3 (a), when a = b = H, attack will more likely take place in the regio-sense shown in Scheme 3 (b). This means that the existing chiral centre in the reagent and the potential one in the product are in any case not adjacent and the likelihood of asymmetric induction is less.

As a consequence, there are very few Type III$_{r.c.}$ reactions, analogous to

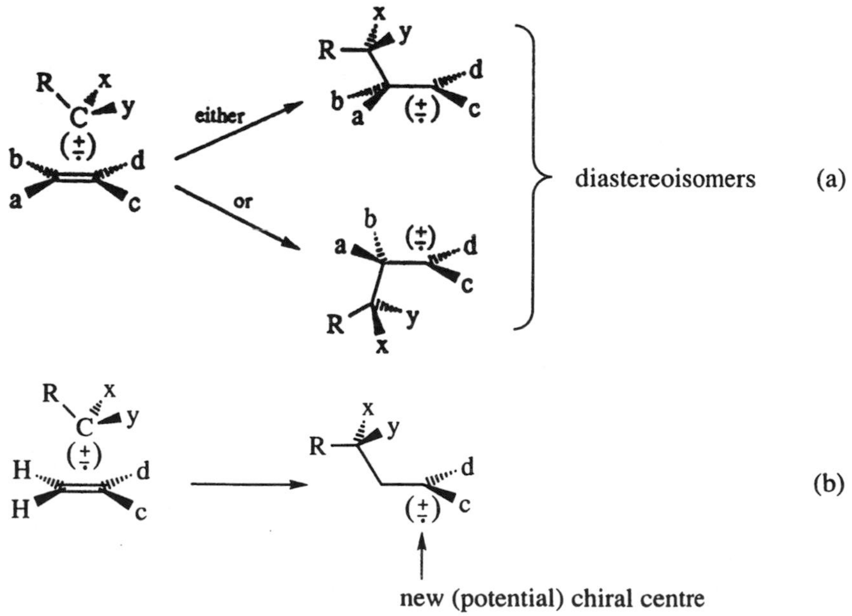

diastereoisomers   (a)

new (potential) chiral centre

Scheme 3

the numerous Type III$_{s.c.}$ reactions in which a chiral centre *adjacent* to a double bond in a substrate induces high diastereoselectivity in addition reactions to that double bond (see Chapters 9 and 10). There are a small number of carbanions having sp^3 hybridisation (see Chapter 4) that can be obtained in enantiopure form and, in principle, could be used to bring about asymmetric induction according to Scheme 3 (a) or by addition to a prochiral carbonyl group with retention of the carbanion configuration, but few reactions of this type have been reported.

Of course, it is always possible to locate the inducing chiral centre at a site in the reagent other than at the atom to which a bond is being formed. Just because this chiral centre is further removed does not mean that the degree of asymmetric induction will *necessarily* be inferior: in the *transition state* for the reaction, this more remote chiral centre may bring about asymmetric induction *via* a well-defined (often cyclic) geometry. Reactions in this category whose stereochemistry can be more easily rationalised are usually Type II/III and will be considered in Chapter 13.

Advantages of Type III$_{r.c.}$ over Type III$_{s.c.}$ reactions

One advantage in synthesis which Type III$_{r.c.}$ reactions have over Type III$_{s.c.}$ reactions is that, in principle, it is possible for a single enantiopure reagent to convert a *range* of achiral alkenes into their respective addition products completely diastereoselectively and thus to prepare single enantiomers of these products. To prepare a corresponding range of enantiopure products using Type III$_{s.c.}$ reactions, it is necessary to prepare each chiral alkene in enantiopure form and then to react with the achiral reagent (completely diastereoselectively). Since enantiopure alkenes are, in general, less readily available than achiral alkenes, so are Type III$_{r.c.}$ reactions advantageous.

Another advantage of Type III$_{r.c.}$ reactions is that they can more easily become Type III$_{r.c./s.c.}$ (by using an alkene substrate containing a chiral centre) to take advantage of double asymmetric induction (see Chapter 15). The conversion of Type III$_{s.c.}$ reactions into Type III$_{r.c/s.c.}$ by the use of a chiral rather than an achiral reagent is, in general, more problematic.

Sub-classification of Type III$_{r.c.}$ reactions

Our sub-classification of Type III$_{r.c.}$ given in Chapter 9 is reproduced in Figure 1.

In many *overall enantioselective* Type III$_{r.c.}$ reactions, the reagents are, or include, metals/metalloids bearing chiral ligands. In these reactions, the chiral ligands, along with the metal/metalloids, are eliminated from the *initial* product in the work-up of the reaction as exemplified in the Type III$_{r.c.}$ hydroboration in Scheme 4.

Type III$_{r.c.}$
(reagent controlled) attack
of chiral reagent on one
enantioface

──────────────

or one enantiotopic atom
or group of substrate

→ chiral centre(s) of reagent retained in the
product which also contains newly created
chiral centre(s)

→ *overall enantioselective*: chiral centre(s) of
reagent *not* retained in the product after work-
up. Reagent used in molar equiv. quantity and
enantiopure form and the product, containing
only new chiral centre(s) isolated in enantio-
enriched form

→ *catalytic enantioselective (cat. enant.)*: as above
but reaction is mediated by an enantiopure
catalyst, i.e. active enantiopure reagent is
generated *in situ* from achiral reagent +
enantiopure catalyst

Figure 1   Sub-classification of Type III$_{r.c.}$ reactions.

In Scheme 4, the initial borane addition product (**4**) is obtained (ideally) as
a single diastereoisomer but hydroboration as a route to alcohols (**5**), by
conversion of the C—B bond into a C—O bond with retention of
configuration at carbon, is invariably carried out without isolation of (**4**). The
isolated alcohol (**5**), therefore, will contain only the created chiral centre; if (**4**)
is indeed a single diastereoisomer, (**5**) will be enantiopure. In general, in these
Type III$_{r.c.}$ reactions, in which the chiral centres of the reagent are not retained
in the isolated product, the diastereoselectivity (d.e.) is equated with the
enantioselectivity (e.e.) in the formation of this isolated product.

The fact that in many Type III$_{r.c.}$ reactions the chiral centres of the reagent
are not retained in the product means that examples in our first sub-category
(see Figure 1) are not common. Even when the intermediate diastereoisomer
which retains the chiral centre(s) of the reagent *is* isolated, it is often only
because it is practicable to do so before elimination of the chiral centre(s)
deriving from the reagent as in the conversion in Scheme 5.[1]

In Scheme 5, the sulphoxide-stabilised carbanion (**6**) adds highly
diastereoselectively to propiophenone to give (**7**). Removal of the sulphur
with Raney nickel gives (**8**), which is a single enantiomer. [Considered
overall, carbanion (**6**) is the synthetic equivalent of a chiral CH$_3^-$ synthon.] It
is likely that the Type III$_{r.c.}$ reaction of lithium carbanion (**6**) and

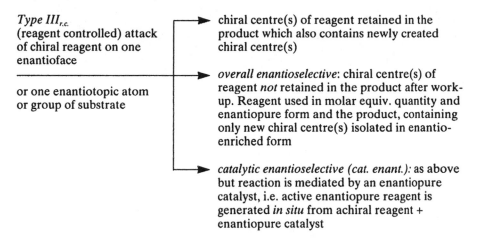

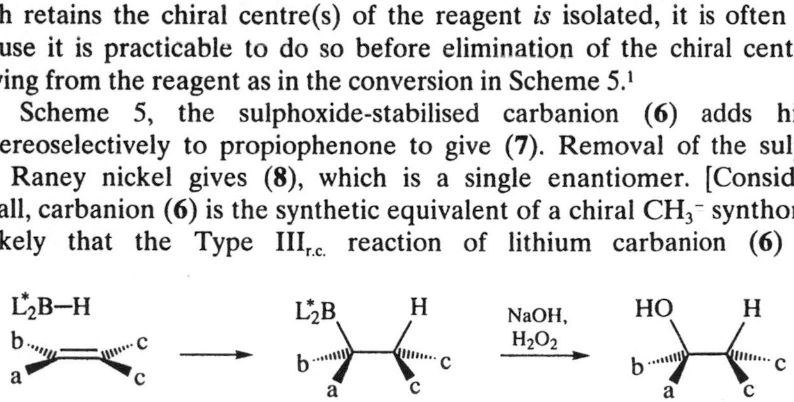

L* enantiopure ligand   (**4**)   (**5**)

Scheme 4

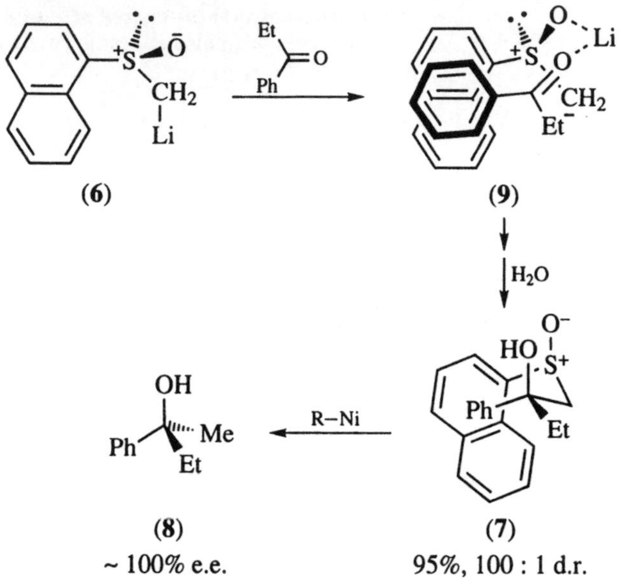

Scheme 5

propiophenone takes place within a lithium complex (9) of the sulphoxide and the carbonyl compound which is stabilised by π–π interaction between the naphthalene and phenyl rings (diastereoselectivity in this addition is poor when the phenyl group is absent). Within this complex, the two faces of the carbonyl group are clearly differentiated and attack on the rear one, as shown in (9), is preferred. If this mechanism is correct, it resembles that in Scheme 1 (c) and differs from that in Scheme 1 (b) in which enantiofaces become diastereofaces only in the transition state of an intermolecular reaction.

Some examples of Type II/III$_{r.c.}$ reactions in this first sub-category are given in Chapter 13. In the remainder of this chapter, we shall consider *overall enantioselective* Type III$_{r.c.}$ reactions (see Figure 1); *catalytic enantioselective* Type III$_{r.c.}$ reactions will be considered in Chapter 14. Type III$_{r.c.}$ reactions in which substitution of one enantiotopic atom or group occurs are considered in Chapter 16.

Overall enantioselective Type III$_{r.c.}$ reactions (those which use at least stoichiometric amounts of the chiral reagent)

Reactions in this category include oxidation, reduction and those using reagents containing metals/metalloids bearing chiral ligands. The majority involve a preliminary complexation of the reagent with the substrate [Scheme 1 (c) or 2] followed by intramolecular diastereoselective reaction. In most of these reactions also, the chiral ligands, along with the metal/metalloid, are eliminated from the initial product in the work-up of the reaction and the isolated product, therefore, contains only the new chiral centre(s) formed (cf. Scheme 4). In such reactions, therefore, the diastereoselectivity (d.e.) in formation of the initial product is identical with the enantiopurity (e.e.) of the isolated product.

REDUCTION OF KETONES

Since diborane, sodium tetrahydroborate, lithium aluminium hydride and their derivatives are such convenient and versatile reagents for the reduction of carbonyl groups, it was natural that chirally modified versions of these were among the first to be examined as enantioselective reducing agents.

Whereas in the 1970s enantioselectivities in the reduction of prochiral ketones were only exceptionally as high as 90%, this figure is routinely achieved with the latest generation of chirally modified reagents. (Summaries of the earlier work have been given ApSimon and co-workers[2].)

Brown *et al.*[3] have undertaken a comparative study of the enantioselectivities achieved in reduction of ten classes of prochiral ketone with six of the most promising chiral borane or alane reagents. Table 1 shows the most enantioselective reagent in each case and it can be seen that at present no one reagent is superior for all classes of ketones.

Reductions using trialkylboranes, e.g. B-Ipc-9-BBN, or dialkyl-monochloroboranes, in Table 1 take place by transfer of a hydride from the β-position of one of the alkyl groups (Scheme 6). It has been suggested[4] that formation of the initial addition product (**10**) is favoured because the 'diaxial' interaction between one of the pinane methyl groups and R_L in (**11**) is greater than the corresponding interactions with R_S in (**12**). Cleavage of

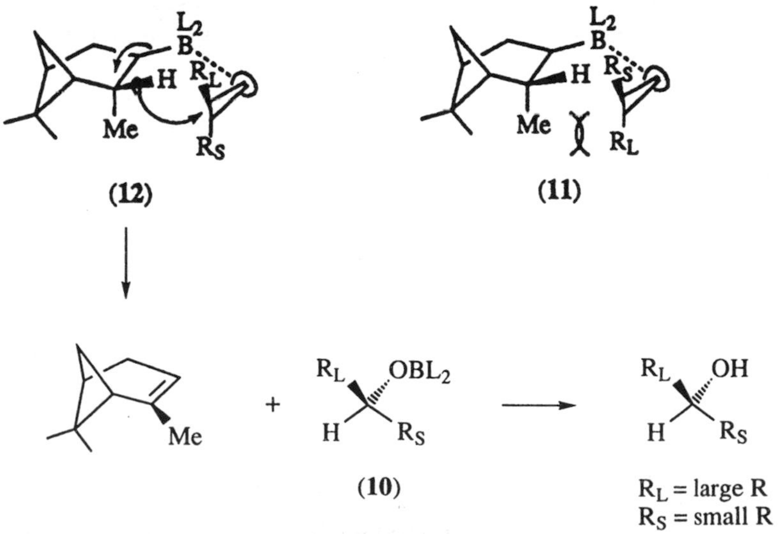

(**12**)   (**11**)

(**10**)

R_L = large R
R_S = small R

Scheme 6

Table 1 Examples of enantioselective reductions of ketones

Ketone type	Structural formula	% e.e.	Preferred reagent	
			Name / conditions	Structural formula
Acyclic		100	(R,R)-2,5-dimethylborolane	
Cyclic		98	Ipc₂BCl, THF	
Aralkyl		100	B-Ipc-9-BBN, 6 kbar, THF (BBN = 9-borabicydo[3.3.1]nonyl)	
Heterocyclic		100	B–Ipc–9–BBN, 6 kbar, THF	
α-Halo		96	B–Ipc–9–BBN, neat	
α-Keto ester		92	K Glucoride	
β-Keto ester		84–92	LiBH₄–ᵗBuOH (R,R')–N,N'–dibenzoylcystine	
Acyclic enone		70	BINAL–H [(R) or (S)]	
Ynone		89	BINAL–H [(R) or (S)]	
Cycloenone		100	LiAlH₄–H	

the boron and accompanying chiral centres of the reagent takes place in the work up using, e.g., ethanolamine.

HYDROBORATION OF ALKENES

The work of Brown and co-workers over more than three decades has established organoboranes as invaluable reagents in a variety of functional group interconversions.[5] Much of this work has as its basis (a) the ability of the boron–hydrogen bond to add *syn* to an alkene double bond, (b) addition of the boron atom to the less sterically hindered carbon of this double bond and (c) replacement of the carbon–boron bond by a carbon–oxygen bond, using alkaline hydrogen peroxide, with retention of configuration at carbon (see Chapter 4).[6]

Enantioselective hydration of prochiral alkenes via diastereoselective hydroboration, first demonstrated as early as 1961, uses enantiopure organoboranes, themselves prepared by hydroboration of enantiopure alkenes, usually readily available terpenes or sesquiterpenes such as α-pinene, limonene or longifolene. Two of the most extensively used chiral hydroborating agents are those derived from (+)- and (−)-α-pinene (Scheme 7) in which one or two molecules of α-pinene are added to borane. In this reaction the boron becomes attached not only to the less hindered end of the double bond but also to the less hindered face. The resulting mono- or diisopinocampheylboranes [abbreviated to IpcBH$_2$ (13) and (Ipc$_2$BH (14)] are obtained in 99% e.e. and are highly diastereoselective in their additions to certain prochiral alkenes (Scheme 8).[7] The diastereoselective use of acetaldehyde in Scheme 8 rather than the usual alkaline H$_2$O$_2$ in the work-up allows the recovery of the α-

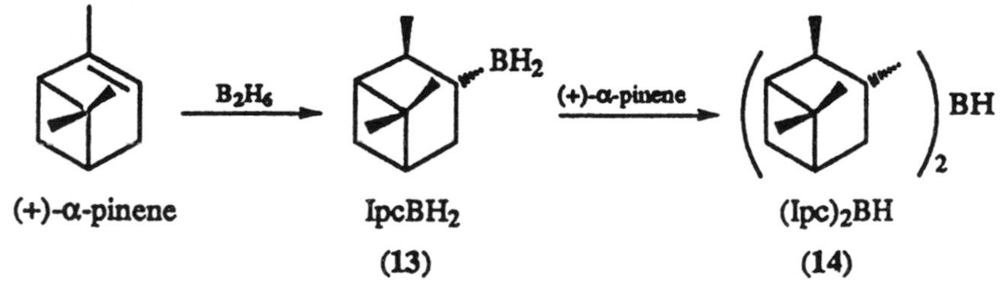

(+)-α-pinene   IpcBH$_2$ (Ipc)$_2$BH

(13)   (14)

Scheme 7

1. (−)-Ipc$_2$BH
2. MeCHO
3. NaOH, H$_2$O

80%, 100% e.e.

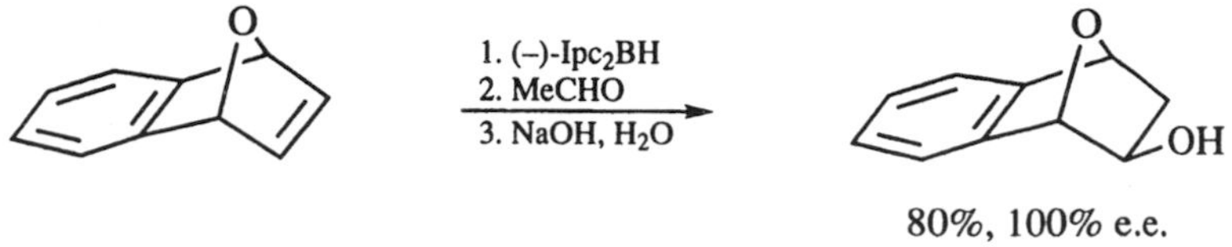

Scheme 8

pinene, it being eliminated by transfer of the β-hydrogen of the adduct to the aldehyde carbonyl group via a six-membered cyclic transition state (Scheme 9; c.f. Scheme 6).[8]

Masemune *et al.*[9] have used chiral *trans*-2,5-dimethylborolanes (**15**) (Scheme 10), which, although generally more enantioselective than (**13**) or (**14**) in their hydrations of simple *cis*- or *trans*-di or -trisubstituted alkenes, are at present less readily available. The proposed transition-state geometry (**16**),

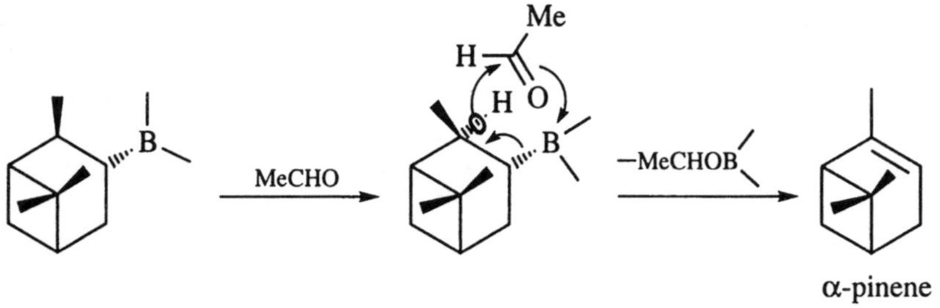

Scheme 9

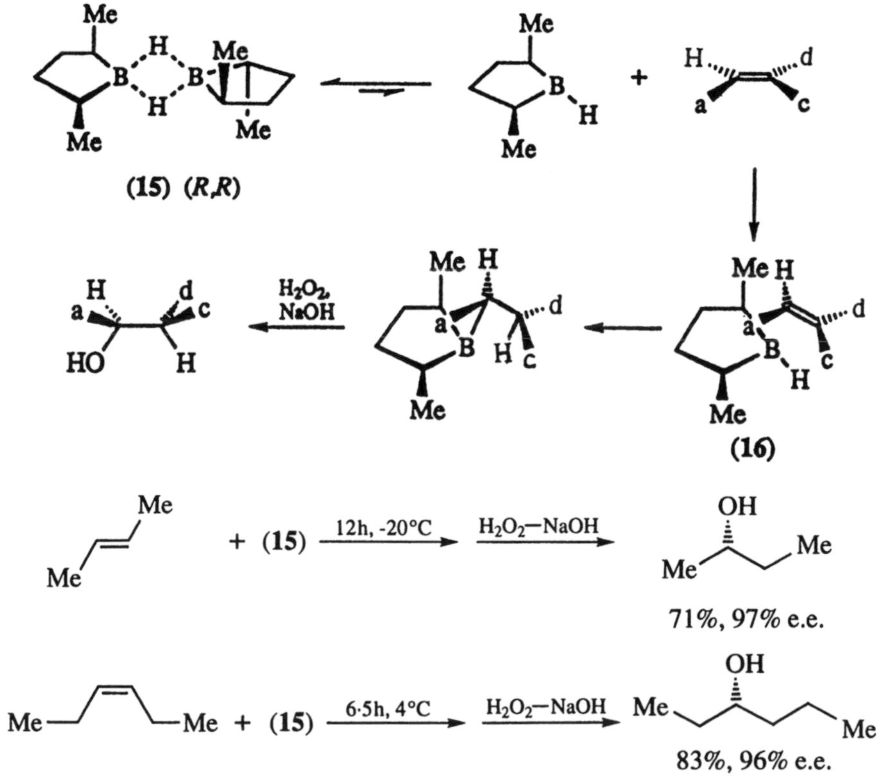

Scheme 10

through which 2,5-dimethylborolane is thought to react with the alkene, accounts for the low enantioselectivity obtained in its application to the hydration of 1,1-disubstituted alkenes.

ALLYL ADDITION TO ALDEHYDES

The simple diastereoselectivity which arises from Type II reactions of achiral allylboron derivatives with aldehydes has been discussed previously (Chapter 8). The preferred chair transition state means that the *cis/trans* relative configuration in the homoallylic product can be manipulated by choice of double bond configuration in the substituted allylboron.

Considerable progress has been made in bringing about this conversion enantioselectively by incorporating chiral centres into allylboronates. Thus Racherla and Brown[10] have achieved almost complete enantioselectivity in the allyl addition to aldehydes using a number of allylboranes including allylbis(2-isocaranyl)borane (**17**) (Scheme 11). Preparation of allylborane (**17**) uses an allylic Grignard reagent; when care is taken to remove the residual magnesium salts, the allylation reactions can be carried out at temperatures as low as $-100\,°C$ with increased overall enantioselectivity.

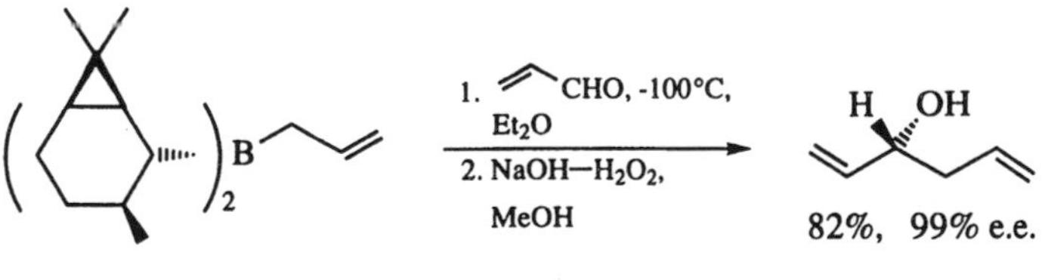

(17)

1. $\diagup\!\!\!\!\diagdown\text{CHO}$, -100°C, Et$_2$O
2. NaOH—H$_2$O$_2$, MeOH

82%, 99% e.e.

Scheme 11

Similar reactions have been effected with high enantioselectivity using a variety of enantiopure allylboron derivatives, including (**18**)–(**23**).[11]

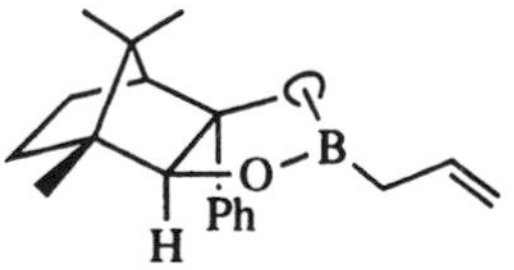

(**18**) Hoffmann 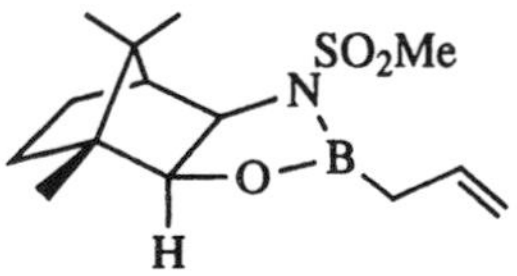   (**19**) Reetz 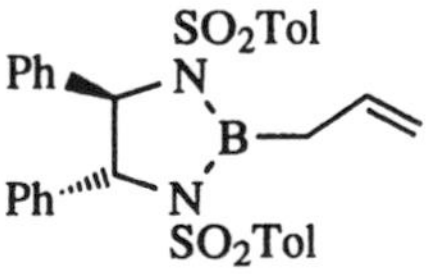   (**20**) Corey

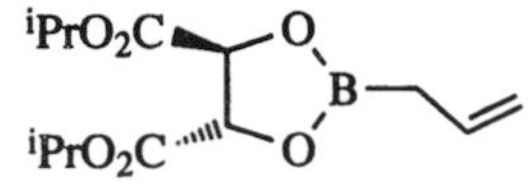

(**21**) Roush 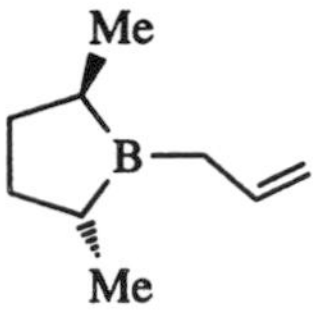   (**22**) Masemune 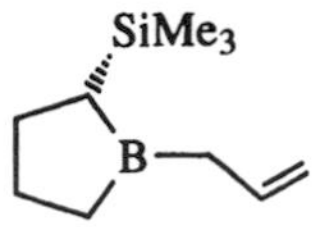   (**23**) Masemune

ALKYNYL AND ALLENYL ADDITIONS TO ALDEHYDES

The (R,R)-B-bromoborane (24) (Scheme 12) has been converted into the corresponding B-propargyl- and B-allenylboranes (25) and (26) by treatment with the allenyl- and propargylstannane derivatives (27) and (28), respectively. These reagents convert aldehydes into allenic and propargylic alcohols, respectively, highly enantioselectively.[12]

Both enantiomers of the B-bromoborane are available from the corresponding 1,2-diphenyl-1,2-diaminoethane enantiomers and so both enantiomeric allenyl and propargyl alcohol products are accessible. The sense of asymmetric induction is thought to arise from a transition-state assembly resembling (29) for addition of (25) to benzaldehyde.

EPOXIDATION OF ALKENES

Oxaziridines (30) are strained rings which resemble dioxiranes (31) and some metal peroxides (32) in their ability to transfer an oxygen atom to alkenes (and other substrates).[13]

| (30) | (31) | (32) |

Transfer of the oxaziridine oxygen to the alkene appears to take place in a single step, with a parallel biplanar geometry (33) (Scheme 13) slightly preferred over the spiro geometry (34), according to calculations.[14]

N-Sulphonyl-substituted oxaziridines, e.g. (35) [Scheme 14 (a)], efficiently transfer oxygen to alkenes at rates comparable to those of peroxyacids. Also, as with peroxyacids, the oxygen transfer is *syn* (Type II) with retention of the alkene configuration in the product (stereospecificity; inherent stereoselectivity). Enantioselectivities of 65% e.e. in these epoxidations have been obtained using the 3-pentafluorophenyl-2-sulphamoyl-substituted oxaziridine (36) [Scheme 14 (b)].

If the mechanism for this oxygen transfer is that given in Scheme 13, then this reaction is one of the few in which the two faces of an alkene are differentiated (i.e. become diastereofaces) only in the transition state for the addition; there is no preliminary complexation of the double bond substituents with the chiral reagent. This reaction is also unusual in that the product, containing only the newly created chiral centre(s), is formed directly and does not require cleavage of the chiral elements of the reagent, i.e. the reaction is directly enantioselective instead of overall enantioselective as in previous examples of Type II$_{r.c.}$ reactions in this chapter.

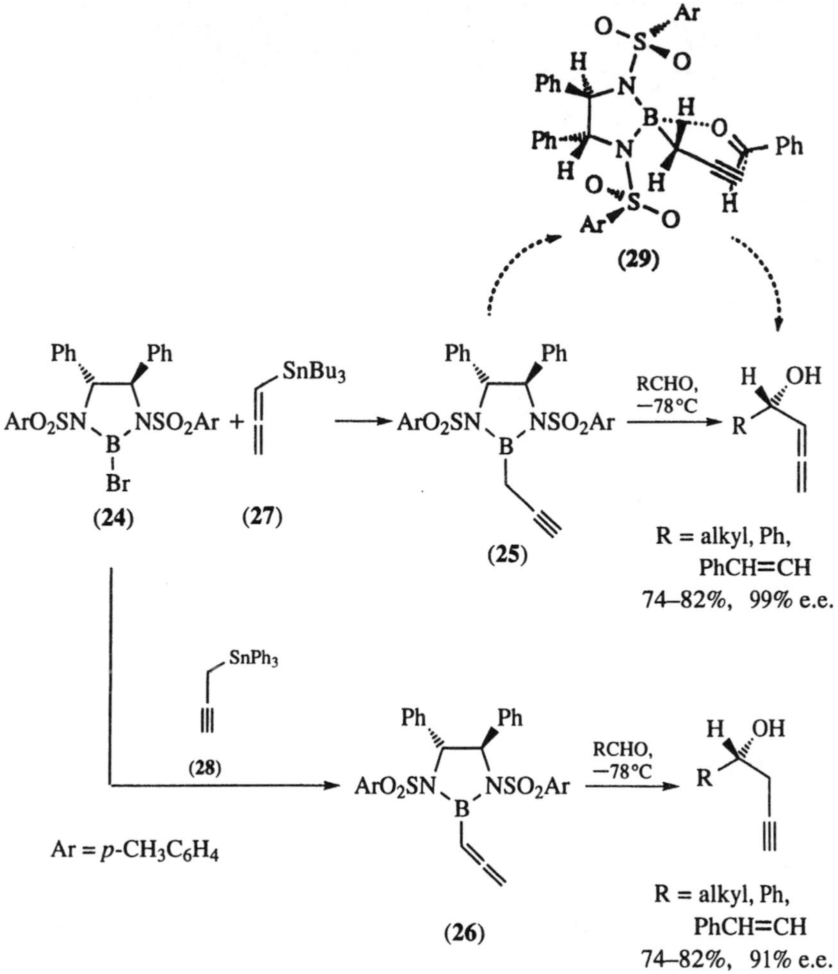

(29)

(24)   (27)   (25)

R = alkyl, Ph,
PhCH=CH
74–82%, 99% e.e.

Ar = p-CH₃C₆H₄

(28)   (26)

R = alkyl, Ph,
PhCH=CH
74–82%, 91% e.e.

Scheme 12

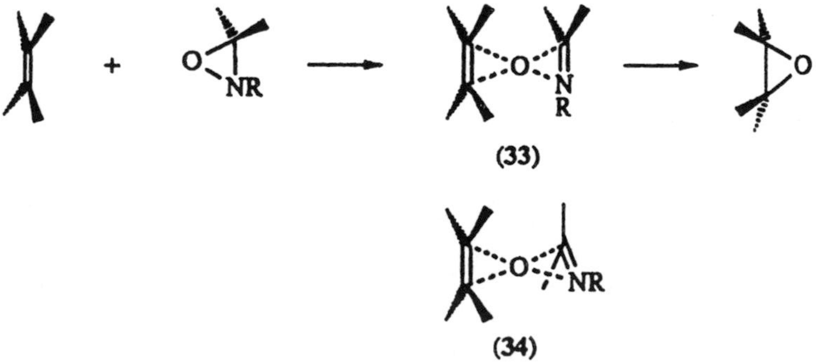

(33)

(34)

Scheme 13

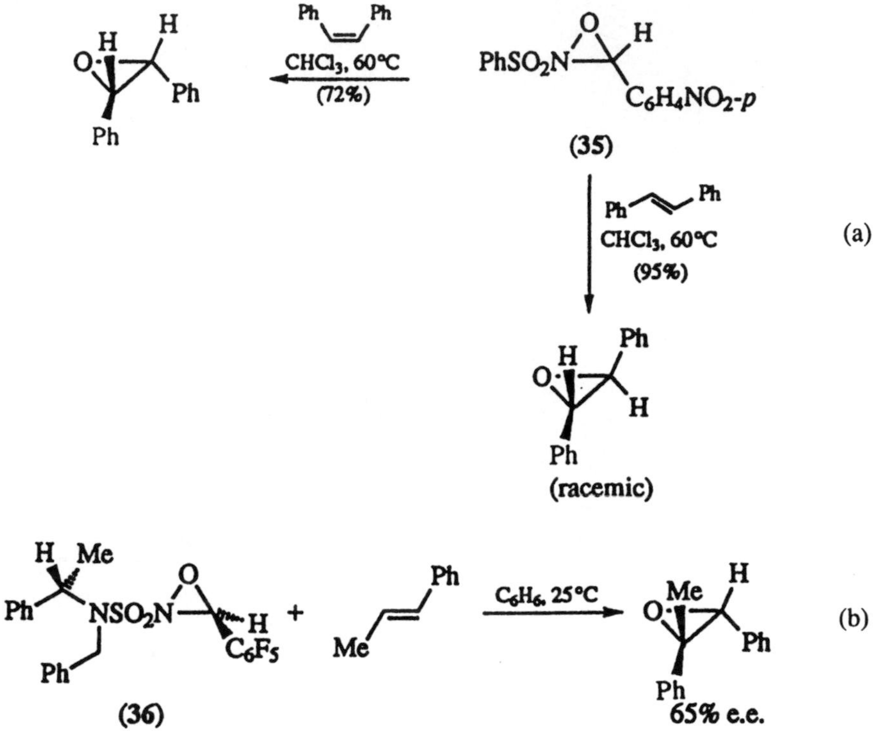

Scheme 14

HYDROXYLATION OF ENOLATES

The *N*-sulphonyloxaziridines above also oxidise enolates to α-hydroxycarbonyl compounds (Scheme 15). Impressive levels of enantioselectivity can be achieved using (8,8-dichlorocamphoryl)sulphonyloxaziridine (37). In this case an intermediate (38) is thought to be involved.[15]

MICHAEL ADDITION

Conjugate addition of organocuprates to cyclic α,β-unsaturated ketones, e.g. cyclohexenones (Scheme 16), has been carried out enantioselectively using a chiral (enantiopure) ligand on the copper.[16] (*R*)-(−)Muscone (39), an important perfume ingredient, has been prepared in 89% e.e. by this route.[17]

Using the ephedrine-derived amino alcohol (40) (Scheme 17) as the chiral ligand, delivery of alkylcopper to the *re*-face of cyclohexenone is proposed as shown in (41).[18] One lithium is chelated, with the oxyanion of

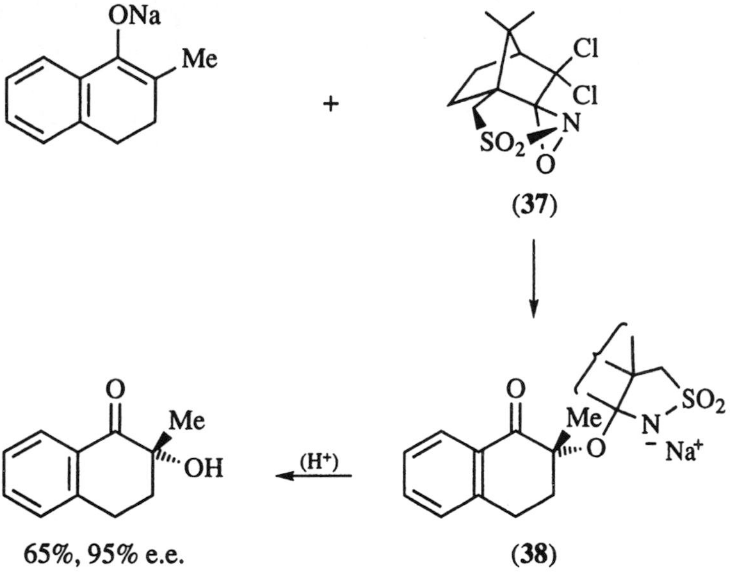

Scheme 15

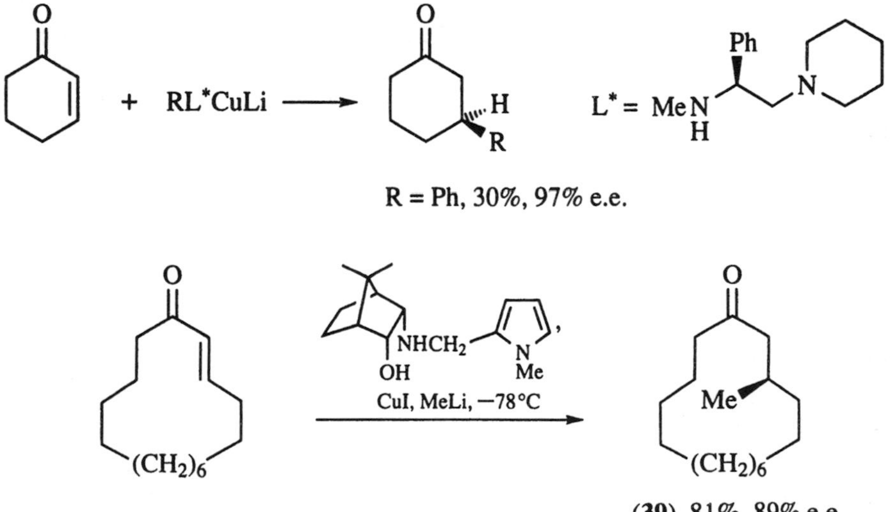

R = Ph, 30%, 97% e.e.

(39) 81%, 89% e.e.
(R)-muscone

Scheme 16

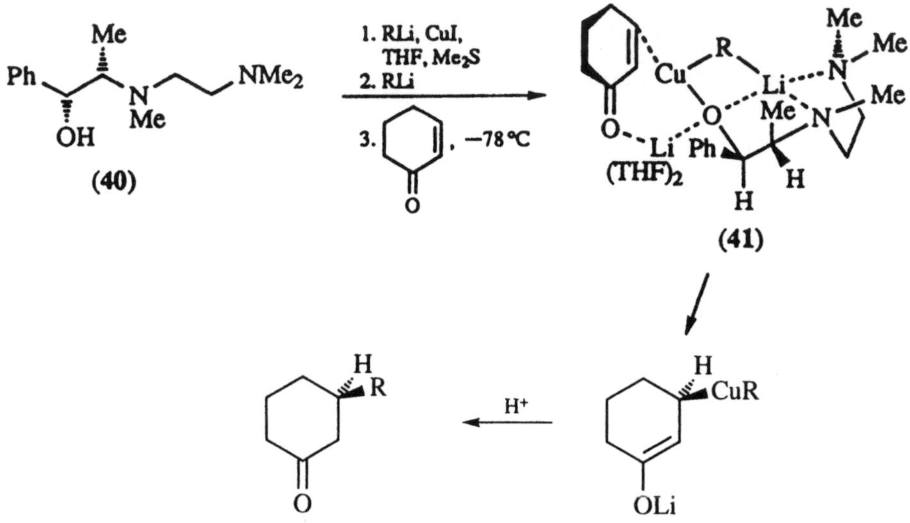

R = Bu, 90%, 89% e.e.

Scheme 17

(40) acting as a tridentate ligand and associated with the alkylcopper as shown. In this transition state, Corey *et al.*[18] envisage preliminary $d\pi^*$ complex formation between the copper and the alkene with the carbonyl carbon made more electrophilic by chelation of the oxygen to the lithium as shown. Conversion of the β-cuprio adduct to the product then occurs by reductive elimination with retention of configuration.

Michael addition of an organolithium reagent to an achiral α,β-unsaturated imine, e.g. (42) or (44), using a molar equivalent of a C_2-symmetric enantiopure catalyst (43) gives the product with high enantioselectivity (Scheme 18).[19]

The sense of induction is rationalised by, for example, a transition state (45) for imine (44) in which the lithium is complexed to the diether (43) and to the nitrogen of the imine. The plane of the imine as drawn in (45) has the cyclohexyl ring tilted (towards the viewer as shown) to minimise its interaction with the O^2-methyl group whose orientation is controlled by the adjacent phenyl group. Intramolecular delivery of the reagent's phenyl group then takes place at the front face of the α,β-unsaturated imine.

PROTONATION

Formation of an enolate from the racemate of a carbonyl compound such as (46) (Scheme 19), followed by completely enantioselective protonation,

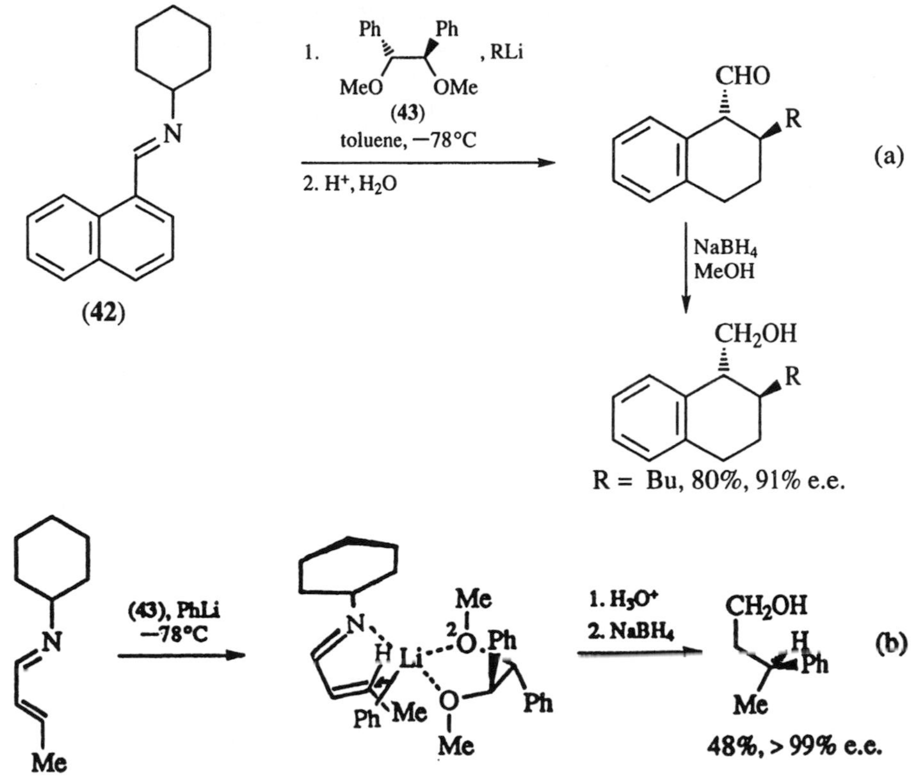

Scheme 18

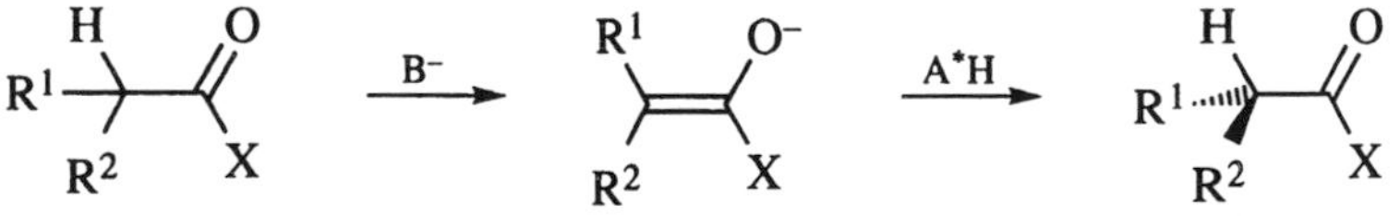

Scheme 19

would result in recovery of the starting material (**46′**) as a single enantiomer (deracemisation). Ideally, the chiral acid A*H used to protonate the enolate would be available in either enantiomeric form, thus making available either enantiomer of the carbonyl compound.

In spite of fairly intensive study, a general method for completely enantioselective enolate protonation has not emerged, although high enantioselectivities have been achieved in particular cases (Scheme 20).[20-22] Weakly acidic and/or hindered enantiopure acids are used to encourage maximum diastereoface discrimination in the transition state for protonation, with the transferred proton as close as possible to chiral elements in the acid. In example (c), the presence and protonation of both enolate diastereoisomers is likely to reduce the overall enantioselectivity. Another problem is that the lithium enolates are present in many cases as aggregates of unknown structure and deaggregation may be important in the protonation step.

The synthesis in Scheme 21 of (S)-α-damascenone, an important perfume ingredient, incorporates an enantioselective protonation step.[23]

Irradiation of α,β-unsaturated esters, e.g. (47) (Scheme 22), at low temperatures generates enol intermediates which are protonated enantioselectively by chiral amino alcohol (48) to give the deconjugated ester, e.g. (49).[24]

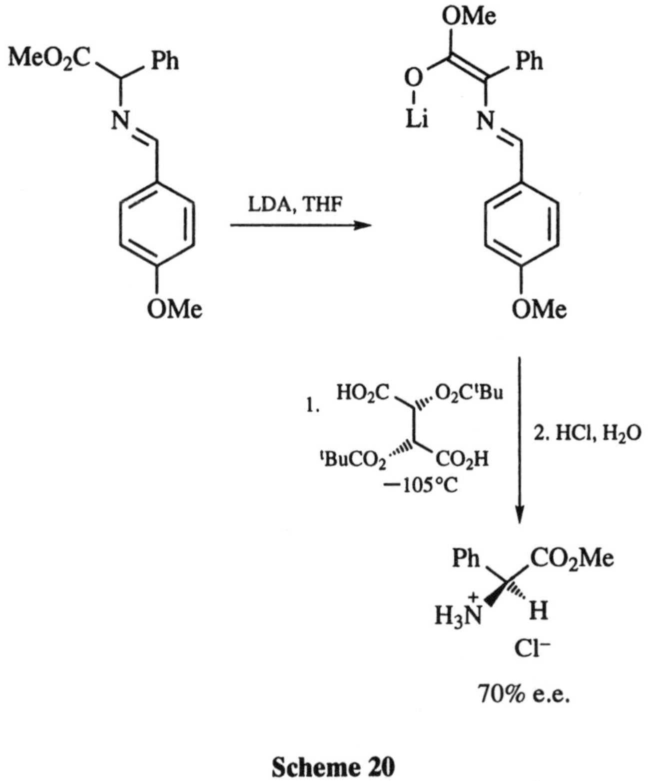

Scheme 20   (continued)

Scheme 20 *(continued)*

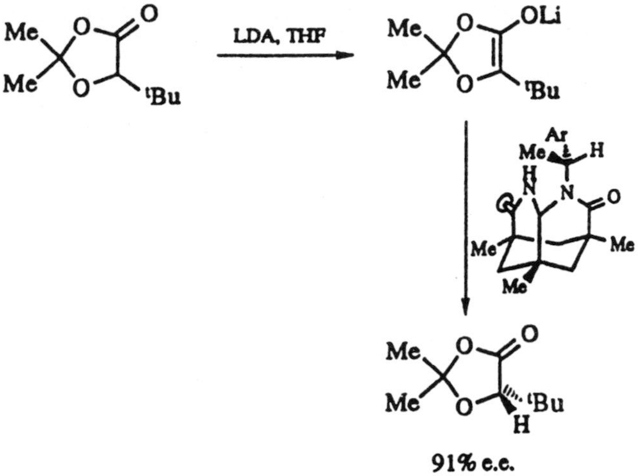

(b)

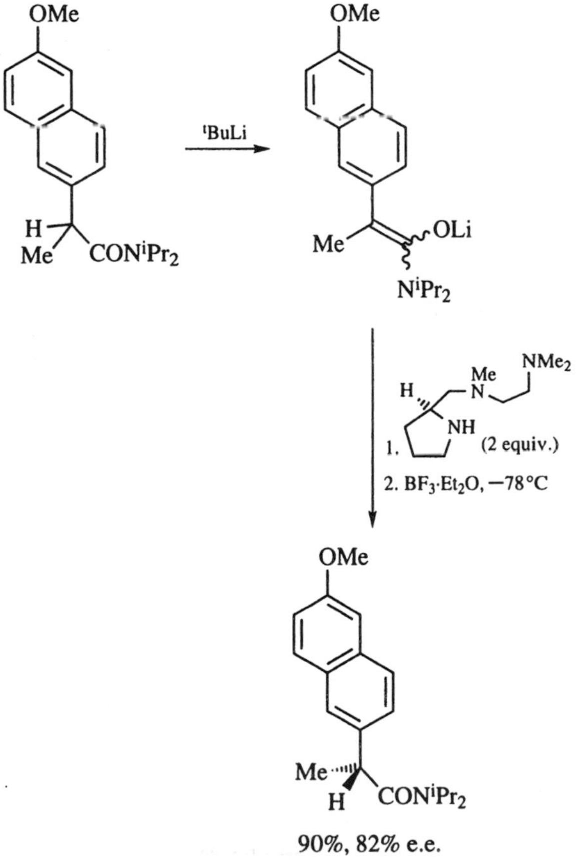

91% e.e.

(c)

90%, 82% e.e.

Scheme 20

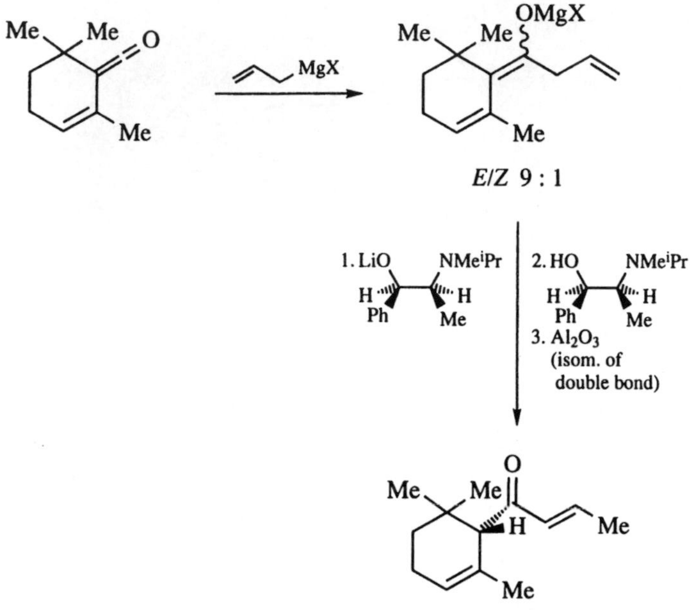

E/Z 9 : 1

(S)-α-damascenone
84% e.e.
(98% e.e. after crystallisation)

Scheme 21

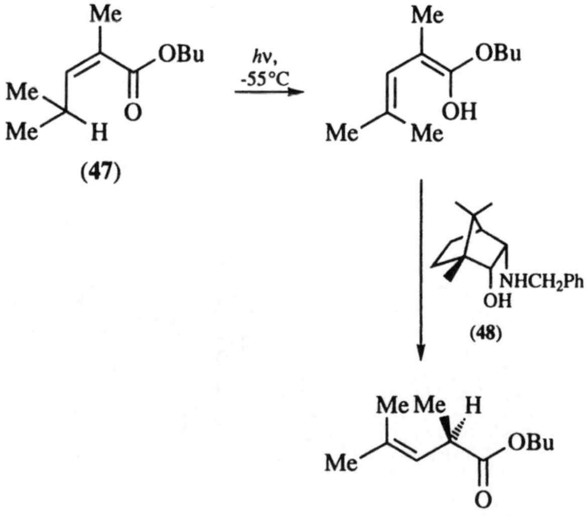

(47)

(48)

(49) 91% e.e.

Scheme 22

Summary

Although there are some highly stereoselective Type III$_{r.c.}$ reactions in which it appears the enantiofaces of the substrate became diastereofaces only in the transition state, there are many more in which preliminary complexation takes place between the substrate and the reagent. As a result of this complexation, the two enantiofaces of the substrate have already become diastereofaces when one of them is attacked in the product-forming step. When the *mechanism* of many of these more common Type III$_{r.c.}$ reactions is considered, therefore, it is clear that diastereoselectivity is brought about in a manner similar to that of intramolecular Type III$_{s.c.}$ reactions with the advantages in stereoselectivity that intramolecularity often confers (see Chapter 11). In overall enantioselective Type III$_{r.c.}$ reactions, the enantiopure reagent, containing chiral substituents or ligands must be used in at least stoichiometric amount.

Reduction of ketones, allyl, alkynyl and allenyl additions of achiral but prochiral aldehydes are all overall enantioselective reactions which involve preliminary complexing of the aldehyde or ketone with the enantiopure reagent. Many Michael additions proceed likewise. Epoxidation of alkenes using enantiopure oxaziridines appears to be an example in which enantioselection takes place in the single transition state in which the oxygen is transferred.

Deracemisation of a carbonyl compound that has a chiral C-centre at the α-position and which is enolisable towards this carbon can be accomplished in particular cases by the use of enantiopure acids to protonate the enolate.

References

1. H. Sakuraba and S. Ushiki, *Tetrahedron Lett.*, 1990, **31**, 5349.
2. J.W. ApSimon and R.P. Seguin, *Tetrahedron*, 1979, **35**, 2797; J.W. ApSimon and T.L. Collier, *Tetrahedron*, 1986, **42**, 5157.
3. H.C. Brown, W.S. Park, B.T. Cho and P.V. Ramachandran, *J. Org. Chem.*, 1987, **52**, 5406.
4. M.M. Midland and J.I. McLoughlin, *J. Org. Chem.*, 1984, **49**, 1316.
5. H.C. Brown, G.W. Kramer, A.B. Levy and M.M. Midland, *Organic Synthesis via Boranes*, Wiley–Interscience, New York, 1975.
6. H.C. Brown and B. Singaram, *Acc. Chem. Res.*, 1988, **21**, 287; H.C. Brown, V.K. Mahindroo, N.G. Bhat and B. Singaram, *J. Org. Chem.*, 1991, **56**, 1500, and references cited therein.
7. H.C. Brown, M.C. Desai and P.K. Jadhav, *J. Org. Chem.*, 1982, **47**, 5065; H.C. Brown and J.V.N.V. Prasad, *J. Am. Chem. Soc.*, 1986, **108**, 2049; H.C. Brown, P.K. Jadhav and A.K. Mandal, *J. Org. Chem.*, 1982, **47**, 5074.
8. H.C. Brown, P.K. Jadhav and M.C. Desai, *Tetrahedron*, 1984, **40**, 1325.
9. S. Masamune, B.M. Kim, J.S. Petersen, T. Sato, S.J. Veenstra and T. Imai, *J. Am. Chem. Soc.*, 1985, **107**, 4549.
10. U.S. Racherla and H.C. Brown, *J. Org. Chem.*, 1991, **56**, 401.
11. Ref. 7 cited in ref. 10.

12. E.J. Corey, C.-M. Yu and D.-H. Lee, *J. Am. Chem. Soc.*, 1990, **112**, 878.
13. F.A. Davis and A.C. Sheppard, *Tetrahedron*, 1989, **45**, 5703.
14. R.D. Bach and G.J. Wolber, *J. Am. Chem. Soc.*, 1984, **106**, 1410.
15. F.A. Davis, R.T. Reddy, W. Han and P.J. Carroll, *J. Am. Chem. Soc.*, 1992, **114**, 1428.
16. B.E. Rossiter and M. Eguchi, *Tetrahedron Lett.*, 1990, **31**, 965; B.E. Rossiter *et al.*, *Tetrahedron Lett.*, 1991, **32**, 3973.
17. K. Tanaka, H. Ushio and H. Suzuki, *Chem. Commun.*, 1990, 795, and see further references to this area therein.
18. E. J. Corey, R. Naef and F.J. Hannon, *J. Am. Chem. Soc.*, 1986, **108**, 7114.
19. K. Tomioka, M. Shindo and K. Koga, *J. Am. Chem. Soc.*, 1989, **111**, 8266.
20. L. Duhamel, P. Duhamel, S. Fouquay, J.J. Eddine, O. Peschard, J.C. Plaquevent, A. Ravard, R. Solliard, J.-V. Valnot and H. Vincens, *Tetrahedron*, 1988, **44**, 5495.
21. D. Potin, K. Williams and J. Rebek, *Angew. Chem. Int. Ed. Engl.*, 1990, **29**, 1420.
22. E. Vedejs and N. Lee, *J. Am. Chem. Soc.*, 1991, **113**, 5483.
23. C. Fehr and J. Galindo, *J. Am. Chem. Soc.*, 1988, **110**, 6909.
24. O. Piva and J.-P. Pete, *Tetrahedron Lett.*, 1990, **31**, 5157.

13 TYPE II/III REACTIONS AND THOSE INVOLVING CHIRALITY TRANSFER

Type II/III reactions (reactions showing simple diastereoselectivity and high levels of asymmetric induction)

In Chapters 6–8 we introduced Type II reactions. The so-called simple diastereoselectivity resulting from these Type II reactions was described as either inherent or occasional. Thus in the Diels–Alder reaction in Scheme 1, the diene and dienophile are prochiral and the *trans* relationship of x and y in the dienophile is retained in both diastereoisomeric products (**1**) and (**2**). This inherent diastereoselectivity (stereospecificity) arises from the (suprafacial) *syn* addition of the diene to the dienophile. Likewise, the *cis* relationship of a and b in (**1**) and (**2**) results from the (suprafacial) *syn* addition of the dienophile to the diene.

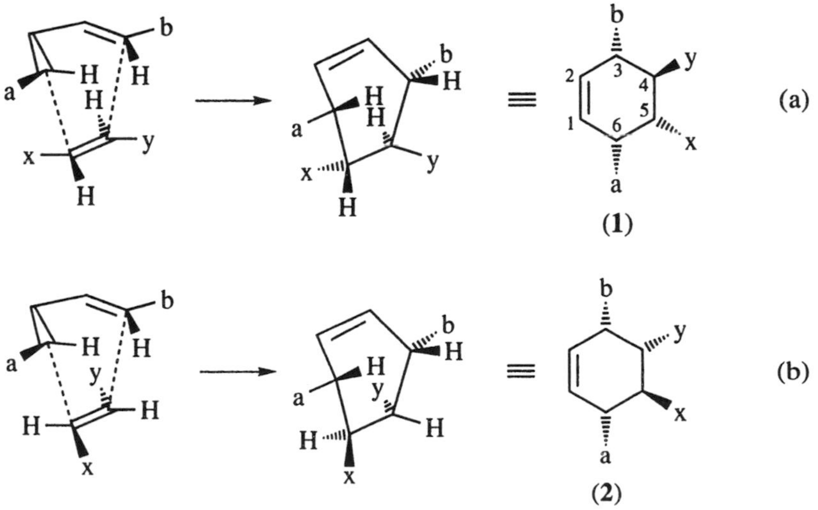

(a)

(**1**)

(b)

(**2**)

Scheme 1

Diastereoisomers (1) and (2) (*exo/endo* diastereoisomers) differ in the relative configuration at C-3 and C-4 and at C-5 and C-6. In a highly diastereoselective Type II reaction, (1) is produced in large excess over (2) or vice versa. Complete diastereoselectivity in this sense is distinguishable from the inherent diastereoselectivity referred to above. It is only occasionally found in the totality of Diels–Alder reactions whereas the inherent diastereoselectivity is found in all of them.

For many Type II reactions, the resulting diastereoselectivity is wholly inherent as, for example, in the *syn* addition of oxygen to an alkene in epoxidation using a peroxy acid. For Some Type II reactions, e.g. Michael additions, the diastereoselectivity is wholly occasional. With other Type II reactions, e.g. aldol condensations, the diastereoselectivity is also occasional but with certain combinations of carbonyl and metal enolate, particular transition state geometries may be preferred, leading to predictable diastereoselectivity.

Although (1) and (2) in Scheme 1 are drawn as single enantiomers, the product from a Type II reaction is always racemic. Scheme 2 represents a Diels–Alder addition closely resembling that in Scheme 1 (a); when the group x is chiral (x*) and if the dienophile is enantiopure, the two products (3) and (4) from this reaction, are now diastereoisomeric since they have the

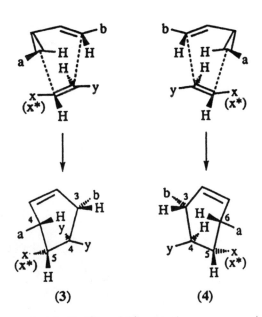

x (achiral) (3) and (4) enantiomers
x* (chiral) (3) and (4) diastereoisomers

Scheme 2

same configuration at x* but opposite configurations at the four other chiral centres created in the cycloaddition. The two transition states by which (**3**) and (**4**) are formed will generally not be isoenergetic and if they differ sufficiently in energy, (**3**) may be formed to the exclusion of (**4**), or vice versa.

The situation here closely resembles that which characterises substrate-controlled Type III reactions: the dienophile has diastereofaces by virtue of the presence of the chiral centre x* and diastereoisomers result from attack on these two faces by the achiral reagent (the diene). The reaction in Scheme 2 differs from the majority of Type III$_{s.c.}$ reactions considered previously (Chapters 9–11), however, because the reagent in the present case is prochiral and gives rise to chiral centres in the product at C-3 and C-6 in addition to those resulting from the dienophile at C-4 and C-5; in Type III$_{s.c.}$ reactions the achiral reagent is not prochiral. In the present case, therefore, we have elements of reagent control (in attack on the prochiral diene) and substrate control (in attack on the prochiral dienophile).

The Diels–Alder reaction in Scheme 2 (x = x*) is in fact a *Type II/III*:† the *relative* configuration at the newly created centres in the product is controlled by the Type II inherent and occasional diastereoselectivity also present in Scheme 1 (a) but their *absolute* configuration, i.e. whether (**3**) or (**4**) is produced in excess, is controlled by a Type III reaction mediated by the chirality of x*.

In the most useful versions of this Type II/III reaction, the dienophile bearing x* is used, as above, in enantiopure form. However, it is important to appreciate that, as in a Type III$_{s.c.}$ reaction, a completely diastereoselective Type II/III reaction does not demand this. If the dienophile were used in racemic form then one enantiomer would give rise to the absolute configuration of the newly created centres as illustrated in Scheme 2 but the other enantiomer of this dienophile would produce the *same* diastereoisomer in which every newly created chiral centre had the opposite configuration to that shown, i.e. the product would be racemic. Thus the product of a Type II/III reaction can be racemic or have any degree of enantiopurity depending on the enantiopurity of the existing chiral centre.

We have illustrated Type II/III by reference to a Diels–Alder reaction containing a chiral substituent on the dienophile, but the effect of introducing a chiral centre into any of the reactants (including any catalyst) of a Type II reaction is to convert it into a Type II/III.

Thus the two products from the [3,3] sigmatropic (Cope) rearrangement

---

†The Type III component of this reaction has Type III$_{s.c.}$ and Type III$_{r.c.}$ elements (see previous paragraph), but this will be abbreviated to Type II/III in this chapter.

in Scheme 3 or the electrocyclic ring closure in Scheme 4 are also diastereoisomeric as a result of introduction of the chiral centre x*. Aldol reactions in which either the aldehyde/ketone component or the enolate contains a chiral centre (see below) are Type II/III whether the

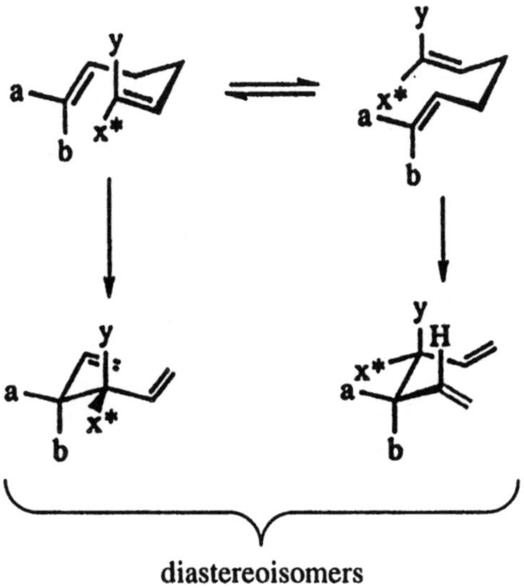

diastereoisomers

Scheme 3

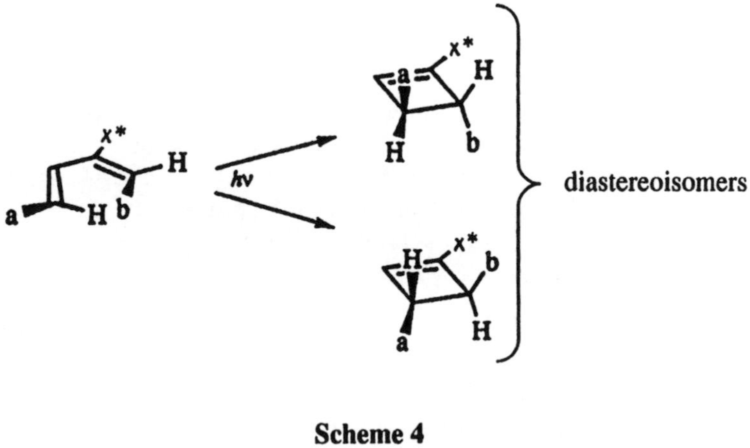

diastereoisomers

Scheme 4

reaction proceeds via a closed (cyclic) transition state or an open one (see Chapter 8).

To account for the stereochemistry of Type II/III reactions, therefore, it is necessary to identify the component Type II reaction and to decide which of the two enantiomeric transition states for this Type II reaction is favoured by involvement of the chiral centre—the Type III component. This procedure succeeds because the factors which bring about inherent Type II diastereoselectivity are not affected by inclusion of a chiral centre in the reaction ensemble. A change in occasional Type II diastereoselectivity may result from the inclusion of a chiral centre in one of the reactants or the catalyst in the reaction, but this is not usually a result of the chirality *per se*, i.e. the same changes may be brought about by similar changes in substitution which do not involve formation of a chiral centre and for which allowance can be made.

How does the chirality of x* in Schemes 2–4 allow differentiation between two otherwise enantiomeric transition states of a Type II? This is much the same question as we have attempted to answer for the Type III$_{s.c.}$ reactions discussed in Chapter 9–11. Indeed, in those chapters, a number of examples were in fact Type II/III. These included epoxidations and reactions of alkenes proceeding via 'onium ions, resulting in the formation of two additional chiral centres: diastereoface selection in Diels–Alder reaction mediated by σ–π interaction coupled with steric effects from substituents in the allylic position of the diene (Chapter 10, Schemes 9 and 10), and some intramolecular reactions, including those in which the two components become appended, *in situ*, during the reaction (Chapter 11).

Briefly, the diastereoselectivity of these Type III$_{s.c.}$ reactions arises from a combination of steric and electronic effects resulting from the presence of one or more existing chiral centres and mediated by conformation control. In diastereoselective Type II/III reactions, incorporation of the Type II component imposes additional conformational restraint on the transition state geometry of the reaction which can often facilitate high levels of asymmetric induction.

The Type III element of the diastereoselective Diels–Alder reaction in Scheme 5 is assumed to be the result of simple steric interaction between the *tert*-butyl group on the diene and the chiral group on the nitrogen of the maleimide dienophile.[1] In particular it is assumed that a phenyl group is larger than methyl and occupies the sterically least encumbered position: (**5**) is preferred to (**6**) because the H–^{t}Bu interaction in (**5**) is less than the Me–^{t}Bu interaction in (**6**).

Note that in (**5**), the *endo* transition state which maleimide normally prefers is apparently retained in spite of the steric interaction noted above. We can deduce that this is so because no diastereoselectivity would be likely in the corresponding *exo* transition state.

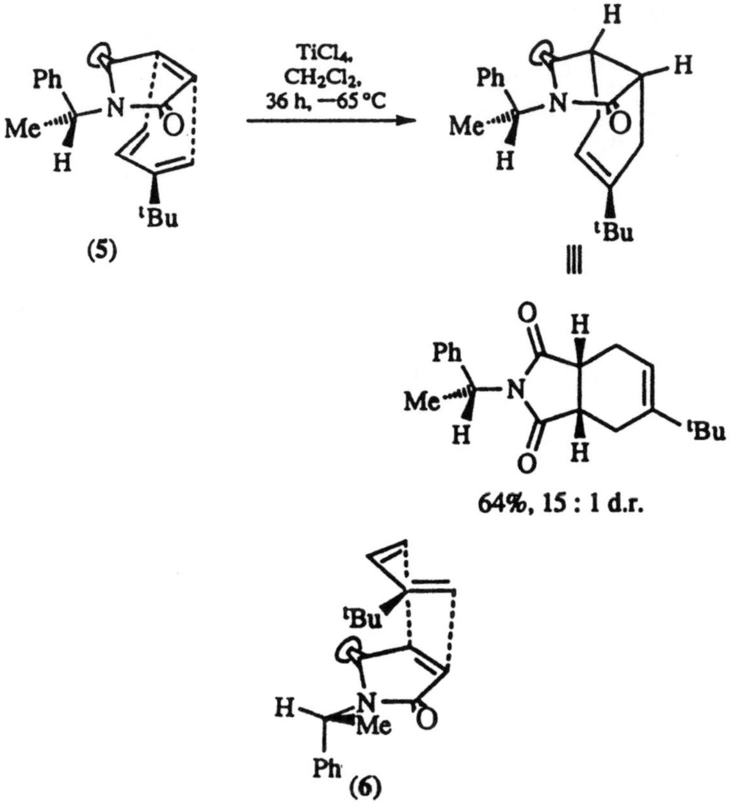

Scheme 5

In the *endo* transition state in Scheme 5, the proximity of the *tert*-butyl group in the diene to the chiral centre in the dienophile is inevitable. For most Diels–Alder reactions, however, some additional conformational restriction in some part of the substituent containing the chiral centre on the diene or dienophile is required before (steric or electronic) effects which mediate the Type III element of the reaction are brought into play. For example, in Scheme 6, cycloaddition of the 2-aza-1,3-diene and a chiral acylnitroso compound as the dienophile has been used in a synthesis of enantiopure α-amino acids.[2]

The transition state depicted in (7) accounts for the sense of the cycloaddition diastereoselectivity in terms of the lesser steric interaction between R and a methoxymethylene substituent of the auxiliary than is the case in the alternative transition state (8). However, this differentiation

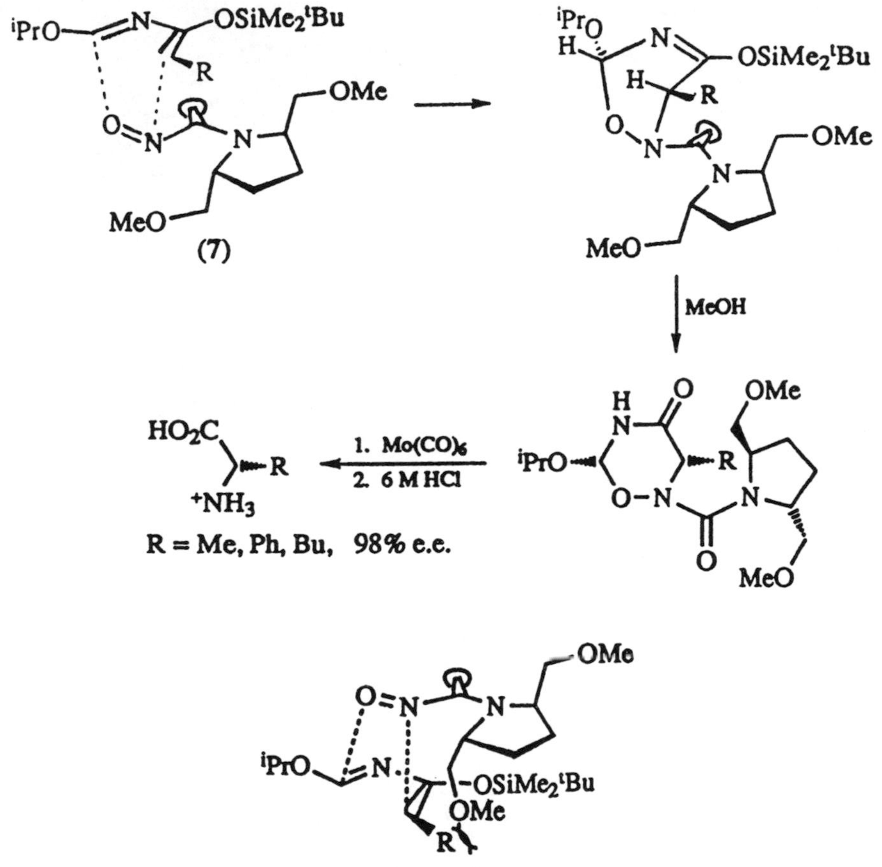

Scheme 6

depends on a number of conformational factors including (a) coplanarity of nitroso and carbonyl groups, (b) *endo* orientation for the carbonyl group and (c) amide resonance which orientates the pyrrolidine ring, and hence its two methoxymethylene substituents, appropriately. Note that because of the C_{2v} symmetry of the chiral pyrrolidine auxiliary, a 180° rotation around the (pyrrolidine) N—CO bond does not change the transition state geometry.

The diene (9) (Trost's diene) in Scheme 7 has been found to add to dienophiles e.g. acrolein to give (10) diastereoselectivity using Lewis acid catalysis. This selectivity is believed to arise from a conformational preference in which the phenyl group shields one face of the diene as shown.[3]

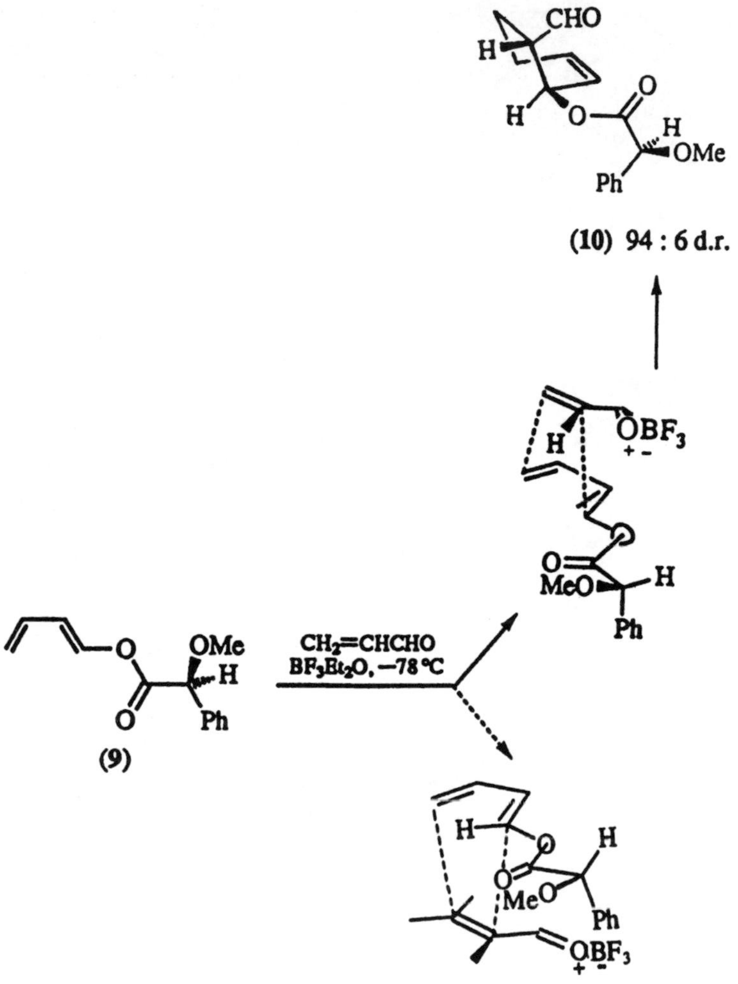

(10) 94 : 6 d.r.

Scheme 7

Here also a number of conformational requirements for the diene substituent are necessary before the phenyl group becomes orientated such that one particular diastereoface of the diene is shielded; they include an *s-cis* conformation for the ester and a *syn*-periplanar arrangement of the methoxy and ester carbonyl groups. The regioselectivity of the reaction, with the BF_3-coordinated acrolein carbonyl group vicinal to the ester in the transition state, will exacerbate the unfavourable interactions associated with attack from the more hindered face of the diene as shown.

In most cases, the supplementary conformational requirements are well known; in some cases their existence can only be inferred from the diastereoselectivity of the reaction! Calculations may help in predicting or confirming the existence of a previously unknown conformational preference as was the case in Scheme 6.

TYPE II/III 1,3-DIPOLAR CYCLOADDITIONS

The Type III 1,3-dipolar cycloaddition of nitrile oxides to monosubstituted alkenes bearing an allylic chiral centre have been studied by Houk *et al.*[4] The major diastereoisomer produced in the reaction shown in Scheme 8 is consistent with addition via the transition state (11) [= (13)] in which the larger group L (${}^tBu$) is *anti* to the 1,3-dipole and the medium-sized group M (MeO) is in the plane of the double bond and 'inside,' i.e. in what might appear a more hindered position. This appears to result in part from a greater repulsive interaction between the oxygen of the nitrile oxide and that of the alkoxy group when the latter is in the 'outside' position, as in (12).

In addition, the ${}^tBu$—C bond in Scheme 8 is a better σ-donor than the

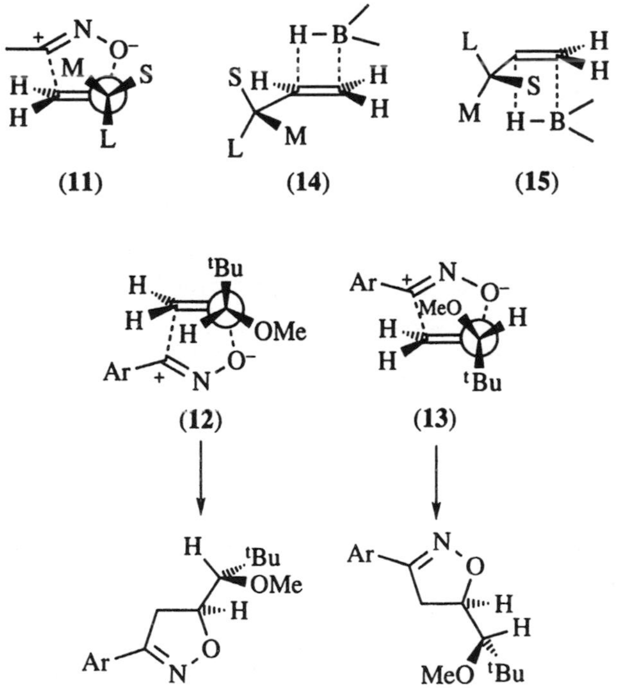

(11)   (14)   (15)

(12)   (13)

64%, > 93 : < 3

Ar = p-NO$_2$C$_6$H$_4$

Scheme 8

O—C bond from the C—OMe group. Since the cycloaddition is likely to be HOMO(alkene)–LUMO(1,3-dipole) controlled, it will be facilitated by location of ^{t}Bu—C bond in the *anti* position where its σ-donor effect raises the alkene HOMO level as much as possible. This accords with the conformation present in (13). The preference for the 'inside methoxy' in (13) is reminiscent of the transition state model for hydroboration of allylsilanes having a chiral centre adjacent to the double bond (see Chapter 10, Scheme 6); in Scheme 8 addition via (14) is sometimes favoured over (15) because the location of M in or close to the plane of the double bond, as in (14), facilitates attack on the face of the double bond which is shielded only by the smallest substituent S. [(14) and (15) are drawn with the M—C or S—C bond eclipsing the C=C bond but there may well be a small angle between the two bonds as in (11)].

In Scheme 8, both of the electronic effects referred to above (the σ–π interaction and the alkoxy–oxide repulsion) *and* the steric effect (*anti* location of ^{t}Bu) can all act in concert in (13). This will not always be the case, and a balance between the conflicting effects will be struck. For example, the Type II/III intramolecular cycloaddition of the nitrone and *cis*-alkene in Scheme 9 gives (19) highly diastereoselectively.[5] There are three identifiable effects in this cycloaddition, not all of which can be optimised in a single transition-state geometry: (a) putting the smallest group H_A 'inside' as in (16) or (17) would minimise $A_{1,3}$-strain with the *cis*-methylene group on the double bond; (b) any repulsion between the nitrone oxygen and the allylic alkoxy groups (cf. Scheme 8) would favour (16) or (18) over (17); (c) assuming that this cycloaddition, like that in Scheme 8, is HOMO(alkene)–LUMO(nitrone) controlled, it will be assisted by a σ-donor (CH_2O) bond in the allyl position as in (17) or (18). Since the product is (19), either (16) or (18) is the preferred transition-state geometry for the reaction; conceivably both are involved.

Many intramolecular cycloaddition reactions are controlled by the preferred location of a substituent on a chiral centre in the connecting tether. Examples of Type II/III reactions in this category can be found in Chapter 11. A further example is the thermal cyclisation of the 1,3-dipole (21) generated, *in situ*, by migration of the silyl group in (22), a reaction used in an early stage of a synthesis of phorbol (Scheme 10).[6] Here the Type III diastereoselectivity is the result of a chair motif for the tether with the methyl group at the existing chiral centre in an equatorial position. Note that this intramolecular addition proceeds with high Type II occasional *exo* selectivity (H_A *syn* to the carbonyl of the pyrone residue); the alternative *endo* transition state (23) would certainly involve severe interactions between the tether and the pyrone ring substituents. We have seen previously in Chapter 6 that intramolecularity in a Type II reaction can increase the likelihood of occasional diastereoselectivity as well as enforcing regioselectivity.

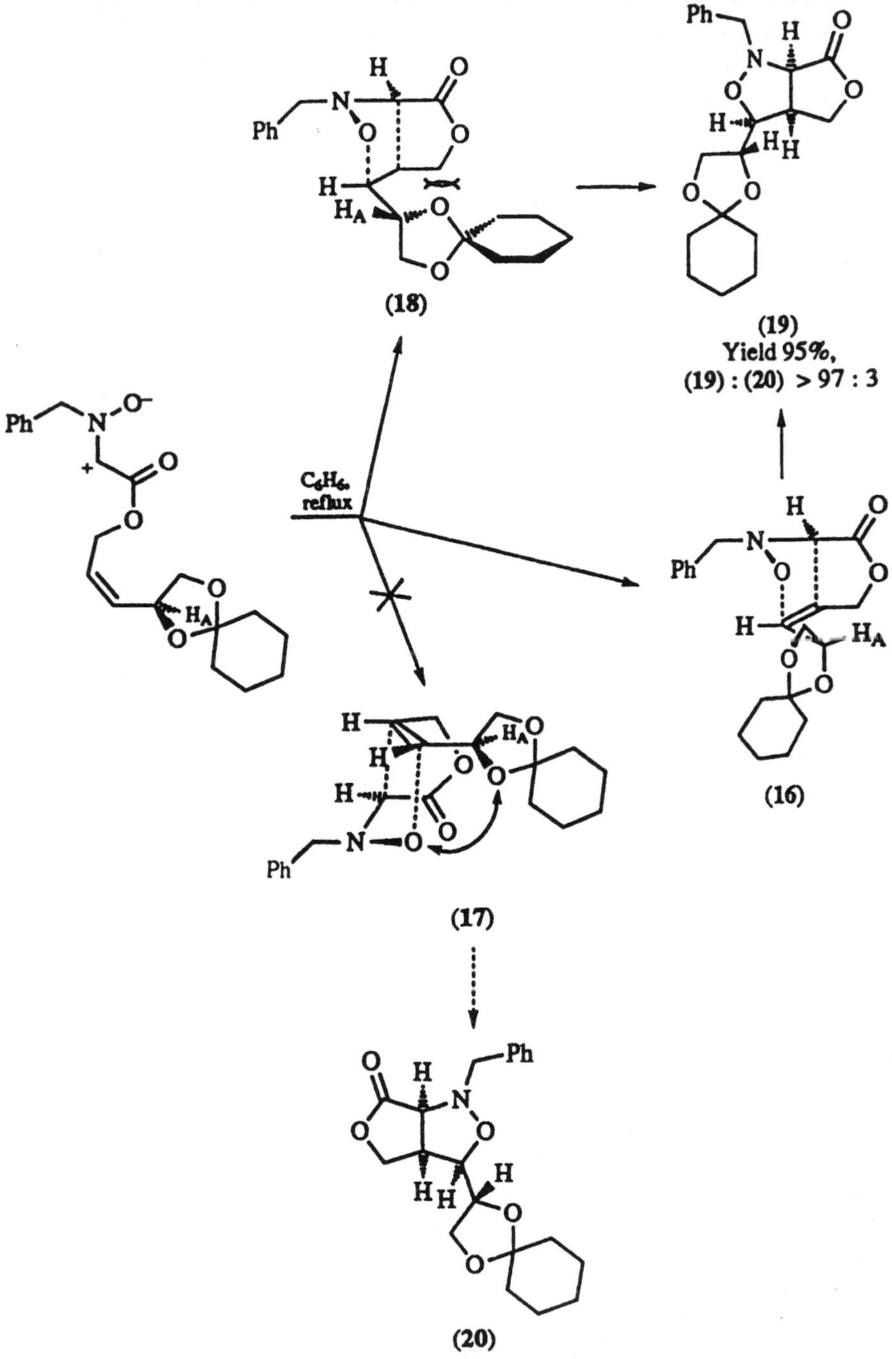

(18)

(19)
Yield 95%,
(19) : **(20)** > 97 : 3

C₆H₆,
reflux

(16)

(17)

(20)

Scheme 9

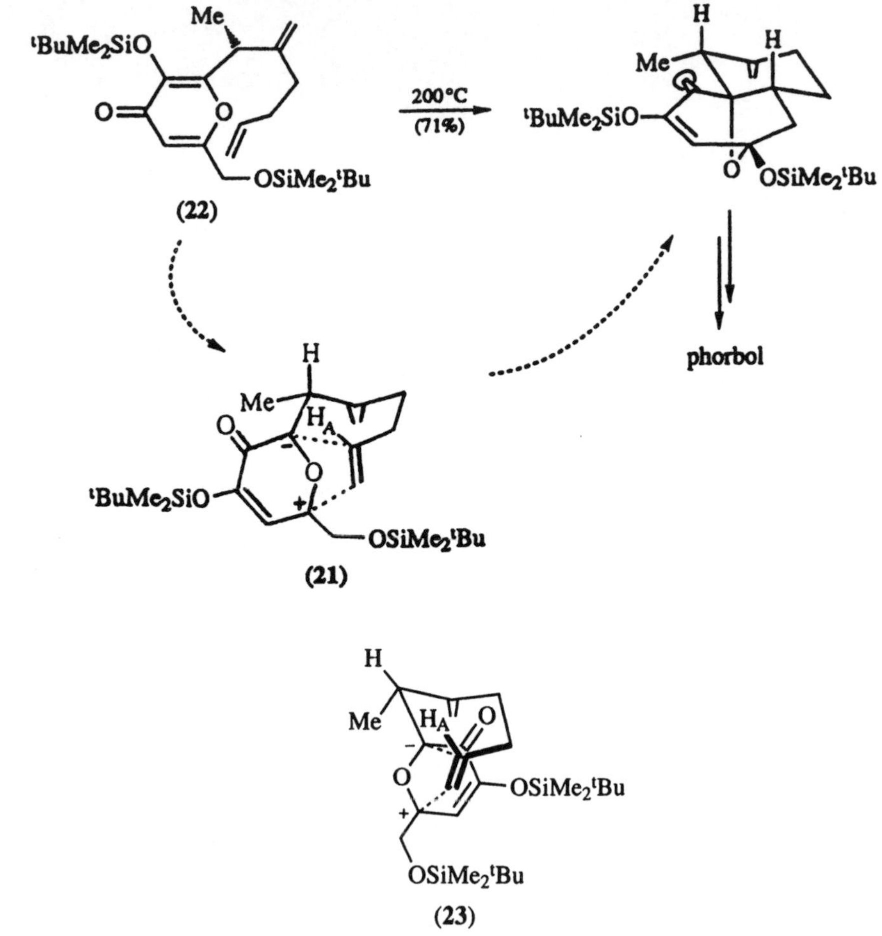

(22)

200 °C
(71%)

(21)

phorbol

(23)

Scheme 10

OTHER TYPE II/III REACTIONS

Scheme 11 illustrates an ene reaction in which chelation of the α-benzyloxy aldehyde with tin tetrachloride leads to a clear preference for attack on one face of the enophile carbonyl group.[7]

In this reaction, with R = H the product is assumed to arise from an *exo*-like transition state (24). However, because the ene component is not prochiral, the *endo*-type transition state (25) (R = H) would lead to the same product (26). When the ene component is prochiral (R = Me), the reaction becomes Type II/III; the occasional diastereoselectivity of the Type II with the expected preference for *exo* transition state (24) is manifest in the relative configuration at the addition chiral centre in (27).

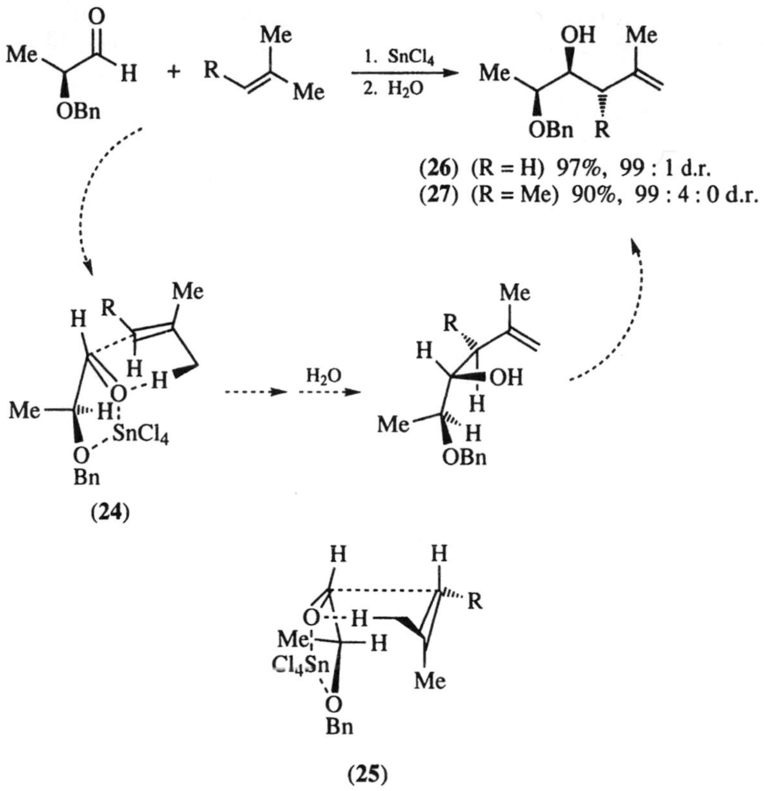
1. $SnCl_4$
2. H_2O
(26) (R = H) 97%, 99 : 1 d.r.
(27) (R = Me) 90%, 99 : 4 : 0 d.r.
H_2O
(24)
(25)

Scheme 11

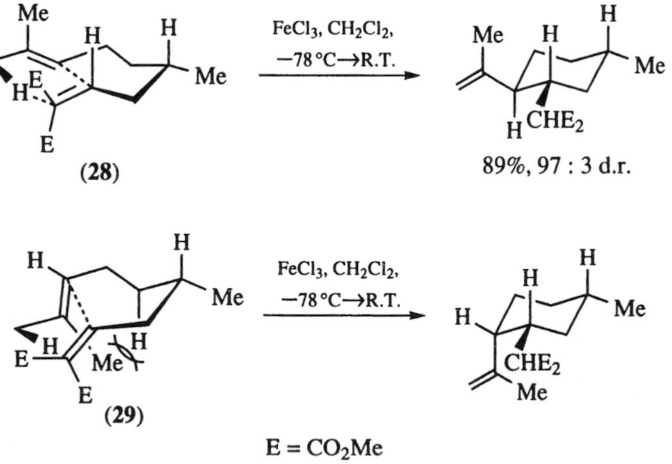
$FeCl_3$, CH_2Cl_2,
−78 °C→R.T.
(28)
89%, 97 : 3 d.r.
$FeCl_3$, CH_2Cl_2,
−78 °C→R.T.
(29)
E = CO_2Me

Scheme 12

In the Type II/III intramolecular ene reaction of the enantiopure diene (28) (Scheme 12), the absolute configurations of the two new chiral centres are controlled by the methyl group at the existing chiral centre, which occupies an equatorial position on a chair-type tether.[8]

In this reaction also, the occasional diastereoselectivity of the Type II element reflects in the preference for transition state (28) over (29) because of steric interaction of the $A_{1,3}$-type indicated in the latter.

For the related intramolecular ene reaction in Scheme 13 (illustrated for one enantiomer), the three-atom tether ensures that of the four possible

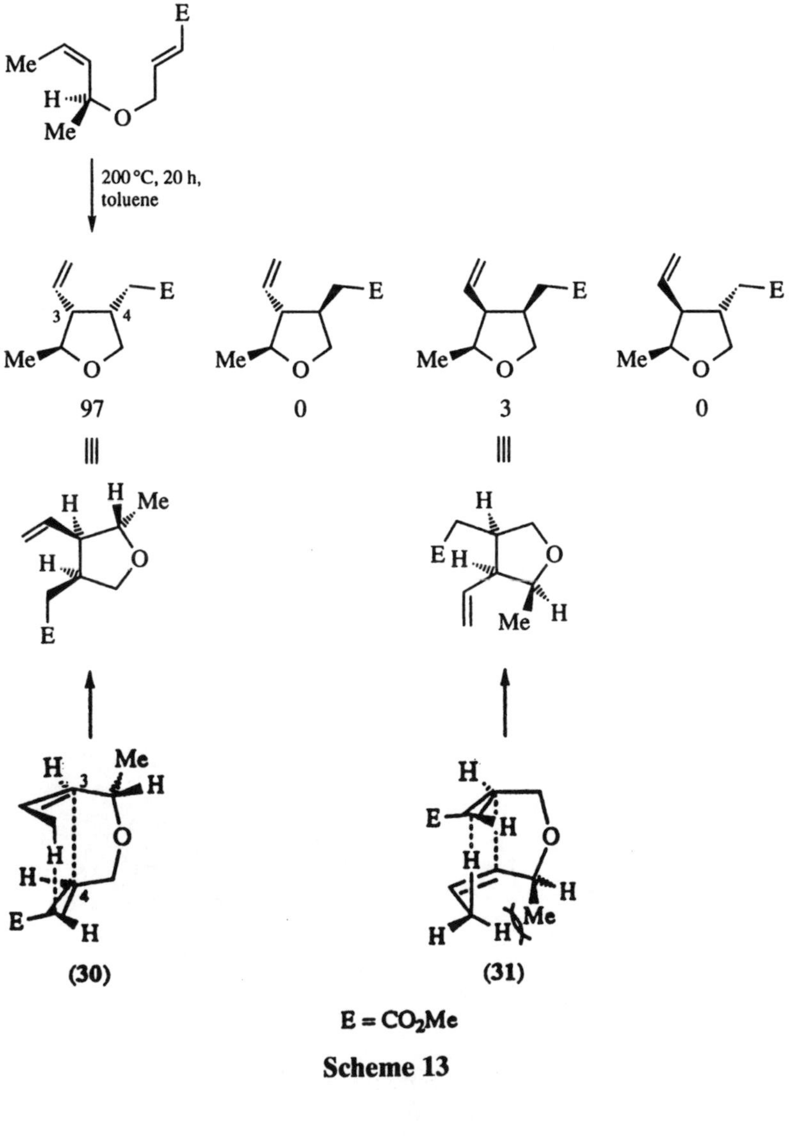

$E = CO_2Me$

Scheme 13

diastereoisomeric products, only those in which the groups on C-3 and C-4 are *cis* (occasional Type II diastereoselectivity) are formed. This is because the transition state energy for formation of the corresponding 3,4-*trans*-substituted diastereoisomers is raised considerably by the limited length of the tether. Of the two transition states in contention, (30) is favoured because it lacks the $A_{1,3}$-strain present in (31). Transition state (30) also accounts for the observed insensitivity of the diastereoselectivity to the configuration of the ester-substituted double bond.[9]

TYPE II/III ALDOL REACTIONS

Aldol reactions of chiral enolates with prochiral aldehydes fall into this category. Likewise, the reaction of an aldehyde containing a chiral centre (usually the α-position) with a prochiral enolate may also show diastereoselectivity arising both from the Type II and Type III elements of the reaction.

Type II/III aldols via closed transition states

Heathcock and co-workers[10] have shown (Scheme 14) that reaction of the enantiopure (Z)-enolate (32) with aldehydes gives the product (33) with high

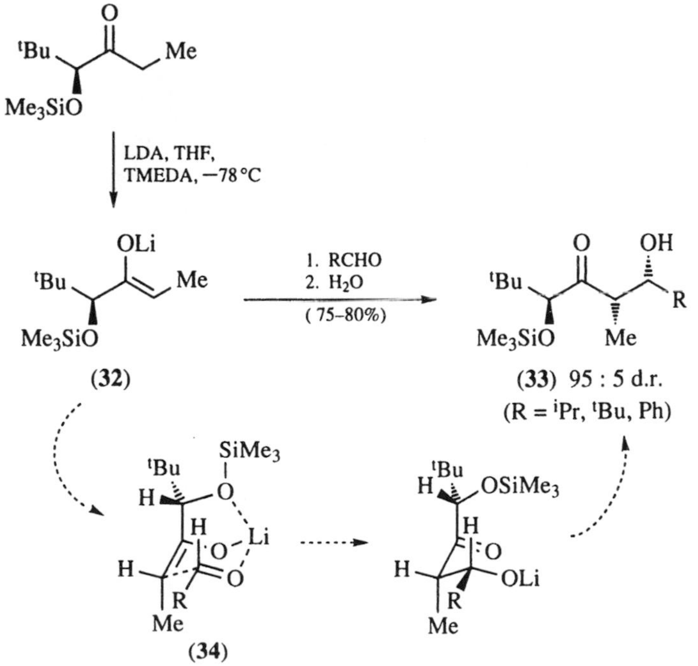

Scheme 14   *(continued)*

Scheme 14 *(continued)*

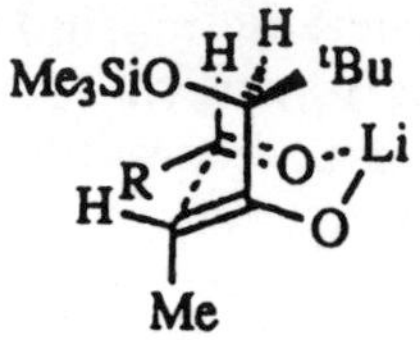

Me₃SiO H H ᵗBu
       O Li
   H R   O
      Me

(35)

Scheme 14

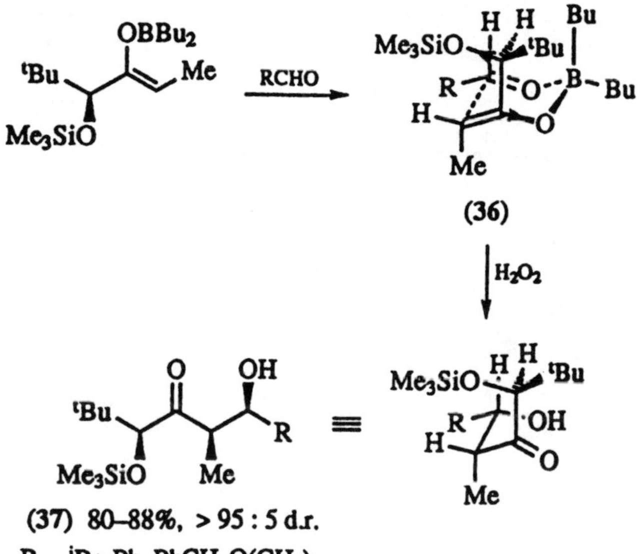

'Bu   OBBu₂
        Me   RCHO
Me₃SiO

Me₃SiO H H Bu
          ᵗBu Bu
   R   O B Bu
   H   O
      Me

(36)

H₂O₂

'Bu   O   OH
            R   ≡
Me₃SiO Me

(37) 80–88%, > 95 : 5 d.r.
R = ᵗPr, Ph, PhCH₂O(CH₂)₂

Me₃SiO H H ᵗBu
          OH
   R   O
   H
      Me

ᵗBu SiMe₃
H   O   Bu
    H   Bu
   H   O B Bu
       O
   R
   Me

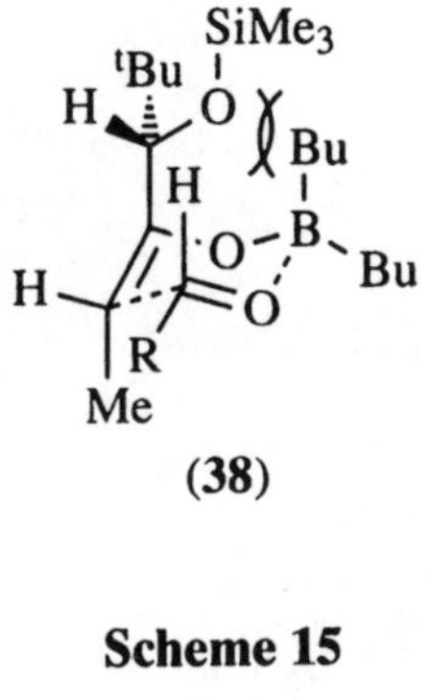

(38)

Scheme 15

diastereoselectivity. The diastereoselectivity can be dissected into its Type II and III elements. The relative configuration at the two created chiral centres (Type II) arises from a chair transition state in which the aldehyde group R occupies an equatorial position. The absolute configuration at those same centres (Type III) arises from the preferred attack on the less hindered enolate diastereoface; chelation of the lithium with the oxygen of the silyloxy group is assumed, as shown in (34). Transition state (35) would give rise to the same *relative* configuration at the two created chiral centres as (34) (Type II), but attack on this diastereoface of the enolate would not permit chelation given the (absolute) configuration of the existing chiral centre and the steric demands of the *tert*-butyl group.

The same reaction (Scheme 15), but with a boron enolate in place of the lithium enolate, shows the same Type II sense of diastereoselectivity but the opposite diastereoface of the prochiral enolate is attacked and a different diastereoisomer (37) is formed.[10]

The boron atom here is coordinatively saturated and therefore incapable of chelation with the silyloxy group, as was the case with the lithium enolate in (34). Instead, a butyl substituent on the sp^3-hybridised boron atom will destabilise transition state (38). Any dipolar repulsion between the two C—O bonds of the enolate in the preferred transition state (36) is also minimised in the conformation shown.

It is known that for aldol reactions proceeding via cyclic transition states, a change in configuration of the enolate double bond brings about a change in the sense of Type II diastereoselection. Thus the use of magnesium and titanium (E)-enolates (Scheme 16) in reactions analogous to those in Schemes 14 and 15 affords access to the two remaining diastereoisomers (40) and (41), respectively [contrast (33) and (37) above]. Chelation like that in (34) (Scheme 14) is believed to be present using the magnesium enolate but not using the triisopropoxy-substituted titanium enolate which reacts via a transition state resembling (36) in Scheme 15.

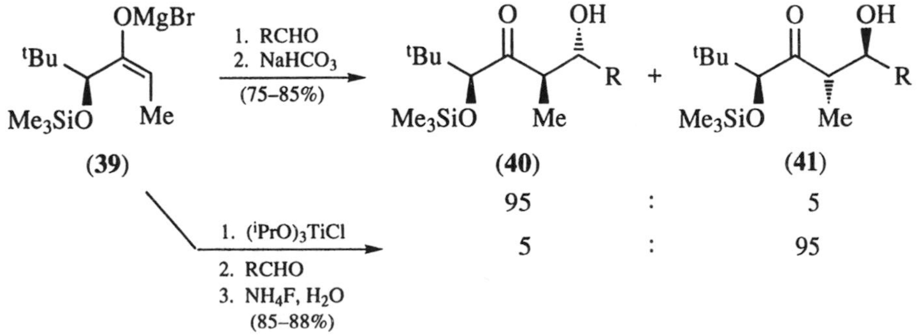

Scheme 16

Formation of the magnesium (E)-enolate (39) by deprotonation using the magnesium salt of 2,2,6,6-tetramethylpiperidine is thought to be the result of severe non-bonded interactions in the transition state leading to the (Z)-enolate (42) (Scheme 17) by comparison with that leading to the (E)-enolate (39).

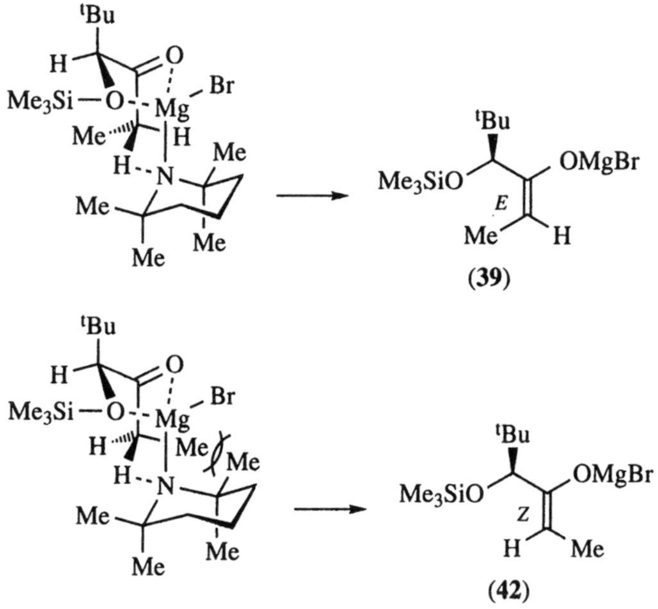

Scheme 17

Thus, starting with a single enantiopure ketone, each of the four possible diastereoisomeric aldol products can be obtained by permutation of the enolate double bond configuration and the diastereoface of the enolate which is attacked (Schemes 14–16).

Reaction as in (43) Scheme 18(a) of a boron enolate, bearing an Evans' chiral auxiliary, with a prochiral aldehyde [Scheme 18 (a)] is a Type II/III reaction. The sense of asymmetric induction here arises from the steric effect of the isopropyl substituent, with the preferred enamide conformation within the (Z)-enolate being as shown.[11]

Note that the opposite sense of induction would arise [Scheme 18 (b)] were the conformation around the N–C bond in the enamide as shown in (44). Nerz-Stormes and Thornton[12] have shown that, using the *titanium* enolate bearing the same auxiliary [Scheme 18 (c)], the sense of diastereoselectivity in the formation of the major product corresponds to that which would arise from (44). It is believed that the triisopropoxy-substituted titanium is able to chelate with the amide carbonyl of the auxiliary, leading to a preference for this alternative enamide conformation as in (45).

The transition-state structures in Scheme 18 do not satisfactorily account for the much reduced level of diastereoselectivity in either (a) or (c) when the propionamide-derived enolate is replaced by the acetamide-derived

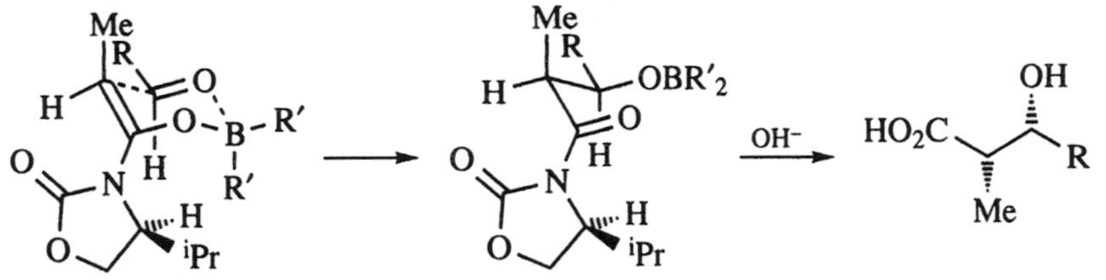

(43)   (a)

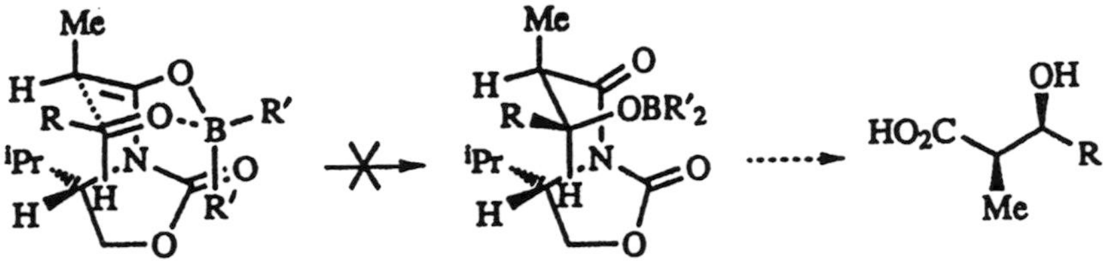

(44)   (b)

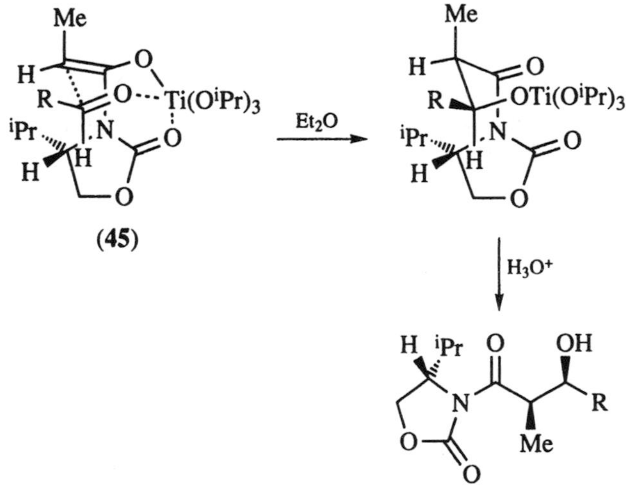

(45)

(c)

R = Ph, 75% (major diastereoisomers)
d.r. 92 : 5 : 3

Scheme 18   (continued)

Scheme 18 *(continued)*

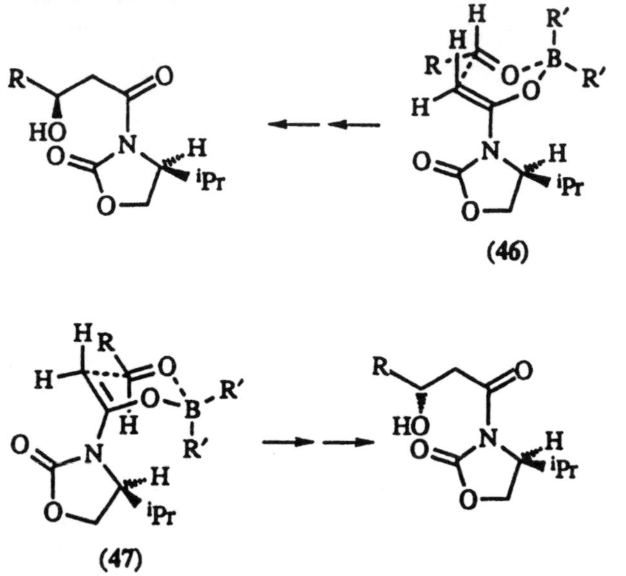

(46)

(47)

(d)

Scheme 18

enolate, i.e. the methyl group in (43) and (45) is replaced by hydrogen. The difference in energy between chair and boat transition states in these cyclic transition states is less than that between the chair and (skew) boat conformations of cyclohexane and it is possible that, in the absence of this methyl group, a (twist) boat transition state e.g. (46) (Scheme 18(d)) becomes comparable in energy to that of the chair (47), leading to an erosion of the diastereoselectivity.

Evans *et al.*[13] have prepared titanium (Z)-enolates (48) from (secondary) alkyl ethyl ketones by the action of titanium tetrachloride and tertiary amine bases and have reacted these with isobutyraldehyde (Scheme 19).

In this Type II/III reaction the Type III diastereoselectivity arises from a transition state (49) in which the largest substituent R_L is in the least hindered position and the distance between the medium-sized group R_M and the substituents on titanium is maximised. Type II (simple) diastereoselectivity arises as usual from the chair transition state with the isopropyl substituent in the equatorial position.

Turning to a Type II/III aldol reaction in which the chiral centre is the α-position of the aldehyde as opposed to contained in the enolate, Roush[14] analysed the reactions of boron (Z)- and (E)-enolates with chiral α-methylaldehydes (Scheme 20).

Attack by the enolate on the aldehyde via transition state (50) would

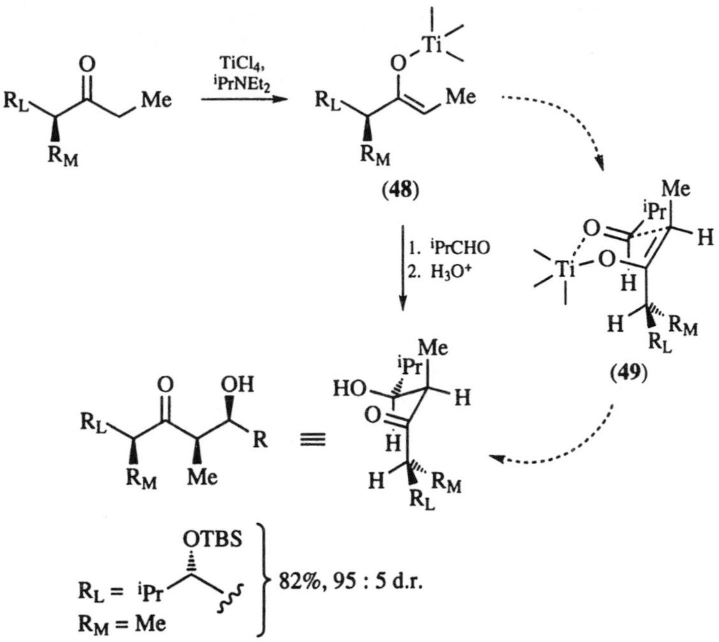

Scheme 19

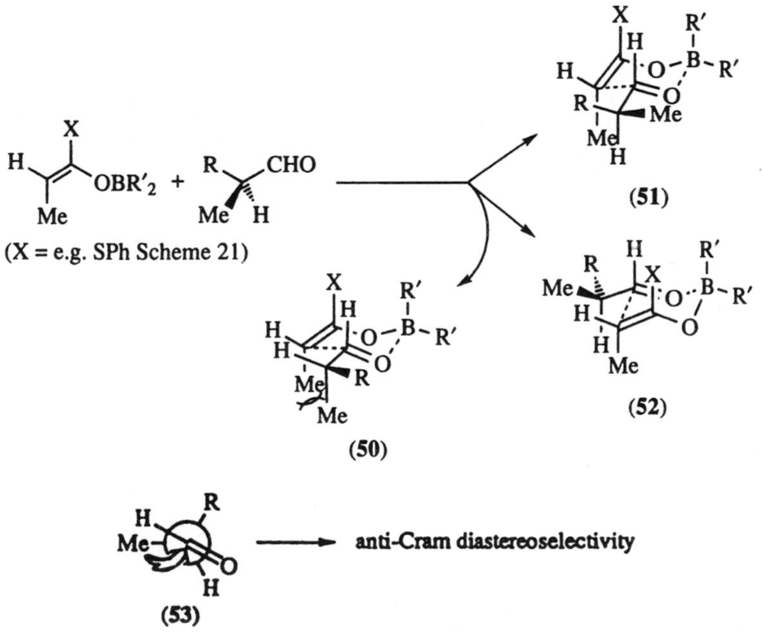

Scheme 20

involve a Me–Me (*syn*-pentane) interaction similar to that between two *cis*-1,3-diaxial methyl groups on a cyclohexane ring. In the transition state (51), this Me–Me interaction is replaced by the less severe R–H interactions. Transition state (52) will be preferred over (51) to the extent that R is larger than Me and the resulting sense diastereoselectivity is *anti*-Cram [cf. (53)] as is found in the reaction in Scheme 21.[15]

The situation using the *trans*-Me/OLi-configured enolate (54) (Scheme 22) is more straightforward and Cram diastereoselectivity prevails because the Me–Me interaction that was present in (50) (Scheme 20) is absent in (55).[14]

Using boron enolates, Type II/III aldol reactions can be carried out with the chiral centres contained in ligands on the boron. Corey's chiral controller group, the substituted *N,N*-disulphonyldiazaborolidine in (56) (Scheme 23), has been used to bring about the reaction of a ketone, as its boron enolate, with aldehydes, e.g. via (57), in the synthesis of the corn weevil aggregation pheromone sitophilure.[16]

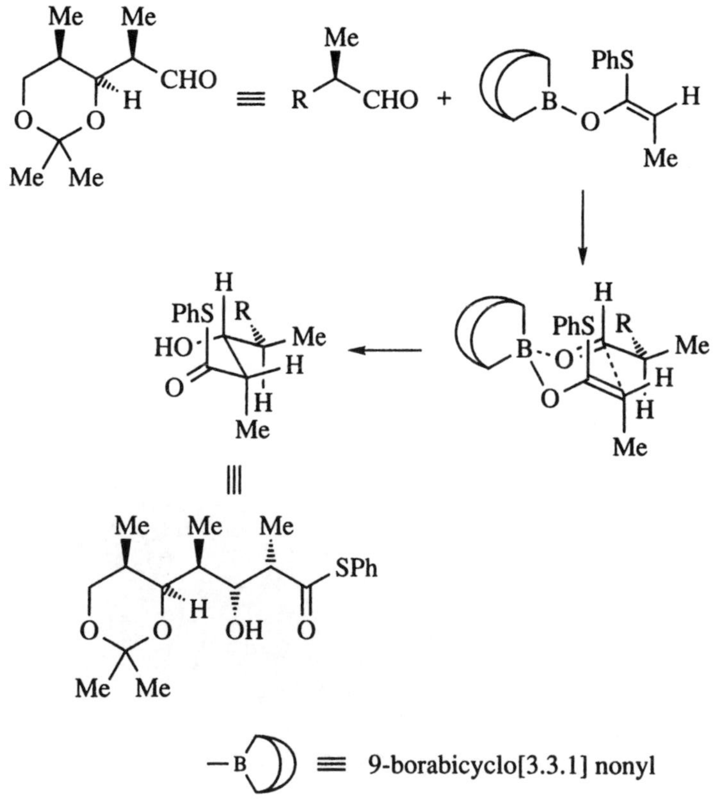

Scheme 21

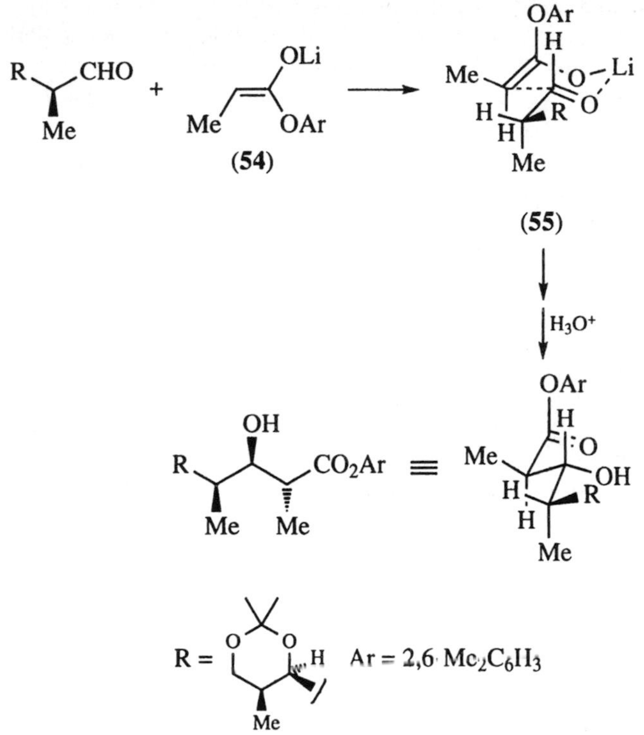

Scheme22

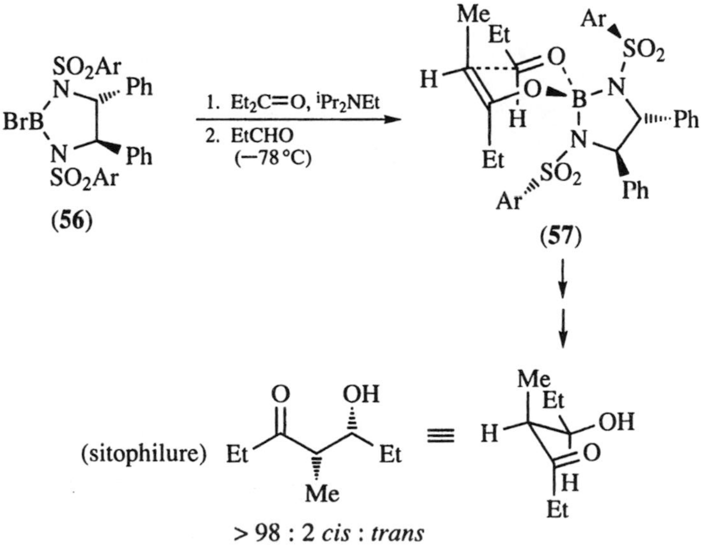

$> 98 : 2$ *cis* : *trans*
cis $> 98\%$ e.e.

Scheme 23

The stereochemistry of this aldolisation is assumed to result from a relayed steric effect from the phenyl rings on the diazaborolidine (cf. Chapter 12, Scheme 12); these direct the orientation of the neighbouring arenesulphonyl groups and hence the placement of enolate and aldehyde in the transition state (57) in Scheme 23. Since the boron, along with its chiral ligands, is lost in the work-up of the reaction, this is an overall highly enantioselective reaction (see Chapter 12) which also incorporates a Type II element; the diastereoselectivity resulting from the latter is also almost complete (98:2 *cis:trans*).

Type II/III reactions of alkenylmetals with aldehydes via closed transition states

Stereochemically, Type II reactions of many allyboronates and other allylmetallics with aldehydes and ketones closely resemble aldol condensations, as has been pointed out in Chapter 8. Likewise, Type II/III allylboronate reactions resemble their aldol analogues. In fact, the foregoing rationalisation of the *anti*-Cram Type II/III diastereoselectivity in Scheme 20 is analogous to

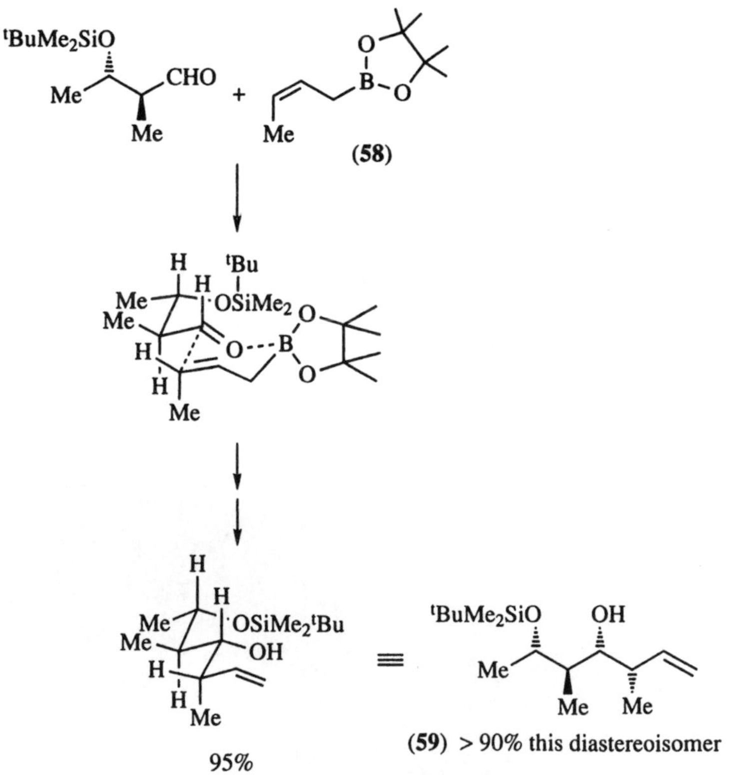

Scheme 24

that proposed by Evans *et al.*[17] to account for the stereochemistry of the reactions of (Z)-crotylboronates with aldehydes observed by Hoffmann and co-workers.[18] The latter workers used (Z)-crotylboronic esters, e.g. (58) (Scheme 24), and found that addition to α-chirally-substituted aldehydes goes with efficient asymmetric induction to give (59); this is *anti*-Cram, analogous to the Type II/III diastereoselectivity in Scheme 21.

Normally the reactions of prochiral ketones with alkenylmetals are not highly diastereoselective. In the intramolecular Sakurai reaction of the hydroxy ketone in Scheme 25 (a), Type II diastereoselectivity arises from a chair-transition state with a bridging covalent silicon–oxygen bond.[19] It is clear that the Type II/III version of this reaction shown will prefer to have the phenyl located in the *exo* position in the transition state as in (60), leading to predictable configurations at the new chiral centres in the product.

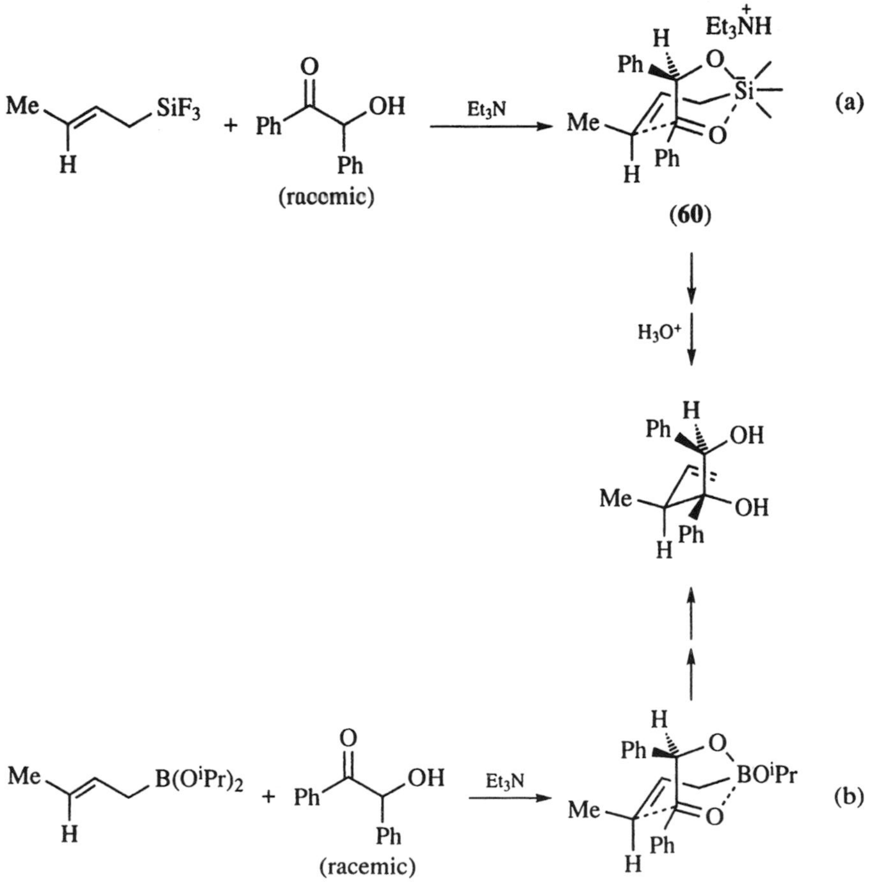

Scheme 25

A similar effect operates in the α-methallylation of α-hydroxy ketones using boranes [Scheme 25 (b)].[20]

Using allyl or crotylboron derivatives, chirality at the boron-substituted allyl position (a C atom) is possible as in (61) (Scheme 26), where it is not in boron enolates (allylic O atom). The Z-configured pentenylboronic ester (61) reacts with aldehydes via (62) rather than (63) because of the $A_{1,3}$-methyl–methyl interaction in the latter.[21]

Likewise crotyltitanium (64), prepared in enantiopurified form (Scheme 27) by virtue of the configurational stability of the oxygen-substituted sp^3-hybridised carbanion (see Chapter 4), reacts with isobutyraldehyde to give (65) with high diastereoselectivity and conservation of the enantiopurity of the crotyltitanium.[22] Although inclusion of the chiral centre within the chair, transition states in Schemes 25 and 26 means that these reactions no longer have a Type II component but are Type $III_{r.c.}$, nevertheless they clearly have similar features to the Type II/III reactions described previously.

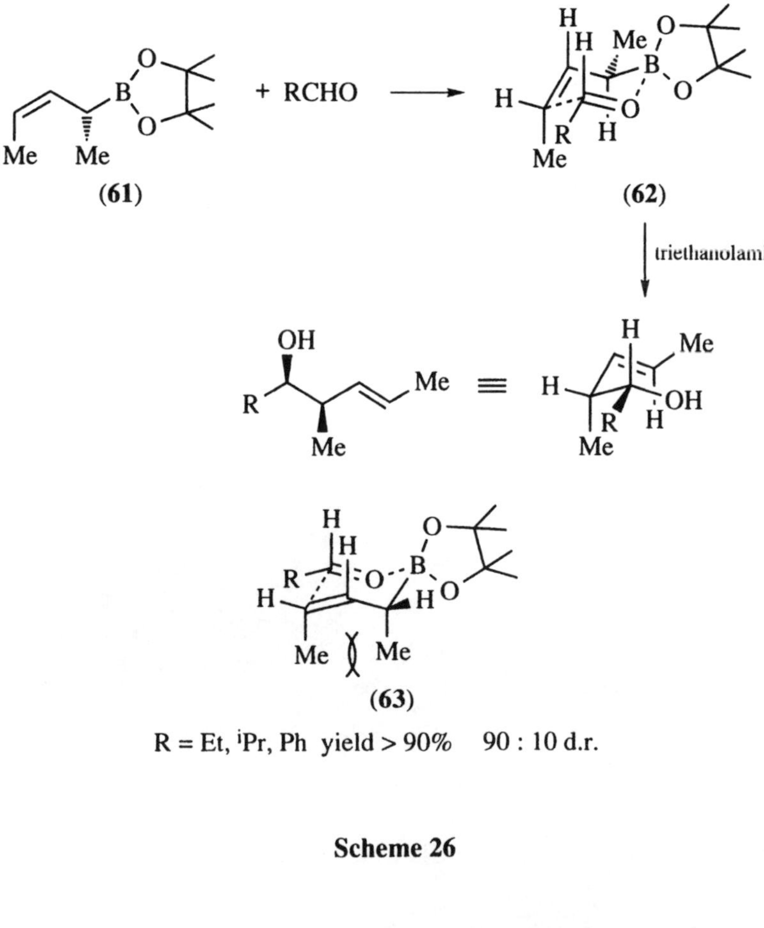

R = Et, ^iPr, Ph yield > 90%   90 : 10 d.r.

Scheme 26

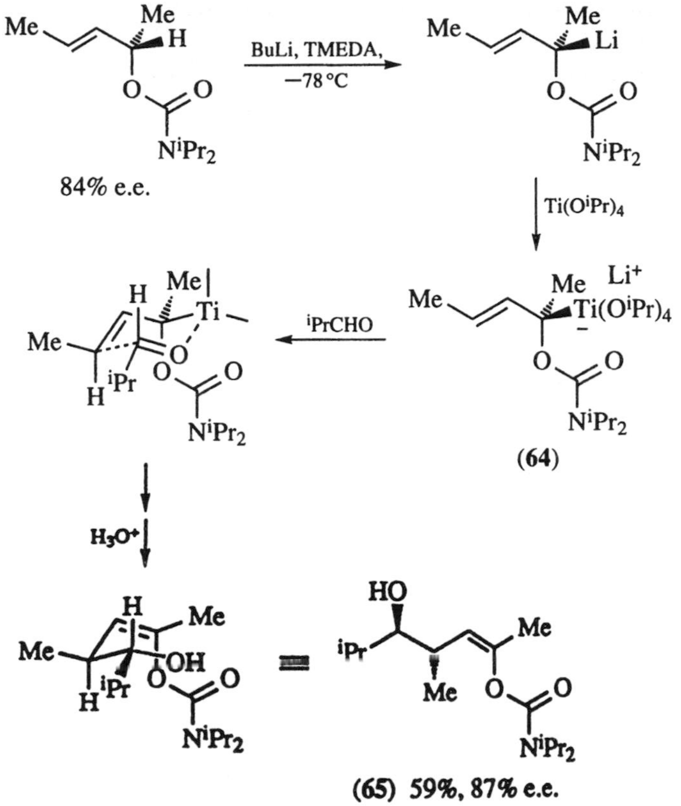

Scheme 27

Type II/III alkenylmetal reactions with aldehydes and aldol-type reactions: open transition states

Kumada and co-workers[23] have shown that the reactions in Scheme 28 proceed with high enantioselectivity using both (E)- and (Z)-crotylsilanes and also with high diastereoselectivity with the former. This is a Type II/III$_{s.c.}$ reaction. The enantioselectivity in both reactions arises from Type III$_{s.c.}$ chirality transfer (see Chapter 5 and below)—the requirement for *anti* addition to the C—Si bond in the S_E2' reaction with the phenyl group and hydrogen on the existing chiral centre occupying the sites shown in (66)–(69). The high (Type II) diastereoselectivity in reaction of the (E)-crotylsilane is a result of the preference for (66) over (67) in the open extended transition state, reminiscent of the Mukaiyama reaction (Chapter 8). By contrast, Type II diastereoselectivity is absent using the (Z)-crotylsilane because of (comparable) adverse interactions in both (68) and (69).

Note that the reaction in Scheme 28 produces two chiral centres in the product which are adjacent. It was pointed out in Chapter 12 that this is not easily accomplished using Type III$_{r.c.}$ addition of a chiral reagent $R_{xy}C^{\ddagger}$ to a double bond because of the lack of configurational stability of such species. The Type II/III reactions like that in Scheme 28 (and also the Type III$_{r.c.}$ reactions in Schemes 26 and 27) provide a partial solution to this problem.

Similar open transition states are implicated in silyl enol ether and crotylsilane addition to oxonium ions generated by treatment of aromatic aldehydes with boron trifluoride or treatment of their acetals with trimethylsilyl triflate (Chapter 8).

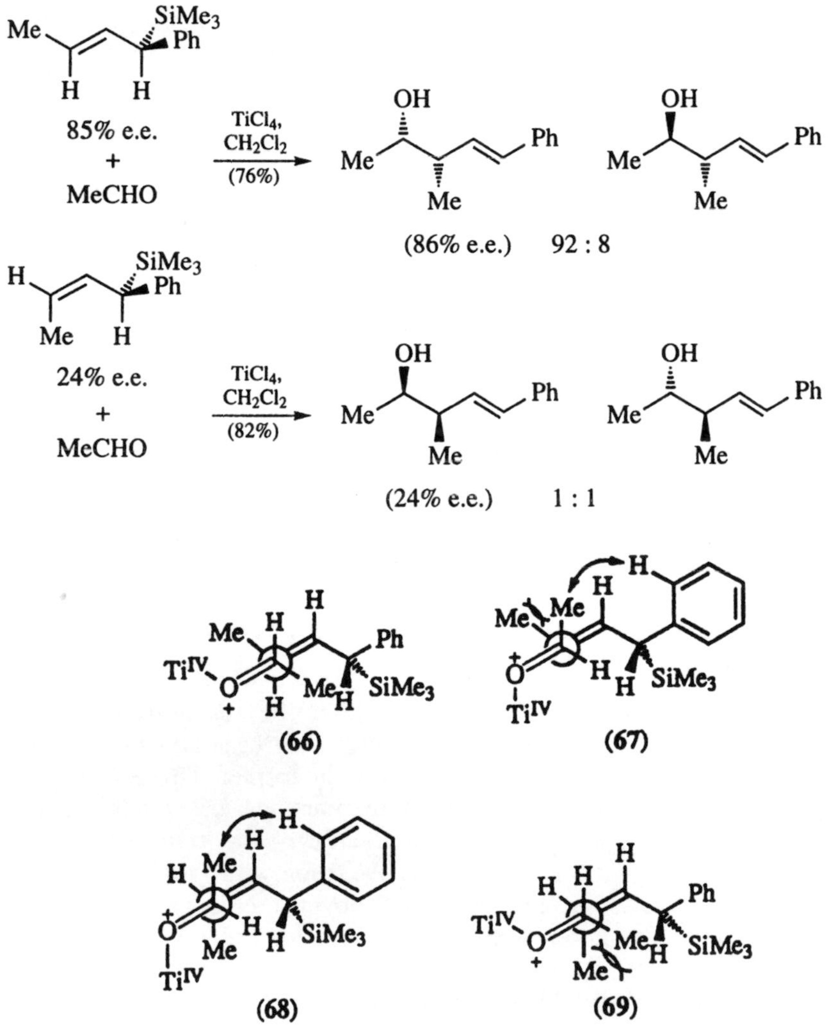

Scheme 28

Addition of the enantiopurified crotyl-substituted silane (70) to acetal-derived cation (71) is assumed to take place as in Scheme 29.[24] The simple Type II diastereoselectivity is controlled, as expected (Chapter 8), by the preference for an *anti* relationship between the aromatic ring and the methyl group of the crotyl-substituted silane as in (72), rather than a *gauche* one as in (73). In both transition states, the stereoelectronic requirement of the S_E2' reaction (see Chapter 5) is satisfied.

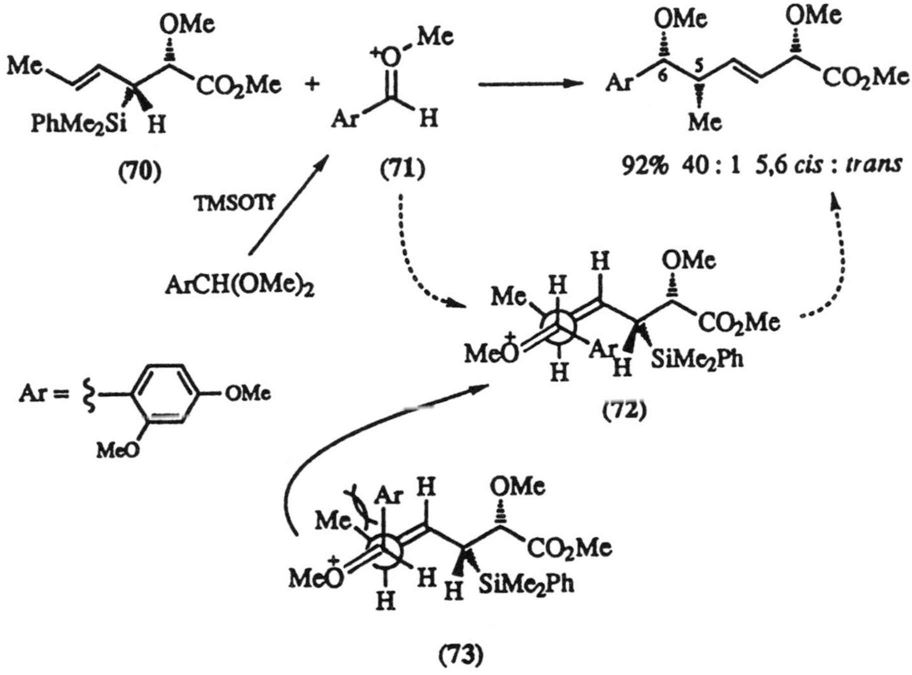

Scheme 29

Work by Reetz *et al.*[25] has shown that α-chiral centre substituted aldehydes, e.g. (74) [Scheme 30 (a)], that bear an oxygen as one of the α-substituents, react diastereoselectively with the prochiral (Z)-enol silane (75) in the presence of Lewis acids ($SnCl_4$, $TiCl_4$) in a Mukaiyama reaction, to give (76). Support for the open transition state shown comes from the fact that the same diastereoisomer (76) is the major product when the (E)-enol silane corresponding to (75) is used; stereoconvergence is a feature of a number of reactions in which open transition states are involved (see Chapter 8).

It is noteworthy that the reaction of the (Z)-enolsilane (75) with an achiral aldehyde lacking chelatable α-substituent, e.g. propanal, is not highly diastereoselective [Scheme 30 (b)], i.e. the chelation present using (74) leads to an increase in diastereoselectivity resulting from the Type II component.

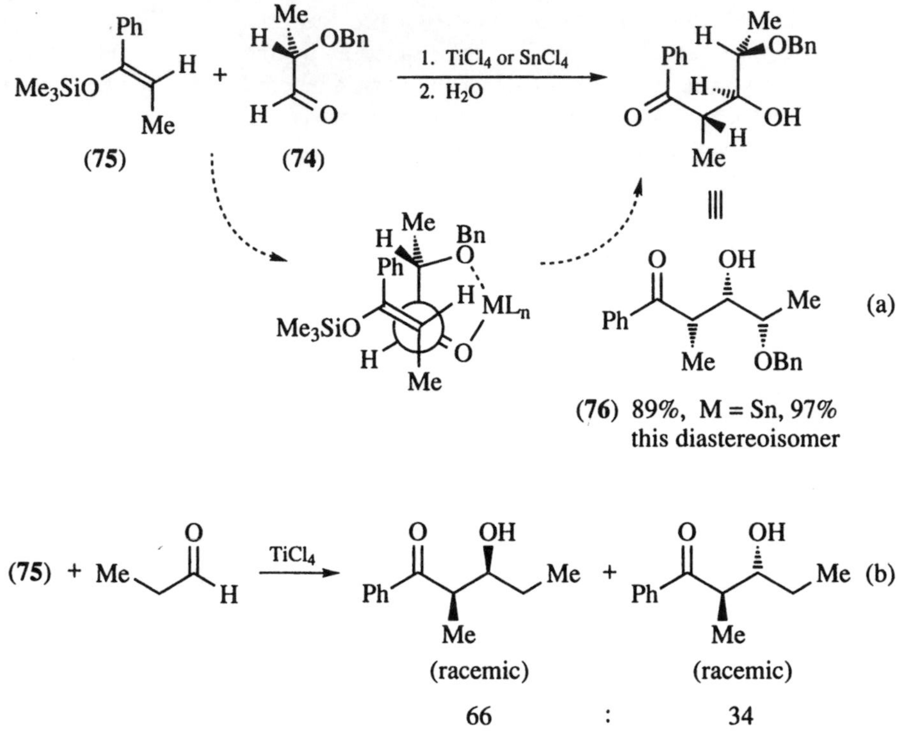

Me₃SiO—C(=CH...)... (75) + (74) → 1. TiCl₄ or SnCl₄ 2. H₂O →

(76) 89%, M = Sn, 97%
this diastereoisomer

(a)

(75) + Me...CHO —TiCl₄→ ...Me (racemic) + ...Me (b) (racemic)

66   :   34

Scheme 30

Reactions involving non-simple chirality transfer

In Chapter 5, simple chirality-transfer reactions, such as the [3,3] sigmatropic rearrangements in Scheme 31, were considered. In such reactions, a single chiral centre is lost and a new chiral centre created at another position, i.e. there is a transfer of chirality from one position to another. Such reactions are of Type III$_{s.c.}$ in the sense that a new chiral centre is created under the influence of an existing chiral centre. They differ from other Type III$_{s.c.}$ reactions in that there is no net increase in the number of chiral centres. Because the original chiral centre is lost in the reaction, these reactions have previously been referred to as being self-immolative of chirality.[26]

Unlike Type III$_{s.c.}$ reactions, which can only be diastereoselective, the Cope rearrangement in Scheme 31 can be not only diastereoselective [rearrangement via (a) rather (b) or vice versa] but also enantioselective. Thus rearrangement via the boat transition state (although less likely) as in (c) would give the enantiomer of (77).

Simple chirality-transfer reactions involve one chiral centre and one

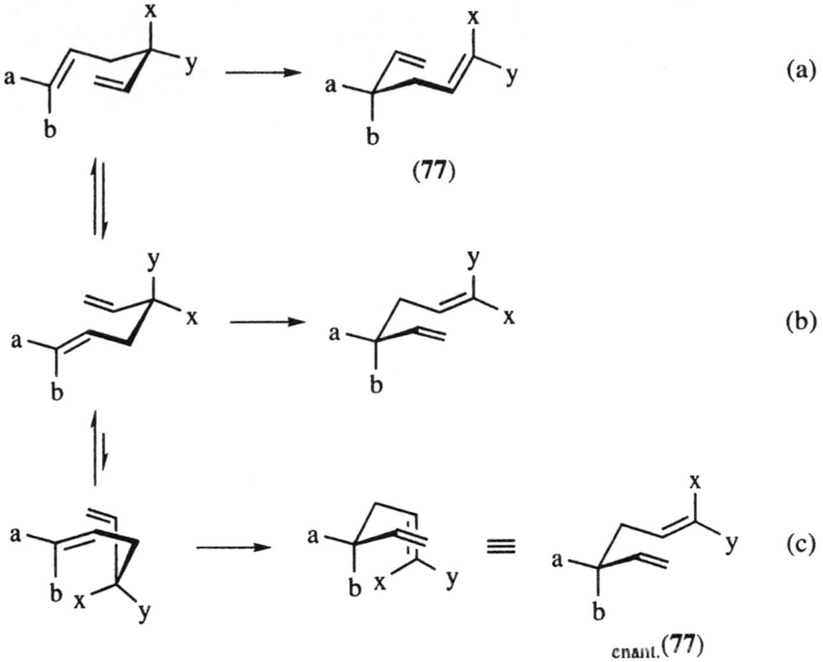

Scheme 31

configured double bond. In Type $III_{s.c.}$ chirality-transfer reactions other than simple ones, more than one chiral centre and one configured double bond are involved. In such cases there may even be a net loss in the number of chiral centres as in the reaction in Scheme 32. In this reaction, the loss of one chiral centre[†] is compensated for by the creation of one additional configured double bond, i.e. overall the number of chiral centres plus configured double bonds remains unchanged; this is a characteristic of this reaction type.

The enantiopure crotylsilane (70) used in Scheme 29 was prepared

Scheme 32

[†]Formally this is still a Type $III_{s.c.}$ reaction since a new chiral centre is formed under the influence of the existing chiral centres.

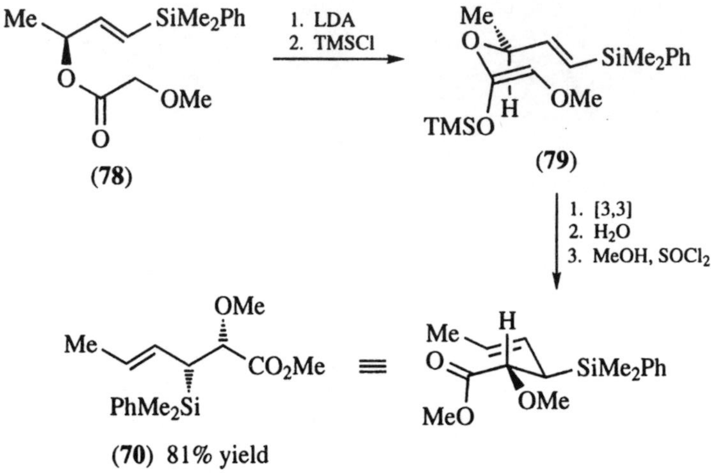

Scheme 33

(Scheme 33) by [3,3] sigmatropic (Ireland–Claisen) rearrangement of the silyl ketene acetal (79), itself prepared from ester (78).[24] Here the inducing chiral centre is lost in the sigmatropic rearrangement but two additional chiral centres are created.

The sense of diastereoselectivity of the reaction is a consequence of the equatorial preference of the methyl group in a chair transition state, together with the Z-configuration of the double bond of the silyl ketene acetal in (79).

Geometrical imperatives may force [3,3] sigmatropic rearrangement to proceed via a boat transition state. In Scheme 34 the Ireland–Claisen

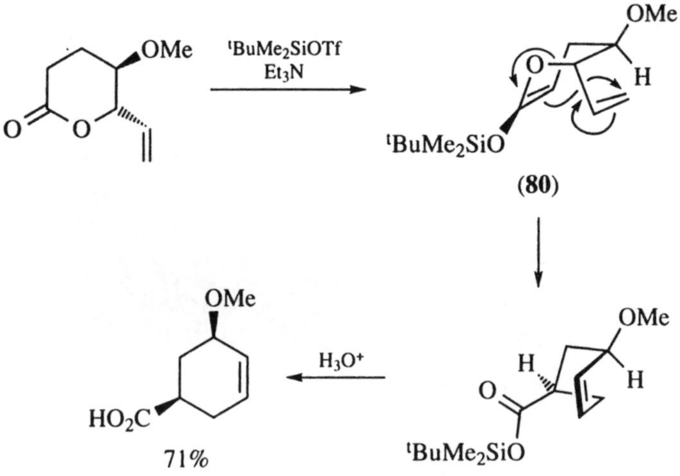

Scheme 34

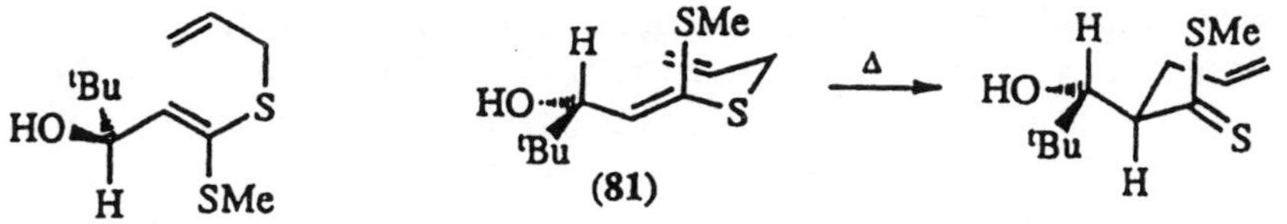

Scheme 35

42%, > 24 : 1 d.r.

(86) (88%)

(82)

Δ, toluene, H⁺

(83)

(85)

Δ, toluene, H⁺

(84)

(89)

(87)

(89′)

(88)

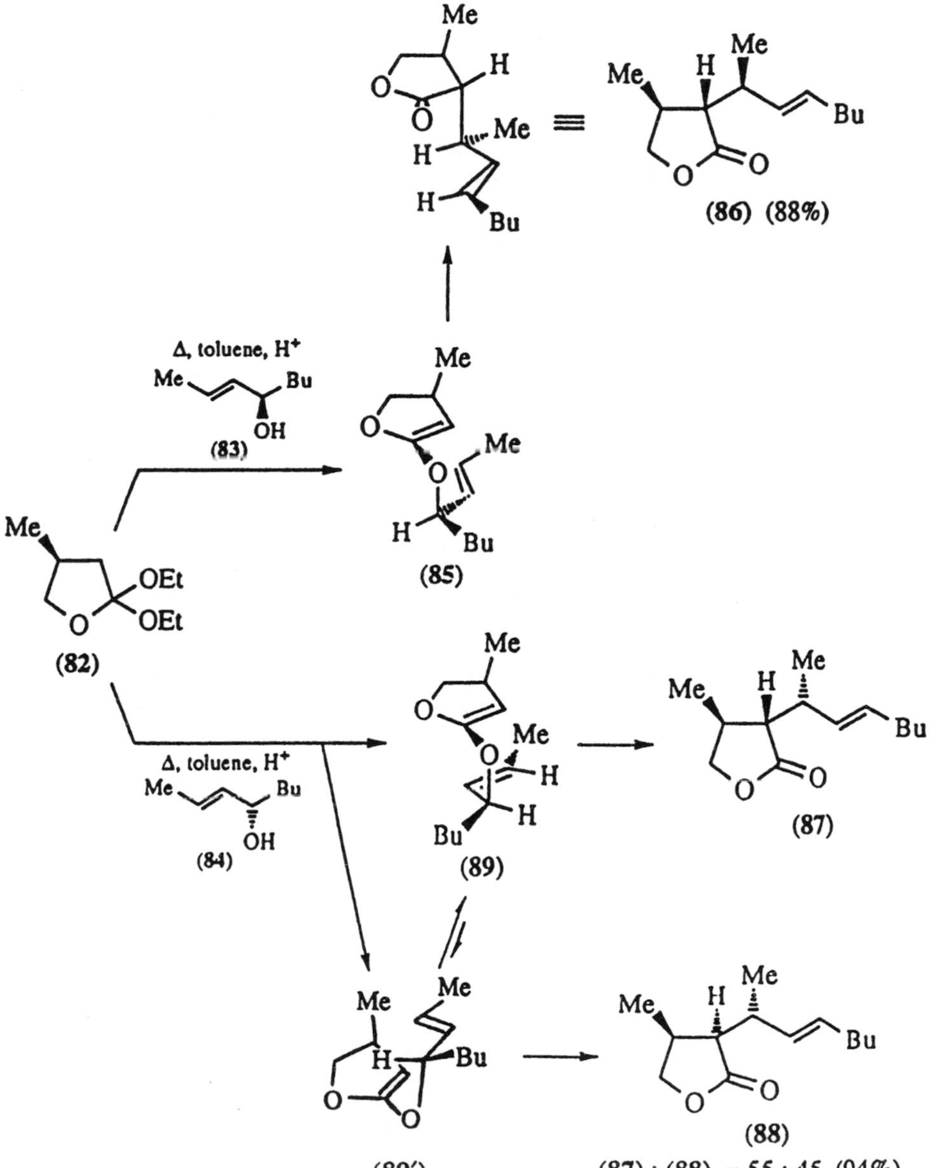

(87) : (88) = 55 : 45 (94%)

Scheme 36

rearrangement of (80) is an example of this[27] (see also Chapter 5).

In the thio-Claisen rearrangement in Scheme 35, the inducing chiral centre is not part of the pericyclic array and is retained in the product; the reaction is therefore a Type $III_{s.c.}$[28] Here the ^tBu group can be assumed to occupy the least hindered site in (81), and the hydrogen on the chiral centre in the position indicated minimises $A_{1,3}$-strain.

The effect of having one chiral centre within *and* one outside the pericyclic array is illustrated in Scheme 36. Here the diastereoisomeric ketene acetals (85) and (89)/(89') are derived *in situ* by heating enantiopure ortholactone (82) with (R)- or (S)-(E)-allylic alcohols (83) and (84), respectively.[29]

In effect these reactions are subject to double asymmetric induction from a pair of matched or mismatched chiral centres (see Chapter 15). In the matching situation, proceeding via (85), the chair transition state has the Bu

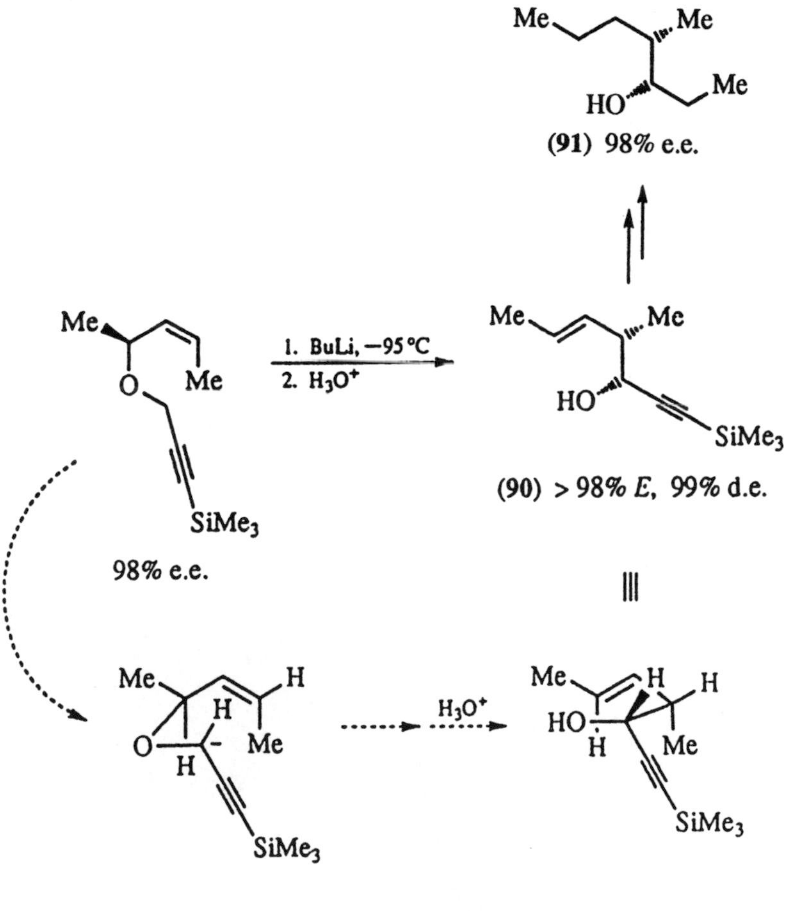

Scheme 37

group equatorial and the face of the ring opposite to the methyl group under attack, leading to a single diastereoisomer (86). By contrast, the senses of induction are mismatched in the rearrangement of the diastereoisomeric ketene acetal obtained from (82) and (84). A 55:45 ratio of (87) to (88) is formed, probably via the boat and chair transition states (89) and (89'), respectively.

A highly diastereoselective chirality-transfer reaction is seen in the [2,3] Wittig rearrangement in Scheme 37. Diastereoselectivity in the Type II [2,3] Wittig rearrangement is particularly high when the oxymethylene carbon bears an alkynyl substituent (Chapter 5). This appears to be the case also in Scheme 37. The product (90) has been converted into (91), an aggregation pheromone of the European elm bark beetle; this shows the completeness of the diastereoselectivity in the formation of (90).[30]

On the other hand, in the [2,3]-sigmatropic rearrangement in Scheme 38, the diastereoselectivity is Type III$_{s.c.}$ with a stereoelectronic effect (π^*–σ^* interaction; see Chapter 10) bringing about attack on the face of the double bond in (92) anti to the electron-withdrawing C—O bond and with minimisation of $A_{1,3}$-strain.[31]

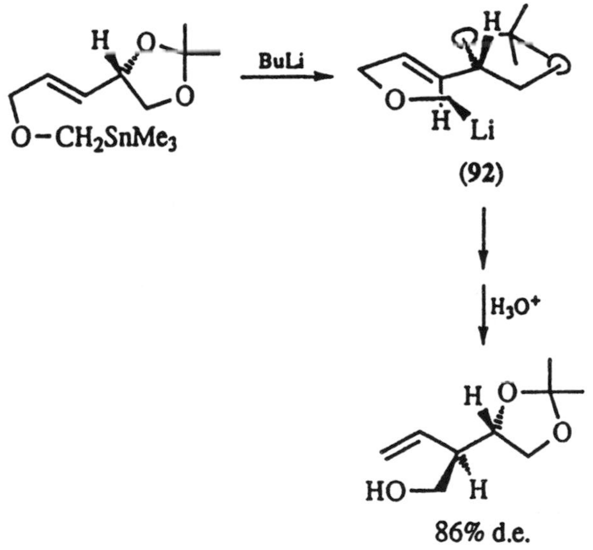

Scheme 38

Summary

A Type II reaction in which one of the components contains at least one chiral centre becomes a Type II/III.

The inherent diastereoselectivity (stereospecificity) of the Type II reaction is unaffected and the occasional diastereoselectivity is usually unaffected by the presence of the chiral centre. Consequently, the diastereoselectivity of the Type II/III can be deduced by considering which of the two enantiomeric Type II transition states will be disfavoured by introduction of the chiral centre.

The means by which this chiral centre brings about discrimination between the two transition states, i.e. brings about asymmetric induction, is much the same as in Type III$_{s.c.}$ or Type III$_{r.c.}$ reactions. Thus, superimposed on the well defined cyclic transition states that characterise most Type II reactions is an interplay between steric, electronic and conformational factors, local to the chiral centre which allows differentiation between two otherwise isoenergetic transition states.

The same devices which bring about diastereoselectivity in Type III$_{s.c.}$ reactions can also operate in Type II/III. For example, a cycloaddition between two functional groups at each end of a tether consisting of four (carbon) atoms can be diastereoselective as a result of the preference of a substituent on this tether for an equatorial position in a chair transition state.

In Type III$_{s.c.}$ chirality transfer reactions other than simple ones (see Chapter 5), more than a single chiral centre and configured double bond undergo changes. In [3,3] sigmatropic rearrangements, the number of chiral centres plus configured double bonds remain unchanged. Highly diastereoselective Type II/III [2,3] sigmatropic rearrangements are known in which an additional chiral centre is created, along with chirality transfer as in the Wittig reaction with ethynyl groups on the oxymethylene carbon.

References

1. S.W. Baldwin, P. Greenspan, C. Alaimo and A.T. McPhail, *Tetrahedron Lett.*, 1991, **32**, 5877.
2. V. Gouverneur and L. Ghosez, *Tetrahedron Lett.*, 1991, **32**, 5349.
3. C. Siegel and E.R. Thornton, *Tetrahedron Lett.*, 1988, **29**, 5225; R. Tripathy, P.J. Carroll and E.R. Thornton, *J. Am. Chem. Soc.*, 1991, **113**, 7630.
4. K.N. Houk, H.-Y. Duh, Y.-D. Wu and S.R. Moses, *J. Am. Chem. Soc.*, 1986, **108**, 2754.
5. R. Annunziata, M. Cinquini, F. Cozzi and L. Raimondi, *J. Org. Chem.*, 1990, **55**, 1901.
6. P.A. Wender and F.E. McDonald, *J. Am. Chem. Soc.*, 1990, **112**, 4956.
7. K. Mikami, T.-P. Loh and T. Nakai, *Tetrahedron: Asymmetry*, 1990, **1**, 13.
8. E.J. Corey and P. Carpino, *Tetrahedron Lett.*, 1990, **31**, 3857. see also L.F. Tietze, U. Beifuss and M. Ruther, *J. Org. Chem.*, 1989, **54**, 3120.
9. K. Mikami, K. Takahashi and T. Nakai, *Tetrahedron Lett.*, 1989, **30**, 357.
10. N.A. Van Draanen, S. Arseniyadis, M.T. Crimmins and C.H. Heathcock, *J. Org. Chem.*, 1991, **56**, 2499.
11. D.A. Evans, J. Bartroli and T.L. Shih, *J. Am. Chem. Soc.*, 1981, **103**, 2127.
12. M. Nerz-Stormes and E.R. Thornton, *J. Org. Chem.*, 1991, **56**, 2489.
13. D.A. Evans, D.L. Rieger, M.T. Bilodeau and F. Urpi, *J. Am. Chem. Soc.*, 1991, **113**, 1047.
14. W.R. Roush, *J. Org. Chem.*, 1991, **56**, 4151.

15. D.V. Patel, F. VanMiddlesworth, J. Donaubauer, P. Gannett and C.J. Sih, *J. Am. Chem. Soc.*, 1986, **108**, 4603.
16. E.J. Corey, R. Imwinkelried, S. Pikul and Y.B. Xiang, *J. Am. Chem. Soc.*, 1989, **111**, 5493; see also E.J. Corey, C.P. Decicco and R.C. Newbold, *Tetrahedron Lett.*, 1991, **32**, 5287.
17. D.A. Evans, J.V. Nelson and T.R. Taber, *Top. Stereochem.*, 1982, **13**, 1.
18. R.W. Hoffmann and H.-J. Zeiss, *Angew. Chem., Int. Ed. Engl.*, 1980, **19**, 218; R.W. Hoffmann and U. Weidmann, *Chem. Ber.*, 1985, **118**, 3966.
19. K. Sato, M. Kira and H. Sakurai, *J. Am. Chem. Soc.*, 1989, **111**, 6429.
20. Z. Wang, X.-J. Meng and G.W. Kabalka, *Tetrahedron Lett.*, 1991, **32**, 1945.
21. M.W. Andersen, B. Hildebrandt, G. Köster and R.W. Hoffmann, *Chem. Ber.*, 1989, **122**, 1777, and references cited therein.
22. T. Kramer and D. Hoppe, *Tetrahedron Lett.*, 1987, **28**, 5149.
23. T. Hayashi, M. Konishi and M. Kumada, *J. Am. Chem. Soc.*, 1982, **104**, 4963.
24. J.S. Panek and M. Yang, *J. Am. Chem. Soc.*, 1991, **113**, 6594.
25. M.T. Reetz, K. Kesseler and A. Jung, *Tetrahedron*, 1984, **40**, 4327; for analogous additions of ketene thioacetals to α-chirally substituted aldehydes, see also C. Gennari *et al.*, *Tetrahedron*, 1986, **42**, 893.
26. K. Mislow, *Introduction to Stereochemistry*, Benjamin, NT, 1964.
27. S.L. Schreiber and D.B. Smith, *J. Org. Chem.*, 1989, **54**, 9.
28. P. Beslin and S. Perrio, *Chem. Commun.*, 1989, 414.
29. F.E. Ziegler, *Chem. Rev.*, 1988, **88**, 1423; this review gives many other examples of diastereoselective Claisen rearrangements.
30. T. Nakai and K. Mikami, *Chem. Rev.*, 1986, **86**, 885; this review gives additional examples of diastereoselective Wittig rearrangements.
31. R. Brückner and H. Priepke, *Angew. Chem., Int. Ed. Engl.*, 1988, **27**, 278.

14 CATALYTIC ENANTIOSELECTIVE REACTIONS (TYPE III$_{\text{cat. enant.}}$)

In Chapter 12, reagent-controlled diastereoselective reactions (Type III$_{\text{r.c.}}$) were considered. In these reactions, the chiral centre(s) in the reagent bring about, e.g., diastereoselective addition to a prochiral double bond in the substrate and these chiral centres, being part of the reagent, must be used in at least stoichiometric quantity since they are retained in the first-formed product.

An important modification of these Type III$_{\text{r.c.}}$ reactions, also previously considered in Chapter 12, is where the reagent-contained chiral centre(s) are not retained in the *isolated* product, but are separated from the first-formed product by a reaction carried out *in situ*. When the reagent in this modification is used in enantiopure form and when the first-formed product is formed diastereoselectively, the overall conversion (reactant to isolated product) is enantioselective.

A *catalytic* enantioselective reagent-controlled Type III reaction, Type III$_{\text{cat. enant.}}$, is a further valuable modification of the Type III$_{\text{r.c.}}$ reaction. In this modification, the reagent x—y is achiral but becomes chiral in the reaction when complexed to an enantiopure catalyst z*; the reaction actually involves, therefore, a chirally modified reagent [x—y]z*, present in low concentration; it is enantioselective since only the created chiral centre(s) are found in the product.

As described in Chapter 12 (Scheme 1), addition of stoichiometric quantities of an enantiopure reagent x*—y to a prochiral double bond can take place in two ways which allow Type III$_{\text{r.c.}}$ reaction of only one enantioface. Scheme 1 shows the most commonly encountered pathway for this conversion, now using an achiral reagent x—y together with a catalytic quantity of enantiopure catalyst z*. The major difference from the analogous reaction previously described (Chapter 12, Scheme 2) is that z* is regenerated at the end of the reaction and recombines with more x—y to reform the chiral reagent [x—y]z*.

For the Type III$_{\text{cat. enant.}}$ process in Scheme 1, there is an obvious analogy with Type III$_{\text{s.c.}}$ reactions involving intramolecular delivery from an *in situ*

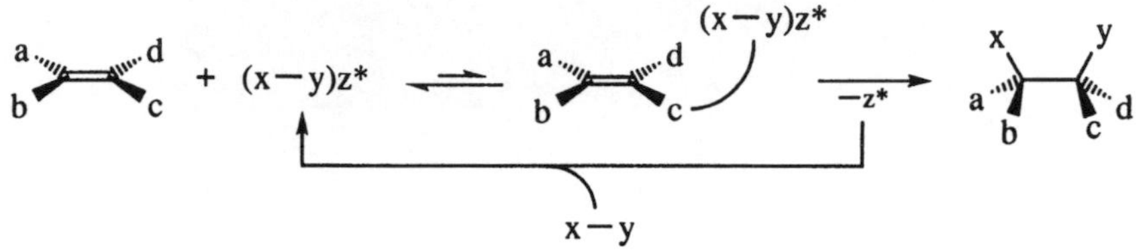

Scheme 1

appended reagent (Chapter 11). For reasons given in Chapter 11, this variation of the Type $III_{s.c.}$ reaction often results in levels of diastereoselectivity superior to those of intermolecular Type $III_{s.c.}$ reactions. It is not surprising, therefore, that this Type $III_{cat.\ enant.}$ variation should also have advantages, in terms of the resulting enantioselectivity, over that in Scheme 2, in which the two faces of the double bond become diastereofaces only in the transition state of the reaction [cf. Chapter 12, Scheme 3 (a)].

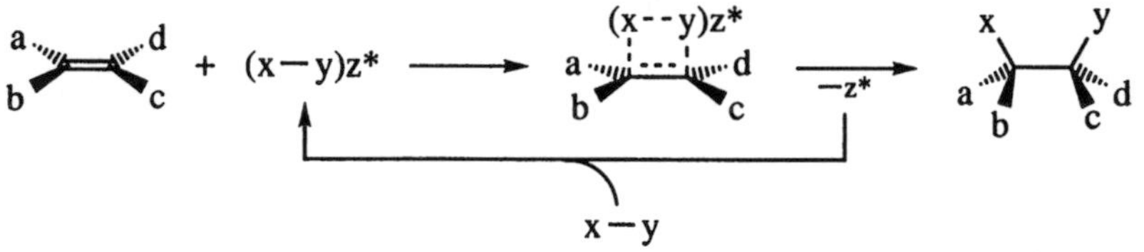

Scheme 2

One major advantage of catalytic and completely enantioselective reactions is that they obviate the need for chiral auxiliaries. The three steps which the use of a chiral auxiliary entails are replaced by one for a catalytic enantioselective addition (Scheme 3). However, the popularity and extensive use of chiral auxiliaries is a reflection of the relative dearth, at present, of methods for bringing about completely enantioselective conversions directly. (The use of chiral auxiliaries may be advantageous in some cases: see Chapter 1).

One advantage of Type $III_{cat.\ enant.}$ reactions relative to overall enantioselective Type $III_{r.c.}$ reactions (see Chapter 12) is a saving in cost, by virtue of the lesser quantity of enantiopure material (catalyst) required. Another advantage is the simplification in product isolation; in the overall enantioselective Type $III_{r.c.}$ reactions discussed in Chapter 12, a molar quantity of the spent reagent containing the inducing chiral centre(s) must be separated from the product. In some Type $III_{cat.\ enant.}$ reactions the chiral catalyst can be immobilised on a solid support, e.g. a polystyrene resin, which can be easily separated and often re-used.

Non-enzymic Type $III_{cat.\ enant.}$ versions of a number of functional group interconversion are known and are considered under the headings below.

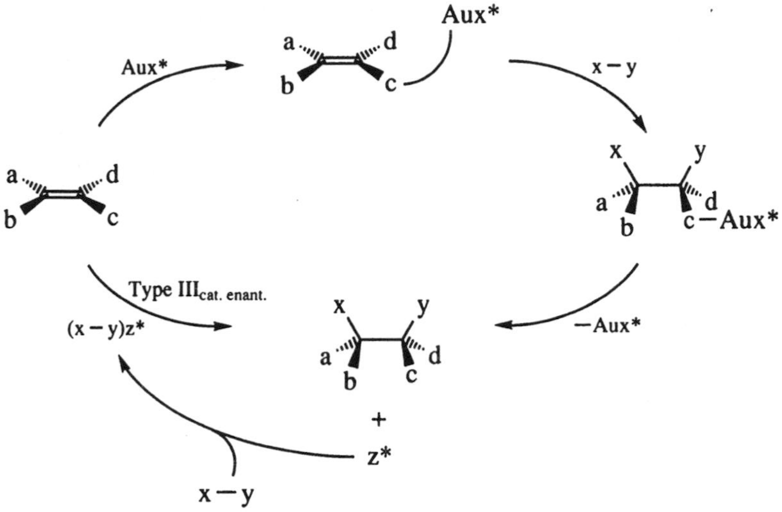

Scheme 3

The Sharpless–Katsuki epoxidation

The Sharpless–Katsuki epoxidation is the conversion of an achiral allylic alcohol into the corresponding epoxide (glycidol) with high enantioselectivity using a single enantiomer of diethyl tartrate in conjunction with *tert*-butyl hydroperoxide and titanium tetraisopropoxide (Scheme 4).[1,2] In the presence of powdered molecular sieve, the diethyl tartrate and titanium isopropoxide can be used in a catalytic amount (0.1 mol equiv.).

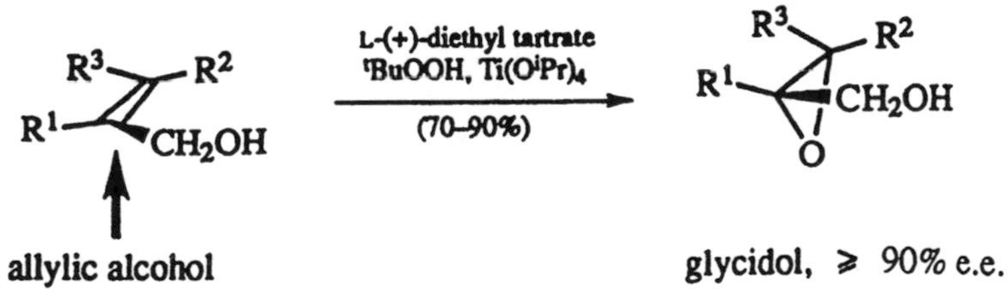

allylic alcohol

glycidol, $\geqslant$ 90% e.e.

Scheme 4

The sense of induction in this epoxidation is remarkably consistent and is given in Scheme 4 as a mnemonic: with the allylic alcohol oriented as illustrated, epoxide formation is on the lower face using L-(+)-diethyl tartrate, but using D-(−)-diethyl tartrate it will be on the top face. Since both enantiomers of diethyl tartrate are commercially available, either glycidol enantiomer can be obtained at will.

Of critical importance for the success of this epoxidation is the ability of titanium(IV) alkoxides to exchange ligands with *tert*-butyl hydroperoxide and with alcohols reversibly. Thus diethyl tartrate, the allylic alcohol and the *tert*-butyl hydroperoxide can be complexed to the same titanium atom and, after the epoxidation event, *tert*-butyl alcohol and the product glycidol can be decomplexed.

Bidentate bonding of the hydroperoxide to the titanium as in (1) is believed to facilitate oxygen transfer to the coordinated allyl alcohol, with attack by the alkene being along the O—O axis as indicated (compare the analogous V^V–*tert*-butyl hydroperoxide-catalysed epoxidation of allylic alcohols (Chapter 11).

(1)   **(2)**

Evidence suggests that the active catalyst has a dimeric structure based on (2) with the substructure in (1) incorporated by ligand exchange of OR and ester groups. Coordination of the allylic alcohol via its hydroxyl to titanium in such a dimeric complex allows discrimination between the two faces of its double bond—now diastereofaces—in the intramolecular epoxidation even if the details are not yet clear.[3]

The problem in establishing the details of the mechanism of this and many other Type III$_{cat. enant.}$ reactions is that the active complex is not usually isolable and its concentration at any time may be exceedingly low, making spectroscopic identification difficult; structural information may therefore only be gained by indirect means.

The impact of the Sharpless–Katsuki epoxidation on organic synthesis over the past decade has been profound. Its use for the kinetic resolution of racemic chiral allylic alcohols (see below) promises to become as valuable as its use for the enantioselective epoxidation of achiral allylic alcohols is now. Three factors, in addition to the routinely high enantioselectivity of the reaction (~90%), are responsible for this impact:

(a) The value of the allylic epoxide products as chiral enantiopure building blocks (chirons: see Chapter 16).

(b) The catholicity of the reaction. Excellent enantioface selectivity is achieved with a wide variety of allylic alcohols, and in this respect it compares favourable with enzyme-catalysed epoxidation, where wide variation from the natural substrate structure is rarely tolerated.

(c) Incorporation of the epoxidation into a synthesis, usually at an early stage, obviates the need for resolution. This factor has influenced the design of many syntheses, such that allylic alcohols have become preferred starting materials or intermediates.

Although the Sharpless–Katsuki epoxidation can also be carried out on homoallylic alcohols, the enantioselectivities that result are in general not as high as those with allylic alcohols. Interestingly, the sense of induction in the major enantiomer from epoxidation of homallylic alcohols is the reverse of that given by the mnemonic in Scheme 4.[4]

KINETIC RESOLUTION

Kinetic resolution is the faster reaction of one enantiomer of a (usually) racemic substrate with an enantiopure reagent. In the most favourable cases only one enantiomer of the racemic substrate reacts, giving an enantiopure product; since a single enantiomer of the starting material remains, a resolution of the original racemic mixture has been achieved. Because, in this favourable case, only one enantiomer of the starting material is consumed, only 0.5 mol equiv. of the reagent needs to be employed.

Kinetic resolution arises as a consequence of double asymmetric induction (see Chapter 15). In practice, for successful kinetic resolution, the rate constant for consumption of the faster reacting enantiomer should be at least 15 times greater than that of the slower reacting enantiomer.

For the majority of Type $III_{cat.\ enant.}$ reactions, there is a preliminary complexing of the enantiopure catalyst-reagent to the substrate (see Scheme 1) and diastereoisomeric complexes can bc formed if the substrate contains a chiral centre. Kinetic resolution, therefore, can result either from different rates of complex formation or from different rates of intramolecular reaction within the diastereoisomeric complexes (or both).

Kinetic resolution using the Sharpless–Katsuki epoxidation

It was the preferential reaction of one enantiomer of racemic isopropyl vinyl carbinol with the now familiar cocktail used for Sharpless–Katsuki epoxidation which led to the discovery of this reaction. Since then, the kinetic resolution has been applied with great success to a variety of racemic allylic alcohols, including those in Scheme 5.[5-8]

In most cases the allylic alcohol and the epoxide are isolated together and must be separated, but in other cases the epoxide is converted into water-soluble by-products in the (aqueous NaOH) work-up, obviating the need for a separation [e.g. (b) in Scheme 5].

Sato and co-workers have used enantiopure iodoallylic alcohols [Scheme 5 (b)] as starting materials in the (Type 0 or Type IV) synthesis of various enantiopure allylic and acetylenic alcohols (Scheme 6)[6,9] and also of lactones and homoallylic alcohols (Scheme 7).[10] Also successfully resolved using this procedure are those substrates containing the allylic alcohol substructure (Scheme 8).[11]

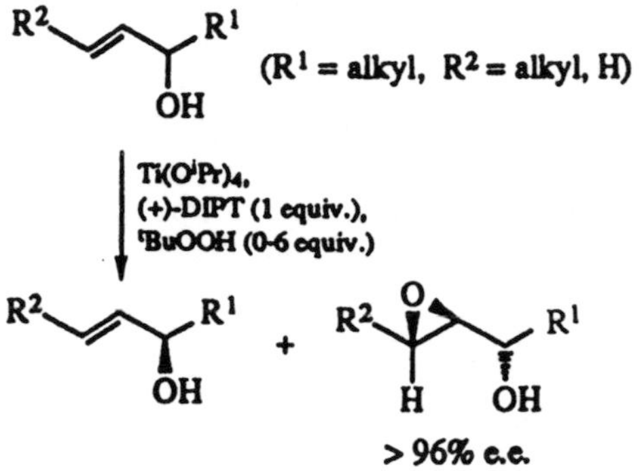

(a)

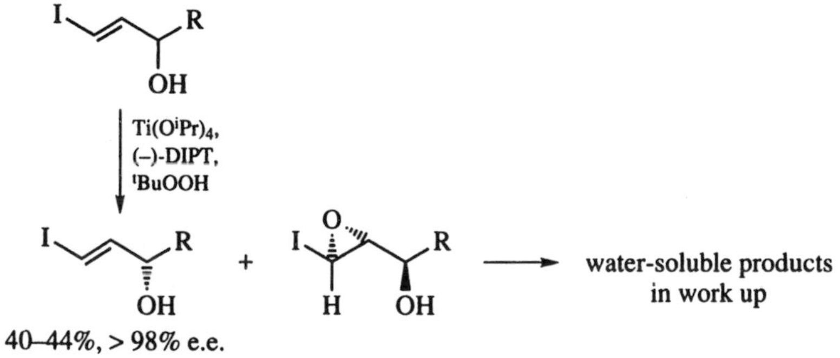

> 96% e.e.

(b)

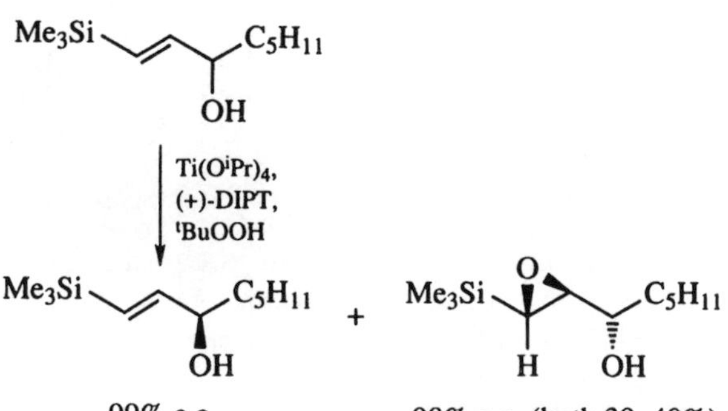

40–44%, > 98% e.e.

water-soluble products in work up

(c)

99% e.e.   98% e.e. (both 39–40%)

Scheme 5   *(continued)*

Scheme 5 *(continued)*

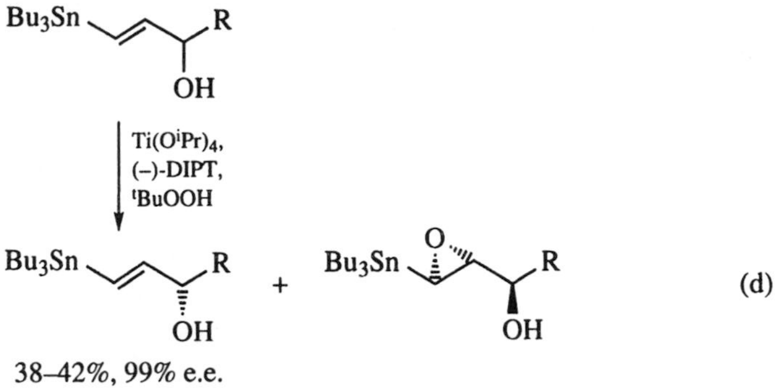

Bu₃Sn‚‚‚‚‚R
OH

Ti(OⁱPr)₄,
(−)-DIPT,
ᵗBuOOH

Bu₃Sn‚‚‚‚‚R   +   Bu₃Sn‚‚‚‚‚R   (d)
OH   OH

38–42%, 99% e.e.

DIPT = diisopropyl tartrate

Scheme 5

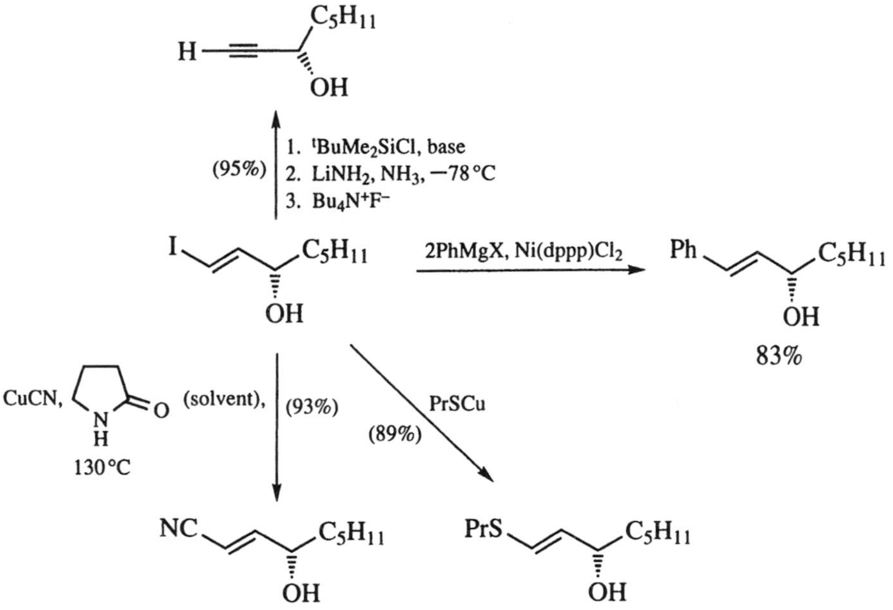

H———C₅H₁₁
OH

1. ᵗBuMe₂SiCl, base
(95%) 2. LiNH₂, NH₃, −78 °C
3. Bu₄N⁺F⁻

I‚‚‚‚‚C₅H₁₁   2PhMgX, Ni(dppp)Cl₂   Ph‚‚‚‚‚C₅H₁₁
OH   OH
                                             83%

CuCN, (pyrrolidinone) (solvent), (93%)   PrSCu
130 °C   (89%)

NC‚‚‚‚‚C₅H₁₁   PrS‚‚‚‚‚C₅H₁₁
OH   OH

dppp = Ph₂P(CH₂)₃PPh₂

Scheme 6

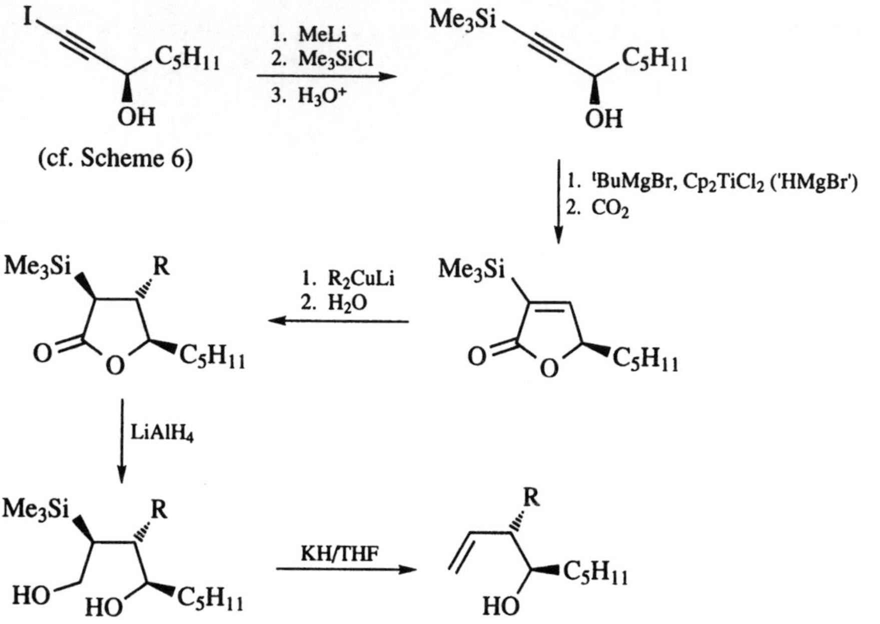

Scheme 7

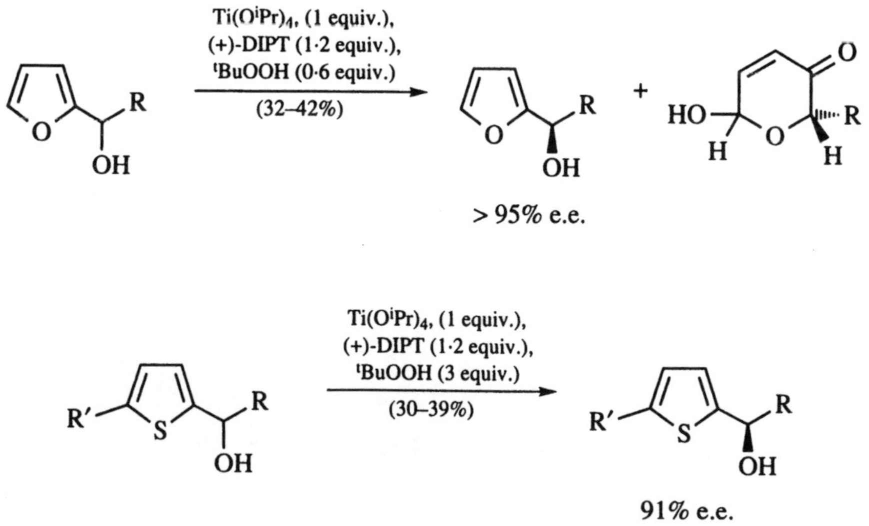

DIPT = diisopropyl tartrate

Scheme 8

The faster reacting enantiomer in the kinetic resolution of a secondary allylic alcohol using the Sharpless–Katsuki epoxidation is given by the mnemonic in Scheme 9; this correctly predicts the absolute configuration of the faster reacting enantiomer in the reactions in Scheme 5.[12]

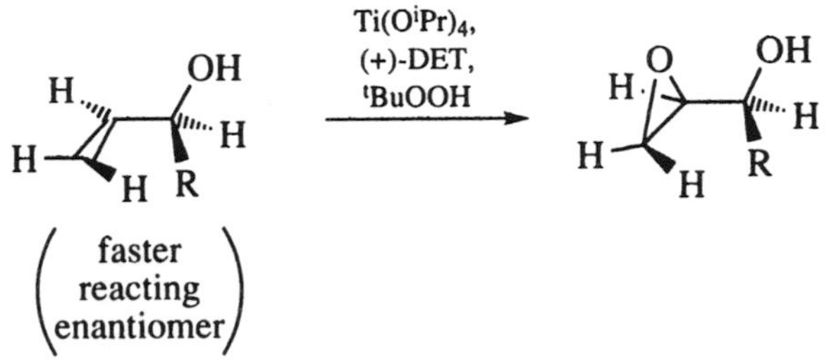

DET = diethyl tartrate

Scheme 9

RING OPENING OF EPOXIDES (GLYCIDOLS)

The major use of epoxides in synthesis is as synthetic relay compounds; their inherent strain makes them prone to ring opening by nucleophiles and electrophiles. Nucleophilic ring opening usually proceeds with inversion of configuration at the carbon undergoing attack [Scheme 10 (a)]. Perhaps surprisingly, ring opening also usually proceeds with inversion when it is electrophile driven, as in treatment with aqueous sulphuric acid [Scheme 10 (b)]. Only if a or b is carbocation-stabilising (and the solvent appropriate) does the carbocation intermediate have a sufficiently long lifetime for attack to take place from both of its diasterofaces.

High *regioselectivity* in the opening of the epoxide ring cannot be taken for

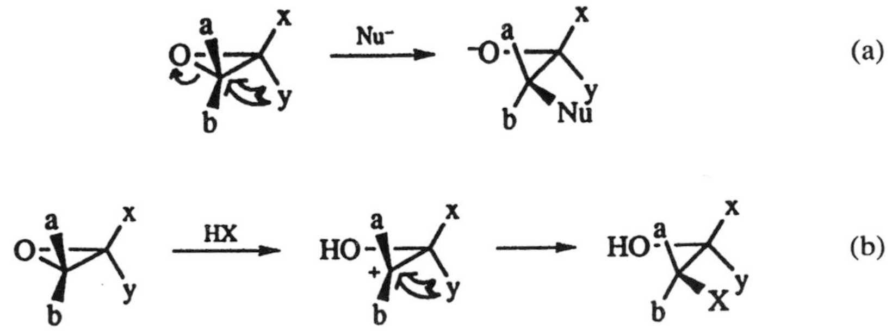

(a)

(b)

Scheme 10

granted. Thus ring opening of epoxide (**3**) [Scheme 11 (a)] with diethylamine
is not very regioselective (and is very inefficient) under normal conditions.[13]

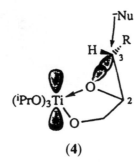

Scheme 11

The advent of the Sharpless–Katsuki epoxidation and with it the
availability of enantiopure or highly enantio-enriched epoxides derived from
allylic alcohols such as (**3**) has stimulated the development of methods for
regioselective ring opening of such unsymmetrically substituted epoxides.

In the presence of 1.5 mol equiv. of titanium tetraisopropoxide, the
reaction in Scheme 11 (b) goes to completion in 5 h at room temperature and
the regioselectivity rises to 20:1.[13] The preferential attack at C-3 by
nucleophiles (including ^-OAc, $^-N_3$ and ^-SPh) in the presence of titanium
tetraisopropoxide is thought to result from a transition state resembling (**4**).
In this transition state, the regioselectivity may arise from the better overlap

of a vacant d-orbital on titanium with a developing orbital from cleavage of the C_3—O bond rather than the C_2—O bond.

Complementary regioselective ring opening at C-2 has been accomplished by intramolecular attack of the nucleophile as in Scheme 12.[14]

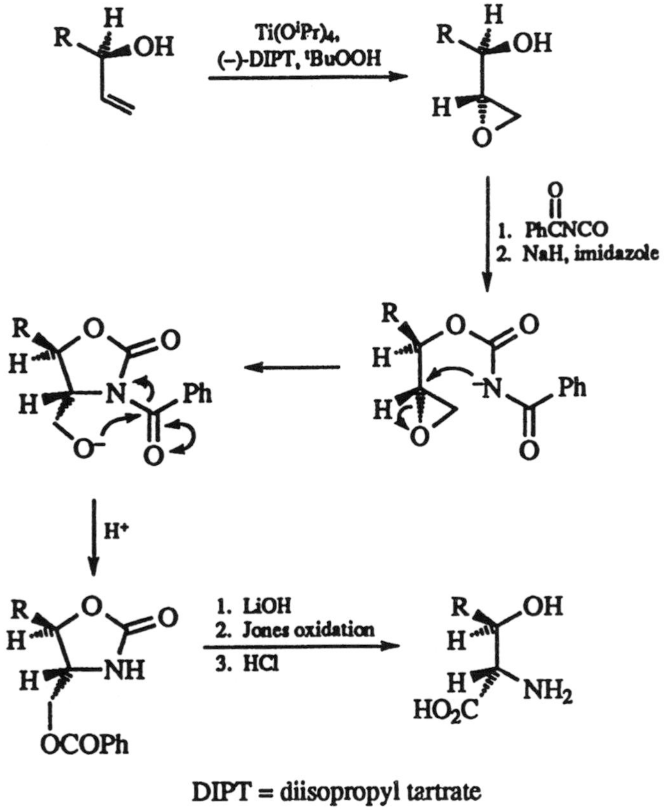

DIPT = diisopropyl tartrate

Scheme 12

A limitation in the kinetic resolution of racemic allyl alcohols using the Sharpless–Katsuki epoxidation is that the maximum yield of resolved allylic alcohol is only 50%, as is the maximum yield of enantiopure glycidol. Reconversion of the enantiopure glycidol back to the allylic alcohol *with inversion at the carbinol chiral centre* would allow quantitative conversion of the original racemic alcohol into a single enantiomer (deracemisation). This has been accomplished by the route in Scheme 13, which takes advantage of the superior nucleophilic properties of the telluride anion.[15]

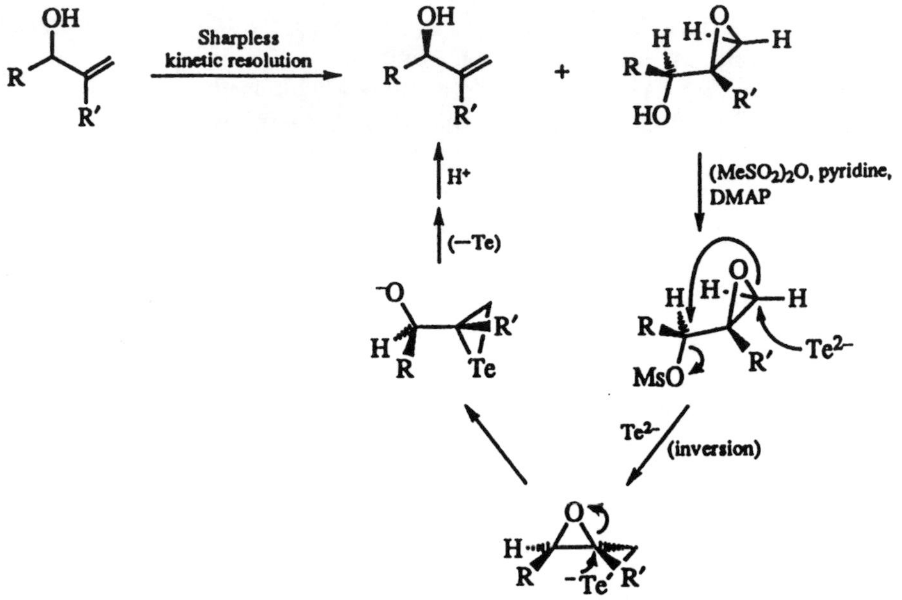

DMAP = 4-dimethylaminopyridine

Scheme 13

Other epoxidation methods

Iron porphyrins are known to catalyse alkane hydroxylation and alkene epoxidation in the presence of oxygen donors such as iodosylbenzene. These reactions have been studied as models for cytochrome P-450 enzymes. In the epoxidation, iron–oxo intermediates are believed to be formed by transfer of the oxygen atom from iodosylbenzene and this oxygen is then transferred to the alkene (Scheme 14).[16]

Enantiopure porphyrins bring about enantioselective formation of epoxides from alkenes using this route. Thus catalytic amounts of the bisbinaphthyl-bridged (vaulted) porphyrin (5) convert (Z)-methylstyrene into its epoxide in 58% e.e. (Scheme 15).[17]

Manganese complexes of enantiopure bases (Mn[III]–salen complexes) also catalyse the epoxidation of alkyl- and aryl-substituted alkenes with high enantioselectivity (Scheme 16).[18]

The sense of enantioselection is in agreement with a side-on and perpendicular approach of the alkene to the manganese–oxo bond (analogous to that shown in Scheme 14), with transfer of the oxo oxygen to the alkene as in (6) (Scheme 16), i.e. on the rear face of the alkene as drawn.

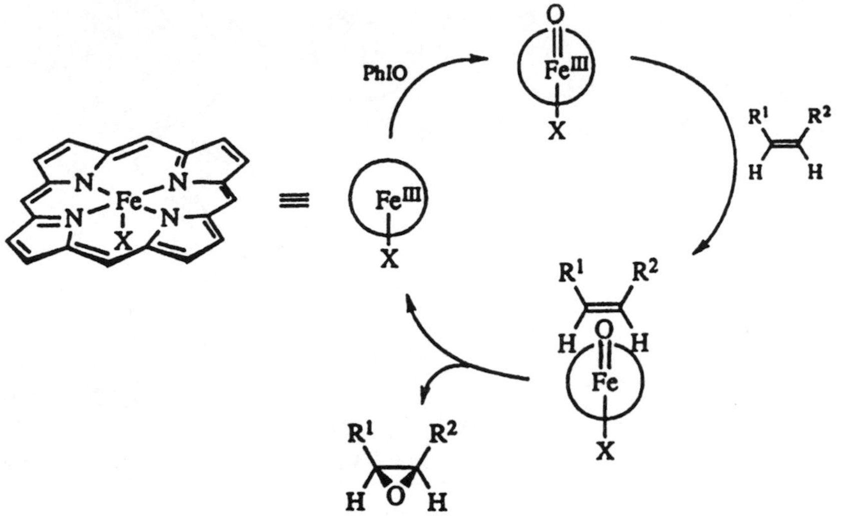

Scheme 14

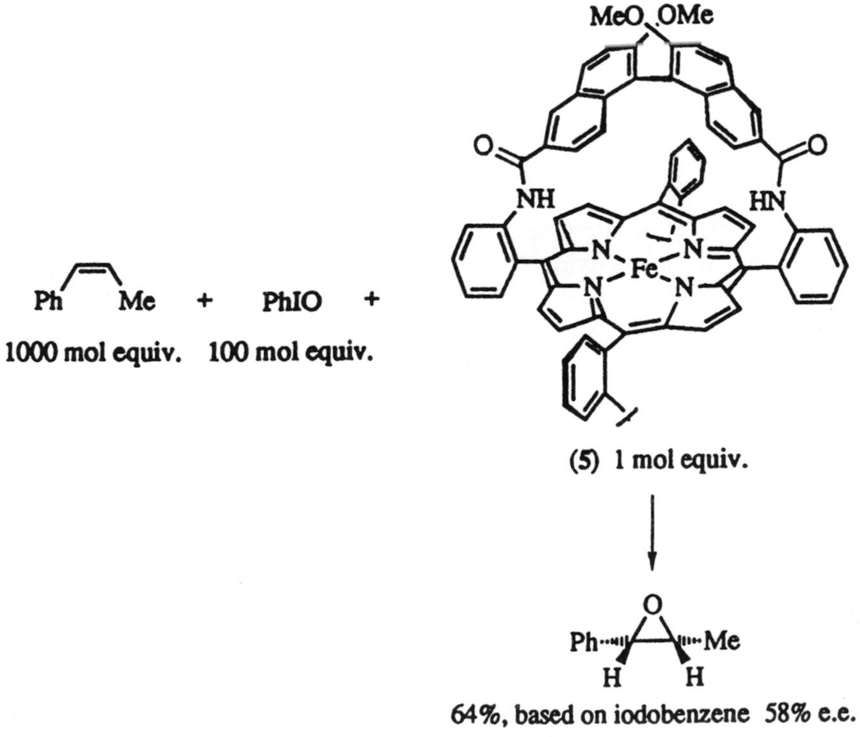

Ph⌒Me + PhIO +

1000 mol equiv.   100 mol equiv.

(5) 1 mol equiv.

Ph‧‧‧⌐○⌐‧‧‧Me
    H   H

64%, based on iodobenzene 58% e.e.

Scheme 15

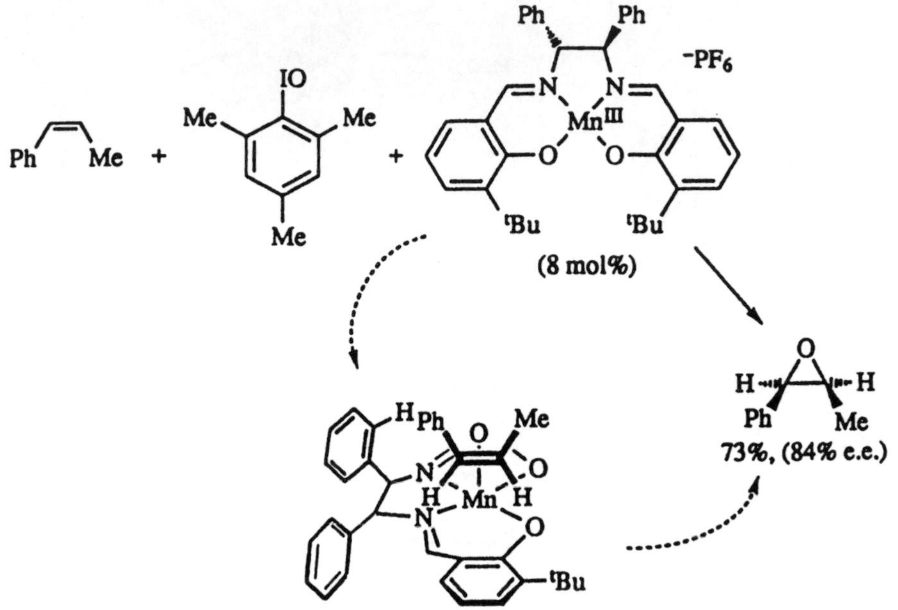

Scheme 16

Attack on the opposite face of the alkene in (6) (keeping the alkene phenyl group distant from the oxidant's *tert*-butyl groups) would give rise to adverse steric interactions between the alkene phenyl group and the rear phenyl group of the oxidant and in particular with the C—H atom shown. This model predicts that *cis*-alkenes will be epoxidised more enantioselectively than *trans*-alkenes, which is found to be the case. Note that in this reaction, unlike the Sharpless–Katsuki epoxidation, no preliminary complex formation is required between the alkene and the epoxidising agent; in principle, therefore, its scope is wider.

Mn[III]–salen-based catalysts have the advantage over porphyrin-based catalysts that the resident chiral centres can be located closer to the metal centre, so better asymmetric induction in the oxygen transfer step is likely. Moreover, the complexes are inexpensive, easy to prepare with a range of substituent groups and robust enough for commercial bleach to be used as the oxidant. Interestingly, using the Mn[III]–salen complex (7) (Scheme 17),[19] the degree of enantioselectivity is dependent on the electronic nature of the substituent X. Since changing the X substituent is unlikely to bring about significant conformational or bond length changes, the increase in

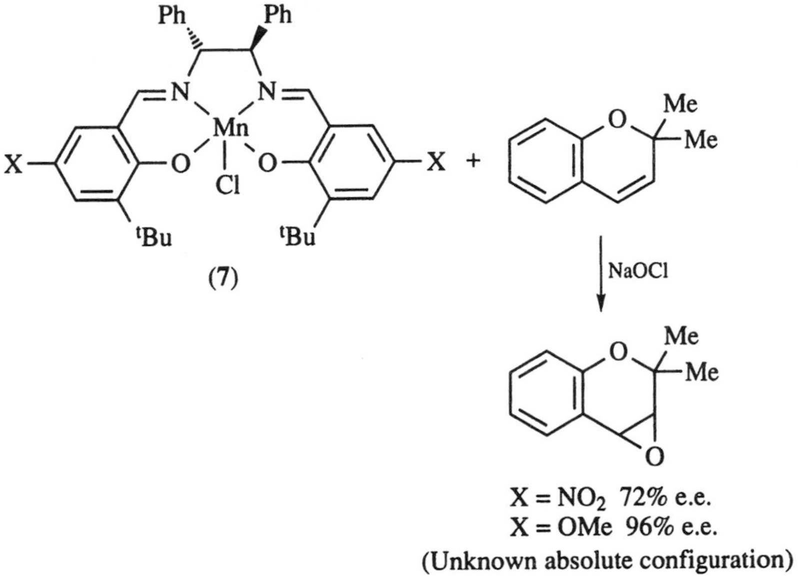
Ph
Ph
N
N
Mn
X
O
O
X
Cl
ᵗBu
ᵗBu
(7)
+
O
Me
Me
NaOCl
O
Me
Me
O
X = NO2 72% e.e.
X = OMe 96% e.e.
(Unknown absolute configuration)

Scheme 17

enantioselectivity when X = MeO may result from a later, product-like, transition state in which non-bonded interactions will be more developed and result in greater discrimination between the two diastereoisomeric transition states.

Reduction of ketones

Methods for overall enantioselective Type $III_{r.c.}$ reduction of ketones have been given in Chapter 12. A highly enantioselective version of this type of reduction which is catalytic in the chiral borane used has been devised by Corey *et al.*,[20] who exploited a reaction first uncovered by Itsuno and co-workers. In this reduction (Scheme 18), the enantiopure borane is an oxazaborolidine (**8**) and the reducing agent is diborane or catecholborane. It is believed that the borane and ketone are complexed to the adjacent nitrogen and boron atoms, respectively, of the oxazaborolidine (**8**). Enantioselectivity arises from intramolecular delivery of hydride from the borane to one diastereoface of the complexed ketone (**9**); subsequent decomplexation regenerates the oxazaborolidine (**8**). Intramolecular delivery of the hydride within any other complexes of ketone, borane and oxazaborolidine that may be formed is apparently slow by comparison with that within (**9**).

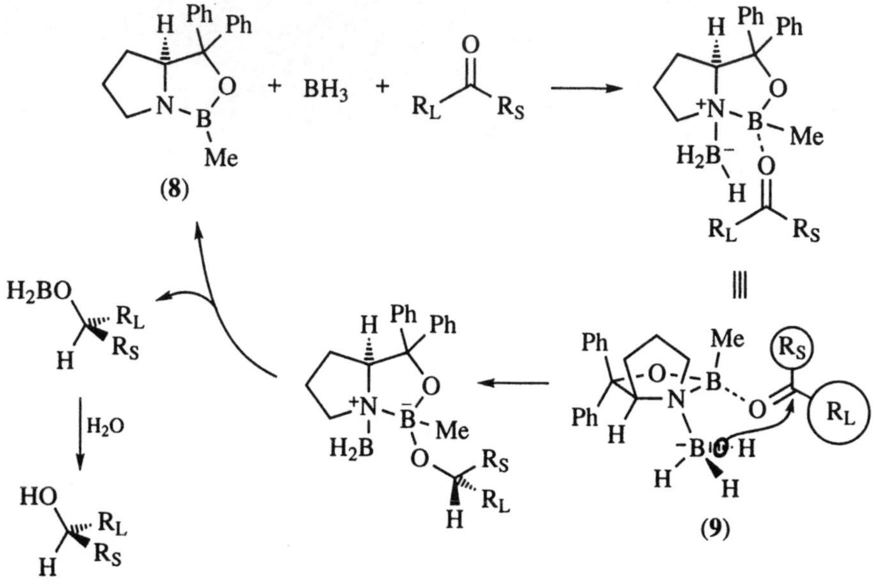

Scheme 18

This CBS reduction is applicable to arylalkyl, dialkyl and α,β-unsaturated ketones with enantioselectivities >90% as illustrated in Scheme 19.[21] The yields in these reactions are good, the experimental procedure simple, the absolute configuration predictable, [cf. Scheme 18, (9)] and the (commercially available) oxazaborolidine recoverable.

This last reaction in Scheme 19 shows that highly enantioenriched primary alcohols, isotopically labelled with deuterium, can be obtained using a dinaphthyl-substituted oxazaborolidine (10).[22]

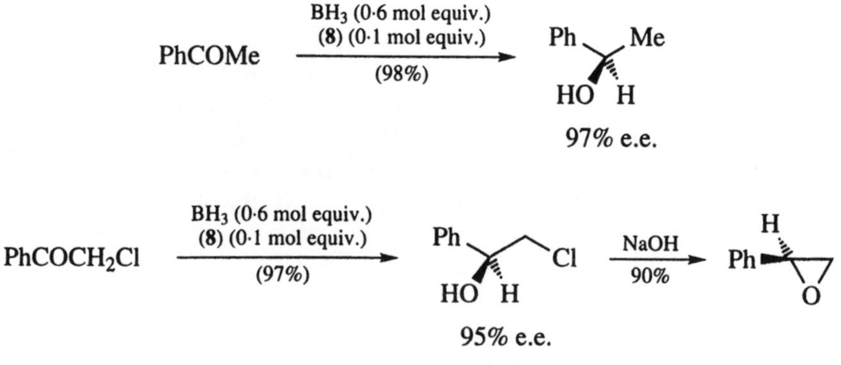

Scheme 19

Scheme 19 *(continued)*

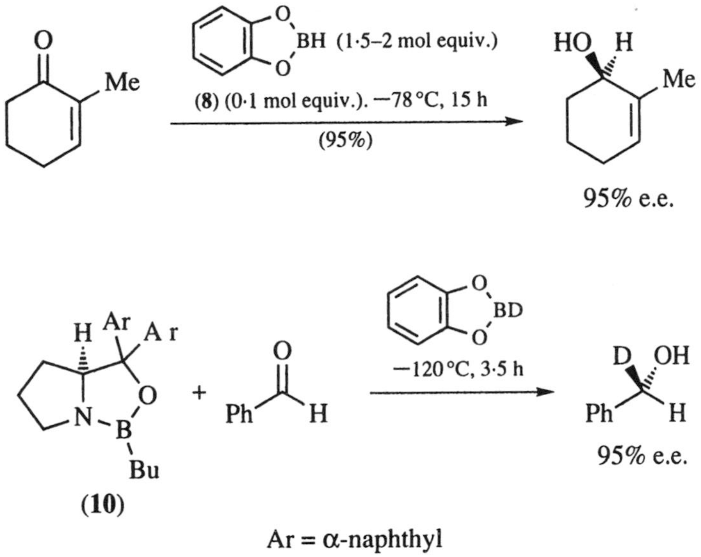

95% e.e.

Ar = α-naphthyl

Scheme 19

Enantioselective homogeneous hydrogenation

Metals are ideal candidates as catalysts for enantioselective hydrogenation since they can be equipped with chiral ligands. These chiral ligands can be tailored to suit the requirements of the reaction in question both in the functional groups they contain and the degree to which they are flexible or rigid. However, as with many other enantioselective reactions, it must be admitted that our understanding of the transition-state geometries or even of the basic mechanisms has not generally advanced to a point where full advantage can be taken of the potential for rational ligand design.

The most productive early studies of homogenous enantioselective hydrogenation used rhodium equipped with chiral phosphine or bisphosphine ligands, and were largely confined to alkenes as substrates.[23] For example, hydrogenation of α-substituted-α,β-unsaturated acids, in particular α-acetamidocinnamic acids and their esters, gave enantio-selectivities >90% (Scheme 20).[24,25] However, with the earlier ligands, only a relatively small range of alkenes could be reduced with such high enantioselectivity. The introduction of binaphthylbisphosphine (BINAP) as a enantiopure ligand by Noyori and co-workers[26] in 1980, and its complexation with ruthenium in particular, has resulted in versatile homogeneous catalysts for reduction of alkenes *and* ketones.

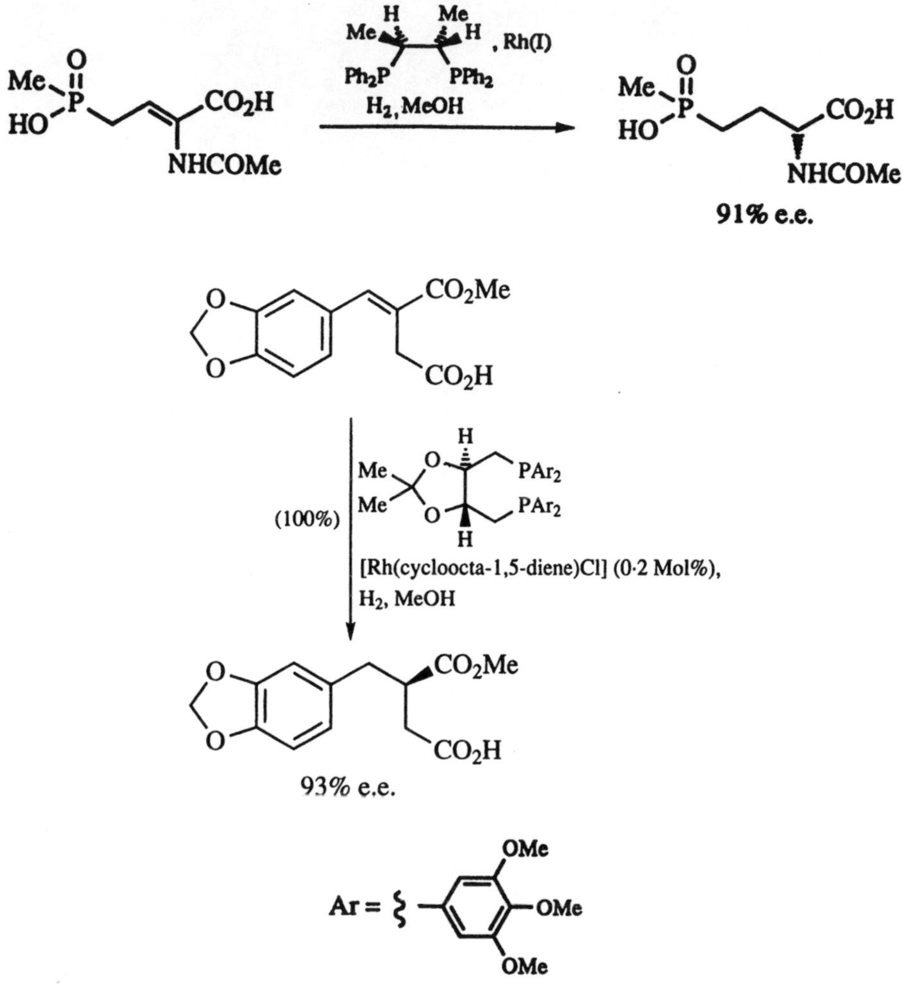

Scheme 20

The chiral ligand BINAP (**11**), by virtue of its C_{2v} symmetry, minimises the number of diastereoisomeric intermediates (and their associated transition states) by which reduction proceeds (a *sine qua non* for stereoselectivity). At the same time it possesses a useful degree of flexibility by virtue of the σ-bond linking the two naphthyl units. Complete rotation around this bond does not occur, of course; if it did the molecule would be racemise. When complexed with rhodium, BINAP is at least as effective in enantioselective

(11)

(S)-(−)-BINAP[2,2′-bis(diphenylphosphino)-1,1′--binaphthyl]

hydrogenation of α-acetamidocinnamic acids etc as the bisphosphines used in the earlier work.

Ruthenium (d⁶) has a preferred coordination number of six whereas rhodium (d⁸) is usually four-coordinated and square planar. In reductions with these metals, rhodium combines with hydrogen to form complexes containing H—Rh—H bonds whereas ruthenium forms complexes in which monohydride Ru—H bonds are present. These differences must contribute to the greater scope of Ru–BINAP in enantioselective reduction. Although it is clear that alkene and hydride are both complexed to Ru–BINAP at some point, the mechanism is not yet known in any detail. Examples of substituted alkenes which have been enantioselectively hydrogenated using Ru–BINAP by Noyori and co-workers[27] are given in Scheme 21. In the last example, the citronellol is produced with a higher enantiopurity than the material isolated from natural sources.

It is in the reduction of ketones that the Ru–BINAP catalysts excel, specifically those having one of a number of functional groups close to the carbonyl group. In Table 1 are examples of such ketones which are reduced highly enantioselectively to the corresponding functionalised alcohols.[28]

Enantioselective reduction of the γ-chloro-β-keto ester (12) (Scheme 22) has been applied to a synthesis of carnitine (13), an unusual amino acid responsible for the transport of long-chain fatty acids through the mitochondrial membrane.[29]

Not only are the enantioselectivities in these ketone reductions high (>90%), but also the yields are good and the substrate/catalyst ratio can be high (200:1). As Table 1 illustrates, various salts of Ru–BINAP have been used.

It is thought that the additional functional group in each of the ketones in Table 1 and Scheme 22 is able to assist in the enantioselective reduction by coordinating with the ruthenium; in this way delivery of hydrogen to one diastereoface of the ketone in the complex is facilitated.

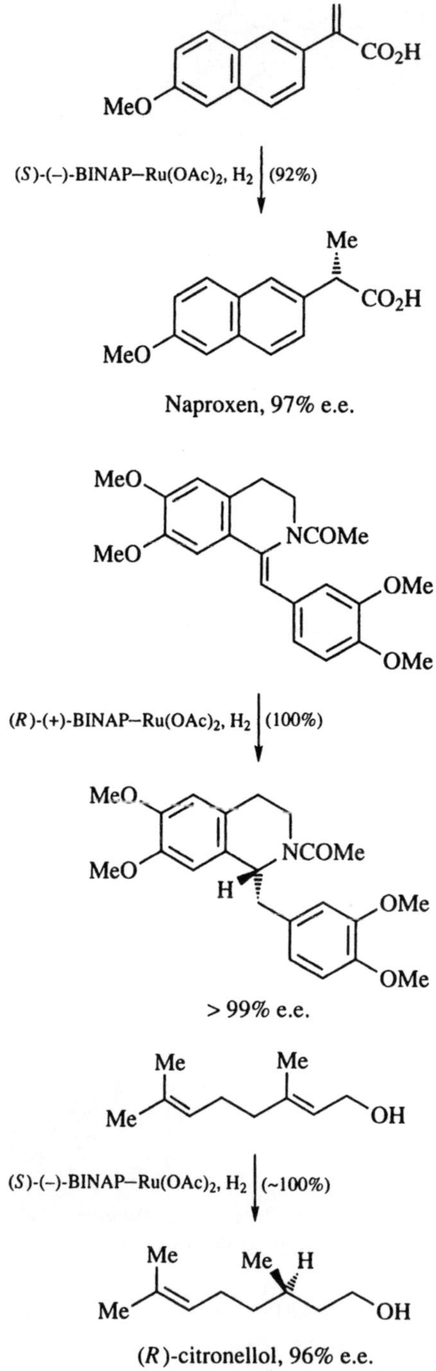
CO2H
MeO
(S)-(–)-BINAP–Ru(OAc)2, H2 (92%)
Me
CO2H
MeO
Naproxen, 97% e.e.
MeO
MeO
NCOMe
OMe
OMe
(R)-(+)-BINAP–Ru(OAc)2, H2 (100%)
MeO
MeO
NCOMe
H
OMe
OMe
> 99% e.e.
Me
Me
Me
OH
(S)-(–)-BINAP–Ru(OAc)2, H2 (~100%)
Me
Me
H
Me
OH
(R)-citronellol, 96% e.e.

Scheme 21

Table 1 Examples of ketones reduced enantioselectively using Ru–BINAP catalysts

Ketone	Alcohol	Yield	e.e. (%)	Catalyst (EtOH solution)
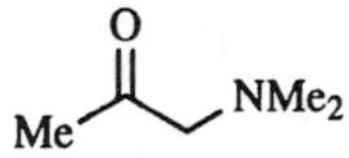 O, Me, NMe$_2$	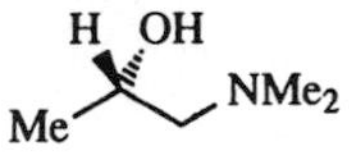 H OH, Me, NMe$_2$	72%	96	Ru(OAc)$_2$[(S)-BINAP]
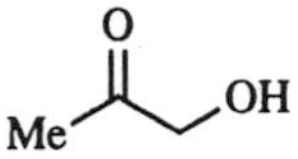 O, Me, OH	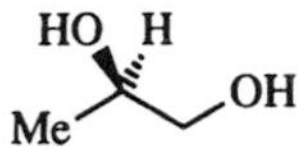 HO H, Me, OH	100%	92	RuCl$_2$[(R)-BINAP]
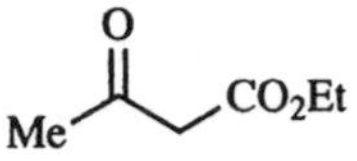 O, Me, CO$_2$Et	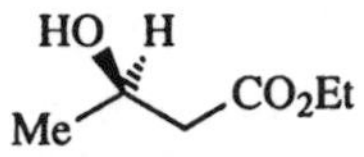 HO H, Me, CO$_2$Et	100%	> 99	RuBr$_2$[(R)-BINAP]
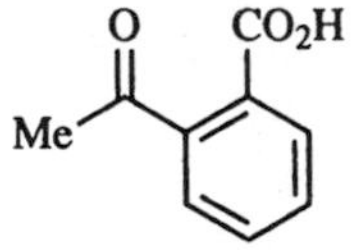 O, CO$_2$H, Me	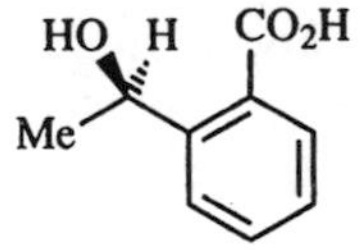 HO H, CO$_2$H, Me	97%	92	Ru$_2$Cl$_4$[(R)-BINAP]$_2$Et$_3$N
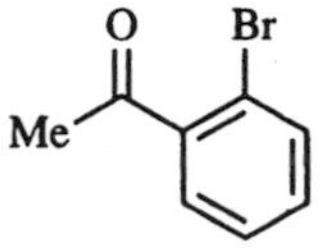 O, Br, Me	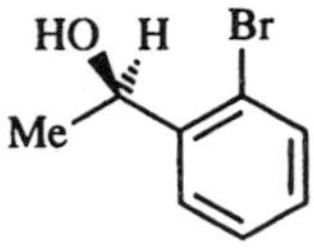 HO H, Br, Me	97%	92	RuBr$_2$[(R)-BINAP]

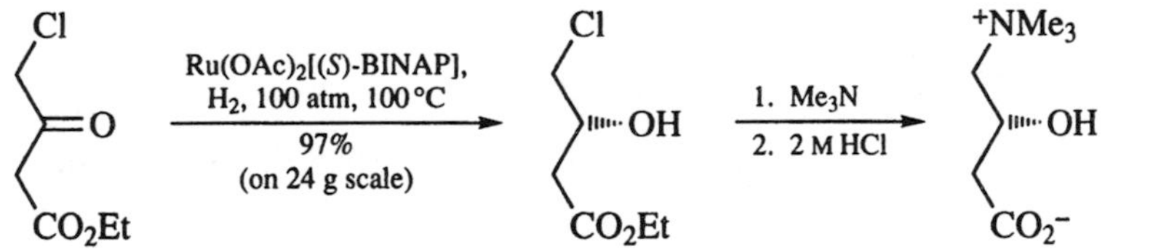

Scheme 22

Enantioselective reduction of the γ-substituted ketone in Scheme 23 requires modification of the ruthenium catalyst by addition of two equivalents of hydrogen chloride. The product (**14**) is a sex pheromone of the female dermistid beetle.[30]

Bakers' yeast is able to accomplish many of the above ketone reductions enantioselectively (see Chapter 16). However, non-enzymic catalytic reductions can be carried out using far higher concentrations of substrate and for larger scale work may be more expedient. Another advantage is that the

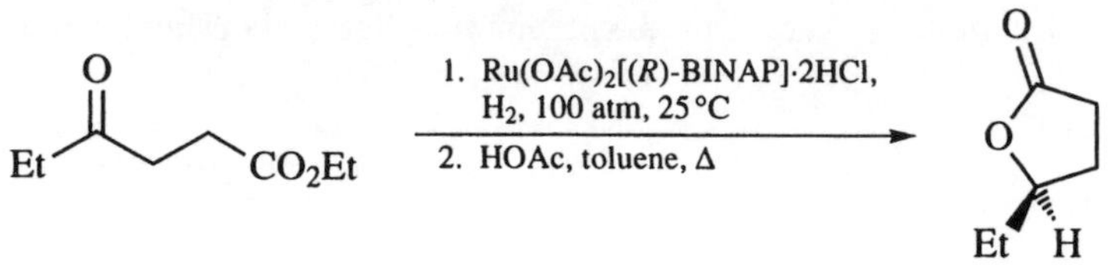

(**14**) 90% e.e.

Scheme 23

product may be obtained in either enantiomeric form, as required, by using (R)- or (S)-BINAP; this facility is not so readily available with bakers' yeast. On the other hand, high-pressure hydrogenation equipment is required for the transition metal-catalysed reactions and greater demands are made on the experimentalist in the manipulation and transfer of air-sensitive catalysts.

Recent research by Noyori and co-workers,[31] who have been responsible for virtually all the results using Ru–BINAP for reduction of ketones, has been directed towards catalysts which are more air-stable and which can be used at low pressures of hydrogen. An air-stable and easily prepared catalyst derived from BINAP and $RuCl_2$(cyclooctadiene)$_2$ has been found by other workers to reduce β-keto esters under ~ 4 atm pressure of hydrogen with an enantioselectivity comparable to that achieved by Noyori and co-workers; a temperature of 80 °C was required, however.[32]

DYNAMIC KINETIC RESOLUTION

A limitation of normal kinetic resolution is that only 50% of each enantiomer is present in the racemic starting material, and this therefore constitutes the maximum theoretical yield of either the enantiopure product or of enantiopure unreacted starting material. If the enantioselective reaction underlying the kinetic resolution is accompanied by racemisation of the slower reacting enantiomer, then in principle *complete* conversion of the racemic starting material to enantiopure product is conceivable.

Noyori *et al.*[33] have accomplished such a conversion using enantioselective reduction of β-keto esters in a medium (CH_2Cl_2) in which racemisation of the starting material is faster than the reduction of the slower reacting enantiomer (Scheme 24). The *sense* of *enantioselective* reduction, i.e. via (**15**) rather than (**16**), is controlled by asymmetric induction from the BINAP residue. The *diastereoselectivity* of product formation, i.e. (**17**) rather than (**18**) has been rationalised by a model in which the ester is ruthenium coordinated and hence *syn* delivery of the hydride takes place as shown in (**15**).

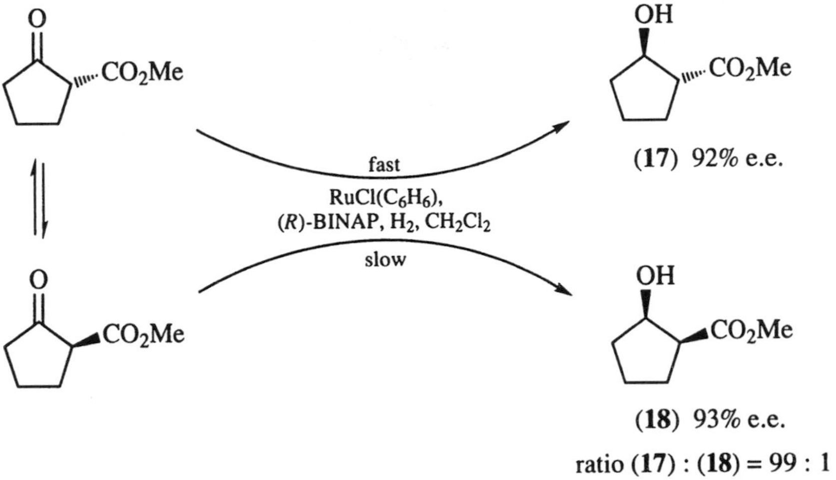

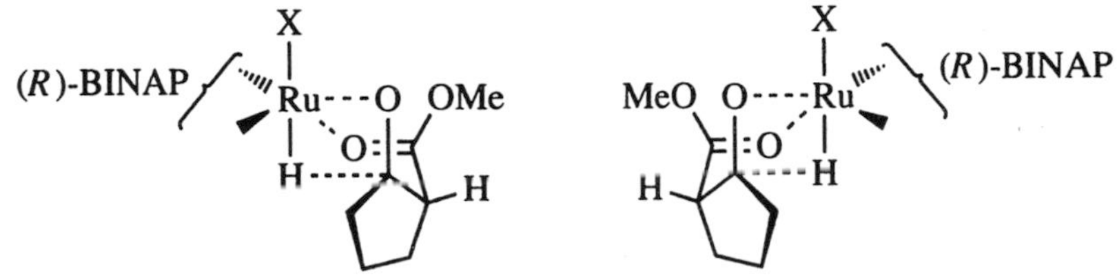

(17) 92% e.e.

(18) 93% e.e.

ratio (17) : (18) = 99 : 1

(15)   (16)

(X = halogen, H, H_2 or solvent)

Scheme 24

Hydroboration

The admirable regioselectivity of the hydroboration of double bonds, with the boron atom adding to the sterically less hindered carbon atom, nevertheless limits the range of organoboranes, and hence alcohols, that can be prepared. Moreover, in overall enantioselective hydration using this method (see Chapter 12), the enantiopure hydroborating reagent must be used in stoichiometric quantity. A seminal discovery by Männig and Noth[34] revealed that the addition of catecholborane to alkenes is catalysed by certain rhodium complexes and that the regioselectivity of addition to styrene is reversed in comparison with that in the uncatalysed addition. The mechanism of the catalysed hydroboration is thought to be as in Scheme 25;[35] Rh–L represents a coordinatively unsaturated rhodium complex.

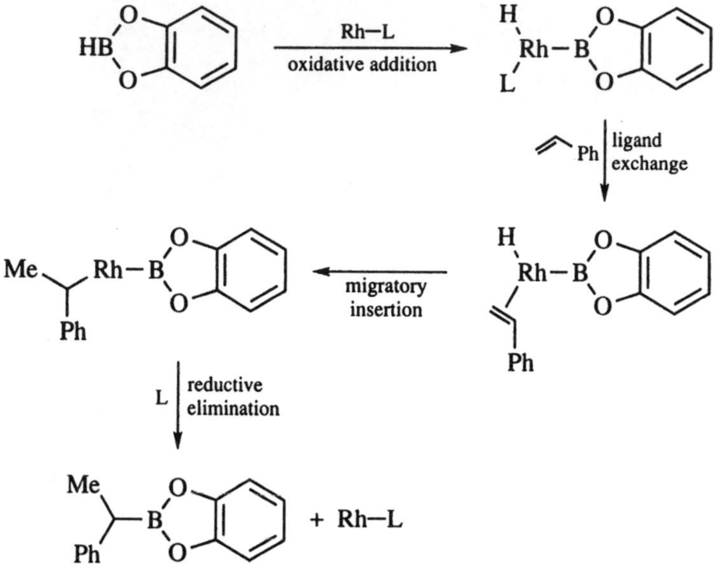

Scheme 25

Using BINAP-complexed rhodium, this catalysed hydroboration has been accomplished with high enantioselectivity using styrene and ring-substituted styrenes (Scheme 26).[35,36]

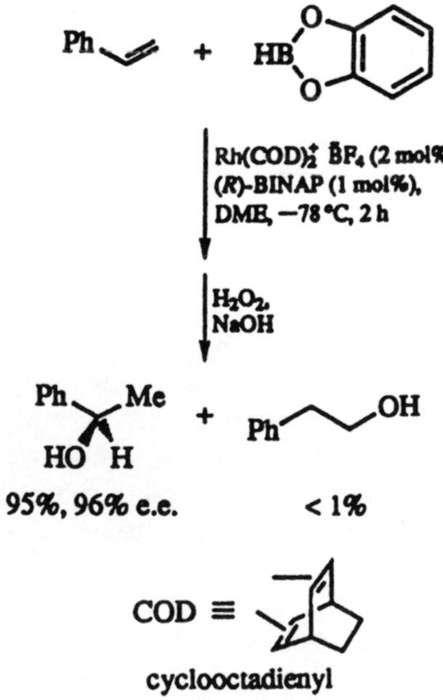

Scheme 26

Cycloadditions

DIELS–ALDER REACTIONS

Diels–Alder reactions using α,β-unsaturated carbonyl derivatives as the dienophile are catalysed by Lewis acids which complex to the carbonyl oxygen atom (Chapter 6). With enantiopure ligands on the Lewis acid (AL_n*), enantioselective Diels–Alder reactions have been successfully mediated.[37]

By complexation of the dienophile with the chiral Lewis acid, its enantiofaces are converted into diastereofaces (Scheme 27); in favourable cases, attack by the achiral diene takes place on only one of these diastereofaces and decomplexation, therefore, lends to a single enantiomer of the adduct.[38,39]

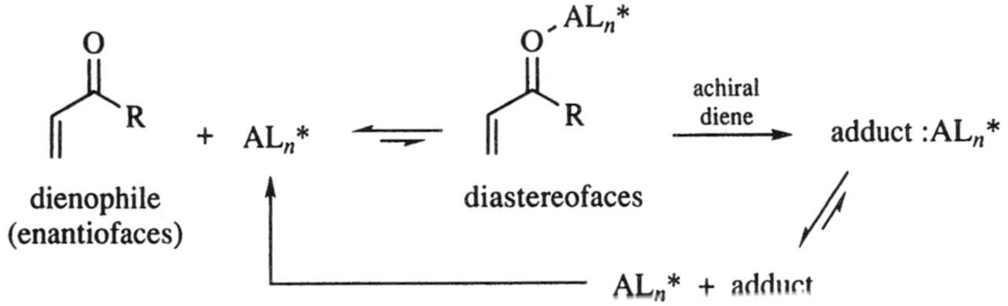

Scheme 27

endo

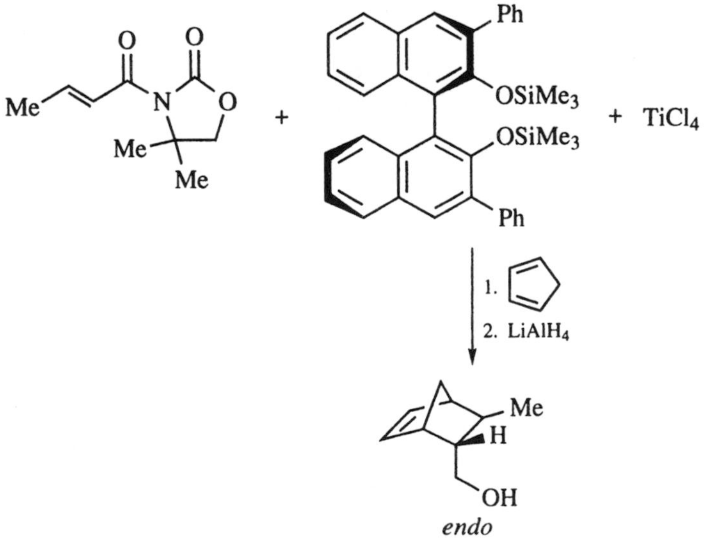

99%, 94 : 6 *endo/exo* (CH_2OH *endo*)
> 98% e.e.

Scheme 28   *(continued)*

Scheme 28 *(continued)*

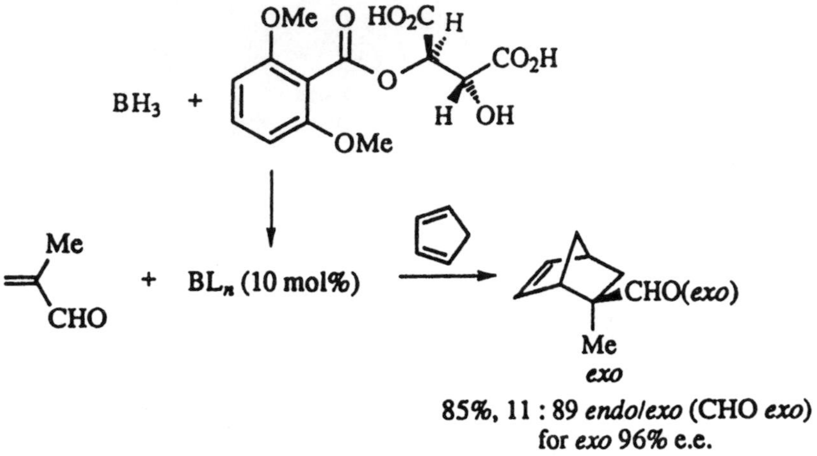
OMe O HO2C H
CO2H
O
BH3 +
H OH
OMe
Me
CHO
+ BLn (10 mol%)
CHO(exo)
Me
exo
85%, 11 : 89 endo/exo (CHO exo)
for exo 96% e.e.

Scheme 28

Attempts have been made to rationalise the sense of asymmetric induction in some cases. Thus, with Corey *et al.*'s enantiopure amidoaluminium alkyl (**19**) (Scheme 29) as a Lewis acid catalyst, the transition state assembly (**20**) is suggested for the Diels–Alder reaction shown.[40]

In this assembly (**20**) a number of conformational restrictions are required to bring about high enantioselectivity in the cycloaddition. Thus, (a) the α,β-unsaturated amide must be in the *s-trans* conformation, (b) complexing to aluminium by the amide carbonyl oxygen lone pair *trans* to the nitrogen is required and (c) the orientation of the CF_3 groups must be such as to shield one diastereoface of the dienophile towards attack by the diene. This orientation has its origin in a preferred location of the CF_3 group *anti* to the adjacent phenyl group in each case.

This Diels–Alder reaction in Scheme 29 is not only enantioselective but also diastereoselective in two occasional Type II senses (see Chapter 6): (i) the diastereoface of the diene which is unsubstituted by the benzyloxymethyl group is that which reacts and (ii) the formation of the *endo* isomer is greatly preferred over the *exo* isomer. The highly enantio-enriched bicyclo[2.2.1]heptene derivative (**21**) was converted into (**22**) (enantiopure after crystallisation), which is an intermediate in a route to prostaglandins previously devised by Corey and Ensley[41] in 1975. This earlier route had as its key step the diastereoselective Type II/III Diels–Alder reaction of an enantiopure acrylate with the same 5-benzyloxymethylcyclopentadiene (Scheme 30), i.e. it made use of a chiral auxiliary in the acrylate.

The Diels–Alder cycloaddition of cyclopentadiene to methyl crotonate in

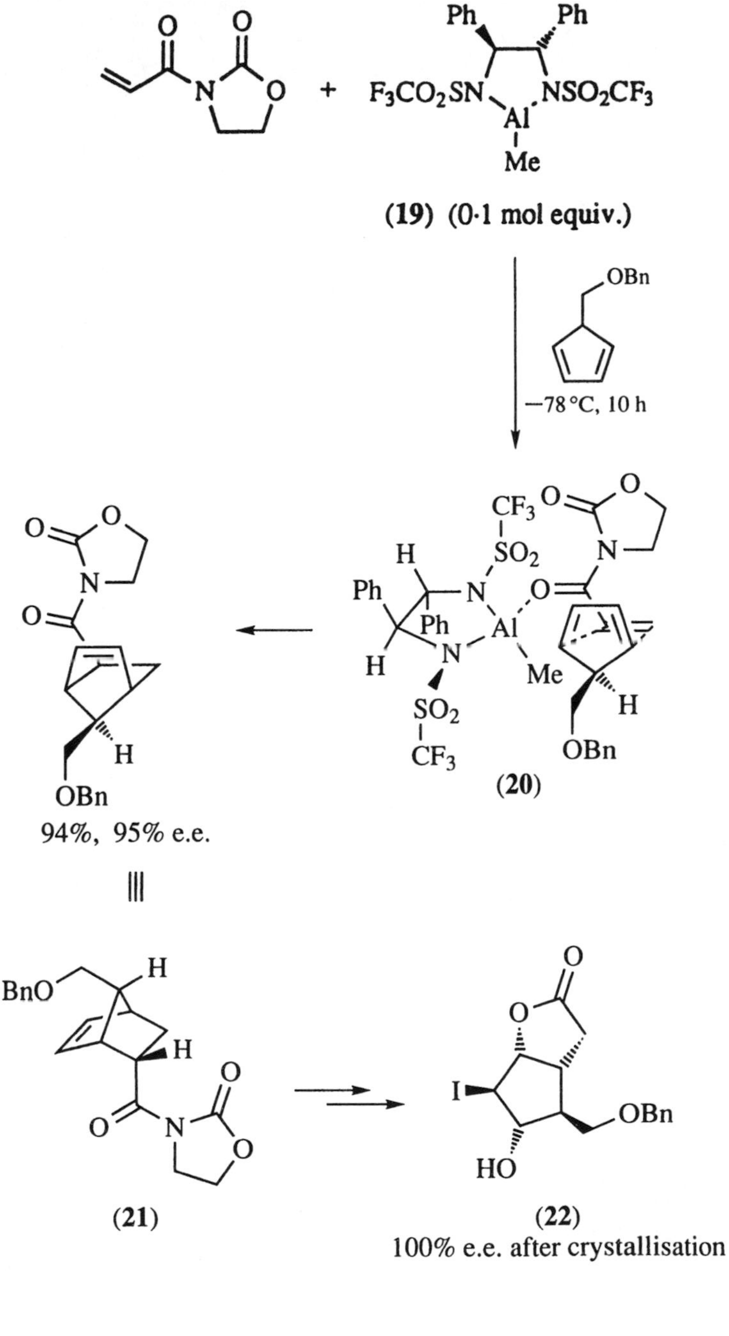
Ph
Ph
F_3CO_2SN
NSO_2CF_3
Al
Me
(19) (0·1 mol equiv.)
OBn
−78 °C, 10 h
CF_3
SO_2
H
Ph
Ph
H
Me
OBn
(20)
94%, 95% e.e.
BnO
(21)
I
HO
OBn
(22)
100% e.e. after crystallisation

Scheme 29

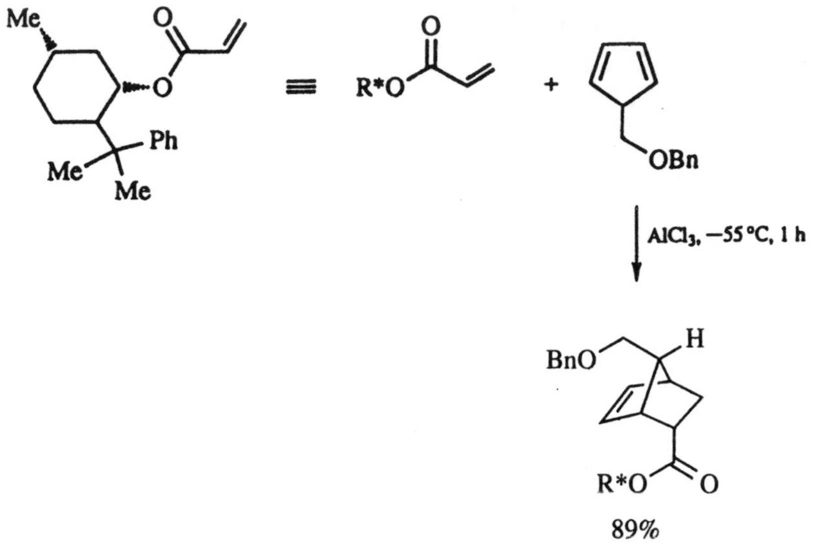

Scheme 30

the presence of the enantiopure dichloroborane (**23**) (Scheme 31) is thought to proceed through a transition state resembling (**24**) in which one face of the dienophile is blocked by the naphthalene ring.[42] This blocking is possibly a result of complexation of the carbonyl oxygen with the (electron-deficient) boron and an attractive electrostatic or dipole–induced–dipole interaction between the (electron-deficient) methoxycarbonyl unit and one of the (electron-rich) rings of the naphthalene. Support for this representation of the transition state comes from examination of the X-ray structure of the crystalline complex (**25**) which the dichloroborane forms with methyl crotonate; the geometry of (**25**) appears from n.m.r. spectroscopy to be maintained in solution.

The work of Bednarski and Danishefsky[43] has shown that hetero Diels–Alder additions of electron-rich dienes with carbonyl or imine groups as dienophiles can be accomplished using various Lewis acid catalysts (Chapter 6). The use of lanthanides substituted with enantiopure ligands, e.g. (+)-Eu(hfc)$_3$, however, gives only modest enantioselectivities with achiral but prochiral dienes (using enantiopure dienes, the diastereoselectivity may be high (double asymmetric induction: see Chapter 15)).

Yamamoto and co-workers[44] have used an enantiopure binaphthyl-complexed aluminium catalyst to bring about similar hetero Diels–Alder additions. Their transition-state assembly for rationalisation of the sense of

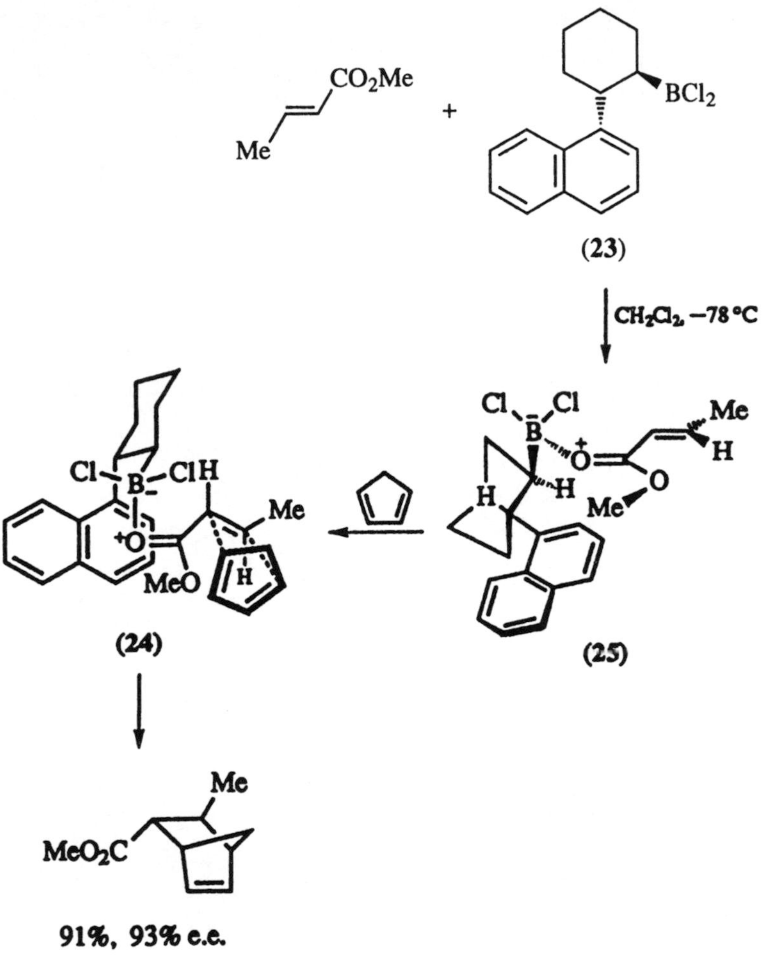

(23)

(24)   (25)

91%, 93% e.e.

Scheme 31

induction is depicted in Scheme 32. In this case the dienophile cannot, of course, exist in *s-cis* and *s-trans* conformations as can α,β-unsaturated ketones; complexation is assumed to be *trans* to the benzene ring of benzaldehyde.[44]

Note that both occasional Type II diastereoselectivity (Me and Ph *cis* or *trans*) and the enantioselectivity are increased by changing the aryl groups on silicon from phenyl to 3,5-dimethylphenyl.

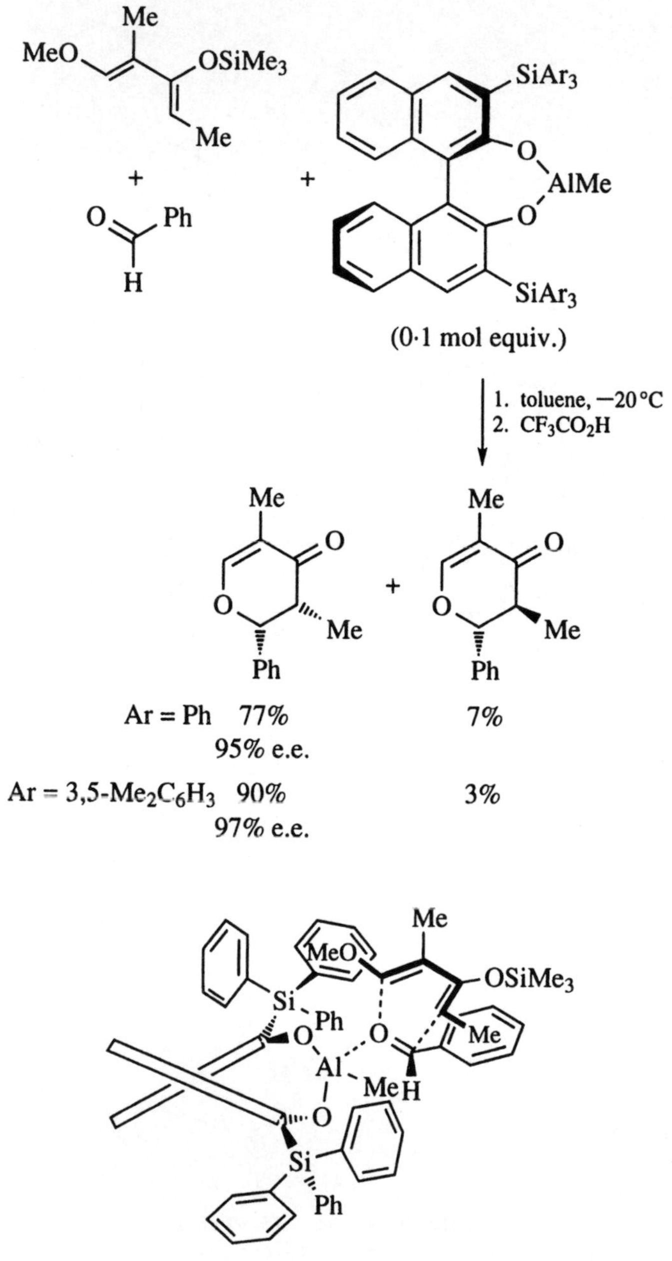
Me
MeO
OSiMe3
Me
SiAr3
O
AlMe
O
SiAr3
O
Ph
H
(0·1 mol equiv.)
1. toluene, −20 °C
2. CF3CO2H
Me
O
O
Me
Ph
Me
O
O
Me
Ph
Ar = Ph 77%
95% e.e.
7%
Ar = 3,5-Me2C6H3 90%
97% e.e.
3%
Me
MeO
OSiMe3
Si
Ph
O
O
Al
Me
MeH
O
Si
Ph
Me

Scheme 32

Ene reactions

The similarity and co-occurrence of the Diels–Alder and ene reactions has been mentioned earlier (Chapter 6). Since the carbonyl group can function as a dienophile or as an enophile in the hetero Diels–Alder and ene reaction, respectively, and since both reactions are catalysed by Lewis acids, it is not surprising that both can occur enantioselectively, and even competitively, in the presence of an enantiopure Lewis acid (Scheme 33).[45]

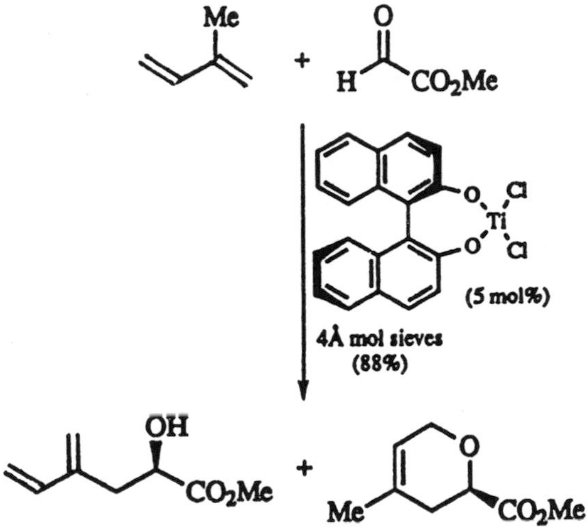

Scheme 33

Competition from Diels–Alder addition can be eliminated by using an ene component that is not part of a diene as in the conversion in Scheme 34.[46] The catalyst in this reaction is prepared *in situ* from binaphthol and

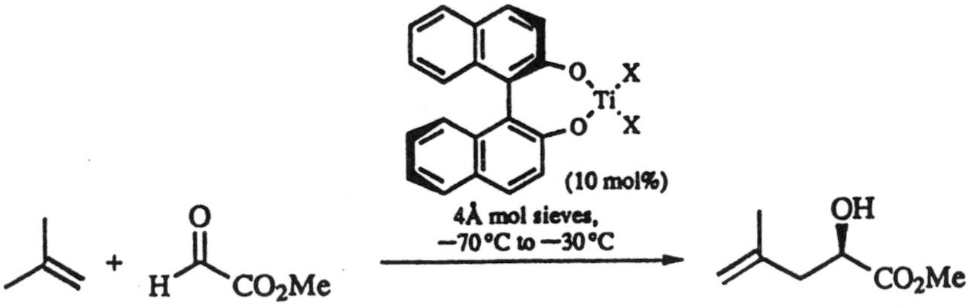

X = Cl 72%, 95% e.e.
X = Br 87%, 94% e.e.

Scheme 34

diisopropoxytitanium dihalide in the presence of 4A molecular sieves; even as little as 1% is effective.

Yamamoto and co-workers[47] have applied the substituted binaphthol-complexed aluminium alkyl (26), as used in the hetero Diels–Alder reaction in Scheme 32, to the enantioselective ene reaction in Scheme 35.[47]

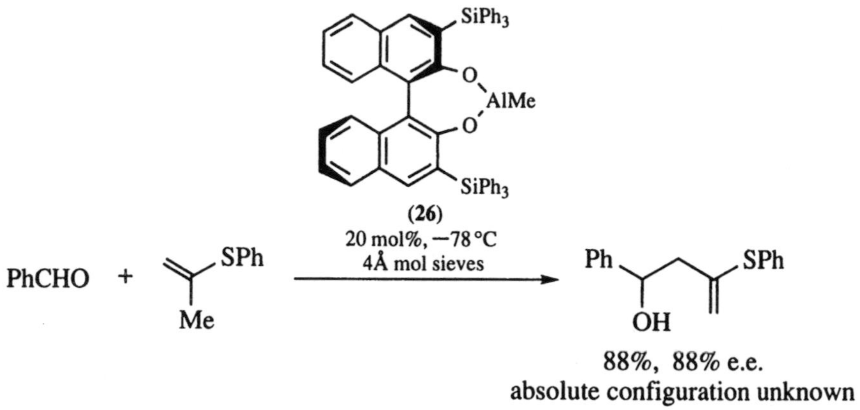

(26)
20 mol%, −78 °C
4Å mol sieves

PhCHO   +   [structure with SPh and Me] → [product: Ph, OH, SPh]

88%, 88% e.e.
absolute configuration unknown

Scheme 35

Hydroxylation (also known as dihydroxylation)

Work by Sharpless and co-workers from the late 1970s onwards showed that enantioselective *syn*-hydroxylation could be accomplished using stoichiometric amounts of osmium tetraoxide with cinchona alkaloids, e.g. dihydroquinine acetate, as chiral ligands (Scheme 36).[48] Since then a variety of other chiral ligands have been used successfully in this reaction, including the diamines (27)–(31).[49]

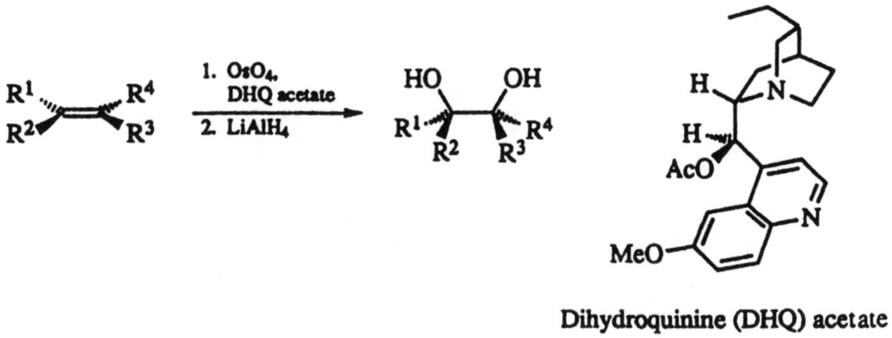

R^1, R^2, R^3, R^4
1. OsO$_4$,
   DHQ acetate
2. LiAlH$_4$

→ HO OH product with R^1, R^2, R^3, R^4

Dihydroquinine (DHQ) acetate

Scheme 36

(27) (R = alkyl)   **(28)**   **(29)**

(30)   **(31)**

The finding by Sharpless and co-workers[50,51] in the late 1980s that this enantioselective hydroxylation could be carried out using *catalytic* amounts of osmium tetraoxide in the presence of N-methylmorpholine N-oxide as reoxidant, was of importance, bearing in mind the expense and toxicity of OsO_4. Enantioselectivities were inferior to those obtained using stoichiometric amounts of OsO_4, but were improved by slow addition of the alkene to a mixture of the other components. This modification of the procedure discourages participation of a second catalytic cycle from which the diol product is enantiomeric with that obtained from the major cycle. It was subsequently found that enantioselectivities comparable to those obtained using stoichiometric quantities of OsO_4 can be achieved by substituting potassium hexacyanoferrate(III) for N-methylmorpholine N-oxide as the reoxidant for osmium(VI).[52]

A further refinement that gives enantioselectivities routinely approaching, and sometimes better than 90% is the use of the 9-O-(4'-methyl-2'-quinolyl) or 9-O-(9'-phenanthryl) ether of dihydroquinidine **(32)** (DHQD) or dihydroquinine **(33)** (DHQ) as the enantiopure ligand on the osmium.[53] Now, of the six possible substitution patterns for alkenes, only the *cis*-di- **(34)** and *tetra*- **(35)** fail to give the corresponding diols with high enantioselectivity.

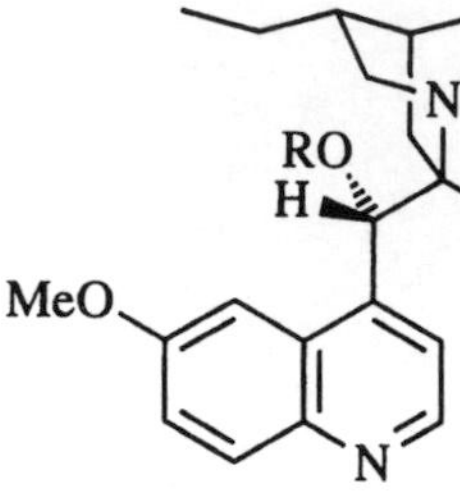

(32) (DHQD) 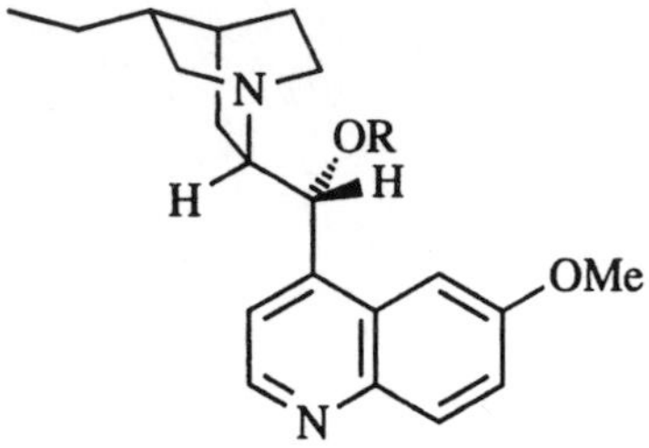   (33) DHQ

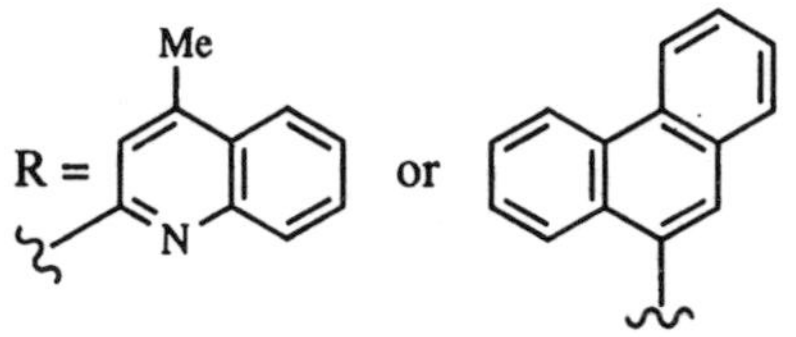

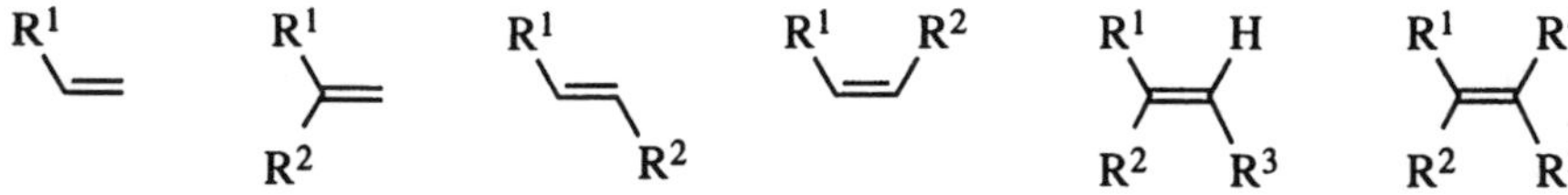

R = or

(34)   (35)

Using the less hazardous involatile solid osmium(VI) salt $K_2OsO_2(OH)_4$ (0.05 mol%) in place of OsO_4 and with an alkaloid ligand concentration of 0.02 M, the four remaining alkene types can be *syn*-hydroxylated to give opposite enantiomers of the diol using (32) or (33), respectively (Scheme 37).[54]

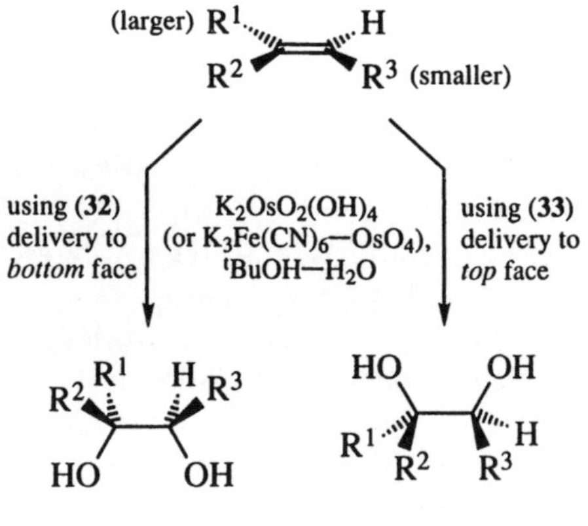

(larger) R^1 H
        R^2 R^3 (smaller)

using (32) $K_2OsO_2(OH)_4$ using (33)
delivery to   (or $K_3Fe(CN)_6$–OsO_4),   delivery to
bottom face ^tBuOH–H_2O *top* face

Scheme 37

The sense of induction is as shown in Scheme 37. For *trans*-disubstituted alkenes (R^2 = H), using dihydroquinidine (32), the hydrogen *cis* to the larger group R′ is placed to the rear right-hand side when viewed with the perspective illustrated and hydroxylation takes place from the bottom face; for trisubstituted alkenes ($R^2 \neq$ H), R^2 must be a small group (CH_3 or CH_2R) for high enantioselectivity. Using dihydroquinine (33) the opposite face of the alkene is attacked and the sense of induction is reversed. In these hydroxylations, dihydroquinidine (32) and dihydroquinine (33) behave as though they were enantiomeric although they are, in fact, diastereoisomeric.

As regards the mechanism of OsO_4 hydroxylation, it is still uncertain whether the osmate ester intermediate is formed (irreversibly) in a [3 + 2] cycloaddition of the alkene to the osmium tetraoxide or whether it arises from a metallooxetane, formed by reversible [2 + 2] addition of an Os=O bond to the C=C π-bond, followed by irreversible rearrangement (Scheme 38).[55]

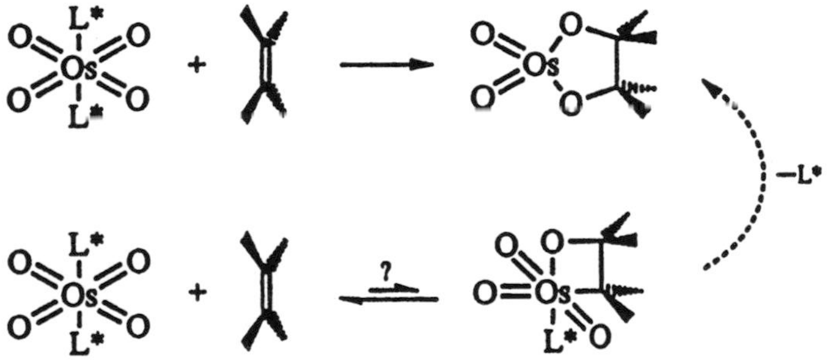

L* = chiral ligand

Scheme 38

The sense of induction in enantioselective hydroxylation of *trans*-stilbene with (30) and (31) as diamine ligands has been interpreted as supporting [2 + 2][49d] and [2 + 3][49e] cycloaddition pathways for the reactions, respectively, although different mechanisms in these cases would not be expected.

It appears that the enantioselective hydroxylations using OsO_4 described above belong to a select group of reactions in which complex formation between the substrate and the reagent prior to addition to the double bond does not take place; the enantiofaces of the starting alkene become diastereofaces only in the transition state for the [3 + 2] or [2 + 2] cycloaddition. (A reversible [2 + 2] cycloaddition, however, would allow for the formation of the thermodynamically preferred metallooxetane.) Enantioselective reactions in this category are inherently of wider scope than those that require a subsidiary functional group for complexing to the reagent, prior to addition to one (diastereo)face of the double bond.

Cyclopropanation[56]

The most successful catalytic enantioselective cyclopropanations have used carbenoids derived from diazoalkanes, and in particular α-diazo esters, with metal catalysts (especially Co, Cu, Rh and Pd) bearing enantiopure ligands. Important developments have included the introduction of soluble copper chelates by Nozaki *et al.*[57] and the discovery by Salomon and Kochi[58] of copper triflate (trifluoromethanesulphonate) as a very active catalyst in the Cu[I] state.

The ability of the ligands on copper to bring about enantioselectivity has usually been assayed using reaction of styrene with ethyl diazoacetate (Scheme 39). In this reaction, enantioselectivity in the formation of the *trans* isomer (36) has been improved from ~10%, obtained (over 25 years ago) by

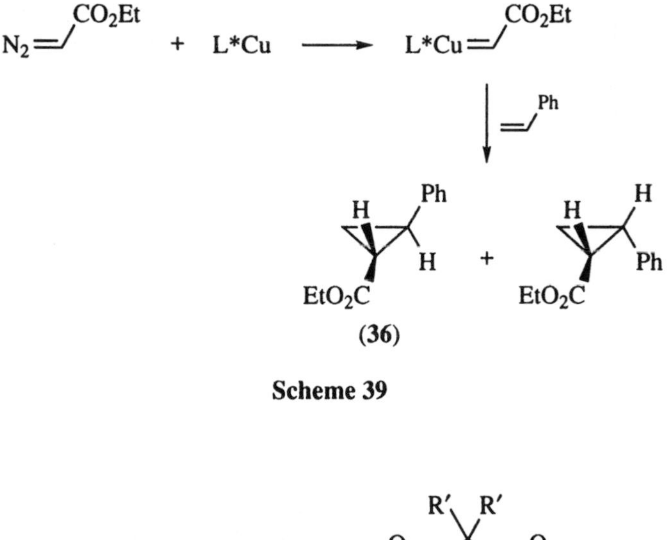

Scheme 39

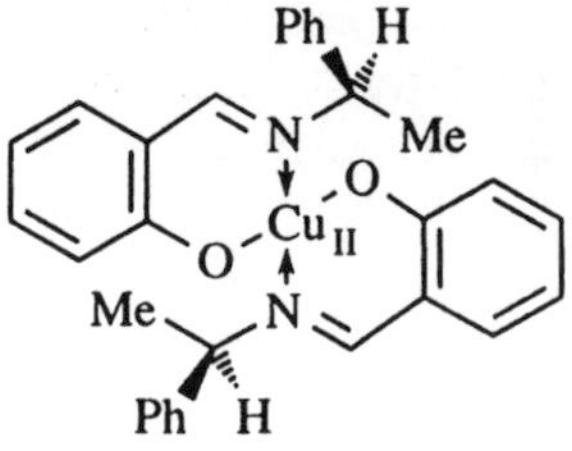

(37) Nozaki

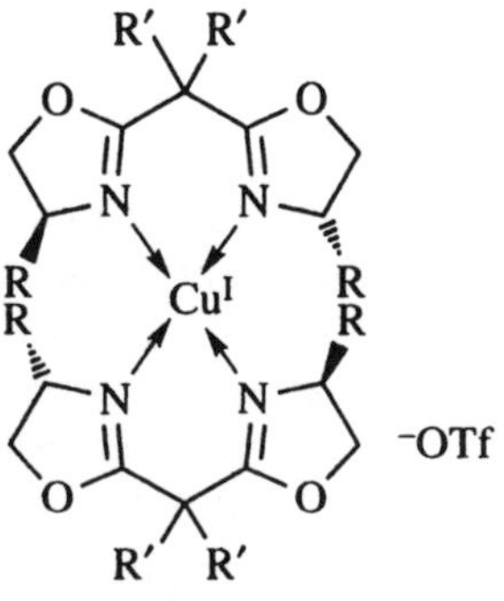

(38) Masamune, Evans

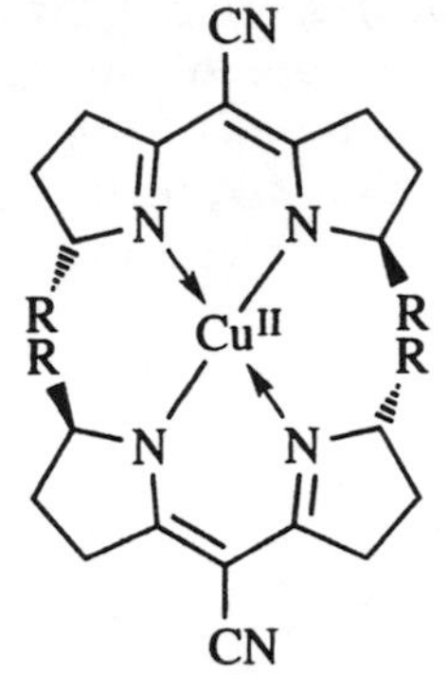

(39) Pfaltz

Nozaki *et al.* using the salicylaldimine ligands (37), to better than 95% using the bisoxazolines (38) of Evans *et al.*[59] and of Masamune and co-workers[60] and the semi-corrin derivatives (39) of Pfaltz and co-workers.[61] Evans *et al.* found that, using (38) and a bulky ester of diazoacetate, not only is there good enantioselectivity but also the *trans:cis* ratio of cyclopropane products can be raised to 94:6 (Scheme 40).

Catalyst design for these enantioselective cyclopropanations has been largely empirical, reflecting the lack of detailed understanding of the means

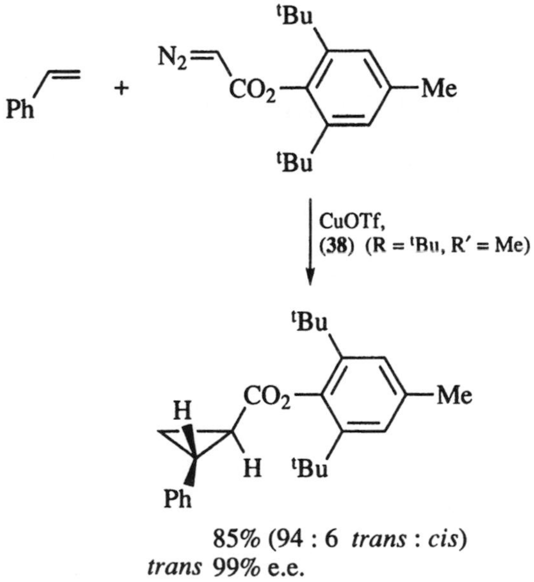

85% (94 : 6 *trans* : *cis*)
trans 99% e.e.

Scheme 40

by which chirality is propagated by the ligands on the metal in the carbenoid intermediate. In the case of the semi-corrin derivatives (39), Pfaltz and co-workers[61] suggested that a copper carbenoid resembling (40) (Scheme 41) is attacked from the rear, by the more nucleophilic carbon of the alkene, because then the R group and ester group move away from each other.

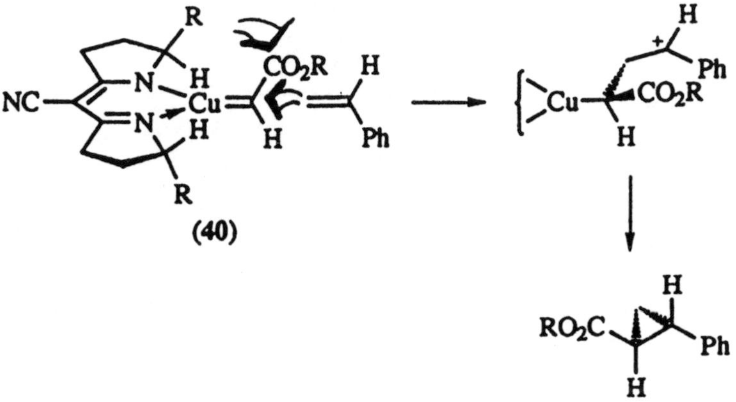

(40)

Scheme 41

This model also accounts for the fact that the *cis:trans* ratio of cyclopropanes is not sensitive to the nature of R in the ligand, because of the distance between it and the (phenyl) group on the alkene in the transition state.

Alkenes other than monosubstituted ones can also be enantioselectively cyclopropanated using (38) and a diazo compound. Evans *et al.*[59] have carried out the conversion in Scheme 42 on a 35 g scale using the enantiomer of (38) with R = ^{t}Bu, R' = Me, at a concentration as low as 0.1%.

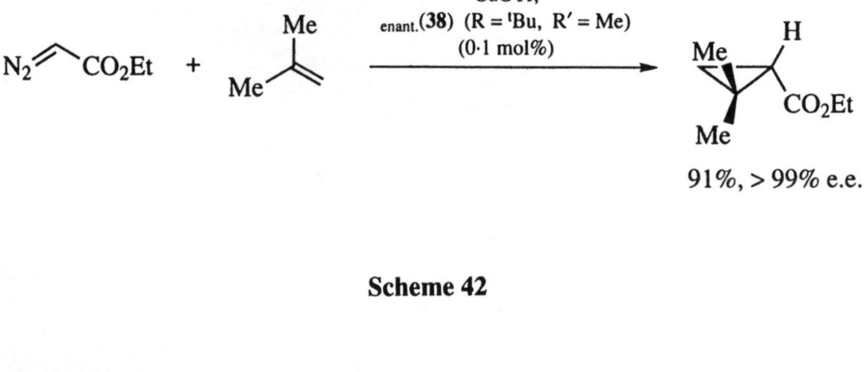

91%, > 99% e.e.

Scheme 42

Enantioselectivity in intramolecular cyclopropanations has been most .successful using rhodium substituted with enantiopure ligands and substrates such as allyl diazoacetates (Scheme 43).[62]

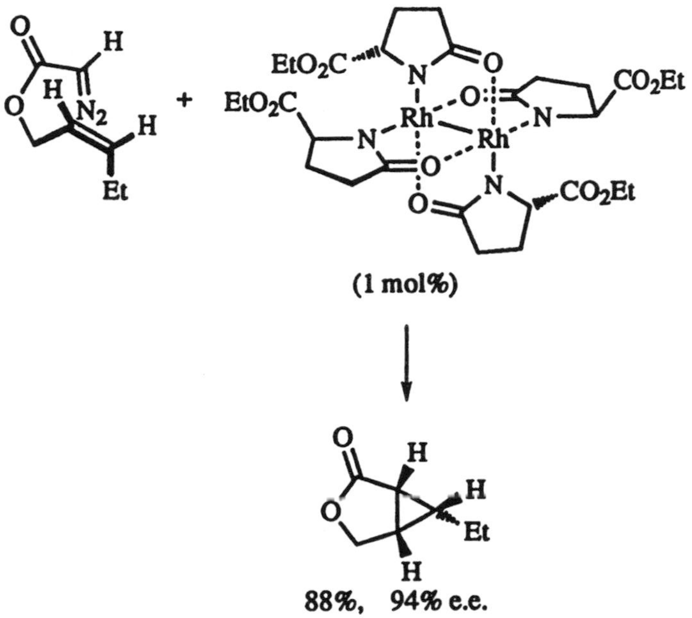

(1 mol%)

88%,   94% e.e.

Scheme 43

Alkyl addition to carbonyl groups

Additions of Grignard and organolithium reagents to aldehydes and ketones are much-used reactions and the development of enantioselective methods for these and other nucleophilic additions to carbonyl groups is desirable.

The use of enantiopure solvents has resulted in only modest levels of enantioselectivity.[63] More success has been achieved by making use of the ability of [1,n]-substituted diamines to complex with the metal, e.g. in an organolithium. Thus Mukaiyama et al.[64] added butyllithium to benzaldehyde enantioselectively in the presence of molar amounts of the proline-derived diamine (41) at very low temperatures (−123 °C) (Scheme 44), and a catalytic version of the reaction was later devised by Mazaleyrat and Cram.[65]

HOOKE

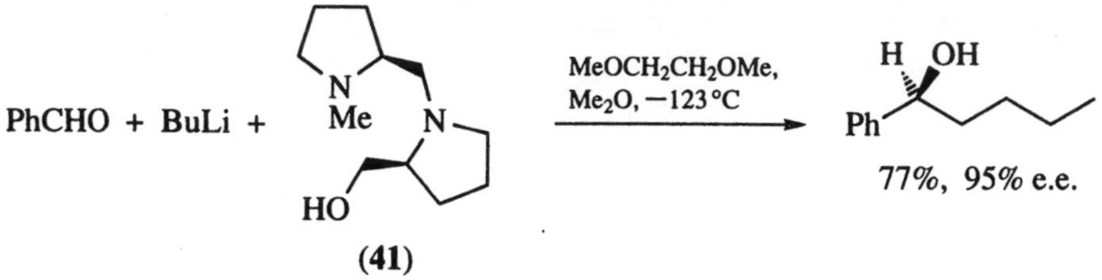
PhCHO + BuLi +
N
Me
N
HO
(41)
MeOCH₂CH₂OMe,
Me₂O, −123 °C
H OH
Ph
77%, 95% e.e.

PhCHO : BuLi : (**41**)
1 : 6·7 : 4

Scheme 44

In these alkylations there are potential advantages in the use of reagents, e.g. organotitanium compounds, in which the chiral substituents can be covalently linked to the metal in a way not possible for lithium. Enantioselectivities >85% have been accomplished using molar equivalent amounts or more of the enantiopure organotitanium reagents (**42**) and (**43**) in additions to aromatic aldehydes (Scheme 45).[66]

In most of the earlier experiments it was necessary to use the enantiopure reagents in at least stoichiometric quantity. More recently, efforts have focused on the development of methods in which only a catalytic quantity of enantiopure co-reagent is used (with the advantages already alluded to).[63,67]

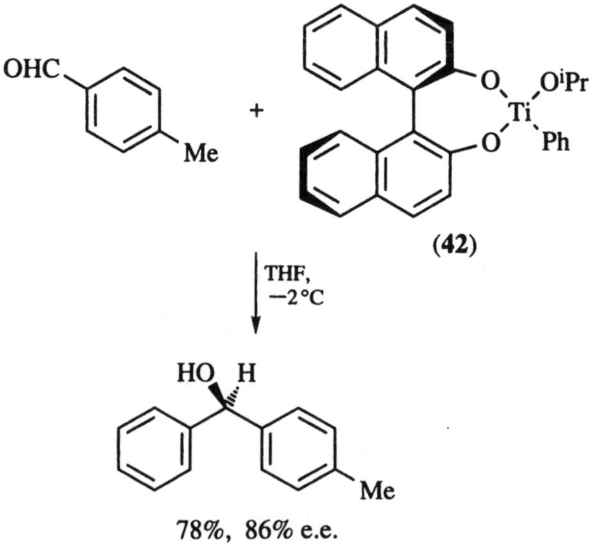
OHC
Me
+
O
OⁱPr
Ti
O
Ph
(42)
THF,
−2 °C
HO H
Me
78%, 86% e.e.

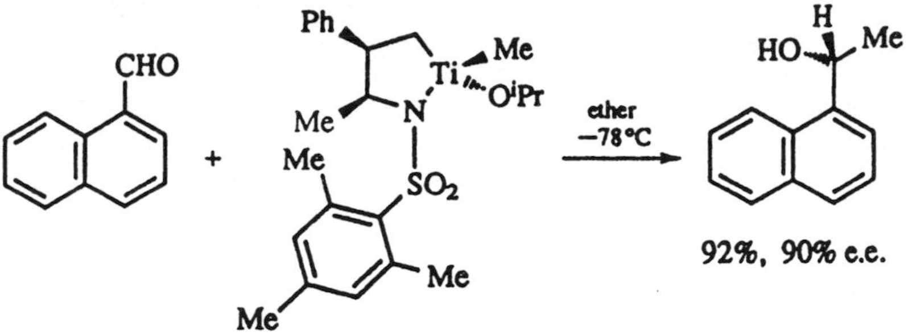

(43)

Scheme 45

Scheme 46 shows the general case in which an organometallic compound R_2M adds enantioselectively to a prochiral carbonyl group via an intermediate RMX* in which X* is an enantiopure catalyst. If a catalytic quantity of X* is to suffice for conversion of all of the carbonyl compound and R_2M into product, then some means must be found for reconstitution of RMX* from the first-formed (**44**) by reaction with R_2M.

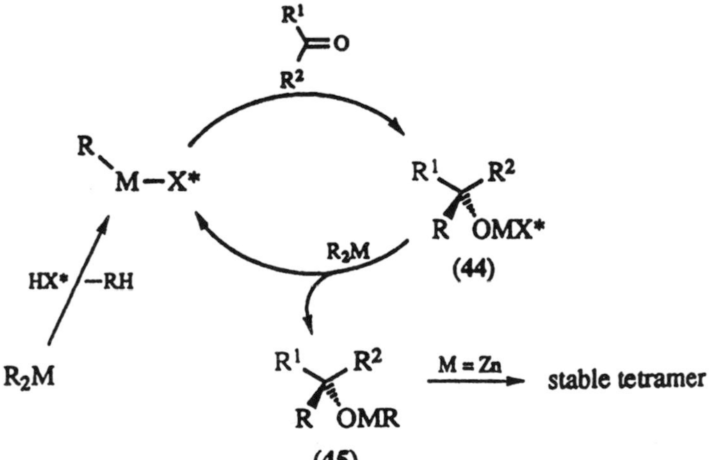

X* = enantiopure ligand

Scheme 46

Apart from the necessity for diastereoselection in addition of RMX* to the carbonyl group, it is clear that the (catalysed) addition of RMX* must be faster than the (uncatalysed) addition of R_2M.

Considerable progress has been made in catalytic enantioselective addition using zinc dialkyls as the alkyl donor R_2M in Scheme 46. Monomeric dialkylzincs are linear and fairly inert towards carbonyl compounds. Coordination of the zinc atom with an oxygen or nitrogen enhances the nucleophilic reactivity of the alkyl groups on the zinc. Most importantly, for alkyl zincates, e.g. (45) (Scheme 46), the formation of stable tetramers in hydrocarbon solvents provides a driving force for the required exchange of the chiral substituent X* and the reconstitution of RMX* referred to above. Following on an initial observation by Ogun and Omi[68] that the reaction of diethylzinc and benzaldehyde is catalysed by chiral 1,2-amino alcohols and the product 1-phenylpropanol is optically active [49% e.e. for (S)-leucinol], Noyori and co-workers[69] screened a variety of 2-amino alcohols for their catalytic (2 mol%) effect on this reaction. Particularly enhanced rates were obtained using $(-)$-3-exo-(dimethylamino)isoborneol [$(-)$-DIAB] (46) and the product from the reaction of Et_2Zn with benzaldehyde was obtained in 98% e.e. (Scheme 47).

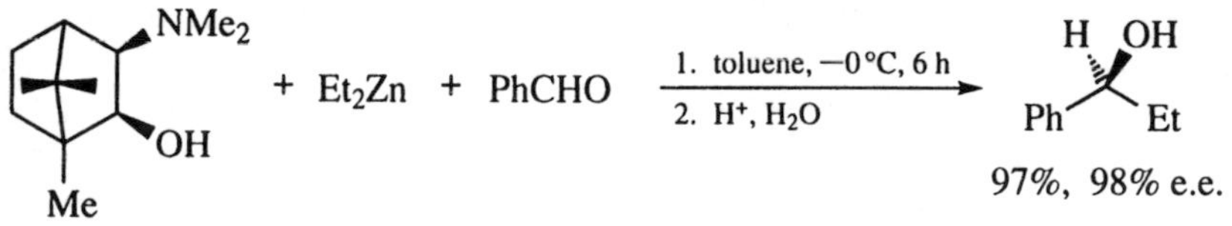

(46) $(-)$-DAIB (2 mol%)

Scheme 47

This enantioselective alkyl group addition could be extended to a range of dialkylzincs and aldehydes. The selectivity is thought to arise from a dinuclear zinc complex of type (47) or (48) (Scheme 48) within which the bridging R group is transferred to the carbonyl carbon. Whether (47) or (48) is preferred is determined by the substitution at the α- and β-positions and at the nitrogen.[70]

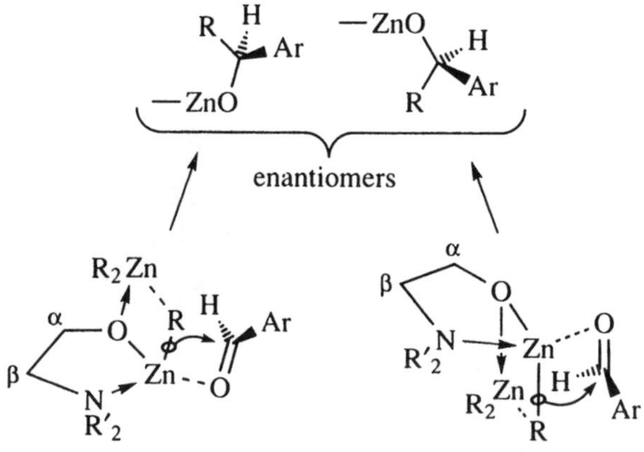

Scheme 48

Using $(-)$-DAIB as ligand, the bicyclic skeleton, including the *gem*-dimethyl bridge, will favour (49) (Scheme 49). The preferred orientation of the benzaldehyde molecule within the dinuclear complex will be as shown in (49) rather than (50) because of the non-bonded interaction between R and

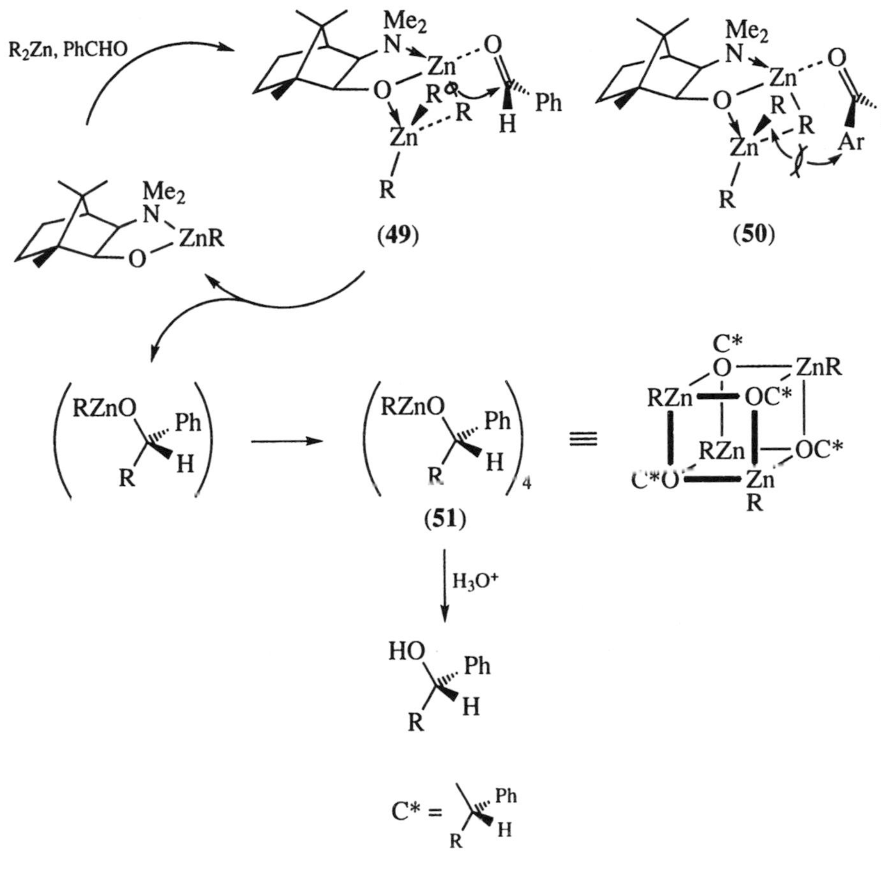

Scheme 49

Ph in the latter. Consequently, the sense of induction arises from transfer of the bridging alkyl group to the carbonyl carbon as illustrated in (49). The alkyl addition product is then eliminated from (49) as its alkylzinc salt and forms the stable tetramer (51) from which the free alcohol is liberated by treatment with acid; the zinc-complexed catalyst is regenerated.

CHIRAL AMPLIFICATION[71]

The enantiopurity of the product from an enantioselective reaction can sometimes be higher than that of the chiral agent responsible for its optical activity. This is the case in the above (−)-DAIB-catalysed reactions of zinc dialkyls with aldehydes. Thus, in Scheme 50, (−)-DAIB of only 15% e.e. results in 1-phenylpropanol of 95% e.e.

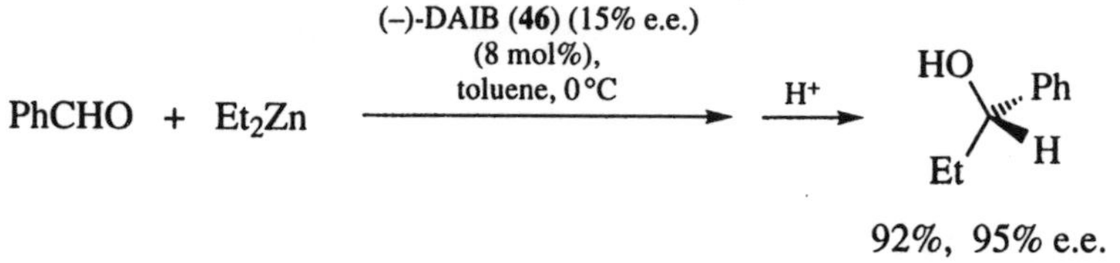

$$\text{PhCHO} + \text{Et}_2\text{Zn} \xrightarrow[\text{toluene, 0°C}]{\substack{\text{(−)-DAIB (46) (15\% e.e.)}\\ \text{(8 mol\%),}}} \xrightarrow{\text{H}^+}$$

92%, 95% e.e.

Scheme 50

This chiral amplification or increase in enantiopurity of the product by comparison with that of the inducing chiral reagent arises from the presence in solution of other (reversibly formed) diastereoisomeric dinuclear zinc complexes besides the reactive one in Scheme 49. Two diastereoisomeric dinuclear complexes (52) and (53) can be formed (Scheme 51) and it appears that (53), comprised of DAIB molecules of the opposite chirality (heterochiral), is more stable than complex (52) which is comprised of molecules of the same chirality (homochiral).

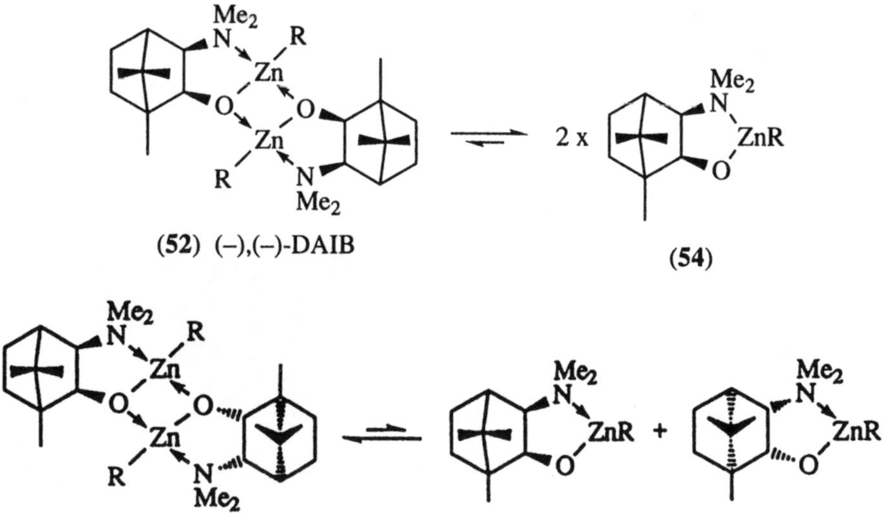

(52) (−),(−)-DAIB

(54)

(53) (−),(+)-DAIB

Scheme 51

Since it is the monomer which is required for formation of the 'productive' dinuclear complex in Scheme 49, reaction takes place essentially only via monomer (54) supplied by (52). The amplified enantiopurity, therefore, results from the effective removal of racemic material from the enantioimpure (−)-DAIB leaving the reaction to proceed with catalysis only by enantiopure (−)-DAIB.

Enantioselective allyl, alkenyl or alkynyl additions to aldehydes can be mediated by other enantiopure amino alcohol derivatives of zinc in which dinuclear complexes similar to those in Scheme 49 are believed to be involved.[72] Schmidt and Seebach[73] have added dialkylzincs to aldehydes in the presence of spiro-titanate complex (55) (Scheme 52), prepared from (R,R)-tartaric acid, and have achieved high enantioselectivities. However, the challenge of a general method for enantioselective addition of organometallic reagents to *ketones* still remains.

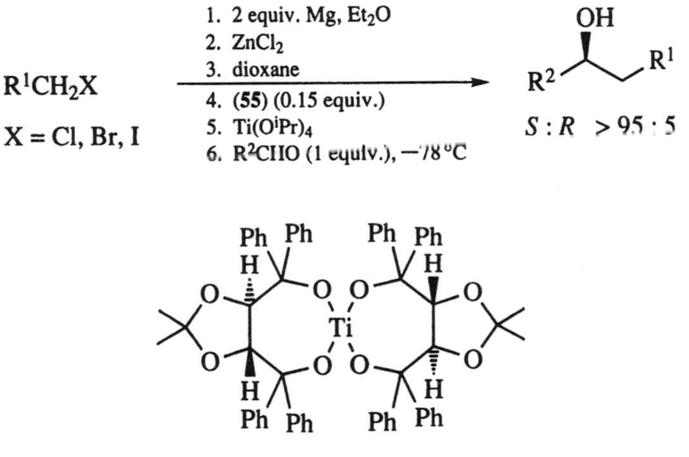

$$R^1CH_2X$$

$$X = Cl, Br, I$$

1. 2 equiv. Mg, Et$_2$O
2. ZnCl$_2$
3. dioxane
4. (55) (0.15 equiv.)
5. Ti(O^{i}Pr)$_4$
6. R^{2}CHO (1 equiv.), −78 °C

$$S : R > 95 : 5$$

(55)

Scheme 52

Aldolisation

Yamamoto and co-workers[74] have used enantiopure (acyloxy)borane complexes (Scheme 53) as catalysts to confer high enantioselectivity in the Mukaiyama reaction.

cis-Aldols result from this reaction irrespective of the silyl enol ether configuration, suggesting that open transition states are involved (see Chapter 8).

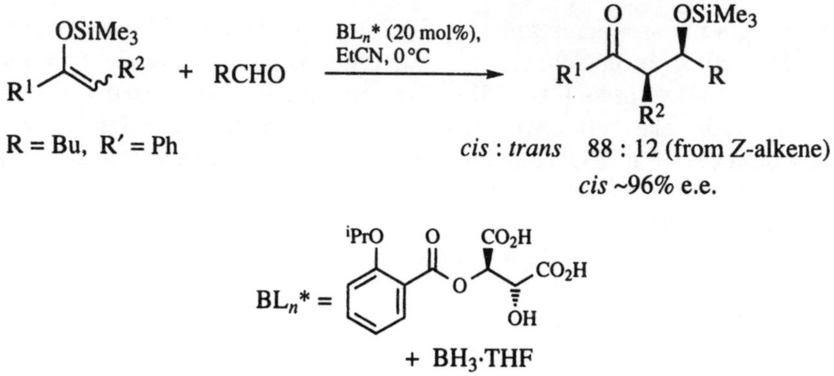

R = Bu, R' = Ph

cis : trans   88 : 12 (from Z-alkene)

cis ~96% e.e.

+ BH$_3$·THF

Scheme 53

The synthesis of oxazolines by the gold(I)-catalysed reaction of α-isocyanoacetates and aldehydes with an enantiopure ferrocenylamine ligand on the gold was shown by Ito and Hayashi[75] to proceed enantioselectively (Scheme 54). High enantioselectivity arises from

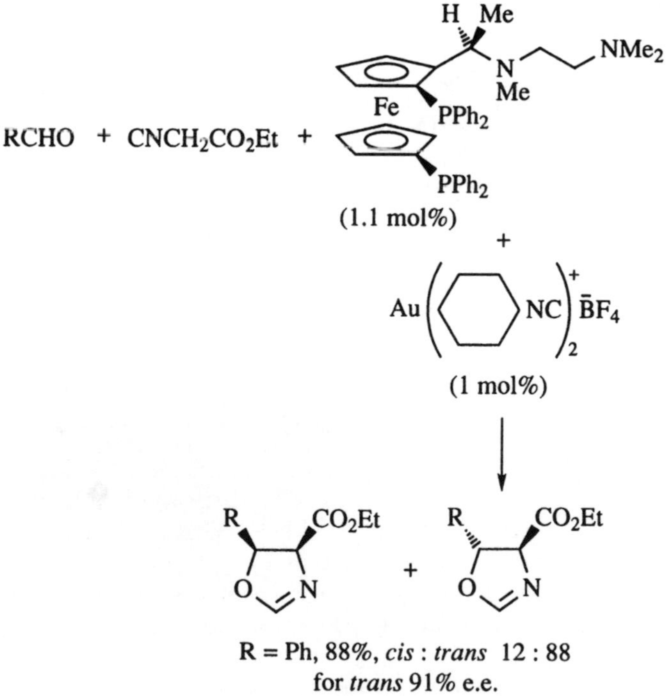

R = Ph, 88%, cis : trans 12 : 88
for trans 91% e.e.

Scheme 54

cooperativity between the chiral centre in the ligand side-chain and the chirality of the ferrocene unit.

One of the most remarkable enantioselective intramolecular aldol condensations is the Hajos reaction (Scheme 55); cyclisation of achiral triketone (56) by a catalytic amount of proline in N,N-dimethylformamide gives the aldol product (57) (two new chiral centres) in excellent yield and enantioselectivity. Although somewhat limited in scope, the reaction can be accomplished on a 50 g scale and the derived α,β-unsaturated ketone (58) is a useful enantiopure building block.[76]

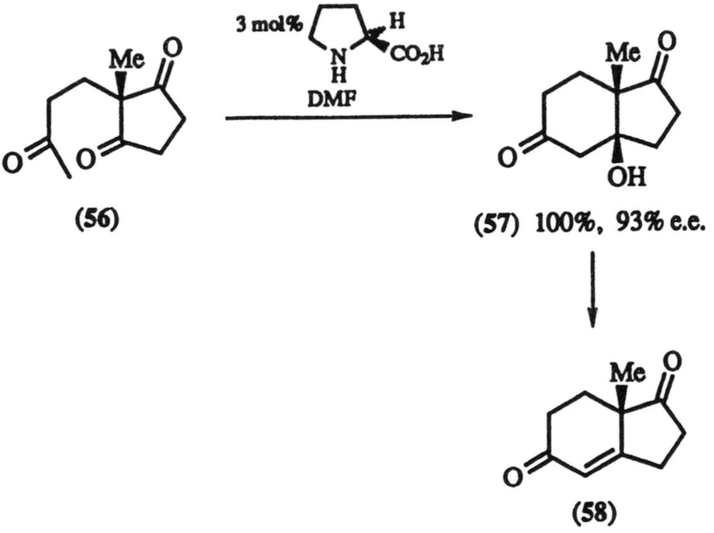

Scheme 55

Michael addition

The Michael addition, like the aldol reaction in Scheme 55, lends itself to Type III$_{cat.\ enant.}$ reactions by the use of enantiopure bases since only a catalytic quantity of the base is used in the normal course of events.

Since the first discovery of enantioselectivity in the base-catalysed Michael addition of the β-keto ester (59) to acrolein by Langstrom and Bergsen,[77] numerous catalysts have been employed in the addition of (59) and other active methylene compounds to Michael acceptors, in particular methyl vinyl ketone. The enantiopurecatalysts have included cinchona alkaloids,[78a] chiral Co^{II} complexes[78b] and the crown ether (60) in Scheme 56.[79]

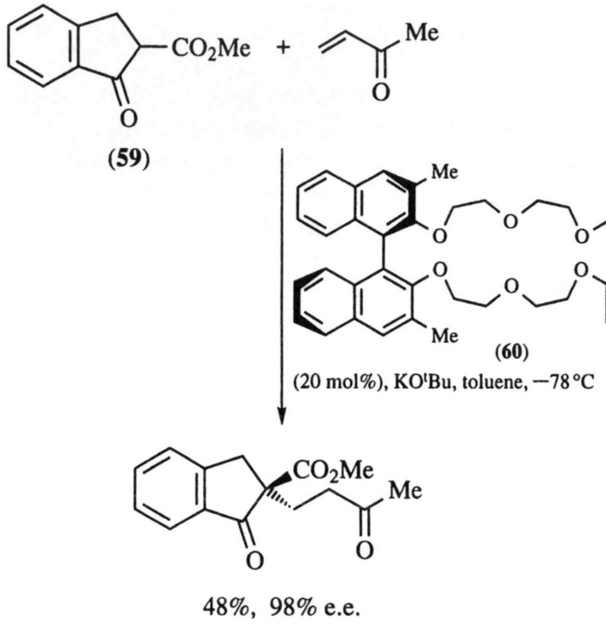

Scheme 56

The anionic polymerisation of a methacrylate ester, which proceeds via serial Michael additions, has been found to give an optically active isotactic polymer when carried out in the presence of Cram's enantiopure crown ether (60), since the chiral centres generated in each successive addition all have the same absolute configuration.[80]

Cross-coupling reactions

Enantioselective formation of substituted allyl derivatives has been accomplished by the cross-coupling of Grignard reagents and vinyl bromides using the enantiopure ferrocenyl-complexed palladium derivative (61) [Scheme 57 (a)].[81]

Enantioselective allylic substitution by Grignard reagents can be accomplished using catalysts by nickel salts complexed with enantiopure phosphines [Scheme 57 (b)].[82]

Isomerisation of alkenes

Rhodium-catalysed enantioselective isomerisation of the achiral allylamine (62) to (R)- or (S)-(E)-enamine (63) is carried out on a large scale (1500 tons/year) by the Takasago company in Japan for the manufacture of highly enantio-enriched citronellal and menthol (Scheme 58).[83]

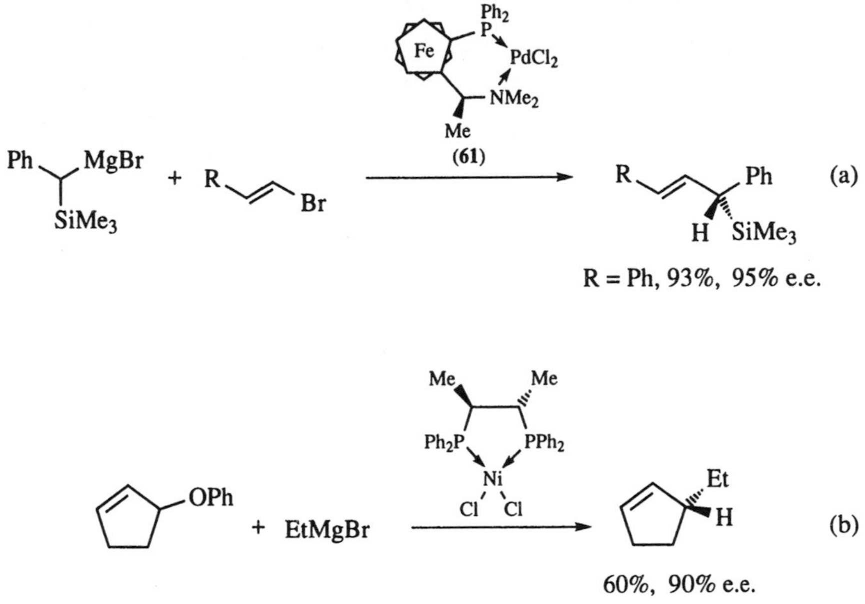

R = Ph, 93%, 95% e.e.   (a)

60%, 90% e.e.   (b)

Scheme 57

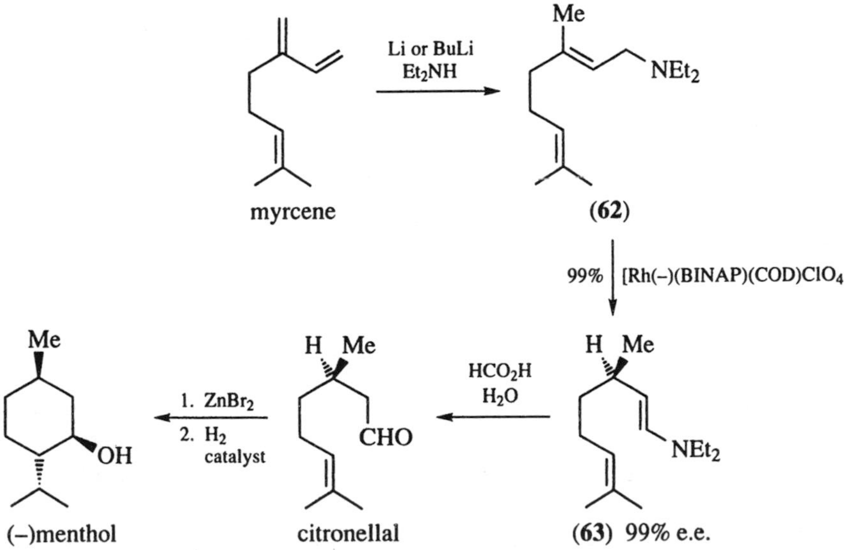

myrcene   (62)

99% | [Rh(−)(BINAP)(COD)ClO$_4$

(−)menthol   citronellal   (63) 99% e.e.

Scheme 58

Hydrocyanation

Addition of the elements of hydrogen cyanide or trimethylsilyl cyanide across a carbonyl group has been accomplished enantioselectively using a variety of catalysts, e.g. Scheme 59. The enantio-enriched cyanohydrin products from these additions, can be hydrolysed to α-hydroxy acids without racemisation.[84]

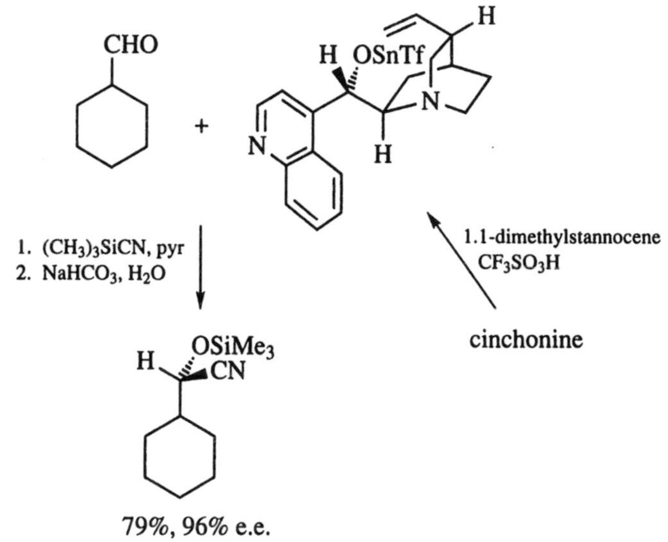

1. (CH₃)₃SiCN, pyr
2. NaHCO₃, H₂O

1.1-dimethylstannocene
CF₃SO₃H

cinchonine

79%, 96% e.e.

Scheme 59

Summary

In catalytic enantioselective additions to double bonds (Type III$_{cat.\ enant.}$), the substrate is achiral and the reagent is chiral (enantiopure) by virtue of being complexed to a chiral (enantiopure)catalyst present in low concentration.

The three steps which are involved in enantioselective addition to a double bond using a chiral auxiliary are telescoped into one using a Type III$_{cat.\ enant.}$ reaction.

Type III$_{cat.\ enant.}$ reactions commonly take place by preliminary bonding, *in situ*, between a functional group in the substrate containing the double bond and the (catalyst-complexed) reagent. The enantiofaces of the substrate are thereby converted into diastereofaces and intramolecular delivery of the reagent to one diastereoface occurs with regeneration of the catalyst.

Less commonly, the double-bond of the substrate reacts directly with the catalyst-complexed reagent, i.e. the enantiofaces of the double bond only become diastereofaces in the transition state for the addition.

The use of Type III$_{cat.\ enant.}$ reactions has advantages relative to the use of overall enantioselective Type III reactions discussed in Chapter 12. Whereas in the latter the (often costly) chiral (enantiopure) elements must be used in molar quantity, in Type III$_{cat.\ enant.}$ reactions, the catalyst concentration is submolar, sometimes as low as 1%. This means that not only is there a saving in cost, but also the isolation of the product is easier because of the absence of molar quantities of the spent reagent and its attached chiral elements.

References

1. T. Katsuki and K.B. Sharpless, *J. Am. Chem. Soc.*, 1980, **102**, 5974; K.B. Sharpless, *Chem. Br.*, 1986, **22**, 38; Y. Gao, R.M. Hanson, J.M. Klunder, S.Y. Ko, H. Masamune and K.B. Sharpless, *J. Am. Chem. Soc.*, 1987, **109**, 5765.
2. K.B. Sharpless and T.R. Verhoeven, *Aldrichim. Acta*, 1979, **12**, 63.
3. M.G. Finn and K.B. Sharpless, *J. Am. Chem. Soc.*, 1991, **113**, 113.
4. B.E. Rossiter and K.B. Sharpless, *J. Org. Chem.*, 1984, **49**, 3707.
5. K.B. Sharpless, C.H. Behrens, T. Katsuki, A.W.M. Lee, V.S. Martin, M. Takatani, S.M. Viti, F.J. Walker and S.S. Woodard, *Pure Appl. Chem.*, 1983, **55**, 589; V.S. Martin, S.S. Woodward, T. Katsuki, Y. Yamada, M. Ikeda and K.B. Sharpless, *J. Am. Chem. Soc.*, 1981, **103**, 6237; P.R. Carlier, W.S. Mungall, G. Schröder and K.B. Sharpless, *J. Am. Chem. Soc.*, 1988, **110**, 2978.
6. Y. Kitano, T. Matsumoto, T. Wakasa, S. Okamoto, T. Shimazaki, Y. Kobayashi, F. Sato, K. Miyaji and K. Arai, *Tetrahedron Lett.*, 1987, **28**, 6351.
7. S. Okamoto, T. Shimazaki, Y. Kobayashi and F. Sato, *Tetrahedron Lett.*, 1987, **28**, 2033.
8. Y. Kitano, T. Matsumoto, S. Okamoto, T. Shimazaki, Y. Kobayashi and F. Sato, *Chem. Lett.*, 1987, 1523.
9. T. Ito, S. Okamoto and F. Sato, *Tetrahedron Lett.*, 1989, **30**, 7083.
10. T. Ito, I. Yamakawa, S. .Okamoto, Y. Kobayashi and F. Sato, *Tetrahedron Lett.*, 1991, **32**, 371.
11. F. Sato *et al.*, *J. Org. Chem.*, 1988, **53**, 1586; 1989, **54**, 994, 2085, 3486.
12. S. Takano, Y. Iwabuchi and K. Ogasawara, *J. Am. Chem. Soc.*, 1991, **113**, 2786.
13. M. Caron and K.B. Sharpless, *J. Org. Chem.*, 1985, **50**, 1557, J.M. Chong and K.B. Sharpless, *J. Org. Chem.*, 1985, **50**, 1560.
14. M.E. Jung and Y.H. Jung, *Tetrahedron Lett.*, 1989, **30**, 6637.
15. R.P. Discordia and D.C. Dittmer, *J. Org. Chem.*, 1990, **55**, 1414.
16. T.J. McMurry and J.T. Groves, in *Cytochrome P-450*, ed. P.R. Ortiz de Montellano, Plenum Press, New York, 1986; J.T. Groves and R.S. Myers, *J. Am. Chem. Soc.*, 1983, **105**, 5791; J.T. Groves ;and P. Viski, *J. Am. Chem. Soc.*, 1989, **111**, 8537.
17. J.T. Groves and P. Viski, *J. Org. Chem.*, 1990, **55**, 3628; for other 'designer porphyrins' which bring enantioselective epoxidation in this way, see Y. Naruta, F. Tani and K. Maruyama, *Chem. Lett.*, 1989, 1269; S. O'Malley and T. Kodadek, *J. Am. Chem. Soc.*, 1989, **111**, 9116, and references cited therein.
18. W. Zhang, J.L. Loebach, S.R. Wilson and E.N. Jacobsen, *J. Am. Chem. Soc.*, 1990, **112**, 2801; R. Irie, K. Noda, Y. Ito and T. Katsuki, *Tetrahedron Lett.*, 1991, **32**, 1055.
19. E.N. Jacobsen, W. Zhang and M.L. Güler, *J. Am. Chem. Soc.*, 1991, **113**, 6703.
20. E.J. Corey, R.K. Bakshi, S. Shibata, C.-P. Chen and V.K. Singh, *J. Am. Chem. Soc.*, 1987, **109**, 7925.

21. E.J. Corey and R.K. Bakshi, *Tetrahedron Lett.*, 1990, **31**, 611; E.J. Corey, S. Shibata and R.K. Bakshi, *J. Org. Chem.*, 1988, **53**, 2861.
22. E.J. Corey and J.O. Link, *Tetrahedron Lett.*, 1989, **30**, 6275.
23. J.M. Brown, *Angew. Chem., Int. Ed. Engl.*, 1987, **26**, 190.
24. H.J. Zeiss, *J. Org. Chem.*, 1991, **56**, 1783.
25. T. Morimoto, M. Chiba and K. Achiwa, *Tetrahedron Lett.*, 1990, **31**, 261.
26. H. Takaya, K. Mashima, K. Koyano, M. Yagi, H. Kumobayashi, T. Taketomi, S. Akutagawa and R. Noyori, *J. Org. Chem.*, 1986, **51**, 629; H. Takaya, S. Akutagawa and R. Noyori, *Org. Synth.*, 1989, **67**, 20; R. Noyori, *Chem. Soc. Rev.*, 1989, **18**, 187.
27. T. Ohta, H. Takaya, M. Kitamura, K. Nagai and R. Noyori, *J. Org. Chem.*, 1987, **52**, 3174; R. Noyori, M. Ohta, Y. Hsiao, M. Kitamura, T. Ohta and H. Takaya, *J. Am. Chem. Soc.*, 1986, **108**, 7117; R. Noyori, *et al.*, *J. Am. Chem. Soc.*, 1987, **109**, 1596.
28. H. Kitamura, T. Ohkuma, S. Inoue, N. Sayo, H. Kumobayashi, S. Akutagawa, T. Ohta, H. Takaya and R. Noyori, *J. Am. Chem. Soc.*, 1988, **110**, 629.
29. M. Kitamura, T. Ohkuma, H. Takaya and R. Noyori, *Tetrahedron Lett.*, 1988, **29**, 1555.
30. T. Ohkuma, M. Kitamura and R. Noyori, *Tetrahedron Lett.*, 1990, **31**, 5509.
31. M. Kitamura, M. Tokunaga, T. Ohkuma and R. Noyori, *Tetrahedron Lett.*, 1991, **32**, 4163.
32. D.F. Taber and L.J. Silverberg, *Tetrahedron Lett.*, 1991, **32**, 4227.
33. R. Noyori, T. Ikeda, T. Ohkuma, M. Widhalm, M. Kitamura, H. Takaya, S. Akutagawa, N. Sayo, T. Saito, T. Taketomi and H. Kumobayashi, *J. Am. Chem. Soc.*, 1989, **111**, 9134.
34. D. Männig and H. Noth, *Angew. Chem., Int. Ed. Engl.*, 1985, **24**, 878.
35. J. Zhang, B. Lou, G. Guo and L. Dai, *J. Org. Chem.*, 1991, **56**, 1670; K. Burgess, W.A. Van der Donk and A.M. Kook, *J. Org. Chem.*, 1991, **56**, 2949, and references cited therein.
36. T. Hayashi, Y. Matsumoto and Y. Ito, *Tetrahedron: Asymmetry*, 1991, **2**, 601.
37. G. Helmchen, R. Karge and J. Weetman, in *Modern Synthetic Methods*, ed. R. Sheffold, Springer, Berlin, 1986, Vol. 4; K. Furuta, S. Shimizu, Y. Miwa and H. Yamamoto, *J. Org. Chem.*, 1989, **54**, 1481 (ref. 2).
38. C. Chapuis and J. Jurczak, *Helv. Chim. Acta*, 1987, **70**, 436.
39. K. Furuta, S. Shimizu, Y. Miwa and H. Yamamoto, *J. Org. Chem.*, 1989, **54**, 1481.
40. E.J. Corey, R. Imwinkelreid, S. Pikul and Y.B. Xiang, *J. Am. Chem. Soc.*, 1989, **111**, 5493; E.J. Corey, N. Imai and S. Pikul, *Tetrahedron Lett.*, 1991, **32**, 7517.
41. E.J. Corey and H.E. Ensley, *J. Am. Chem. Soc.*, 1975, **97**, 6908.
42. J.M. Hawkins and S. Loren, *J. Am. Chem. Soc.*, 1991, **113**, 7794.
43. M. Bednarski and S. Danishefsky, *J. Am. Chem. Soc.*, 1986, **108**, 7060.
44. K. Maruoka, T. Itoh, T. Shirasaka and H. Yamamoto, *J. Am. Chem. Soc.*, 1988, **110**, 310.
45. M. Terada, K. Mikami and T. Nakai, *Tetrahedron Lett.*, 1991, **32**, 935.
46. K. Mikami, M. Terada and T. Nakai, *J. Am. Chem. Soc.*, 1990, **112**, 3949.
47. K. Maruoka, Y. Hoshino, T. Shirasaka and H. Yamamoto, *Tetrahedron Lett.*, 1988, **29**, 3967.
48. S.G. Hentges and K.B. Sharpless, *J. Am. Chem. Soc.*, 1980, **102**, 4263.
49. (a) M. Hirama, T. Oishi and S. Ito, *Chem. Commun.*, 1989, 665; (b) T. Oishi and M. Hirama, *J. Org. Chem.*, 1989, **54**, 5834; (c) T. Yamada and K. Narasaka, *Chem. Lett.*, 1986, 131; (d) K. Tomioka, M. Nakajima, Y. Iitaka and K. Koga, *Tetrahedron Lett.*, 1988, **29**, 573; (e) E.J. Corey, P.D. Jardine, S. Virgil, P.-W. Yuen and R.D. Connell, *J. Am. Chem. Soc.*, 1989, **111**, 9243.

50. J.S.M. Wai, I. Marko, J.S. Svendsen, M.G. Finn, E.N. Jacobsen and K.B. Sharpless, *J. Am. Chem. Soc.*, 1989, **111**, 1123, and references cited therein.
51. B.B. Lohray, T.H. Kalanter, B.M. Kim, C.Y. Park, T. Shibata, J.S.M. Wai and K.B. Sharpless, *Tetrahedron Lett.*, 1989, **30**, 2041.
52. Y. Ogino, H. Chen, H.-L. Kwong and K.B. Sharpless, *Tetrahedron Lett.*, 1991, **32**, 3965.
53. Y. Ogino, H. Chen, E. Manoury, T. Shibata, M. Beller, D. Lübben and K.B. Sharpless, *Tetrahedon Lett.*, 1991, **32**, 5761.
54. K.B. Sharpless, W. Amberg, M. Beller, H. Chen. J. Hartung, Y. Kawanami, D. Lübben, E. Manoury, Y. Ogino, T. Shibata and T. Ukita, *J. Org. Chem.*, 1991, **56**, 4585.
55. K. A. Jorgensen and B. Schiott, *Chem. Rev.*, 1990, **90**, 1483.
56. M.P. Doyle, *Chem. Rev.*, 1986, **86**, 919.
57. H. Nozaki, S. Moriuti, M. Yamabe and R. Noyori, *Tetrahedron Lett.*, 1966, 59.
58. R.G. Salomon and J.K. Kochi, *J. Am. Chem. Soc.*, 1973, **95**, 3300.
59. D.A. Evans, K.A. Woerpel, M.M. Hinman and M.M. Faul, *J. Am. Chem. Soc.*, 1991, **113**, 726.
60. R.E. Lowenthal, A. Abiko and S. Masamune, *Tetrahedron Lett.*, 1990, **81**, 6005.
61. H. Fritschi, U. Leutenegger and A. Pfaltz, *Helv. Chim. Acta*, 1988, **71**, 1553.
62. M.P. Doyle, R.J. Pieters, S.F. Martin, R.E. Austin, C.J. Oalmann and P. Müller, *J. Am. Chem. Soc.*, 1991, **113**, 1423.
63. R. Noyori and M. Kitamura, *Angew. Chem., Int. Ed. Engl.*, 1991, **30**, 49.
64. T. Mukaiyama, K. Soai, T. Sato, H. Shimizu and K. Suzuki, *J. Am. Chem. Soc.*, 1979. **101**, 1455.
65. J. P. Mazaleyrat and D.J. Cram, *J. Am. Chem. Soc.*, 1981, **103**, 4585.
66. D. Seebach, A.K. Beck, S. Roggo and A. Wonnacott, *Chem. Ber.*, 1985, **118**, 3673; M.T. Reetz, T. Kükenhöhner and P. Weinig, *Tetrahedron Lett.*, 1986, **27**, 5711.
67. S. Suga, K. Kawai, S. Okada and R. Noyori, *Pure Appl. Chem.*, 1988, **60**, 1597.
68. N. Oguni and T. Omi, *Tetrahedron Lett.*, 1984, **25**, 2823.
69. M. Kitamura, S. Suga, K. Kawai and R. Noyori, *J. Am. Chem. Soc.*, 1986, **108**, 6071.
70. R. Noyori, S. Suga, K. Kawai, S. Okada, M. Kitamura, N. Oguni, M. Hayashi, T. Kaneko and Y. Matsuda, *J. Organomet. Chem.*, 1990, **382**, 19.
71. M. Kitamura, S. Okada, S. Suga and R. Noyori, *J. Am. Chem. Soc.*, 1989, **111**, 4028.
72. E.J. Corey, P.-W. Yuen, F.J. Hannon and D.A. Wierda, *J. Org. Chem.*, 1990, **55**, 784; N.N. Joshi, M. Srebnik and H.C. Brown, *Tetrahedron Lett.*, 1989, **30**, 5551; W. Oppolzer and R.N. Radinov, *Tetrahedron Lett.*, 1991, **32**, 5777.
73. B. Schmidt and D. Seebach, *Angew. Chem., Int. Ed. Engl.*, 1991, **30**, 99; D. Seebach, L. Behrendt and D. Felix, *Angew. Chem., Int. Ed. Engl.*, 1991, **30**, 1008.
74. K. Furuta. T. Maruyama and H. Yamamoto, *J. Am. Chem. Soc.*, 1991, **113**, 1041; S. Kiyooka, Y. Kaneko, M. Komura, H. Matsuo and M. Nakano, *J. Org. Chem.*, 1991, **56**, 2276.
75. A. Togni and S.D. Pastor, *J. Org. Chem.*, 1990, **55**, 1649, and references cited therein.
76. C. Agami, *Bull. Soc. Chim. Fr.*, 1988, 499, and references cited therein.
77. B. Langstrom and G. Bergson, *Acta Chem. Scand.*, 1973, **27**, 3118.
78. (a) K. Hermann and H. Wynberg, *J. Org. Chem.*, 1979, **44**, 2238; (b) H. Brunner and H. Hammer, *Angew. Chem., Int. Ed. Engl.*, 1984, **23**, 312.
79. D.J. Cram and G.D.Y. Sogah, *Chem. Commun.*, 1981, 625.
80. D.J. Cram and D.Y. Sogah, *J. Am. Chem. Soc.*, 1985, **107**, 8301; see also Y. Okamoto, K. Suzuki, K. Ohta, K. Hatada and H. Yuki, *J. Am. Chem. Soc.*, 1979, **101**, 4763.

81. I. Ojima, N. Clos and C. Bastos, *Tetrahedron*, 1979, **45**, 6901; T. Hayashi, M. Konishi, Y. Okamoto. K. Kabeta and M. Kumada, *J. Org. Chem.*, 1986, **51**, 3772.
82. G. Consiglio and R.M. Waymouth, *Chem. Rev.*, 1989, **89**, 257.
83. S. Otsuka and K. Tani, *Synthesis*, 1991, 665; S. Akutagawa, in *Chirality in Industry*, eds A.N. Collins, G.N. Sheldrake and J. Crosby, Wiley, Chichester, 1992.
84. S. Kobayashi, Y. Tsuchiya and T. Mukaiyama, *Chem. Lett.*, 1991, 537, 541; A. Mori, H. Nitta, M. Kudo and S. Inoue, *Tetrahedron Lett.*, 1991, **32**. 4333.

15 TYPE III$_{s.c./r.c.}$ (DOUBLE ASYMMETRIC INDUCTION)

In Chapters 9–12 and 14, Type III reactions have been considered which were either substrate or reagent controlled. This depended on whether the prochiral element (usually a double bond) and chiral element (usually a chiral centre) were contained in the same (Type III$_{s.c.}$) or different (Type III$_{r.c.}$) molecules.

When both the substrate *and* the reagent contain chiral centres and the substrate contains, e.g., a prochiral double bond which is undergoing addition, both Type III$_{s.c.}$ and Type III$_{r.c.}$ reactions are operative. Asymmetric induction will arise from the combined influence of chiral centres present in the substrate and in the reagent, so-called *double asymmetric induction*.[1]

Consider epoxidation of enantiopure (**1**) (Scheme 1), in which the double bond has diastereofaces, by an achiral reagent [O]. Suppose [Scheme 1 (a)] that this attack takes place to a greater extent on the upper face (Type III$_{s.c.}$: substrate-controlled diastereoselectivity) and hence (**2**) is formed in preference to (**3**).

Now suppose that a similar alkene (**1A**) is epoxidised using enantiopure reagent [O]* from the top face selectively [Scheme 1 (b)] and that attack on this face is independent of the configuration of any chiral centre in R (Type III$_{r.c.}$ reagent-controlled diastereoselectivity).

If now the alkene enantiomer (**1**) in Scheme 1 (a) is reacted with the enantiopure epoxidising agent used in Scheme 1 (b), it is clear that epoxide formation will be directed on to the top face preferentially [Scheme 1 (c)]. The induction brought about by both the existing chiral centre C$_{xyz}$ and the chiral centre(s) contained in the epoxidising agent [O]* are *matched* and a single diastereoisomer (**2**) of the product will predominate.

A corollary of this matching induction is that if, as in Scheme 2(a), the enantiomer of the alkene [$_{enant.}$(**1**)] is reacted with the achiral epoxidising agent [O], the result will be formation of $_{enant.}$ (**2**) and $_{enant.}$ (**3**) in relative amounts equal to those of (**2**) and (**3**) produced in Scheme 1 (Reaction of a racemic substrate with an achiral reagent is bound to produce racemic products). Since the enantiopure epoxidising agent [O]* will now attack (**1A**) as drawn from the bottom face (Scheme 2(b)) the resulting inductions will be

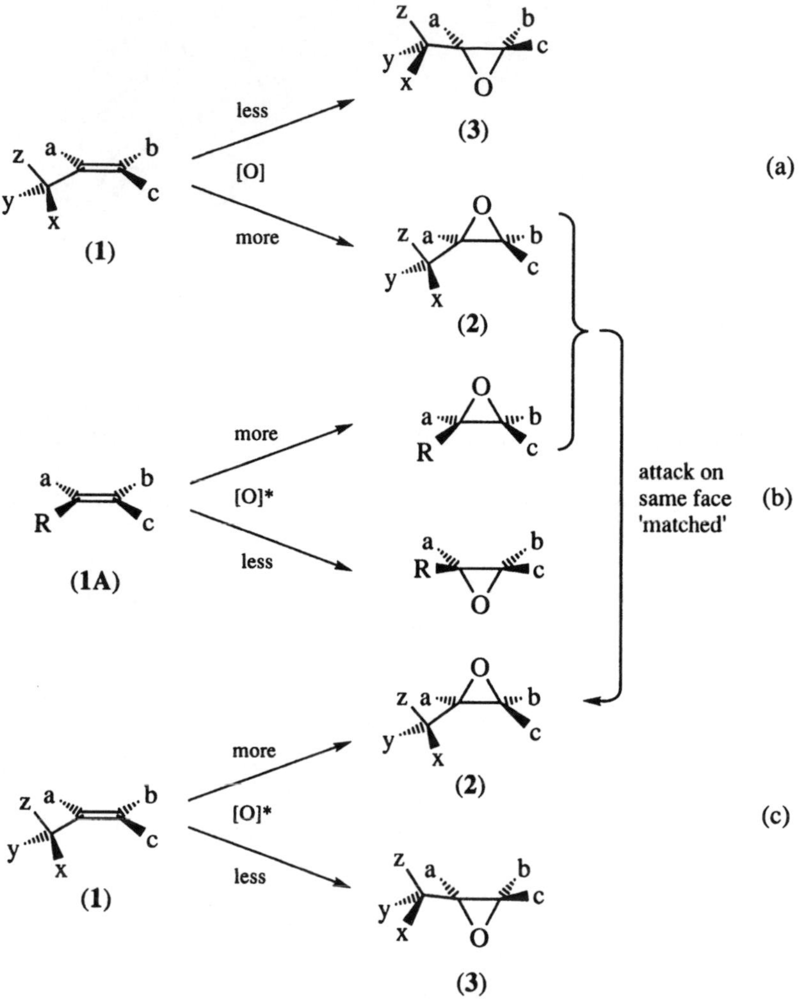

Scheme 1

mismatched when (**1A**) ≡ $_{enant.}$ (1) and a mixture of more comparable amounts of enantiopure diastereoisomers $_{enant.}$ (2) and $_{enant.}$ (3) will result.

Likewise it follows that the inductions will also be mismatched when the alkene enantiomer used in Scheme 1 (a) is reacted with the *enantiomer* of the epoxidising agent $_{enant.}$ [O]* used in Scheme 1 (b); given that in Scheme 1 (b) epoxidation takes place preferentially from the top face, the enantiomeric epoxidising agent is bound to have a preference for the bottom face, i.e. there will also be a mismatching of the face preference in this case.

In Schemes 1 and 2 we have used a reaction in which the chiral centre(s) in the reagent are not retained in the product, but this is not bound to be the

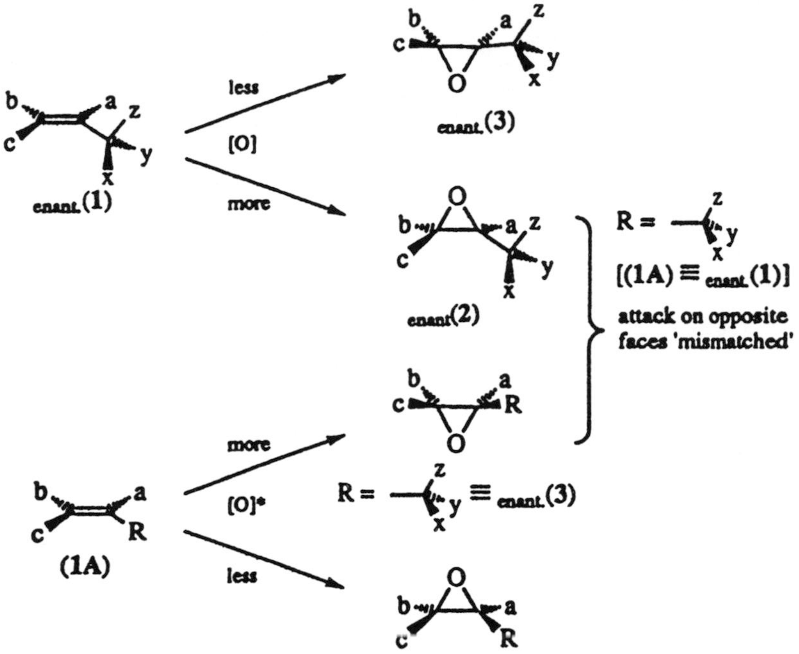

(a)

(b)

Scheme 2

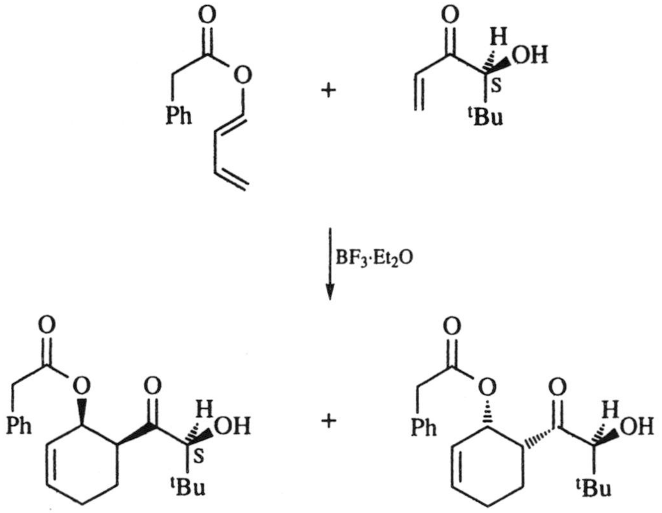

BF$_3$·Et$_2$O

(a)

(4)

> 100 : 1

Scheme 3 *(continued)*

Scheme 3 *(continued)*

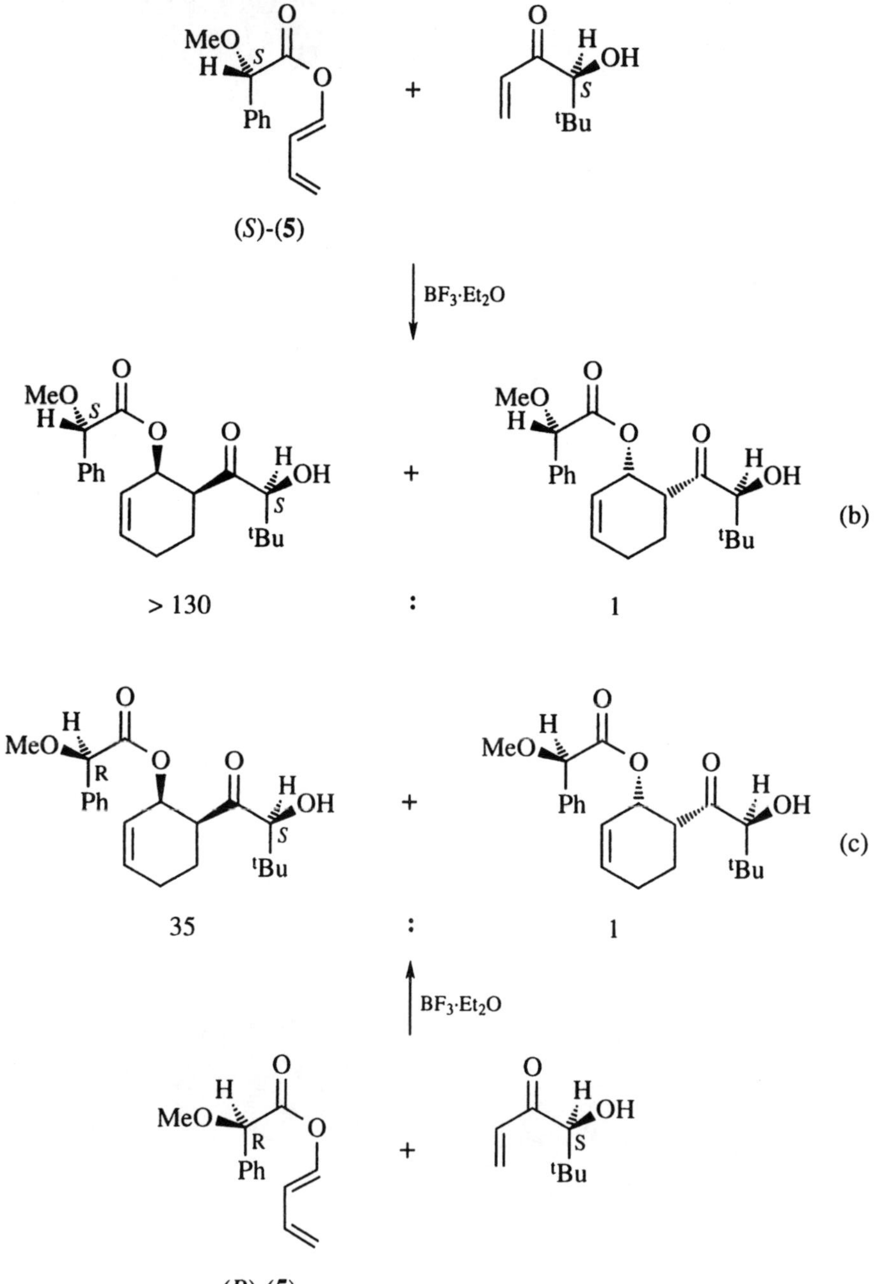

Scheme 3

case.[1] In Scheme 3 (b) and (c) is a Diels–Alder reaction in which both diene and dienophile each contain a chiral centre and both these chiral centres are retained in the product. In Scheme 3 (a), the ratio of diastereoisomers of the product (4) is 100:1 (Type II/III:single asymmetric induction). Clearly, the introduction of a chiral centre into the diene (5) having the S configuration gives rise to a matched pair in (b) whereas the R configuration at this same centre results in a mismatched pair in (c).

Of course, it is always possible that the chiral centre in the substrate or in the reagent is too remote from the centre being formed to influence its configuration; the asymmetric induction will then be simply reagent or substrate controlled. However, it is unwise to assume that a more distant chiral centre will always have less influence on the induction than a more local one. In Scheme 4 is the reaction of a boron enolate (6) with the aldehyde enantiomers (S)-(7) and (R)-(7).[2] The sense of induction using either aldehyde enantiomer is the same, i.e. the created chiral centre has the same absolute configuration in the major diastereoisomers (8) and (9). Complete reagent control, mediated by the chiral substituents on boron, is operating in this reaction in spite of the closer proximity of the chiral centre in the substrate to the nascent chiral centre.

The complete reagent control in this reaction is the result of two factors. First, of course, asymmetric induction depends on the effects that substituents at chiral centres have on the two diastereoisomeric *transition states* in the reaction; substituents which may appear remote in the preferred conformation of the starting material may nevertheless become decisive in distinguishing between two diastereoisomeric transition states. In Scheme 4 it is likely that

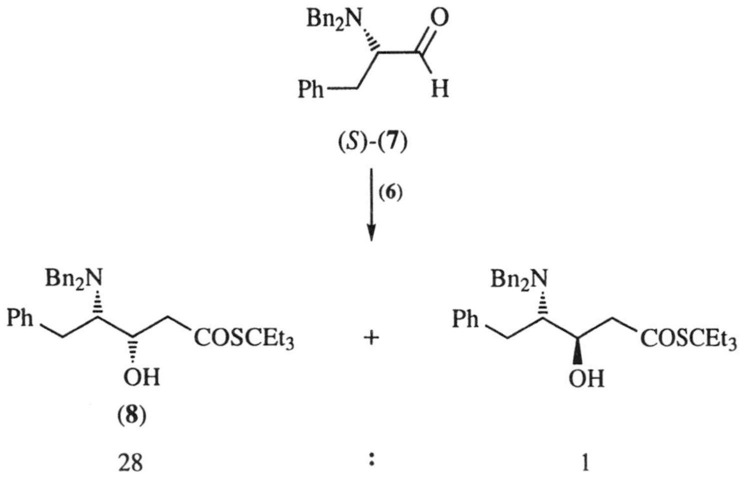

Scheme 4   (*continued*)

Scheme 4 *(continued)*

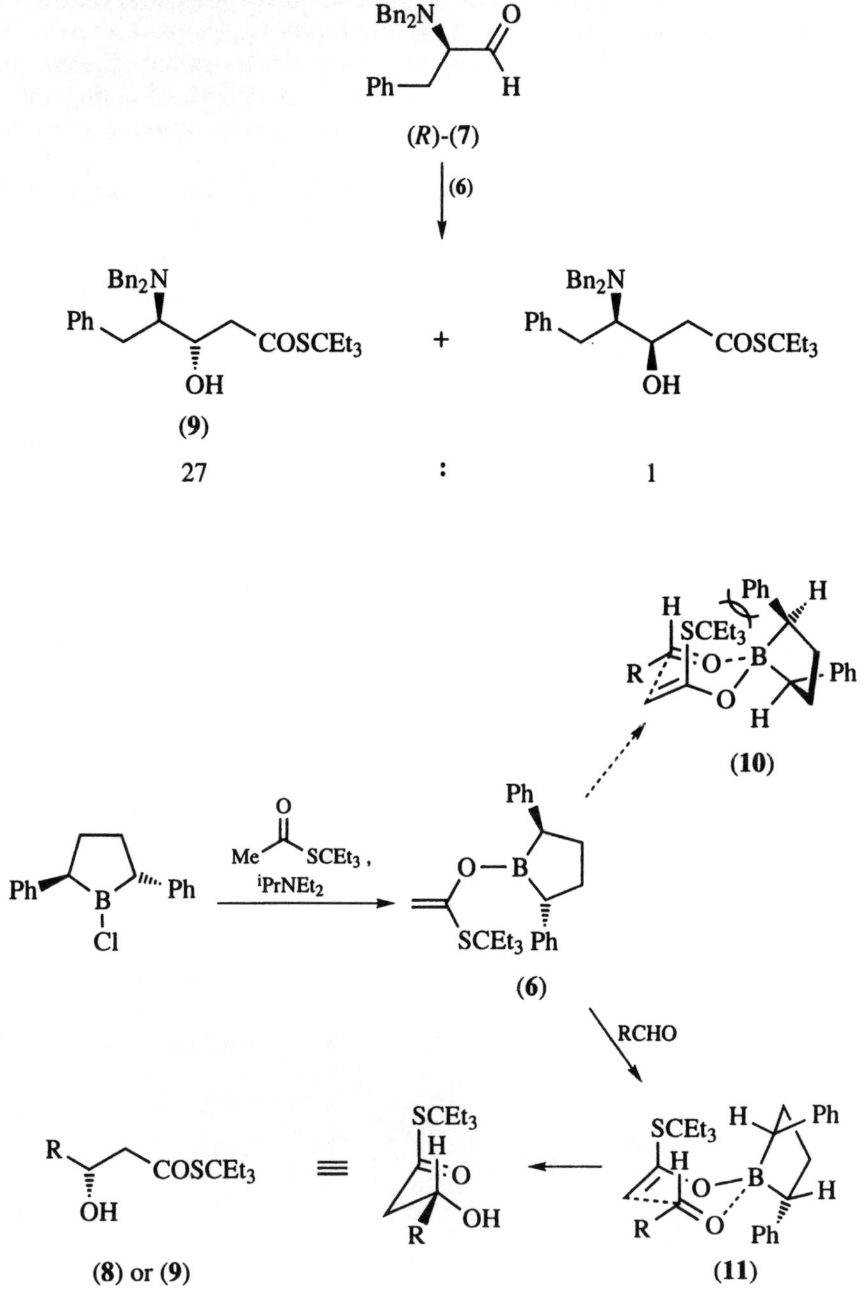

(R)-(7)

(6)

(9)

27 : 1

(10)

(6)

(8) or (9)

(11)

Scheme 4

the two diastereoisomeric transition states can be represented by (10) and (11) with the preference for the latter a result of steric interaction between the not-so-remote phenyl group and the SCEt$_3$ group in (10) as shown.

Second, induction by the α-chiral centres in aldehydes (S)-(7) and (R)-(7) is insignificant, as is revealed by the reaction of (S)-(7) with the achiral boron enolate (12); the ratio of (13) to (14) in this case is 1:1 (Scheme 5). This absence of induction with aldehyde (S)-(7) contrasts with its highly diastereoselective reactions with lithium alkyls and other nucleophiles (Chapter 10).

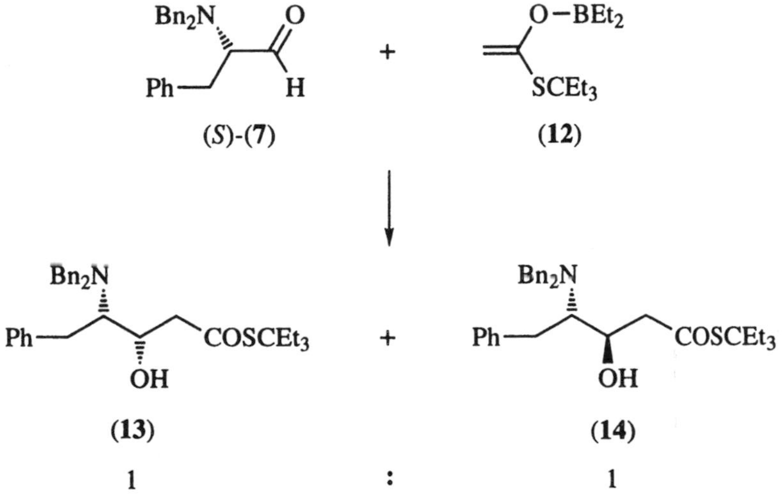

Scheme 5

In both of the previous examples, the double asymmetric induction is dominated by one particular chiral centre. In Scheme 3 it is the chiral centre on the dienophile, as shown by the high diastereoselectivity in the formation of (4) and the same sense of diastereoselectivity irrespective of the configuration of the chiral centre in the diene (5). In Scheme 4 the chiral centre on the aldehyde contributes little if anything towards the asymmetric induction that gives rise to the preferential formation of (8) or (9). In other reactions, however, there may be more balanced contributions from Type III$_{s.c.}$ and Type III$_{r.c.}$ inductions, as revealed by a greater disparity between matched and mismatched ratios. Thus, in Scheme 6, a 73:27 ratio of (15) to (16) in the reaction of the achiral allylborane (17) with aldehyde (18) is raised to 93:7 using the allylborane (R,R)-(19) but is only 36:64 using (S,S)-(19).[3] *Diastereoselectivity in a Type III$_{s.c.}$ reaction can be enhanced by using an*

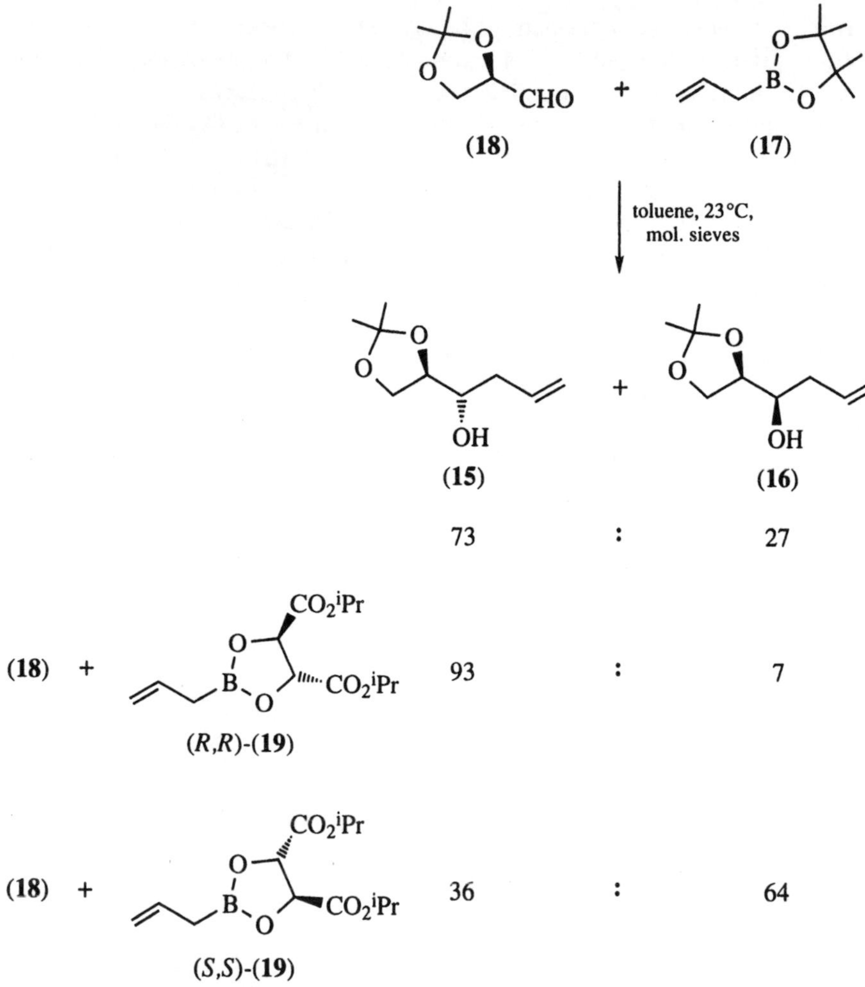

(18)   (17)

toluene, 23 °C,
mol. sieves

(15)   +   (16)

73   :   27

(18) + (R,R)-(19)   93   :   7

(18) + (S,S)-(19)   36   :   64

Scheme 6

*enantiopure reagent and by making use of the matched pair in double
asymmetric induction.*

Reactions in which the chiral centres of the reagent are eliminated as in
Scheme 6 are desirable but even if they are retained in the product, they may
be removed in a subsequent step, leaving only the created chiral centre(s)
with a higher level of diastereoselectivity than is feasible in the analogous
Type III$_{s.c.}$ reaction.

Another example in which matched and mismatched pairs lead to

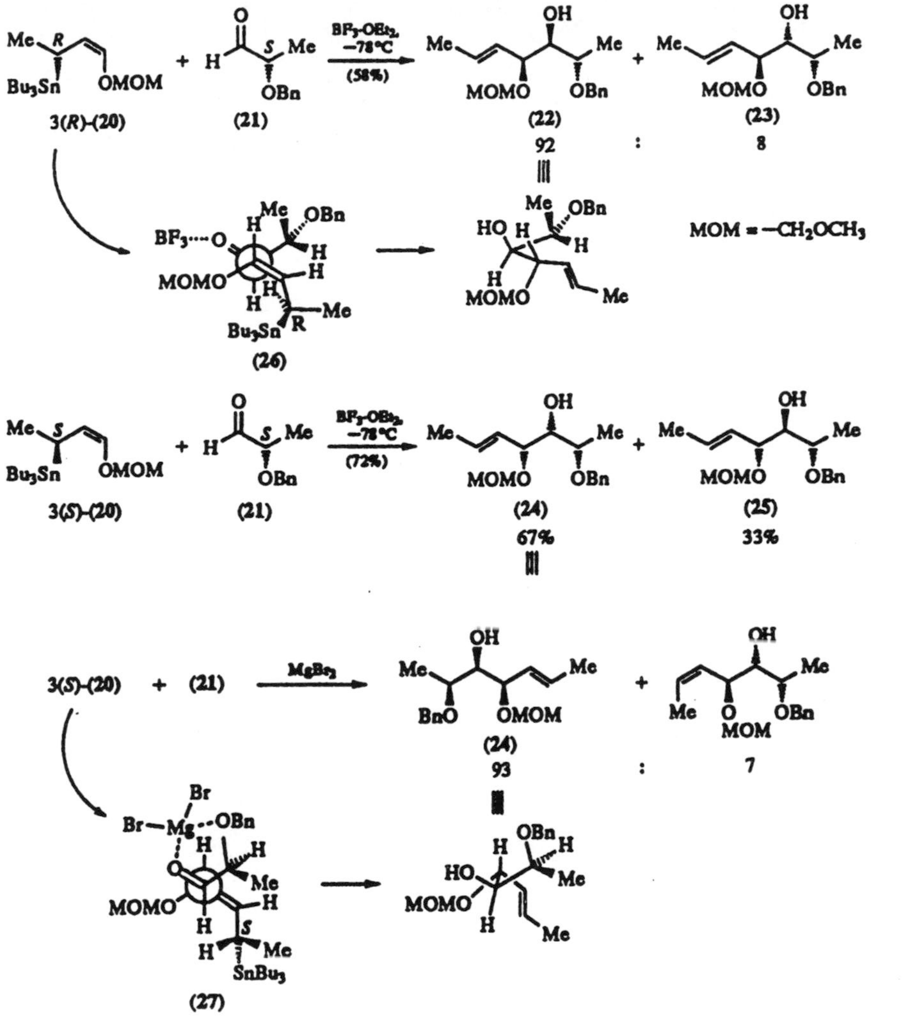

Scheme 7

significantly different ratios of diastereoisomers is that in Scheme 7.[4] Thus the allylstannane (20) having the 3(R) configuration reacts with the 2(S)-aldehyde (21) in the presence of BF$_3$ to give a 92:8 ratio of products (22) and (23) (matched), whereas the 3(S) enantiomer of (20) gives a 67:33 ratio of (24) to (25) (mismatched).

The major diastereoisomer (22) is assumed to be formed via an open transition state (26) in which the usual *anti*-stereoelectronic requirement of an S$_E$2′ and a preference for an *E*-configured incipient double bond are combined with a σ^*–π^* dominated Felkin–Anh attack (see Chapter 10) on the aldehyde having the benzyloxy group *anti* to the C—C bond which is forming.

Interestingly, if the reaction is catalysed by magnesium bromide instead of boron trifluoride, the matched and mismatched pairings are reversed. The predominant diastereoisomer (24) from the matched pair 3(*S*)-(20) + (21) is thought to arise from the chelated aldehyde being attacked in transition state (27).

Marshall and Wang[5] have exploited double asymmetric induction in the S_E2' reaction of enantiopure allenylstannanes with aldehydes bearing an oxygen substituent on the (chiral) α-carbon (Scheme 8). With the matched pair (28) and (29) and using magnesium bromide as catalyst, the chelated aldehyde is attacked via an extended open transition state (30) giving (31) highly diastereoselectively. Note that the allene (29) is chiral although it lacks a chiral centre.

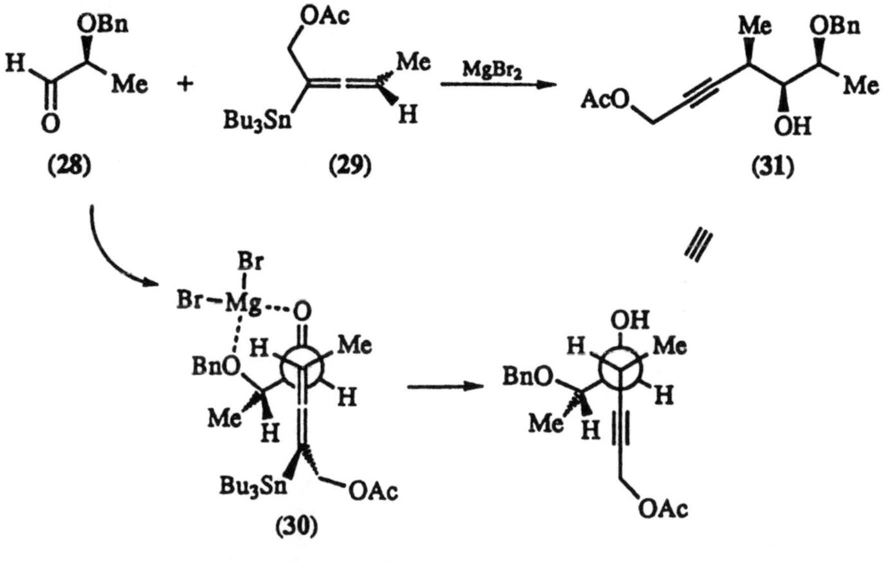

Scheme 8

Double asymmetric induction can also be accomplished using an achiral reagent which becomes complexed to an enantiopure catalyst in the reaction. Thus the Diels Alder cycloaddition of achiral diene (32) [Scheme 9 (a)] to benzaldehyde in the presence of the enantiopure Lewis acid catalyst (+)-Eu(hfc)$_3$ is a Type III$_{cat. enant.}$ reaction (*cf* Chapter 14; Scheme 32) but the enantioselectivity in formation of (33) is low [the ratio of (33) to $_{enant.}$(33) was deduced from the enantiopurity of (34) formed with loss of one of the chiral centres].[6]

Reaction of the diene (35) bearing *1*-menthyloxy as a chiral auxiliary with benzaldehyde in the presence of the *achiral* Eu(fod)$_3$ gave only a slight excess of diastereoisomer (37) over (36) [Scheme 9 (b)].

However, the reaction of the same diene (35) with benzaldehyde in the presence of (+)-Eu(hfc)$_3$ gave (36) highly diastereoselectively [Scheme 9 (c)] [the reaction of $_{enant.}$(35) with benzaldehyde and (+)-Eu(hfc)$_3$ showed little diastereoselectivity].

From the magnitudes of the product ratios in Scheme 9, it appears that

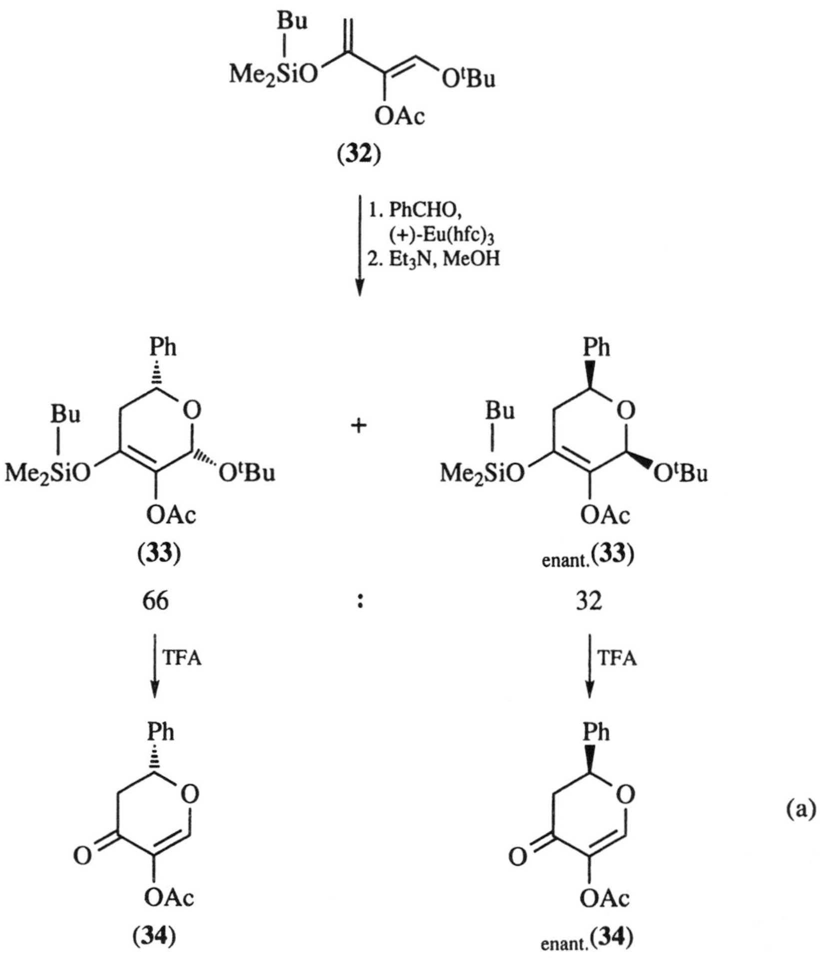

Scheme 9

(continued)

Scheme 9 *(continued)*

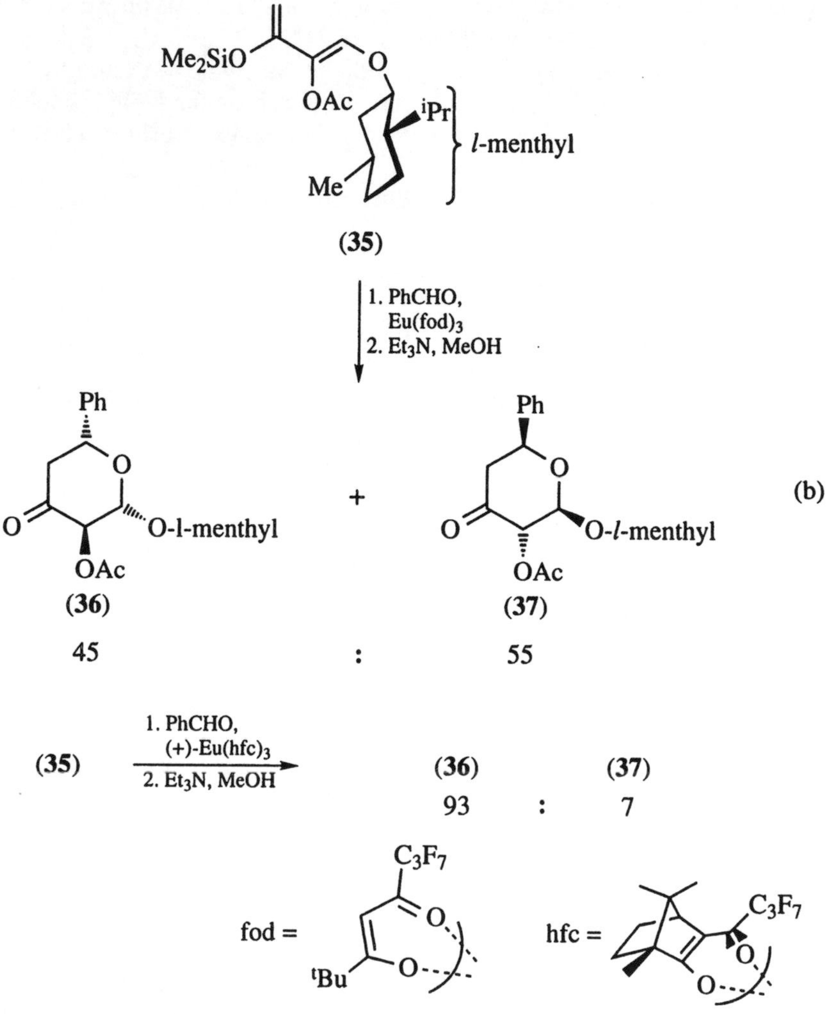

Scheme 9

there is some specific interaction between the chiral centres in the diene (**35**) and those on the catalyst in the double asymmetric induction. In Schemes 1 and 2 it was assumed that the face of the alkene epoxidised by the enantiopure reagent is independent of the existing chiral centre in the substrate, but this will not always be the case.

Bednarski and Danishefsky[6] used the highly diastereoselective reaction of (**38**) with benzaldehyde in the presence of (+)-Eu(hfc)$_3$ to prepare unnatural L-glucose by the route shown in part in Scheme 10.

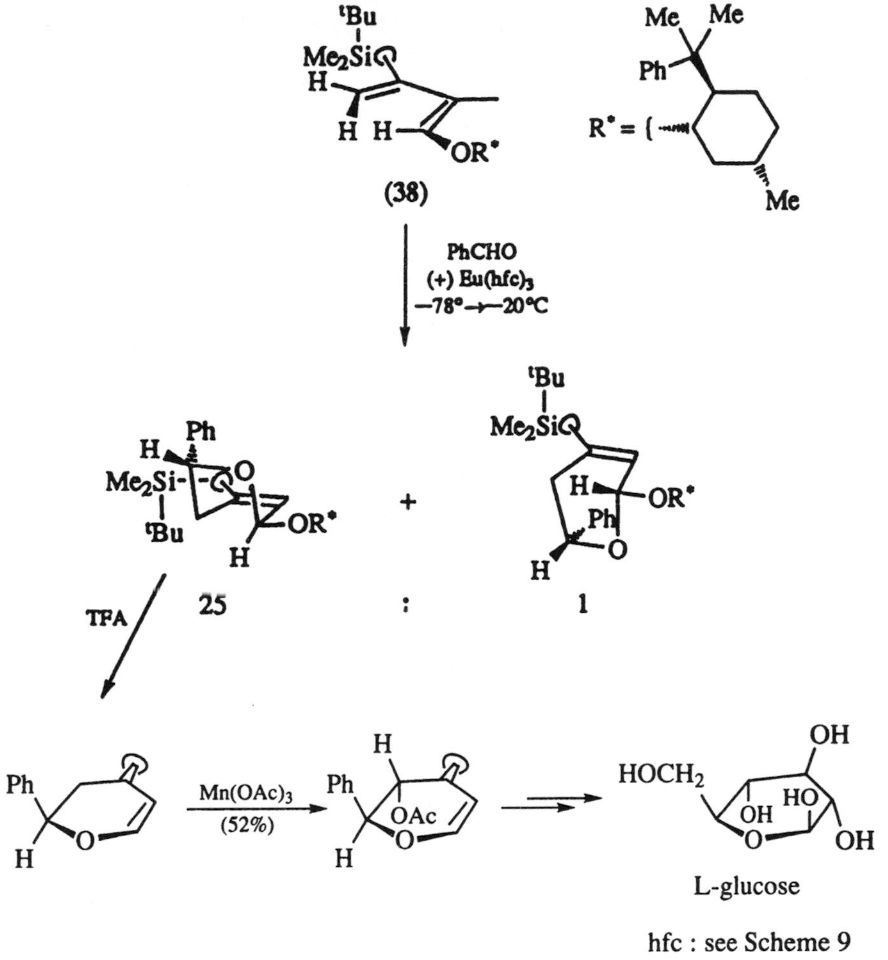

Scheme 10

Further uses of double asymmetric induction

KINETIC RESOLUTION[7]

In epoxidation of the racemic alkene (1) + $_{enant.}$(1) with an enantiopure epoxidising agent in Schemes 1 and 2, we expect the reaction leading to (2) in Scheme 1 to be faster than that leading to $_{enant.}$(2) in Scheme 2 (Scheme 11). This is because in Scheme 1, the chiral centres in *both* the substrate *and* the reagent are contributing to a lowering of the transition state energy, and therefore increasing the rate, for attack on the *same* diastereoface of (1). In Scheme 2, on the other hand, these chiral centres are each lowering the transition state energy for attack on *different* diastereofaces.

If we start with a racemic mixture of (1) and $_{enant.}$(1), therefore, and treat with the enantiopure epoxidising agent [O]* used in Schemes 1 and 2, we would expect the situation in Scheme 11 in which (1) reacts to give one enantiopure diastereoisomer (2) preferentially, whereas $_{enant.}$(1) reacts more slowly to give a mixture of $_{enant.}$(2) and $_{enant.}$(3).

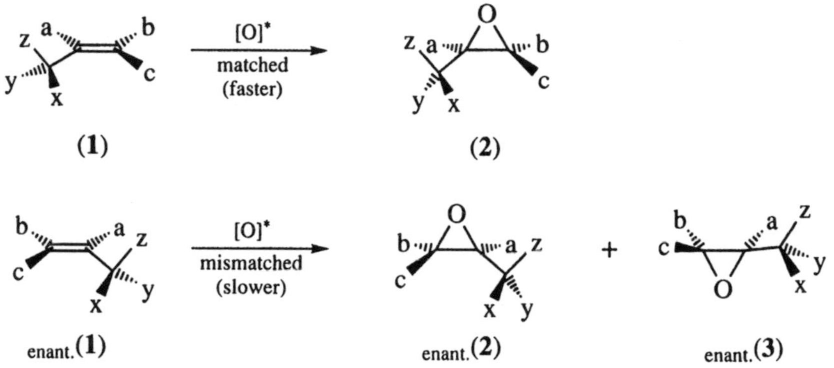

Scheme 11. (Complete) kinetic resolution; (1) + $_{enant.}$(1) react with 0·5 mol equiv. of [O]* (an enantiopure epoxidising agent) and only (1) reacts leaving $_{enant.}$(1) behind.

If the rates of reaction of these two enantiomeric alkenes are sufficiently different and if only 0.5 mol equiv. of enantiopure epoxidising agent is used, it follows that only (1) will react and $_{enant.}$(1) will be left unchanged. In other words, there will have been a resolution of the racemic alkene.

Outstandingly high levels of kinetic resolution are found in the Sharpless–Katsuki epoxidation of racemic allylic alcohols (see Chapter 14).

ENANTIOPURIFICATION BY DOUBLE ASYMMETRIC INDUCTION

Another use to which double asymmetric induction can be put is the enhancement of product enantiopurity relative to that of the starting material(s). Thus Roush et al.[8] have shown that the enantiopure allylboronate (39) (Scheme 12) reacts with the chiral epoxyaldehyde (40) of 90% e.e. to give a product (41) of >98% e.e. highly diastereoselectively.

To understand the origin of the increased enantiopurity of the major product diastereoisomer (41) relative to that of the starting aldehyde (40), we should remember that if reaction with an enantiopure reagent is matched for one enantiomer of the substrate (as it is for the major one of (40) above) then

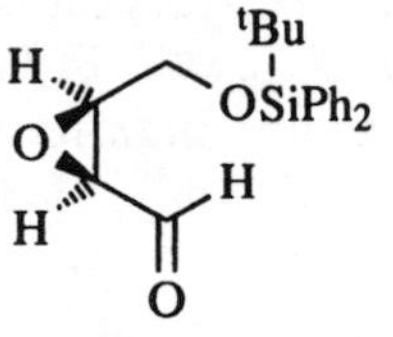

(40) (90% e.e.)

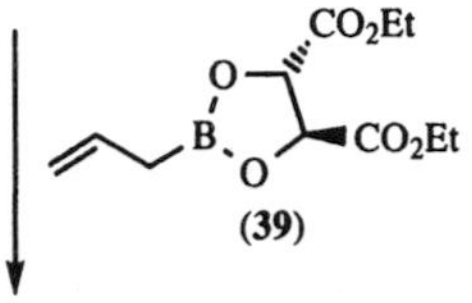

(39)

(41) + **(42)**

>98% e.e.

enant.**(41)** + enant.**(42)**

58% e.e.

$$\text{ratio } \frac{(41) + \text{enant. } (41)}{(42) + \text{enant. } (42)} = \frac{96}{4}$$

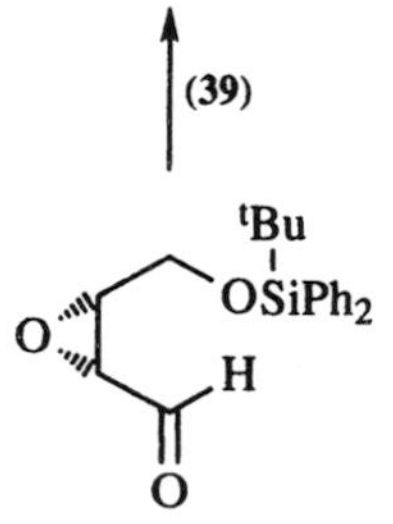

(39)

enant.**(40)**

Scheme 12

it is bound to be mismatched for the minor enantiomer. Consequently, the 15% of the minor enantiomer $_{enant.}$(40) in Scheme 12 is bound to react *less* diastereoselectively than the major enantiomer (40). Thus the ratio of (41) to (42) will be greater than the ratio of $_{enant.}$(41) to $_{enant.}$(42). Consequently, the product (41) will contain *less* than 15% of its enantiomer, leading to an increase in its e.e. The minor diastereoisomer (42), however, will suffer a loss in its enantiopurity by comparison with (40) because a greater proportion of its enantiomer $_{enant.}$(42) will be produced in the mismatched reaction of (39) with $_{enant.}$(40); the major enantiomer of this minor diastereoisomer is $_{enant.}$(42) of 58% e.e.

It should be noted that although the product (41) is enhanced in *enantiopurity* by comparison with the starting material, it is correspondingly less *diastereopure* as a result of the increased amount of $_{enant.}$(42) present. However, in practice this is normally advantageous given that separation of diastereoisomers is usually easier than separation of enantiomers.

Although raising the enantiopurity from 90% to >98% e.e. might not seem of much significance, it is likely to become increasingly so as the requirement for enantiopurity, e.g. in medicine components, becomes more pervasive.

As a general rule, the major diastereoisomer from a reaction arising from a matched double asymmetric induction will always have enhanced enantiopurity relative to the starting material.

References

1. S. Masamune, W. Choy, J.S. Petersen and L.R. Sita, *Angew. Chem., Int. Ed. Engl.*, 1985, **24**, 1.
2. M.T. Reetz, E. Rivadeneira and C. Niemeyer, *Tetrahedron Lett.*, 1990, **31**, 3863.
3. W.R. Roush, L.K. Hoong, M.A.J. Palmer and J.C. Park, *J. Org. Chem.*, 1990, **55**, 4109; W.R. Roush, M.A. Adam, A.E. Walts and D.J. Harris, *J. Am. Chem. Soc.*, 1986, **108**, 3422.
4. J.A. Marshall and G.P. Luke, *J. Org. Chem.*, 1991, **56**, 483.
5. J.A. Marshall and X. Wang, *J. Org. Chem.*, 1991, **56**, 3211.
6. M. Bednarski and S. Danishefsky, *J. Am. Chem. Soc.*, 1986, **108**, 7060.
7. H.B. Kagan and J.C. Fiand, *Top. Stereochem.*, 1988, **18**, 249.
8. W.R. Roush, J.A. Straub and M.S. VanNieuwenhze, *J. Org. Chem.*, 1991, **56**, 1636.

16 TYPE III REACTIONS OF DIASTEREOTOPIC AND ENANTIOTOPIC ATOMS AND GROUPS: ENZYME-CATALYSED REACTIONS

Diastereotopic atom selectivity

Our discussion of stereoselectivity so far has been concerned almost exclusively with addition to diastereofaces/enantiofaces. It was pointed out in Chapter 1, however, that the prochiral element may be a centre rather than a double bond. Typically, two apparently identical atoms or groups a attached to a tetrahedral atom in a molecule containing a chiral centre as in (1) are actually diastereotopic and can react at different rates with achiral reagents. In a molecule such as (2), however, the atoms or groups a are enantiotopic and can be distinguished chemically only by reaction with a chiral reagent, which is generally used in enantiopure form.

$$\qquad (1) \qquad\qquad\qquad (2)$$

A common example of (1) has a = H and removal of one of the diastereotopic hydrogens as a proton is facilitated by the presence of, e.g., a carbonyl group in b as in (3) (Scheme 1). In this case, however, diastereoselectivity in formation of the *product* arises from preferred attack on one of the two *faces* of the enolate (4), i.e. a situation previously discussed in Chapter 9–11.

Likewise, removal of a proton, a hydrogen atom or a hydride is facilitated by an adjacent double bond. The resulting species, which may each react to form a new chiral centre (Scheme 2), is a planar allyl anion, radical or cation. Here also diastereoselectivity in formation of the product is controlled by a

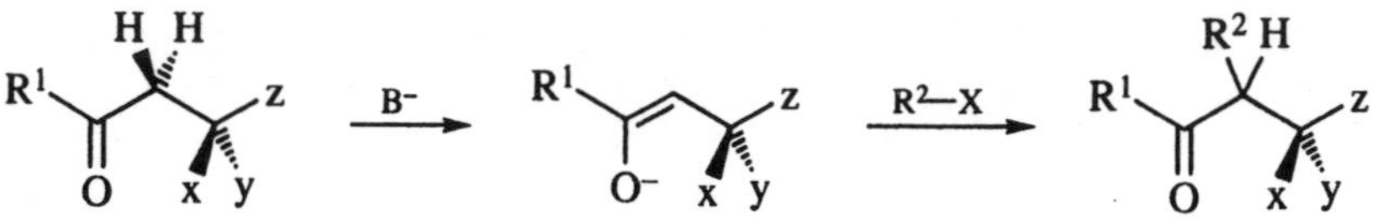

Scheme 1

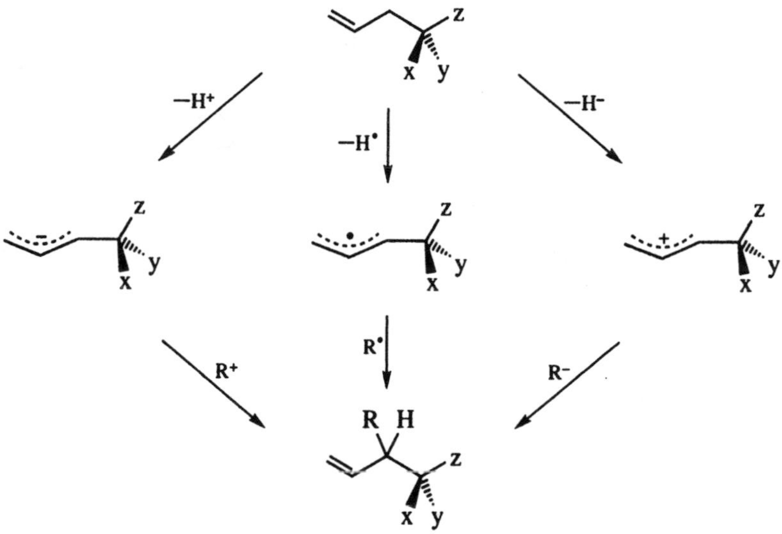

Scheme 2

preference for attack by the anion R⁻, radical R• or cation R⁺ (or their synthetic equivalents) on one of the two diastereofaces of the intermediate. The factors controlling which diastereotopic hydrogen is removed are not necessarily the same as those that determine which diastereoface of the intermediate reacts more readily.

Even when there is no resonance delocalisation, the species formed by cleavage of the C—a bond in (1) is likely to be planar (cation) or inverting rapidly (radical, anion) except in special cases. Any overall diastereoselectivity, therefore, will not necessarily be a reflection of the relative rates of removal of the diastereotopic hydrogens.

Examples in which one of the two diastereotopic atoms a in (1) is selectively removed, the hybridisation of the prochiral centre is retained and reaction occurs with retention of configuration are known in cyclopropane and epoxide chemistry. This is because carbanions in three-membered rings are pyramidal and have rates of inversion which are grossly retarded by comparison with those of other carbanions (see Chapter 4).[1] Thus reaction of the dibromocyclopropane (5) in Scheme 3 with methyllithium followed by deuteriation results in exclusive substitution of the *syn*-bromine.[2]

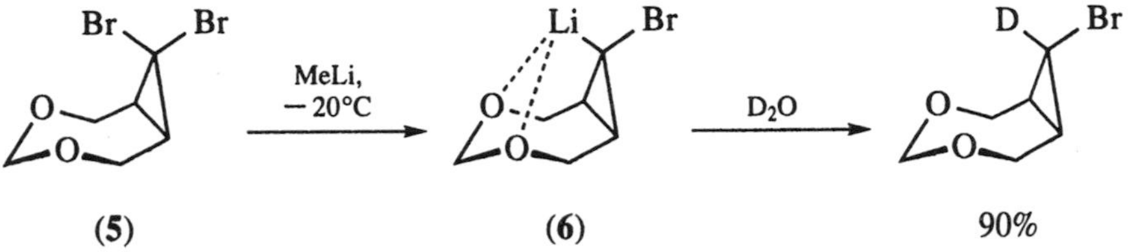

Scheme 3

The greater reactivity of the *syn*-bromine is believed to arise from coordination of the methyllithium with the acetal oxygen(s); a proximity effect then favours attack on this bromine. [Coordination of the acetal oxygens with the lithium appears to confer additional configurational stability on (6).]

In intramolecular reactions, one of two diastereotopic atoms may be more reactive as a result of geometrical factors. Thus carbene insertion within (7) in Scheme 4 takes place into only the *syn* C—H bond since the *anti* C—H bond is less accessible.[3]

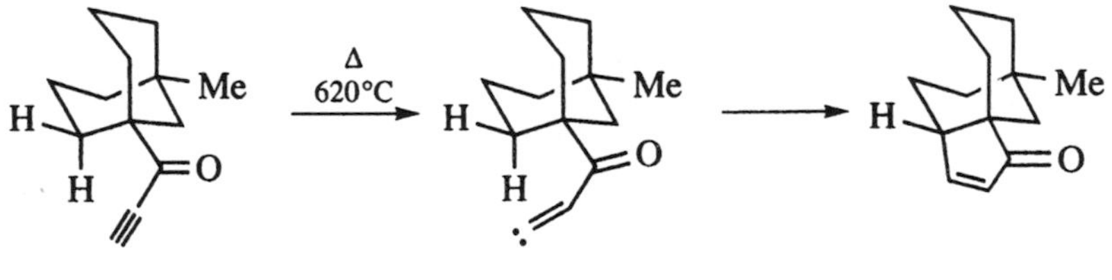

Scheme 4

Diastereotopic group selectivity

In intermolecular reactions using achiral reagents there is frequently a selectivity between diastereotopic groups, particularly when the groups are substituents on a ring and attack on one of them is sterically hindered by comparison with the other. Thus, using a limited quantity of base, there is selective hydrolysis of the ester group *trans* to the phenyl in the diester (8) [Scheme 5 (a)].[4] In Scheme 5 (b), one of the diastereotopic lone pairs on the sulphur atom in (9) is more readily alkylated by allyl triflate (prepared *in situ*) because of the shielding effect of the adjacent isopropyl group.[5] Even unactivated diasterotopic methyl groups can be differentiated as in the bromination of 3-bromocamphor (Scheme 10, Chapter 3), although in that case it is not a direct attack on one of the methyl groups.

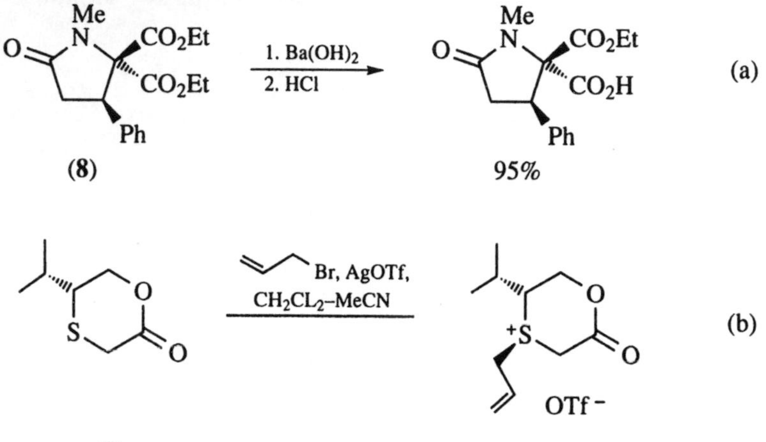

(a)

(b)

Scheme 5

One of the two diastereotopic dichlorophenoxy substituents on phosphorus in (**10**) (Scheme 6) is selectively displaced in an intermolecular nucleophilic substitution by an *n*-butoxide, as is revealed by the enantiopurity of the methanolysed product (**11**);[6] methanolysis is known to proceed with inversion of configuration.

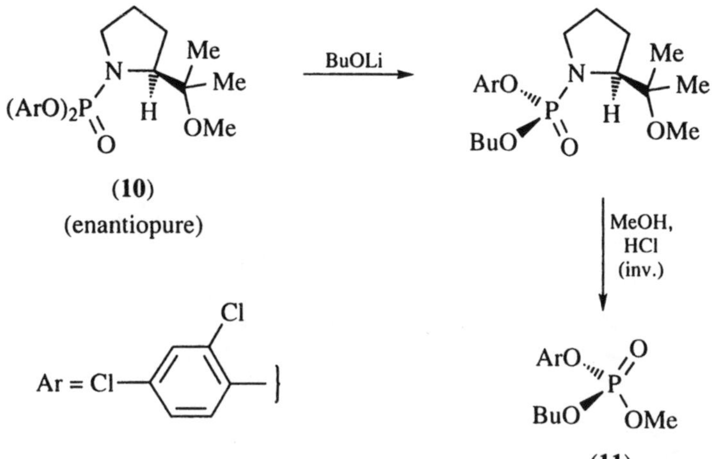

Scheme 6

Distinguishing between two diastereotopic groups that are not substituents on rings is easier in intramolecular reactions. The two hydroxymethylene groups in (12) (Scheme 7) are diastereotopic and one of them acts preferentially in intramolecular attack on the episulphonium intermediate (13) leading to (14).[7]

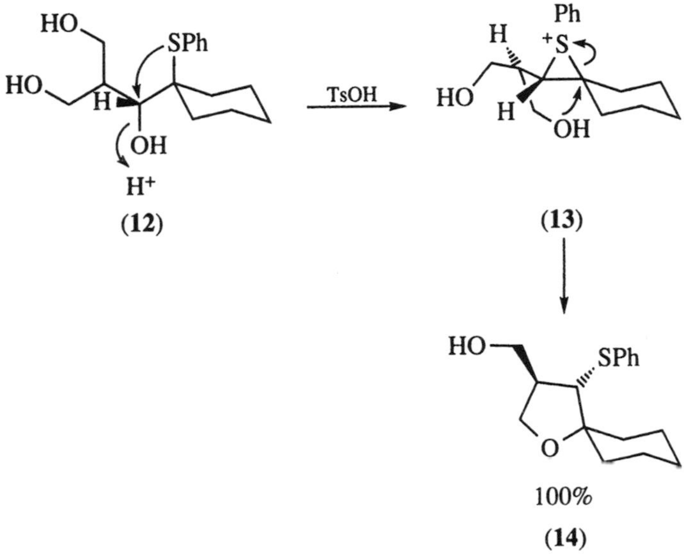

Scheme 7

Distinguishing chemically between enantiotopic atoms and groups (Type III$_{r.c.}$)

Substitution of one of two enantiotopic atoms a in (2) by a second atom or group d will result in an enantiopure product (Scheme 8).

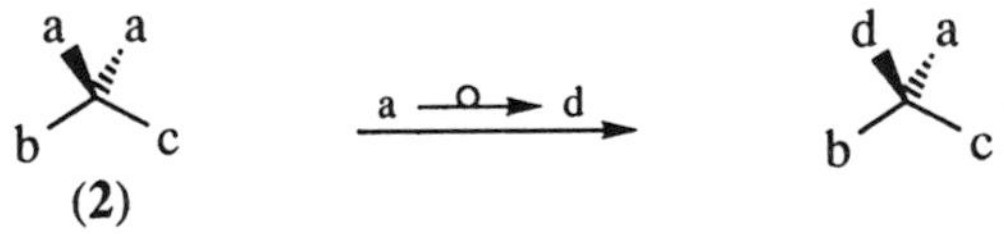

Scheme 8

Apart from enzyme-mediated reactions, few cases are known where enantioselectivity results from the faster reaction of one of two enantiotopic atoms attached to the same prochiral centre. However, as we have seen in Chapter 1, two atoms (or groups) may be enantiotopic but bonded to different atoms as are the two hydrogen atoms H_A and $H_{A'}$ in the *meso* compound (15) [Scheme 9 (a)].

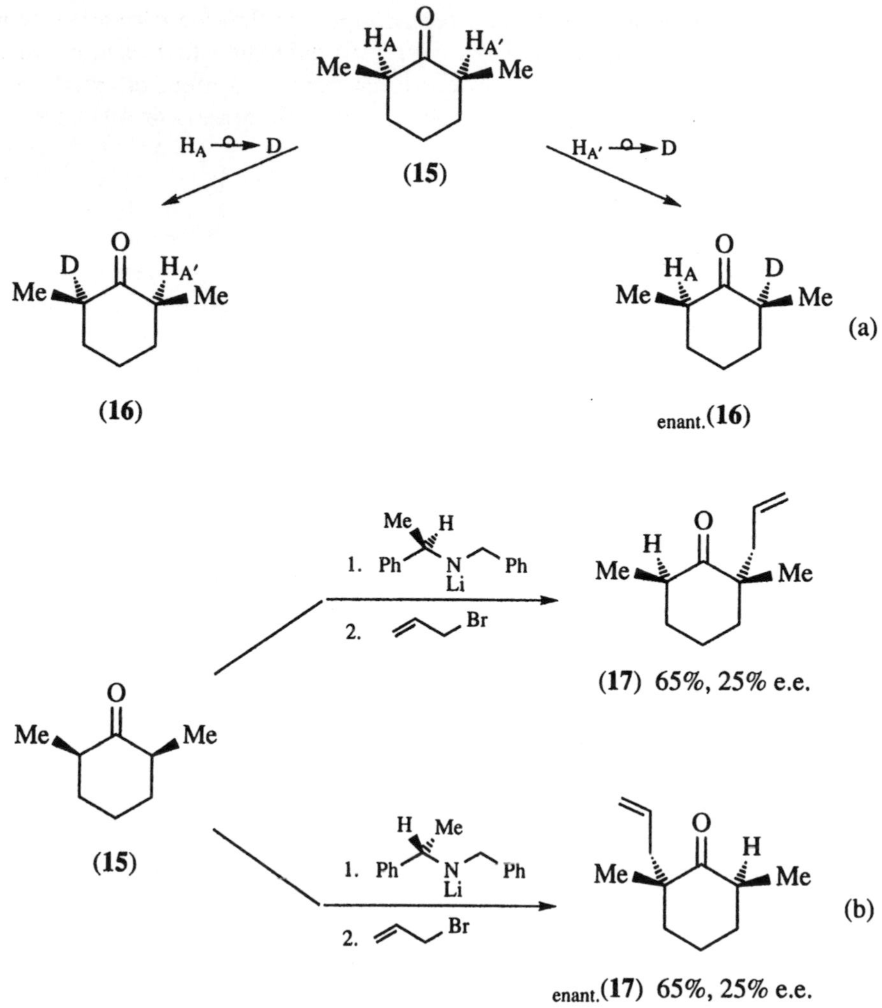

Scheme 9

Replacement of H_A and $H_{A'}$ in turn by deuterium gives (16) and $_{enant.}$(16), respectively, which clearly have an enantiomeric relationship showing that H_A and $H_{A'}$ are indeed enantiotopic. Molecule (15) is achiral but prochiral in a reaction in which substitution of only one of the enantiotopic hydrogen atoms H_A or $H_{A'}$ occurs since the product is chiral.[8] In fact, cis-2,6-dimethylcyclohexanone has been enantioselectively deprotonated using a single enantiomer of a chiral base and the anion (enolate) trapped with allyl bromide to give (17) in 25% e.e. [Scheme 9 (b)]; using the enantiomeric chiral base gave the enantiomeric product $_{enant.}$(17).

It is worth emphasizing that, in the reaction, enantioselectivity arises from the faster removal of one enantiotopic proton over the other. The resulting enolate has diastereofaces and only one diastereoface is attacked by the allyl bromide. However, the product would still be enantio-enriched even if the other diastereoisomer were formed with the methyl groups *trans*. Enantioselectivity, therefore, results from the preferential removal of one enantiotopic proton and diastereoselectivity from the preferential alkylation at one face of the enolate.

Trapping of the enantioselectively produced enolates from (15) and (19) (Scheme 10) by trimethylsilyl chloride already present in the reaction mixture minimises racemisation and gives the silyl enol ethers (18) and (20), with 96% and 97% e.e., respectively.

Conversion of the enolsilane (20) (Scheme 10) into α,β-unsaturated ketone (21) by treatment with palladium acetate then allows introduction of groups

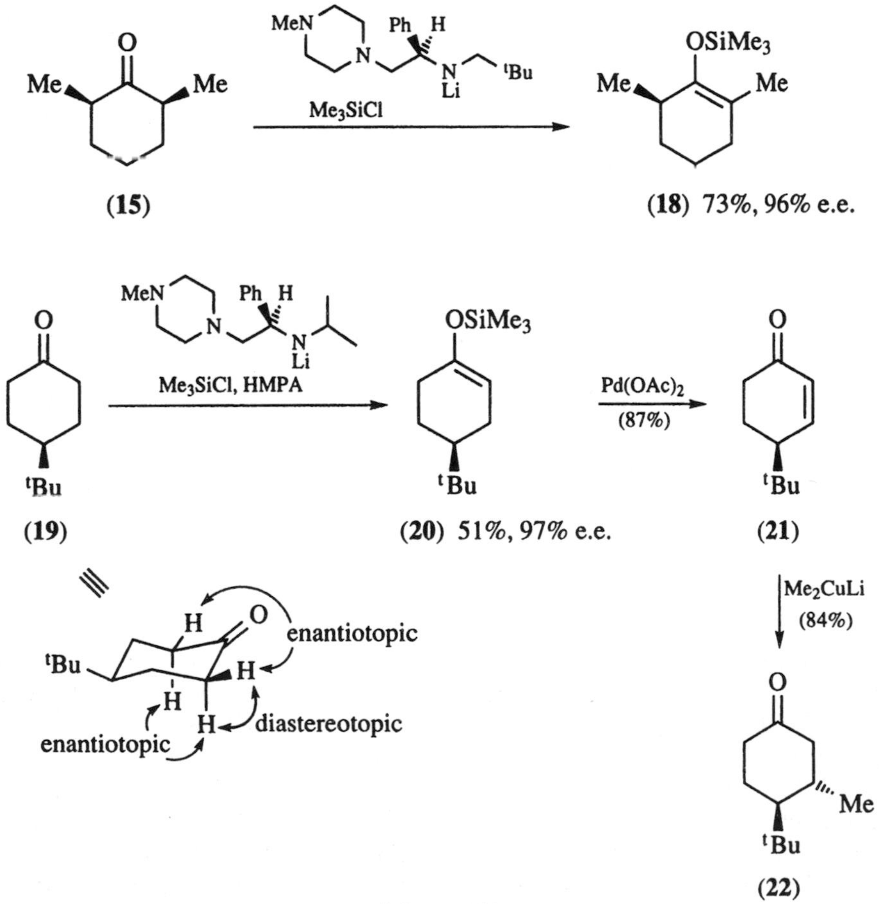

Scheme 10

into the β-position. By this means, the enantiopurity of the ketone (22), a compound whose optical rotation was known, was used to assay the enantioselectivity of the enolsilane formation.[9]

The *principle* which allows for the faster removal of one of the two enantiotopic protons in, e.g., (15) by an enantiopure base is very similar to that which allows an enantiopure reagent to distinguish between the two enantiofaces of a prochiral double bond; the enantiotopic relationship between these two protons in (15) is converted into a diastereotopic one by association of (15) with the chiral base. This association may be transient or, as is likely in the present case, a longer lived complex may be formed from which one of the now diastereotopic protons is preferentially removed (see Chapter 1).

The most successful chiral lithium amides, such as those in Scheme 10, have an additional nitrogen atom in the molecule which can form a five-membered chelate structures. This chelate structure not only brings about conformational control but also may fix the orientation of the nitrogen lone pair responsible for the deprotonation and, as a result maximise the asymmetry induced by the adjacent chiral centre in the deprotonation.

Prochiral molecules containing enantiotopic atoms or groups are often, like (15), *meso* compounds and contain an even number of chiral centres. However, they may, like (19) (Scheme 10), have no chiral centres but still possess pairs of enantiotopic protons.

Meso compounds such as that used in Scheme 9 contain two chiral centres although the molecule is achiral. In our classification, a *meso* compound, with its equal numbers of identically substituted R and S chiral centres, is regarded as having no net chiral centres (cf. Type III$'_{s.c.}$, Chapter 9). Accordingly, the enantioselective reaction in Scheme 9 results in the formation of two net chiral centres.

Enantioselective transformations of other prochiral molecules resulting from preferential attack on one enantiotopic proton are those in Scheme 11.[10,11] These reactions are reagent-controlled Type III since the prochiral element and the chiral element are contained in different molecules. Other reactions in this category, in which two enantiotopic groups are distinguished chemically, are illustrated in Scheme 12.[12–14]

Note that in Scheme 12 (a) epoxidation is both enantio- and diastereoselective, i.e. the two vinyl groups in (23) are enantiotopic and each has diastereofaces. Likewise, in Scheme 12 (b), hydroboration in diastereoselective (attack only from the *exo* face) and enantioselective with the *meso*-bicyclo[3.2.0]heptene (24) behaving as a prochiral molecule having enantiotopic olefinic carbons.

In Scheme 12 (c), attack by the enantiopure amine takes place preferentially on the enantiotopic carbonyl group on the R-centre of the anhydride.

The reactions in Scheme 12 (a) and (b) are, according to our classification in Chapter 9, Type III$_{cat. enant.}$ and overall enantioselective Type III reactions, respectively: that in (c) is a Type III$_{r.c.}$ reaction but the chiral element arising

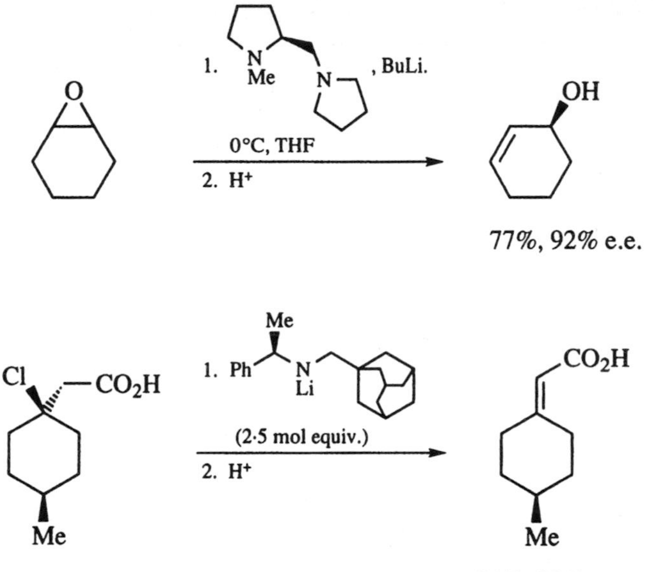

Scheme 11

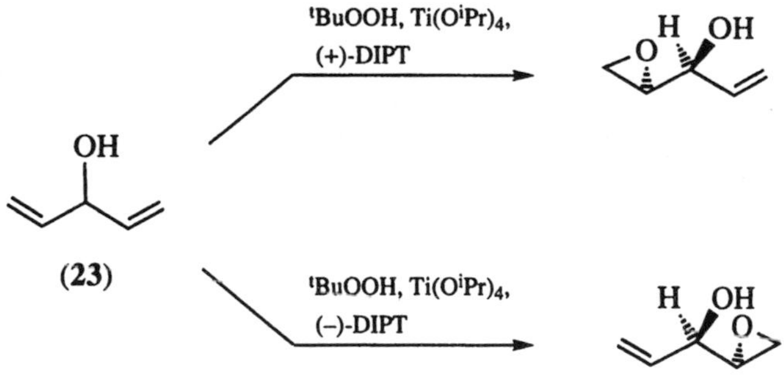

(a)

(23)

DIPT = diisopropyl tartrate

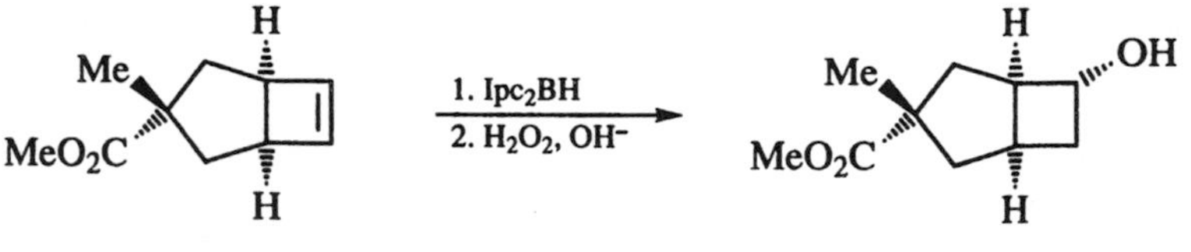

(b)

(24)   73%, ~ 92% e.e.

Ipc = isopinylcampheyl
(see Chapter 12, Scheme 7).

Scheme 12   *(continued)*

Scheme 12 *(continued)*

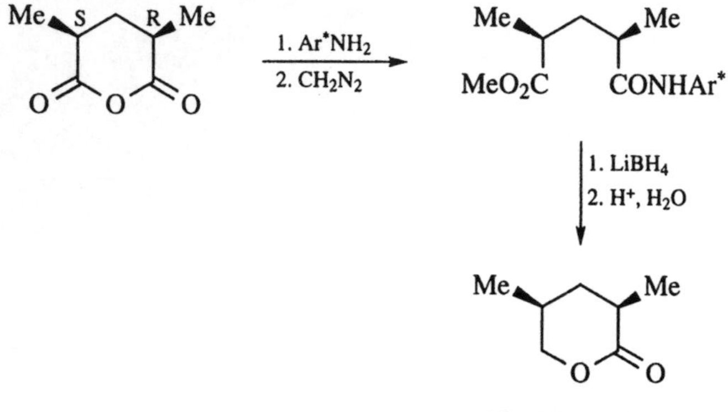

(c)

(25) 77%, 92% e.e.

$$Ar^*NH_2 =$$

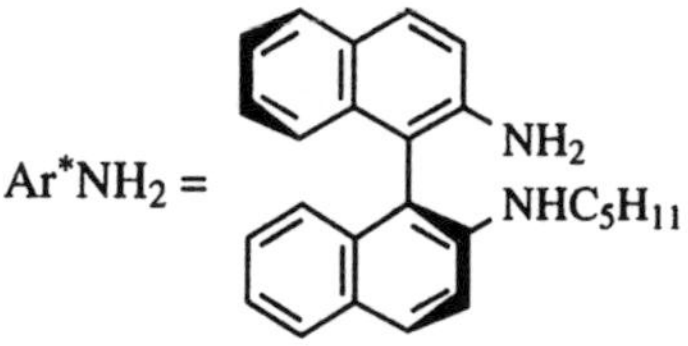

Scheme 12

from the reagent is eliminated in the subsequent step to give (25) in high enantiopurity.

Just as the electron pairs on bivalent sp³-hybridised sulphur are diastereotopic when an additional chiral centre is present, so they are enantiotopic in, e.g., (26). Reaction with the same reagents that bring about enantioselective epoxidation of alkenes (the Sharpless–Katsuki reaction) converts (26) into a sulphoxide enantiomer in 89% e.e. (Scheme 13).[15]

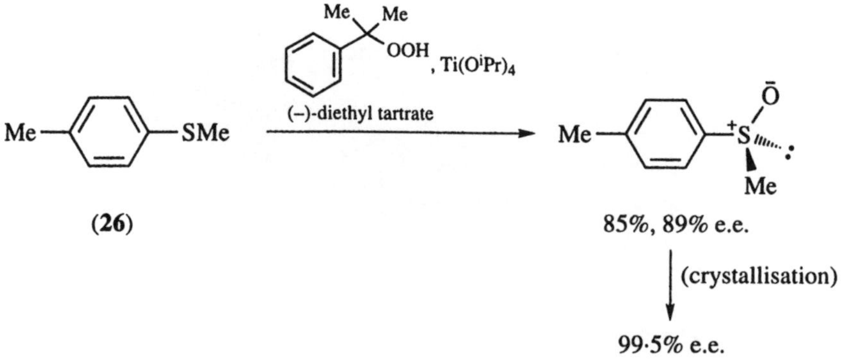

(26)   85%, 89% e.e.

(crystallisation)

99·5% e.e.

Scheme 13

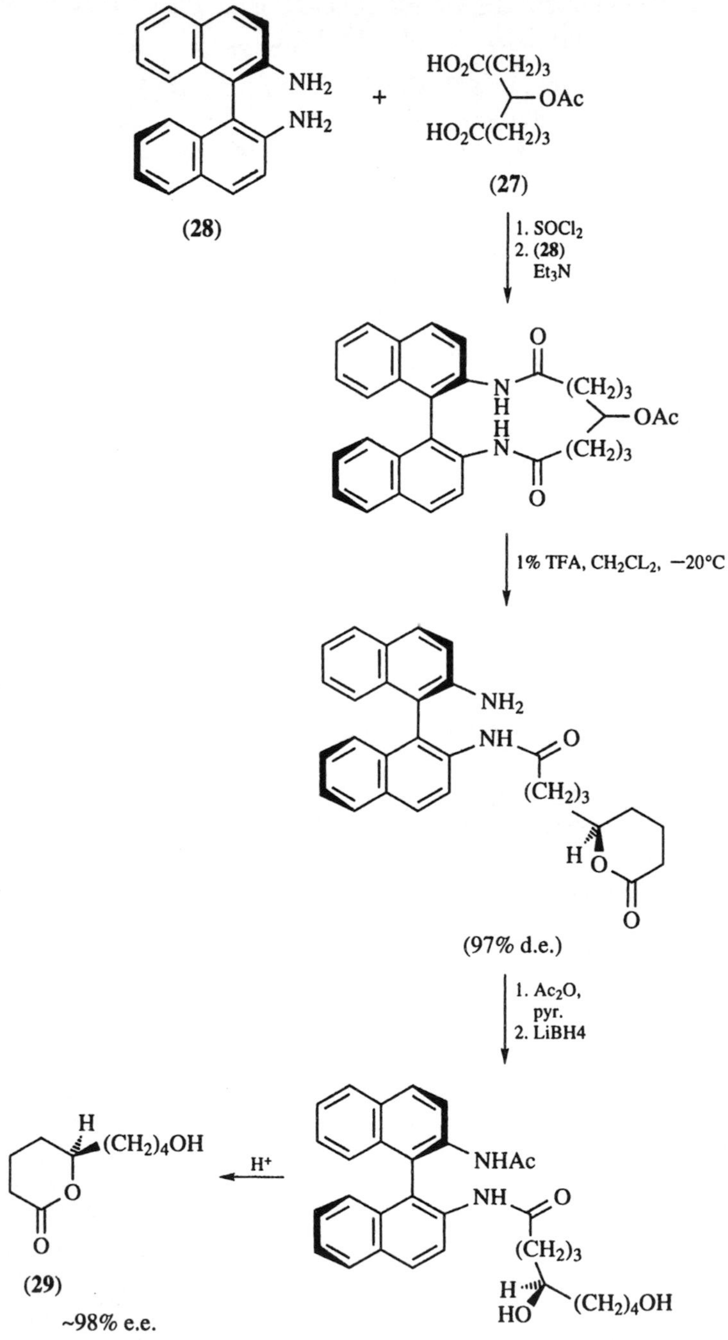
NH2
NH2
(28)
+
HO2C(CH2)3
OAc
HO2C(CH2)3
(27)
1. SOCl2
2. (28)
Et3N
1% TFA, CH2CL2, −20°C
(97% d.e.)
1. Ac2O,
pyr.
2. LiBH4
NHAc
(CH2)4OH
H+
(29)
~98% e.e.

Scheme 14

USE OF CHIRAL AUXILIARIES TO DISTINGUISH BETWEEN ENANTIOTOPIC GROUPS IN A MOLECULE

Just as two enantiofaces of a double bond are converted into diastereofaces by attachment of a chiral auxiliary (Chapter 1), so two enantiotopic groups are rendered diastereotopic by incorporation of a chiral auxiliary into the molecule.

The two carboxyl groups in diacid (27) (Scheme 14) are enantiotopic but diastereotopic amide groups result from their reaction with the binaphthyl diamine (28). Subsequent intramolecular reaction of the acetate-derived hydroxyl group with one of these amide groups followed by elimination of the auxiliary leads to the lactone (29) in high enantiopurity.[16]

Enzymes in stereoselective synthesis[17]

Enzymes are nature's catalysts. Many functional group interconversions that do not involve stereochemical change (Type 0 reactions) can be accomplished enzymatically, but enzymes which bring about regio- or stereoselective changes are those most valued by chemists.

Most enzymes have evolved in plant or animal systems to bring about a single chemical reaction on a defined substrate—their *natural substrate*. However, many enzymes are able to accept substrates which differ from their natural one, and it is this which allows them to be employed in stereoselective synthesis. *Chemo-enzymatic synthesis* is the synthesis of a molecule, usually in enantiopure form, by a route involving a number of chemical transformations at least one of which is mediated by an enzyme. For an enzyme to be widely useful in this way it is desirable that it be effective for a broad range of substrates, and that it brings about a transformation with (ideally) the same complete stereoselectivity as for the natural substrate.

In practice, the enzymes most used in organic synthesis are those that are commercially available, that can be used to process gram quantities of substrate in a reasonable time (hours rather than weeks), and that can be used without specialised biochemical equipment.

Some enzymes are isolable as powdered solids and are *cell-free*; others can carry out their function only inside the cell and the substrate, therefore, must be able to pass through the cell membrane to be processed by the enzyme. Many, but not all, enzymes require *cofactors* or *coenzymes* for their action. In transformations which use growing cultures, addition of cofactors is usually unnecessary because they are produced by the organism itself. However, the use of cell-free enzymes often requires added cofactors and since these are generally expensive, they are therefore used in less than stoichiometric quantity and are regenerated *in situ* by other enzymes specifically added for this purpose. Adenosyl triphosphate (ATP) and nicotinamide adenine dinucleotide (NADH) are two commonly used cofactors (NADPH is NADH phosphate—this may be more often used).

Enzymes are notable for the mildness of the conditions under which they operate most efficiently—usually close to ·neutral pH, at ambient temperatures, and in aqueous solution. The environment and locality of the *active* site of the enzyme, where chemical transformation of the substrate takes place, is fashioned by the folding of the protein chain(s) of the enzyme. This *tertiary structure* is the result of hydrogen bonds, van der Waals and other forces within the enzyme and between its surface and its environment. In water, hydrophobic amino acid side-chains in the protein will be directed towards the interior of the enzyme to avoid unfavourable interactions with the polar aqueous environment. Conversely, hydrophilic amino acid side-chains will be located on the outside where they can interact favourably with water.

Many of the substrates that the chemist wishes to modify using enzymes are insoluble in water, and will react slowly if at all in this medium. The question of what happens to the enzyme when an organic solvent is added to help dissolve the substrate is, therefore, of great importance. When the solvent is changed from water to a hydrocarbon the preferences for the exterior of hydrophilic and hydrophobic amino acids in the enzyme protein are reversed. Consequently, if the tertiary structures permits, there may be changes in the contours of the enzyme surface and in the environment of the active site which is often on or close to the surface.

Lipases are a group of enzymes which catalyse hydrolysis of esters. In fact they transfer acyl groups from one hydroxyl to another (including that of water) and to other receptor groups also. Lipases are more stable than most other enzymes in organic solvents, and in particular in hydrocarbons, and in these solvents they may bring about transesterification with enantioselectivities superior to those obtained in aqueous solution.

It appears that the protein chains of lipases are flexible to a degree not matched by most other enzymes. This flexibility enables them to accommodate the change from aqueous to non-polar solvents, and is probably also responsible for the relatively wide deviation from the natural substrate structure that lipases can cope with.

Most enzymes, however, do not function as efficiently as lipases in hydrocarbons but their stability (and that of lipases) towards addition of water-miscible organic solvents can be enhanced by *immobilisation*. This can be done by covalent linking of the enzyme to a support such as an agarose derivative, cellulose or a synthetic polymer such as polyacrylamide.

ENZYME PURITY

Although some commercially available enzymes are homogeneous, many contain minor amounts of other enzymes. A (unnatural) substrate for a reaction mediated by the major enzyme is often also a substrate for the minor one which may be enantioselective in the opposite sense. There are a number of ways in which competition from other enzyme impurities can be eliminated, without recourse to (often lengthy) enzyme purification procedures. These include changing the pH of the medium, adding chemicals (e.g. detergents or organic solvents), using specific inhibitors or carrying out controlled heating of the enzyme preparation. Alternatively, minor modifications of the *substrate* by derivatisation may reduce its suitability as a substrate for all except one enzyme. Where a number of enzymes are available for carrying out a given transformation, these may be screened for their suitability for the substrate in question.

ENGINEERED ENZYMES

Naturally occurring enzymes have evolved to handle a single substrate or a narrow range of substrates under closely defined conditions of pH, temperature and solvent (water).

Recombinant DNA technology has allowed modification of existing enzyme structures to provide mutants which may be better suited for application to transformations of non-natural substrates and that are more stable in organic solvents or more tolerant of pH and temperature changes. Briefly, this genetic engineering involves locating the gene in the organism which codes for the amino acid sequence of the enzyme, excising this gene using enzymatic methods (restriction enzymes) and, after mutation, transforming it into cells of *Escherichia coli* where protein (enzyme) can now be produced in large amounts (over-expression).

The mutation of the gene can be done in a precise way, via *site-directed mutagenesis* where changes in the nature of one or more selected amino acids in the expressed protein of the enzyme can be accomplished. The effects of changing one or two amino acids at a time in the enzyme on its activity can be determined. This information, together with X-ray crystallography and molecular modelling, can be used to describe the active site of the enzyme and its *modus operandi*. This technology is now being used to modify enzymes so as to optimise the reaction of substrates *in vitro* under particular conditions.[18]

CATALYTIC ANTIBODIES: ABZYMES[19]

All reactions pass through at least one transition state. Enzymes function by binding to both the substrate and the product. According to transition state theory of absolute rates applied to enzyme catalysis, the enzyme structure is designed to bind maximally to the transition state and it is this differential affinity for binding of the transition state over the ground state that lowers the activation energy.

Like enzymes, antibodies are also proteins which bind to other molecules (antigens) thereby earmarking them for destruction by other components of the immune system. Over the past decades it has been found that antibodies can also be induced to catalyse reactions. By raising antibodies to a ground state structure that mimics the transition state (*transition state mimic*) for a desired chamical transformation, a number of these antibodies will fortuitously show the same greater affinity for binding of the transition state over the ground state as enzymes and will therefore function catalytically. With a suitable method of screening these antibodies (abzymes), the efficient catalysts can be selected and exploited. For example, in the lactonisation in Scheme 15, the transition state has tetrahedral character at the ester carbonyl carbon. A transition-state mimic is the phosphonate (**30**) (tetrahedral phosphorus) and monoclonal antibodies raised against this phosphonate catalyse the lactonisation.

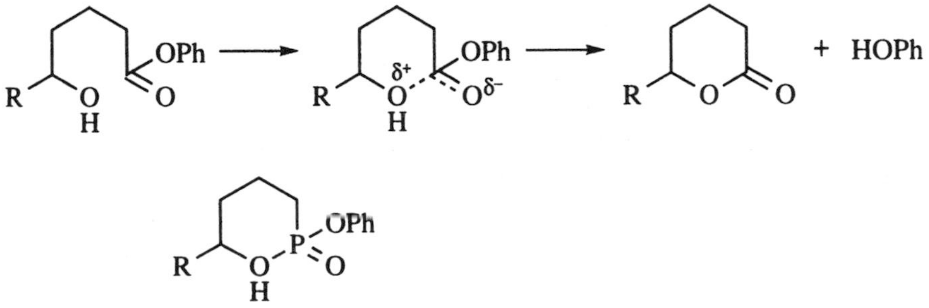

(**30**)

Scheme 15

Catalytic antibodies or *abzymes* are now also known for bimolecular reactions including the Diels–Alder reaction. Like enzymes, abzymes can be immobilised on solid supports and, like enzymes, the reactions that they bring about are often highly stereoselective.

Enzyme-mediated stereoselective transformations

Enzyme reactions are often completely regioselective and diastereoselective in situations where regioisomers or diastereoisomers are conceivable. *syn*-Dehydrogenation of stearoyl-CoA (**31**) to oleyl-CoA (Scheme 16) is such a biological transformation which it would not be possible to carry out by direct chemical means at the present time.[20] However, it is the ability of enzymes also to effect highly enantioselective transformations that is of particular value to the chemist in synthesis. Even when an enzyme reaction is not highly diastereoselective, it may nevertheless be the case that each of the diastereoisomers may be produced with high enantioselectivity (Scheme 17).[21]

In addition to bringing about enantioselective addition to prochiral double

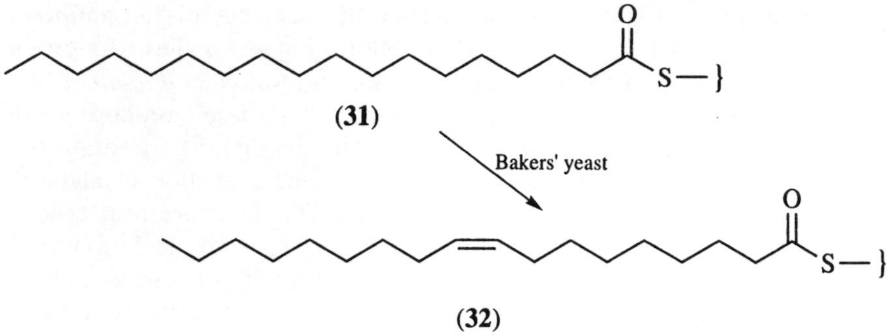
O
S—}
(31)
Bakers' yeast
O
S—}
(32)

Scheme 16

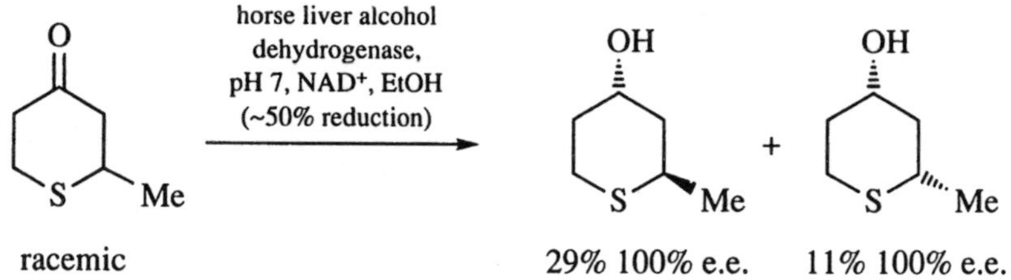
O
S
Me
racemic
horse liver alcohol
dehydrogenase,
pH 7, NAD+, EtOH
(~50% reduction)
OH
S
Me
29% 100% e.e.
+
OH
S
Me
11% 100% e.e.

Scheme 17

bonds, enzymes are often excellent catalysts for the (kinetic) resolution of racemates where one enantiomer reacts and the other is unaffected. Enantioselective reactions brought about by enzymes also include those in which enantiotopic atoms or groups attached to the same atom [Scheme 18 (a)], or to different atoms [Scheme 18 (b)] as in the *meso*-substrate discussed previously (Scheme 9), are selectively transformed.

One advantage in the use of achiral substrates, as in Scheme 18, is that

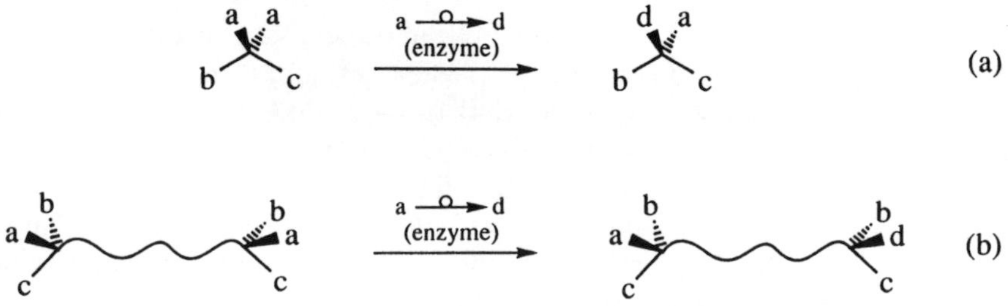
a a
b c
a → d
(enzyme)
d a
b c
(a)
b
a
c
b
a
c
a → d
(enzyme)
b
a
c
b
d
c
(b)

Scheme 18

all of the starting material can be converted into product. By contrast, in the kinetic resolution of a racemate only 50% of the starting material will be utilised, although it may be possible to convert the unreactive enantiomer back into the racemate so that all of the material can be processed.

LIPASES (ACYL TRANSFER ENZYMES)[22]

About 20 lipases are commercially available, many of them inexpensive. As mentioned above, the great advantage in their use is their ability to function in hydrocarbon solvents, including cyclohexane and toluene. Also, they do not require cofactors.

In hydrocarbon solvents, lipases bring about transacylations of a wide variety of substrates with enantioselectivities that are often superior to those achieved in water. Procedures for the conversions shown in Scheme 19 are

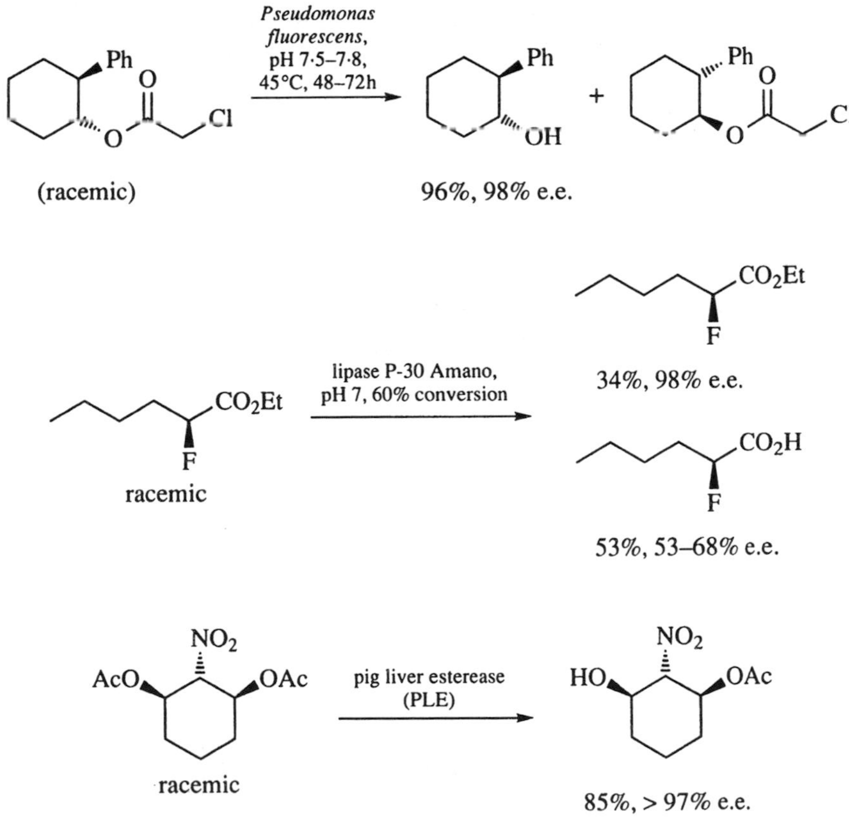

Scheme 19

described in *Organic Synthesis*; they make use of three readily available lipases, and can be carried out on at least a 10g scale.[23]

Lipases can also be used for resolution of racemic acids. Thus *Candida cylindracea* has been applied to the resolution of 2-bromo- and 2-chloropropanoic acids (starting materials for synthesis of herbicides based on phenoxypropanoic acid) and is currently being used to produce these enantiopurified acids on a 100kg scale commercially.[17d]

Scheme 20 illustrates the use of lipase P-30 as a catalyst either for hydrolysis of a diester or for acetylation of a diol.[24] It is expedient to use isopropenyl acetate for lipase-mediated acetylations since the *trans*-acetylation in this case is irreversible [Scheme 20 (b)]. Scheme 20 (a) illustrates a commonly observed situation in which there is a trade-off between the yield of the product and its enantiopurity. The lower yield (36%) after 46h is the result of increased conversion of the monoester to the diol, but the enantiopurity of the monoester is greater because the minor enantiomer [mirror image of (33)] is deacetylated *faster* than (33) itself. A 92% e.e. in this case is, therefore, the result of modestly selective hydrolysis of one of the enantiotopic ester groups followed by kinetic resolution of the enantioimpure monoacetylated product.

That there should be preferential consumption of the minor enantiomer in this kinetic resolution can be understood from consideration of the model for the lipase-mediated conversion shown in Scheme 21. Here the substrate is assumed to be held on the enzyme surface by receptor sites p′ and q′ as a

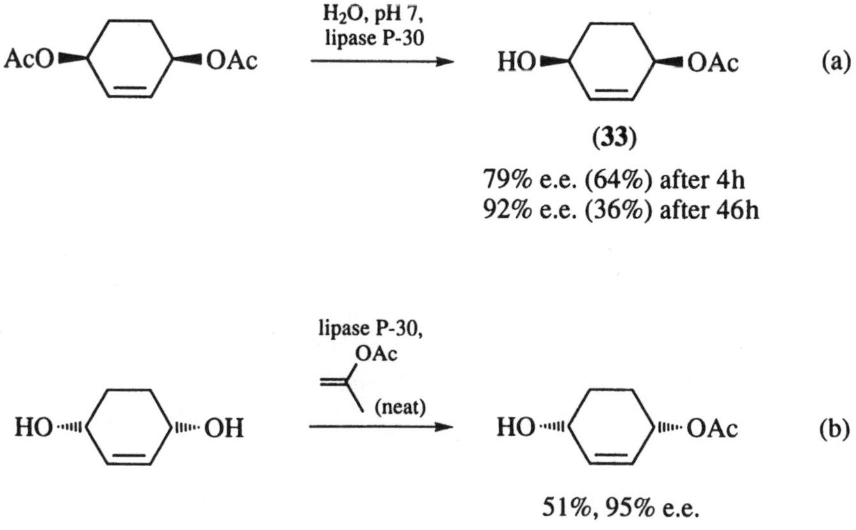

(a)

(33)

79% e.e. (64%) after 4h
92% e.e. (36%) after 46h

(b)

51%, 95% e.e.

Scheme 20

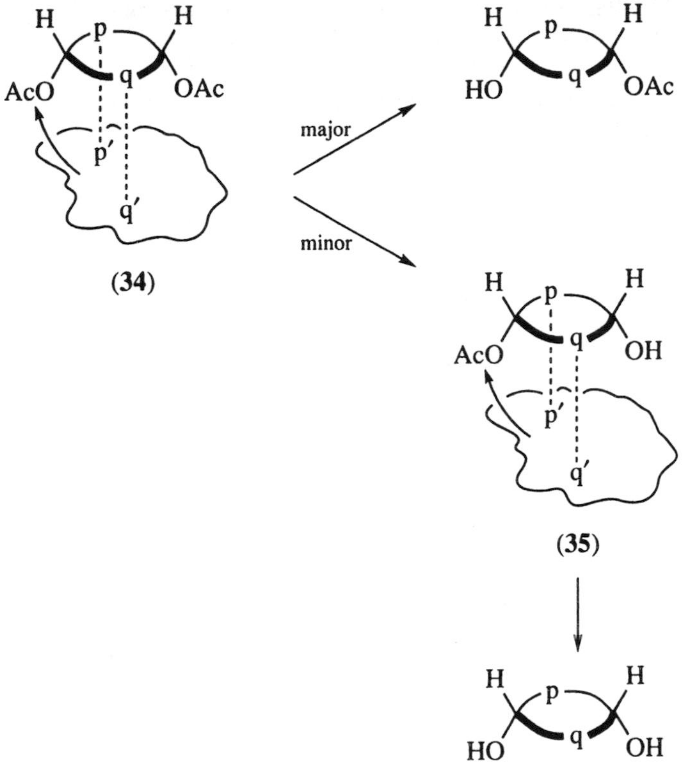

Scheme 21

result of which one of the two now-diastereotopic acetoxy groups is hydrolysed preferentially, as indicated in (34). With the *same* receptor sites and *same* hydrolysis mode, attack on the second acetoxy group is more likely for the *minor* enantiomer, as illustrated in (35), and this leads to enantio-purification of the major monoacetoxy product.

Although these lipase-catalysed acetyl transfers lead (ideally) to a single enantiomer of the 'desymmetrised' product, the application of judicious functional group interconversions to this enantiomer can lead to derivatives of *either* enantiomer, as illustrated in Scheme 22.[25] Thus, although the lipase catalyses the formation of a single enantiomer of acetoxyalcohol (36), either of the two enantiomeric dideoxypyranose derivatives (37) or $_{enant.}$(37) can be obtained by double bond cleavage after acetoxy/hydroxyl group manipulation.

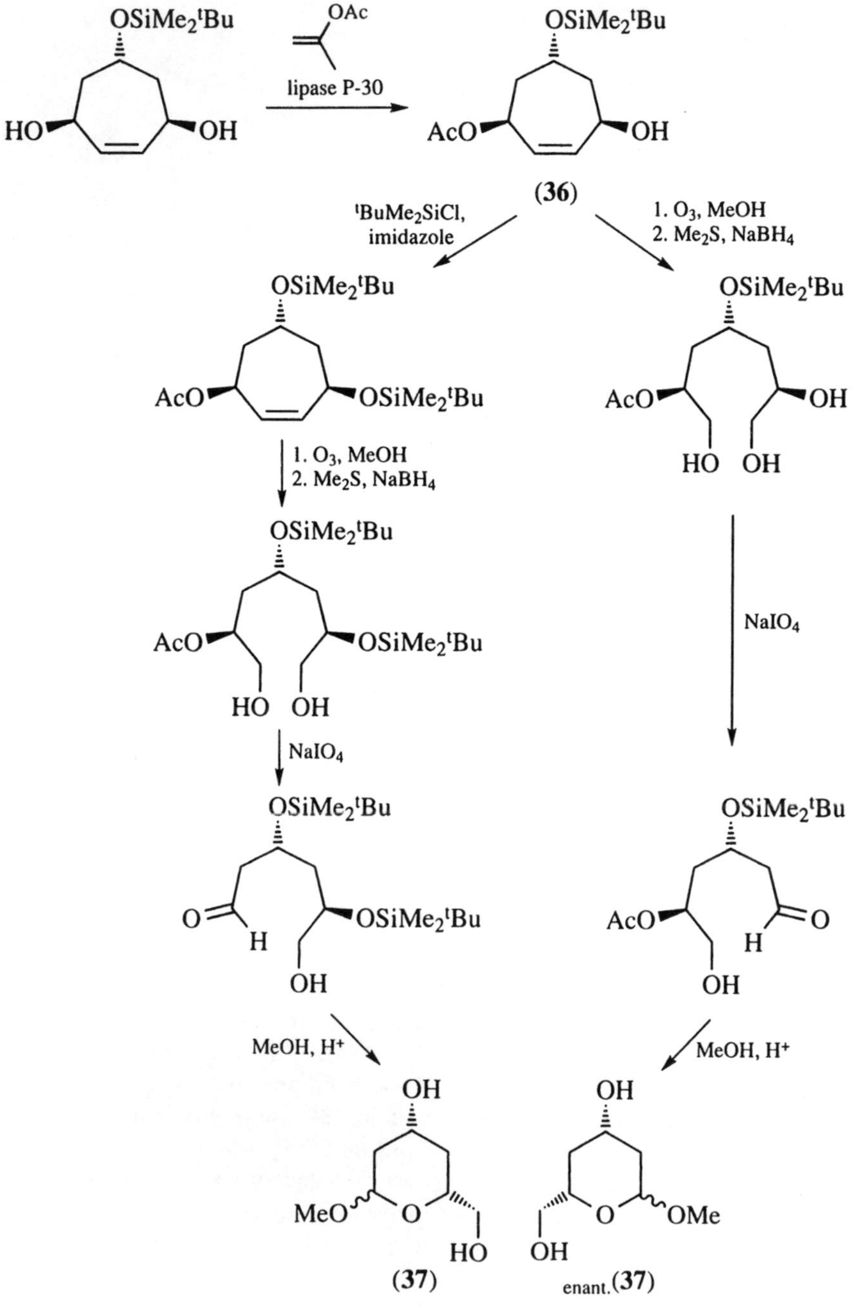

OSiMe2tBu
HO
OH
OAc
lipase P-30
AcO
(36)
tBuMe2SiCl, imidazole
1. O3, MeOH
2. Me2S, NaBH4
NaIO4
MeOH, H+
O
H
MeO
OMe
(37)
enant.(37)

Scheme 22

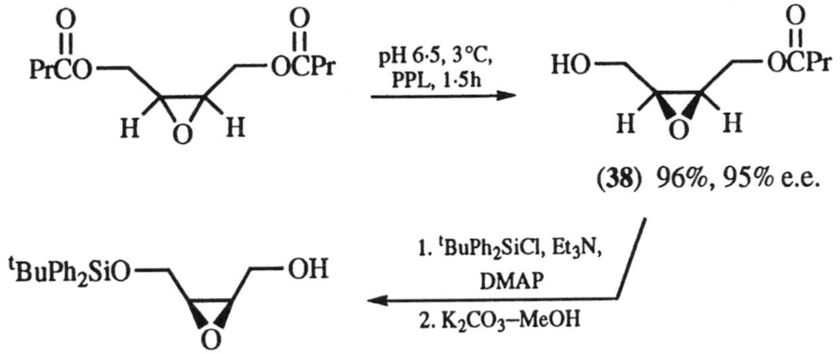

Scheme 23

Another example of this tactic is shown in Scheme 23 using porcine pancreas lipase (PPL).[26] The initial product (38) can be recrystallised to give material of 100% e.e., a common means of increasing enantiopurity. Inspection of (38) and (39) reveals that each of the two enantiotopic substituent groups in the starting material has been individually made available for further reaction by conversion to the hydroxymethylene. Overall, this sequence, which has been carried out on a 50 g scale, provides enzymatic alternative to the Sharpless–Katsuki epoxidation, a reaction which gives only modest enantioselectivities with some cis-allylic alcohols.

Schemes 22 and 23 are just two examples of a large number of acyclic, monocyclic and bicyclic meso compounds which have been desymmetrised highly enantioselectively using lipases.[27] Enantiotopic group differentiation by enzymes is, of course, not limited to lipases.

CHIRONS[28]

A chiron is a molecule containing a small number of chiral centres, of defined absolute configuration, and one or more functional groups which allow it to be used as a precursor for a range of other enantiopure derivatives. The desymmetrised products from lipase-catalysed reactions in Scheme 22 and 23 are chirons. Scheme 24 illustrates part of a synthesis of brefeldin, which uses porcine pancreas lipase (PPL) to produce the chiron (40).[29] Singlet oxygen cycloaddition followed by reductive cleavage and acetylation is a much used route to 1,4-diacetoxy-2-ene-containing substrates which are precursors for compounds of type (40).

The enantiopure β-lactam (41), useful for synthesizing carbapenem antibiotics, can be derived by retrosynthetic analysis from the enantiopure 3-

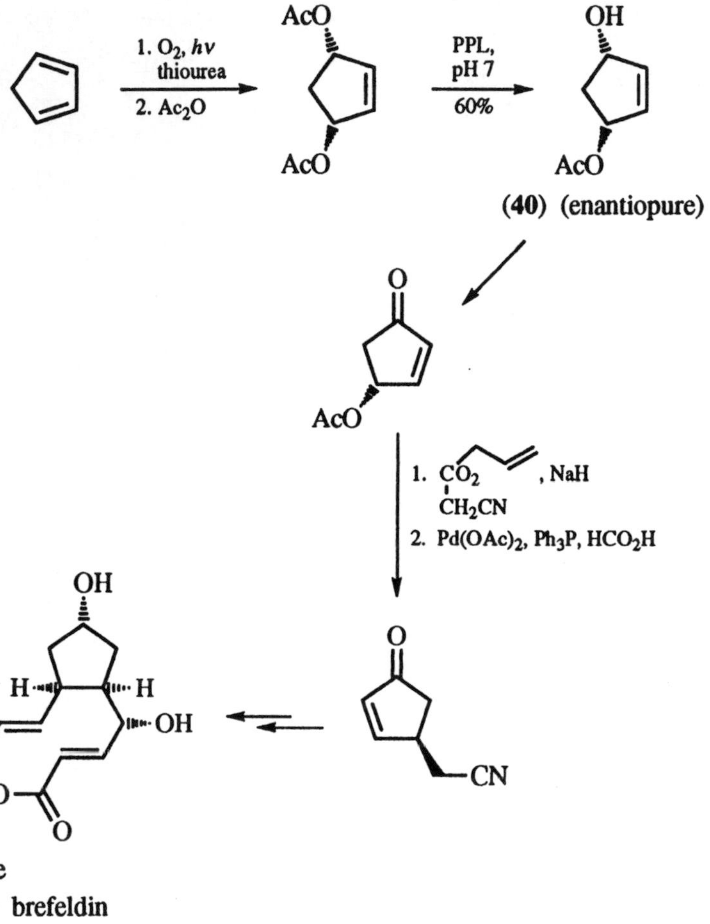

Scheme 24

aminoglutarate monoester (**42**), a chiron obtained by lipase-mediated hydrolysis of the prochiral glutarate diester (**43**) [Scheme 25 (a)].

Efficient synthesis of (**41**; R = Me) was accomplished by protection of the amino group of dimethyl β-aminoglutarate, pig liver esterase (PLE) hydrolysis of the *pro-R*-methyl ester and ring closure [Scheme 25 (b)].[30]

Pig liver esterase also converts the readily available diester (**44**) (Scheme 26) into the chiron (**45**) (10–100 g scale). Appropriate functional group interconversion including C—C→C—N with retention of configuration (Type 1) leads to useful 4,5-*cis*- or *trans*-disubstituted cyclohexene derivatives (**46**)–(**49**).[31]

Chirons are also available by the action of lipases on racemic substrates in the kinetic resolution mode (Scheme 27). Thus treatment of the racemic

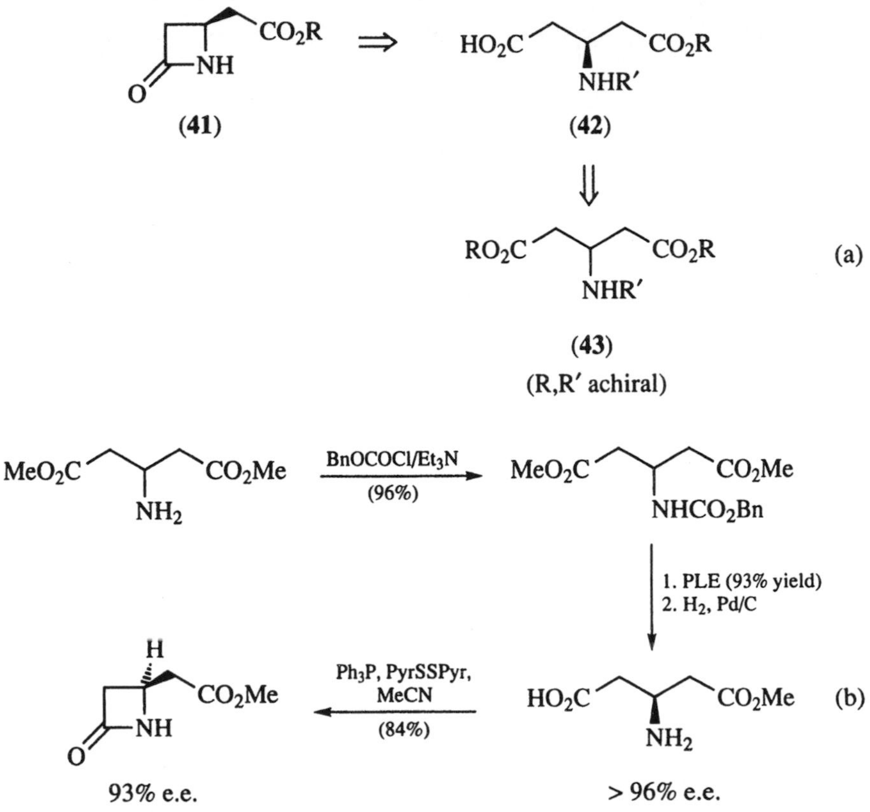

$$(41) \Rightarrow (42) \quad (a)$$

(43)
(R,R' achiral)

93% e.e.   > 96% e.e.

Scheme 25

secondary alcohol-derived acetate (50) with lipase gives, after 50% conversion, the alcohol (51) and the unreacted acetate (50), both in 98% e.e. As illustrated in Scheme 27, these products are themselves sources of other chirons, as in their conversion into acetal-protected epoxyaldehydes (52) and $_{enant.}$(52).[32]

In this discussion of lipase-catalysed reactions we have concentrated on those in which one of two enantiotopic groups is acetylated or hydrolysed. When two groups are diastereotopic, exclusive reaction of one of them usually takes place. A testing case of this is the desymmetrisation of the pentitol (D-arabinitol) tribenzyl derivative (53) (Scheme 28), which can easily be prepared by selective protection of the primary alcohol groups of the corresponding pentol followed by benzylation and then deprotection. The two hydroxymethylene groups in (53) are diastereotopic and only one of them is acetylated by vinyl acetate in the presence of the lipase enzyme.[33]

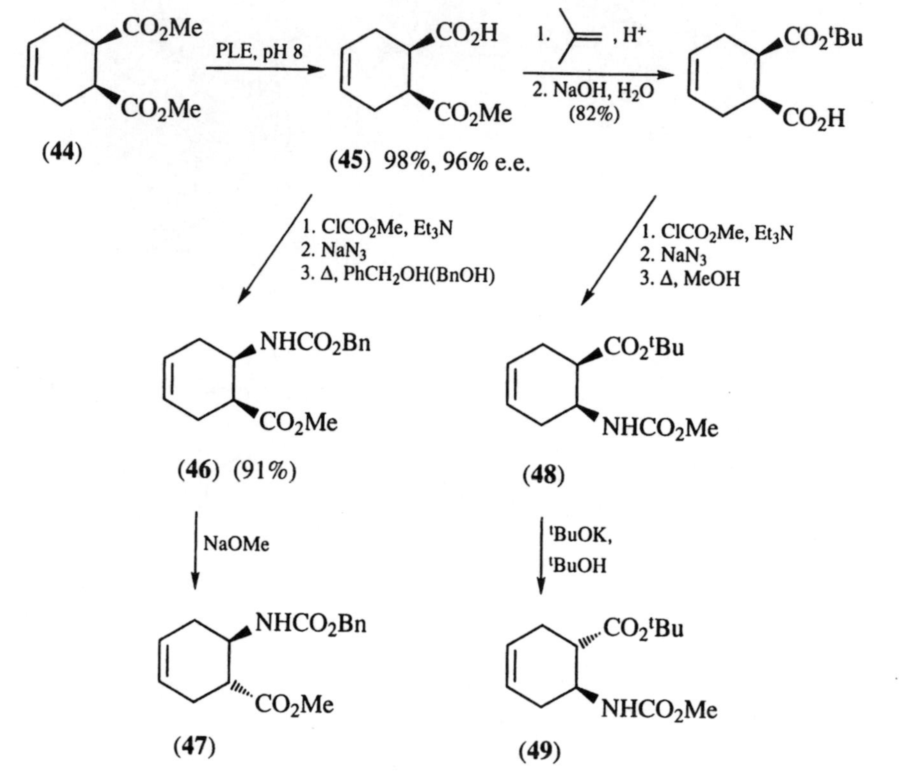

Scheme 26

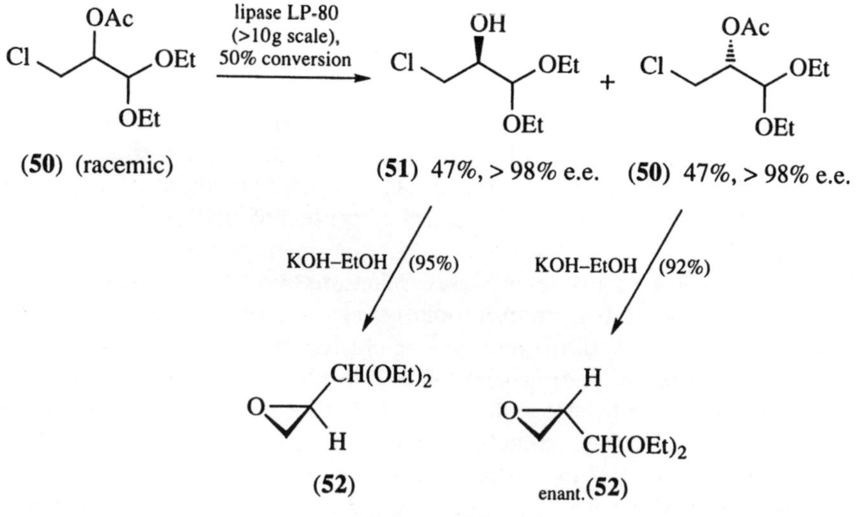

Scheme 27

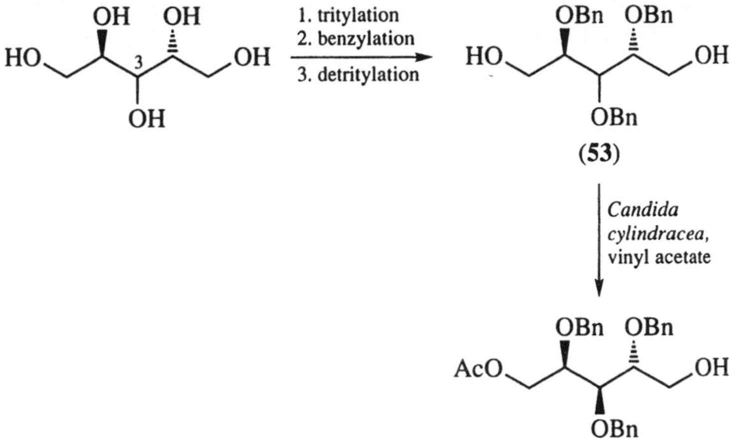

(54) 58%, 95% e.e.

Scheme 28

ACTIVE SITE MODELS

The mechanism of enzyme action is known in detail for very few transformations. Active site models are an attempt to describe the contours of the active site and, in particular, the size and nature (hydrophobic, hydrophilic) of the 'pockets' into which substituents of the substrate are required to fit. This description is arrived at empirically from analysis of the profile of the group of related compounds which are substrates for the enzyme and which are transformed in the same enantio-sense.

Pig liver esterase (PLE) is believed to consist of a number of closely related enzymes (isozymes) which act in a similar way. The arrangements in space of the hydrophobic and hydrophilic 'pockets' at the active site have been assessed from analysis of the PLE-catalysed hydrolysis of about 100 methyl esters; the result has been expressed in terms of a cubic space diagram in which the dimensions of the cubes (pockets) can be estimated.[34]

The purpose of these active-site models is, of course, to be able to predict the feasibility of using the enzyme reaction with a previously untried related substrate and the sense of the enantioselectivity of its transformation.

More readily available are descriptors which allow prediction of the more reactive enantiomer of a racemate, or the more reactive one of a pair of enantiotopic groups, in an enzyme-mediated reaction. Thus the model for porcine pancreatic lipase in Figure 1 allows prediction of the more readily hydrolysed acetate enantiomer, e.g. **(55)**, or the more rapidly hydrolysed enantiotopic acetate of a *meso*-diacetate, e.g. **(56)**.[35]

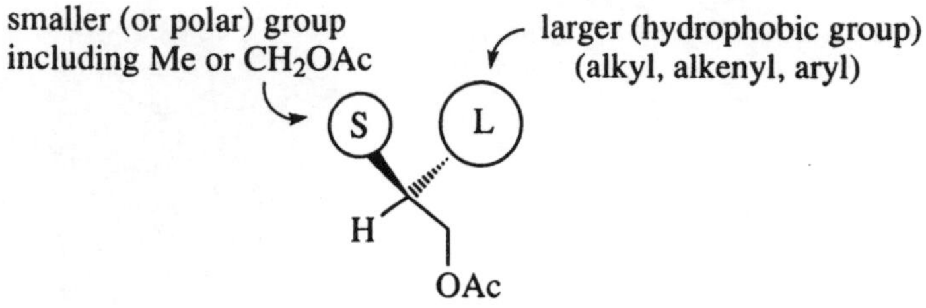
smaller (or polar) group
including Me or CH2OAc
larger (hydrophobic group)
(alkyl, alkenyl, aryl)
S
L
H
OAc

Figure 1

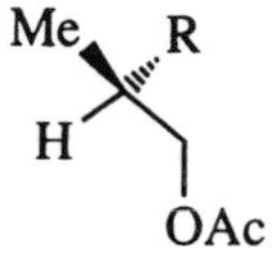
Me
R
H
OAc

(55)

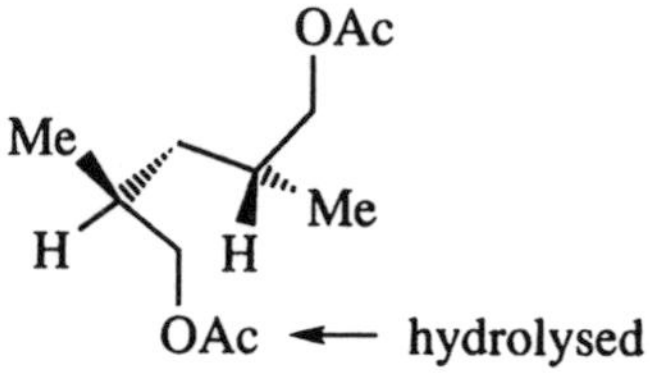
OAc
Me
Me
H
H
OAc
hydrolysed

(56)

OTHER ENZYMES USEFUL IN CHEMO-ENZYMIC SYNTHESIS

Enzyme-mediated addition of the elements of X and Y to a double bond has much in common with the *catalytic* enantioselective reactions discussed in Chapter 14. In these enzymatic additions, both the substrate and X and Y are held at the active site of the enzyme, which then mediates the stereoselective addition of X and Y to the double bond.

The enzymes in this category that have found much use in synthesis are the oxidoreductases, particularly bakers' yeast, which is cheap and easy to use. Reduction of a great variety of ketones and some aldehydes has been accomplished with high enantioselectivity using bakers' yeast.[36] Scheme 29 gives three examples.[37-39] In (a) the alcohol was converted into (*S*)-(+)-(**57**) the enantiomer occurring naturally in the berries of the mountain ash. In (b), enolisation of the (racemic) starting dithioester allows high enantioselective and diastereoselective formation of the *cis* product.

It is known that bakers' yeast accomplishes carbonyl reduction using NADPH (see below). In enantiofacially selective reductions, therefore, X is a hydride ion supplied by the dihydropyridine unit of NADPH and Y is a proton added to the carbonyl oxygen from a suitably located protonated amino group in the enzyme (Scheme 30).

Bakers' yeast contains at least three enzymes which are capable of reducing ketones.[40] Although reduction by each of these enzymes is likely to be completely enantioselective, it is not necessarily so in the same sense in

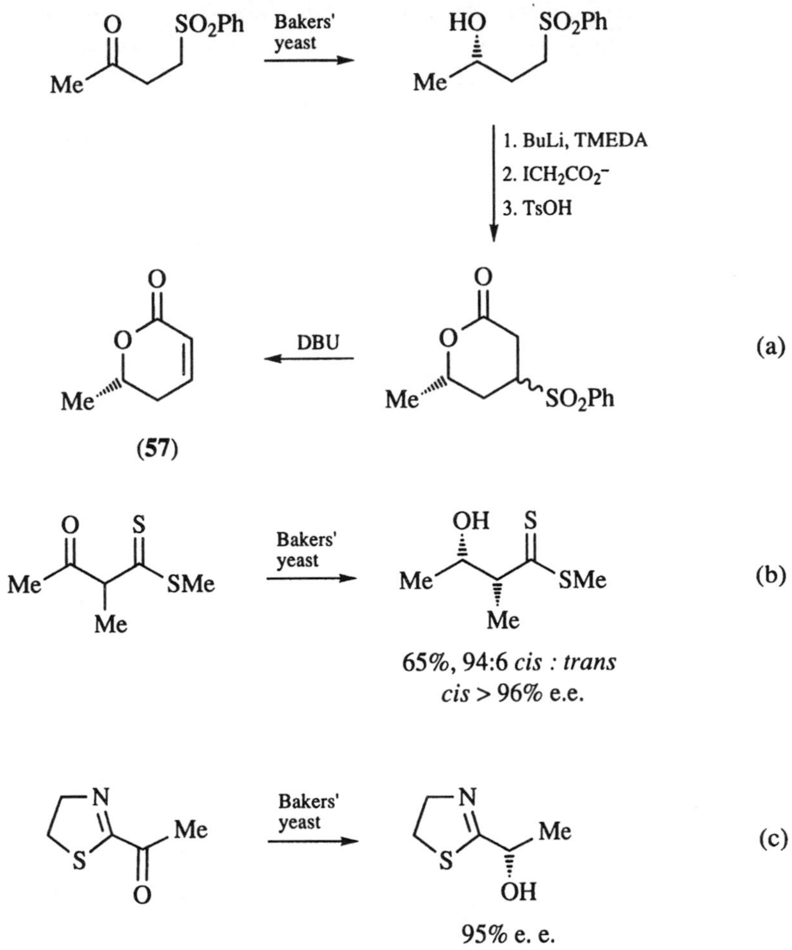

Scheme 29

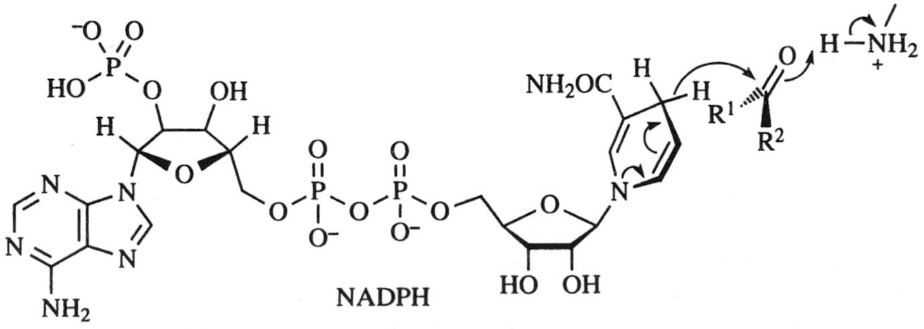

Scheme 30

each case. Consequently, the greater the number of enzymes that participate in the reduction of the substrate, the greater is the likelihood of diminished enantiopurity of the product.

A number of methods have been used to limit the reduction by all except one enzyme, including the addition of inhibitors such as ethyl α-chloroacetate, allyl alcohol or an α,β-unsaturated carbonyl compound. For example, reduction of ethyl 3-oxopentanoate with bakers' yeast gives the (R)-alcohol with only 12% e.e., but when the reducing system is first treated with ethyl α-chloroacetate (Scheme 31), the enantiomeric (S)-alcohol is produced with 91% e.e.[41] Heat treatment of bakers' yeast before use can enhance the stereoselectivity of a reduction.[42] The experimental procedure can also affect the enantioselectivity. For example, slow addition of the substrate to the reducing medium, thereby minimising its concentration, often results in a higher enantioselectivity. Immobilisation of the yeast can also have beneficial effects on the enantioselectivity.[43]

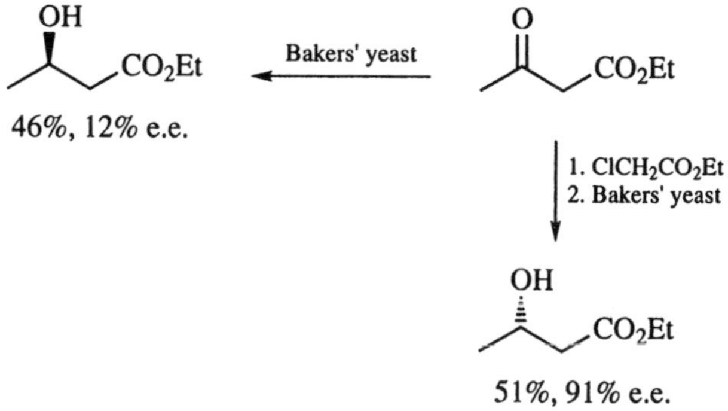

46%, 12% e.e.

1. $ClCH_2CO_2Et$
2. Bakers' yeast

51%, 91% e.e.

Scheme 31

One of the most widely used methods for avoiding multiple enzyme involvement in bakers' yeast reductions is tailoring the substrate so that it becomes less attractive to all except one enzyme. Thus reduction of the γ-chloro-β-keto ethyl ester (**58**) (Scheme 32) gives the alcohol in 55% e.e. but this is raised to 95% e.e. when the n-octyl ester (**59**) is used (with reversal of the sense of enantioselectivity).[44]

Horse liver alcohol dehydrogenase (HLADH) is another commercially available NADPH-dependent oxidoreductase which catalyses $\rangle$CHOH $\rightleftharpoons$ $\rangle$ C=O interconversions. One use is for making specific isotope-labelled alcohol enantiomers which can then be used to probe the stereochemistry of biosynthetic pathways. In Scheme 33, tritiated cyclopentanol is used as a means of regenerating the tritiated dihydropyridine ring of NADPH.[45]

Selective reduction of one of two enantiotopic carbonyl groups can be accomplished as in the reduction of *cis*- and *trans*-decalindiones (**60**) and (**61**)

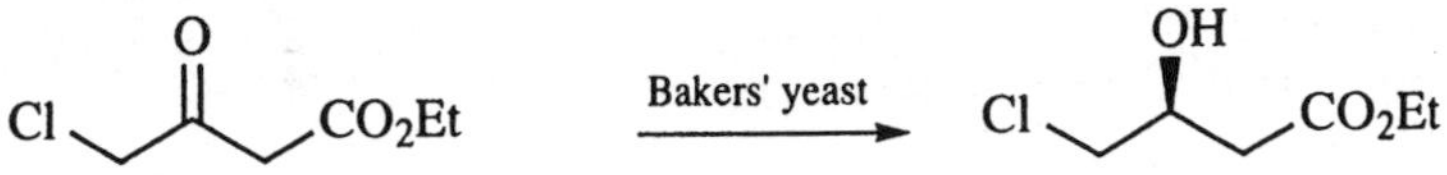

(58)   Bakers' yeast   55% e.e.

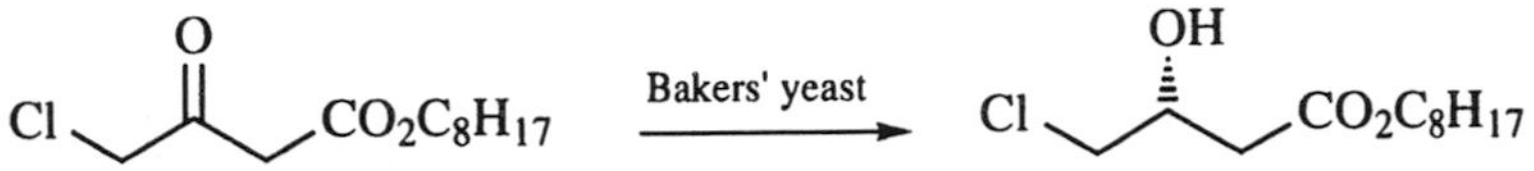

(59)   Bakers' yeast   95% e.e.

Scheme 32

HLADH

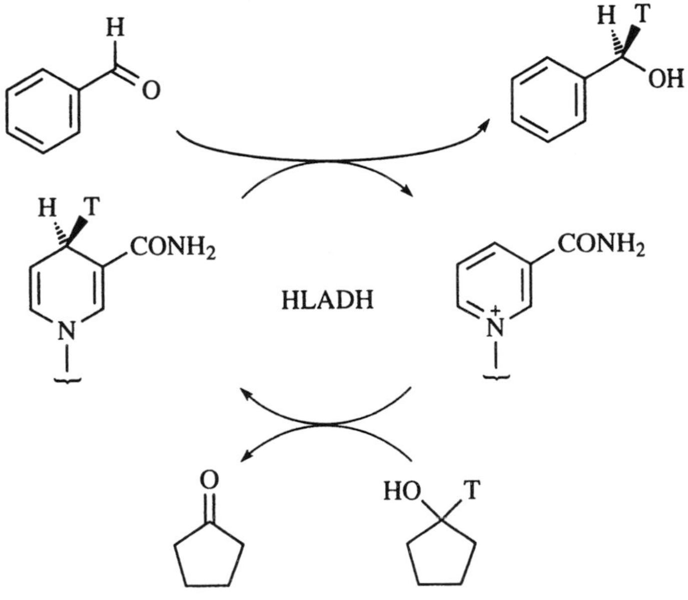

Scheme 33

(Scheme 34); in each case also a single diastereoisomer of the product is produced.[46]

Horse liver alcohol dehydrogenase can be used to distinguish between enantiotopic hydroxyl groups in *meso*-diols (Scheme 35). As with lipases, a broad range of *meso*-diol structures can be desymmetrised, in this case by conversion into the corresponding lactones by further oxidation of intermediate hemiacetals (Scheme 35).[47]

Rabbit muscle aldolase (RAMA) (fructose-1,6-diphosphate aldolase) is

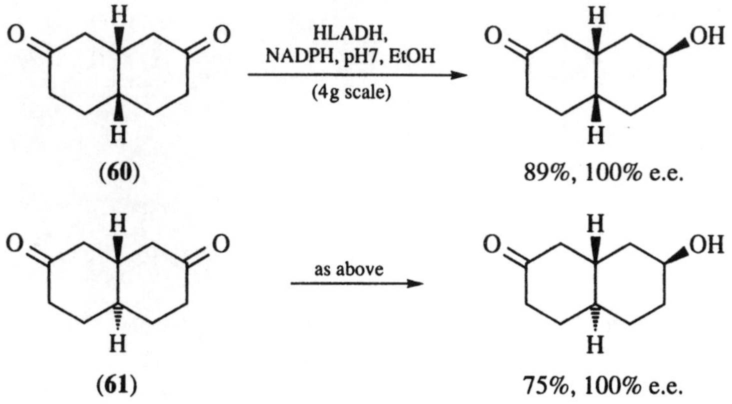

(60)   89%, 100% e.e.

(61)   75%, 100% e.e.

Scheme 34

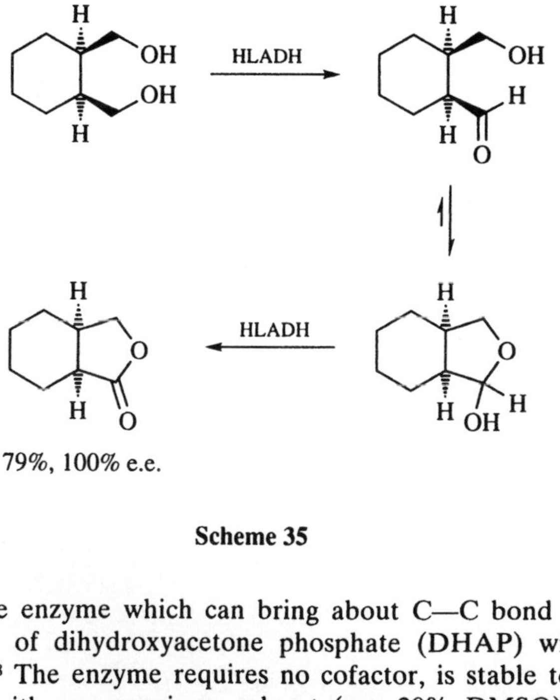

79%, 100% e.e.

Scheme 35

an inexpensive enzyme which can bring about C—C bond formation by aldol reaction of dihydroxyacetone phosphate (DHAP) with aldehydes (Scheme 36).[48] The enzyme requires no cofactor, is stable to oxygen and can be used with an organic co-solvent (e.g. 20% DMSO). A range of aliphatic aldehydes can be used but reaction is retarded by bulky substituents in the α-position. Aromatic aldehydes are unreactive. RAMA-catalysed aldol reactions tolerate very little variation in the nucleophilic component (DHAP).

As indicated in Scheme 36, the reaction is an equilibrium but in most cases it lies well to the right-hand side. The enantioselectivity and

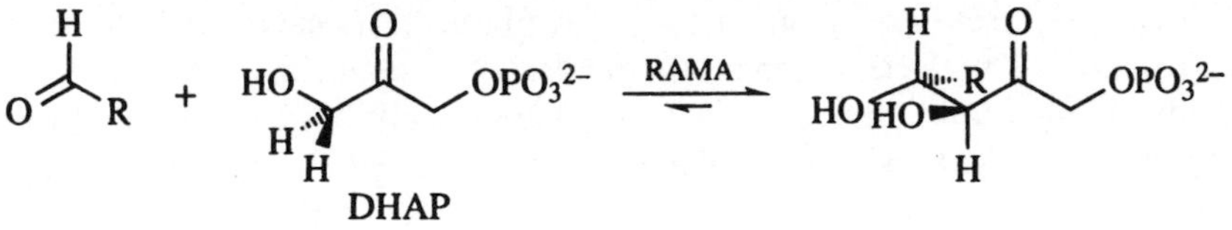

Scheme 36

diastereoselectivity are as shown.

The reaction lends itself to the synthesis of sugars, e.g. D-xylulose (Scheme 37).[49] One obvious advantage over chemical methods of sugar synthesis is that multiple protection–deprotection steps are unnecessary. The enzyme has also been applied to a synthesis of *exo*-brevicomin, a sex pheromone of the western pine beetle (Scheme 38).[50]

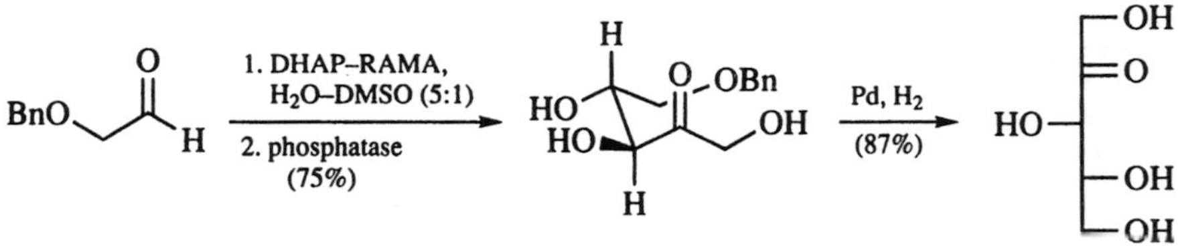

D-xylulose

Scheme 37

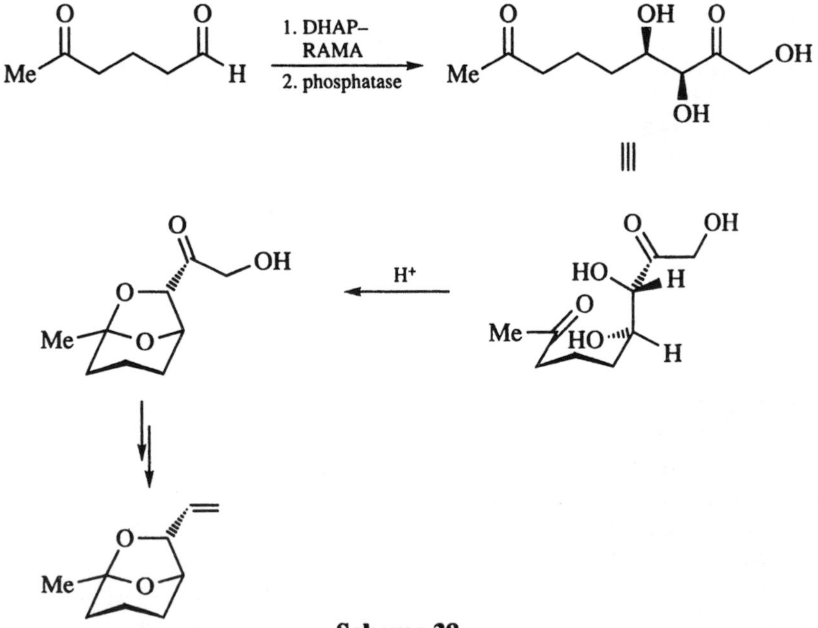

Scheme 38

Other aldolases are commercially available for carrying out similar transformations with complementary diastereo- or enantioselectivity.[51]

There is no doubt that enzymes will increasingly be used in all areas of stereoselective synthesis, particularly for reactions where chemical methods are either not available or lack efficiency in yield and/or stereoselectivity. One of the principal tasks of the synthetic chemist will be to develop stereoselective methods for the many reactions that cannot be catalysed by enzymes.

References

1. G. Boche and H.M. Walborsky, *Cyclopropane-Derived Reactive Intermediates*, Wiley–Interscience, Chichester, 1990.
2. K.G. Taylor and J. Chaney, *J. Am. Chem. Soc.*, 1972, **94**, 8924.
3. J. Ackroyd, M. Karpf and A.S. Dreiding, *Helv. Chim. Acta*, 1984, **67**, 1963.
4. W. Hartwig and L. Born, *J. Org. Chem.*, 1987, **52**, 4352.
5. S.H. Tahir, M.M. Olmstead and M.J. Kurth, *Tetrahedron Lett.*, 1991, **32**, 335.
6. K. Nakayama and W.J. Thompson, *J. Am. Chem. Soc.*, 1990, **112**, 6936.
7. F.H. Sansbury and S. Warren, *Tetrahedron Lett.*, 1992, **33**, 539.
8. N.S. Simpkins, *Chem. Soc. Rev.*, 1990, **19**, 335.
9. R. Shirai, M. Tanaka and K. Koga, *J. Am. Chem. Soc.*, 1986, **108**, 543.
10. R.S. Ward, *Chem. Soc. Rev.*, 1990, **19**, 1.
11. M. Asami, *Chem. Lett.*, 1984, 829.
12. S. Hatakeyama, K. Sakurai and S. Takano, *Chem. Commun.*, 1985, 1759.
13. A.E. Greene, M.-J. Luche and A.A. Serra, *J. Org. Chem.*, 1985, **50**, 3957.
14. Y. Kaurakami, J. Hiratake, Y. Yamamoto and J. Oda, *Chem. Sommun.*, 1984, 779.
15. S.H. Zhao, O. Samuel and H.B. Kagan, *Org. Synth.*, 1990, **68**, 49.
16. A. Sakamoto, Y. Yamamoto and J. Oda, *J. Am. Chem. Soc.*, 1987, **109**, 7188.
17. (a) G.M. Whitesides and C.-H. Wong, *Angew. Chem., Int. Ed. Engl.*, 1985, **24** 617; (b) J.B. Jones, *Tetrahedron*, 1986, **42**, 3351; (c) H. Yamada and S. Shimizu, *Angew. Chem., Int. Ed. Engl.*, 1988, **27**, 622; (d) A.M. Klibanov, *Acc. Chem. Res.*, 1990, **23**, 114; (e) D.H.G. Crout, S.M. Roberts and J.B. Jones, *Tetrahedron: Asymmetry*, 1993, **4**, 757.
18. C.-H. Wong, *et al.*, *J. Am. Chem. Soc.*, 1990, **112**, 945, and references cited therein.
19. D.P. Weiner, *Chem, Ind. (London)*, 1991, 347, and references cited therein.
20. P.H. Buist and D.M. Marecak, *Tetrahedron Lett.*, 1991, **32**, 891.
21. J. Davies and J.B. Jones, *J. Am. Chem. Soc.*, 1979, **101**, 5405.
22. Review: C.-S. Chen and C.J. Sih, *Angew. Chem., Int. Ed. Engl.*, 1989, **28**, 695.
23. *Org. Synth.*, 1990, **69**, 1, 10, 19.
24. K.J. Harris, Q.-M. Gu, Y.-E. Shih, G. Girdaukas and C.J. Sih. *Tetrahedron Lett.*, 1991, **32**, 3931.
25. C.R. Johnson, A. Golebiowski and D.H. Steensma, *Tetrahedron Lett.*, 1991, **32**, 3931; C.R. Johnson *et al.*, *Tetrahedron Lett.*, 1991, **32**, 2597.
26. D. Grandjean, P. Pale and J. Chuche, *Tetrahedron Lett.*, 1991, **32**, 3043.
27. H.-J. Gais, G. Bülow. A. Zatorski, M. Jentsch, P. Maidonis and H. Hemmerle, *J. Org. Chem.*, 1989, **54**, 5115; L.-M. Zhu and M.C. Tedford, *Tetrahedron*, 1990, **46**, 6587.
28. S. Hanessian, *Total Synthesis of Natural Products, the Chiron Approach*,

Pergamon Press, Oxford, 1983.
29. J. Nokami, M. Ohkura, Y. Dan-Oh and Y. Sakamoto, *Tetrahedron Lett.*, 1991, **32**, 2409.
30. M. Ohno, S. Kobayashi, T. Iimori, Y.-F. Wang and T. Izawa, *J. Am. Chem. Soc.*, 1981, **103**, 2405.
31. S. Kobayashi, K. Kamiyama, T. Iimori and M. Ohno, *Tetrahedron Lett.*, 1984, **25**, 2557.
32. R.L. Pederson, K.K.-C. Liu, J.F. Rutan, L. Chen and C.-H. Wong, *J. Org. Chem.*, 1990, **55**, 4897.
33. K. Burgess and I. Henderson, *Tetrahedron Lett.*, 1991, **32**, 5701.
34. E.J. Toone, M.J. Werth and J.B. Jones, *J. Am. Chem. Soc.*, 1990, **112**, 4946.
35. P.G. Hultin and J.B. Jones, *Tetrahedron Lett.*, 1992, **33**, 1399.
36. R. Csuk and B.I. Glänzer, *Chem. Rev.*, 1991, **91**, 49; S. Servi, *Synthesis*, 1990, 1; C.J. Sih and C.-S. Chen, *Angew. Chem., Int. Ed. Engl.*, 1984, **23**, 570.
37. A.S. Gopolan and H.K. Jacobs, *Tetrahedron Lett.*, 1990, **31**, 5575.
38. T. Itoh, Y. Yonekawa, T. Sato and T. Fujisawa, *Tetrahedron Lett.*, 1986, **27**, 5405.
39. A. Dondoni, G. Fantin, M. Fogagnolo, A. Mastellari, A. Medici, E. Nefrini and P. Pedrini, *Gazz. Chim. Ital.*, 1988, **118**, 211.
40. W.-R. Shieh, A.S. Gopolan and C.J. Sih, *J. Am. Chem. Soc.*, 1985, **107**, 2993.
41. K. Nakamura, Y. Kawai and A. Ohno, *Tetrahedron Lett.*, 1990, **31**, 267.
42. K. Nakamura, Y. Kawai and A. Ohno, *Tetrahedron Lett.*, 1991, **32**, 2927.
43. K. Nakamura, Y. Kawai, S. Oka and A. Ohno, *Tetrahedron Lett.*, 1989, **30**, 2245.
44. B. Zhou, A.S. Gopolan, F. VanMiddlesworth, W.-R. Shieh and C.J. Sih, *J. Am. Chem. Soc.*, 1983, **105**, 5925.
45. A.R. Battersby, in *Enzymes in Organic Synthesis*, Ciba Foundation Symposium 111, Pitman, London, 1985, pp. 22–30.
46. D.R. Dodds and J.B. Jones, *Chem. Commun.*, 1982, 1080.
47. I.J. Jakovac, H.B. Goodbrand, K.-P. Lok and J.B. Jones, *J. Am. Chem. Soc.*, 1982, **104**, 4659.
48. E.J. Toone, E.S. Simon, M.D. Bednarski and G.M. Whitesides, *Tetrahedron*, 1989, **45**, 5365; G.M. Whitesides, *et al.*, *J. Am. Chem. Soc.*, 1989, **111**, 627.
49. J.R. Durrwachter and C.-H. Wong, *J. Org. Chem.*, 1988, **53**, 4175.
50. M. Schultz, H. Waldmann, W. Vogt and H. Kunz, *Tetrahedron Lett.*, 1990, **31**, 867.
51. T. Kajimoto, L. Chen, K.K.-C. Liu and C.-H. Wong, *J. Am. Chem. Soc.*, 1991, **113**, 6678.

INDEX

HOOKE LIBRARY
12 AUG 1998
OXFORD